Christine Burik

PRINCIPLES OF PLASMA DISCHARGES AND MATERIALS PROCESSING

PRINCIPLES OF PLASMA DISCHARGES AND MATERIALS PROCESSING

MICHAEL A. LIEBERMAN
ALLAN J. LICHTENBERG

A Wiley-Interscience Publication
JOHN WILEY & SONS, INC.
New York / Chichester / Brisbane / Toronto / Singapore

Copyright © 1994 by John Wiley & Sons, Inc.

Library of Congress Cataloging in Publication Data:
Lieberman, M. A. (Michael A.)
   Principles of plasma discharges and materials processing / Michael
A. Lieberman, Allan J. Lichtenberg.
     p. cm.
  Includes bibliographical references and index.
  ISBN 0-471-00577-0
  1. Plasma dynamics. 2. Thin films—Surfaces. 3. Plasma etching.
4. Plasma chemistry. I. Lichtenberg, Allan J. II. Title.
QC718.5.D9L54 1994
530.4$'$4—dc20   94-28954

Printed in the United States of America

10 9 8 7 6 5 4 3 2

CONTENTS

8   MOLECULAR COLLISIONS   217

9   CHEMICAL KINETICS AND SURFACE PROCESSES   265

B   THE COLLISION INTEGRAL AND KINETIC THEORY   **547**

C   STOCHASTIC HEATING IN AN INDUCTIVE DISCHARGE   **555**

PREFACE

This book discusses the fundamental principles of partially ionized, chemically re-active plasma discharges and their use in thin-film processing. Plasma processing is a high-technology discipline born out of the need to access a parameter space in materials processing unattainable by strictly chemical methods. The field is in-terdisciplinary, combining the areas of plasma physics, surface science, gas-phase chemistry, and atomic and molecular physics. The common theme is the creation and use of plasmas to activate a chain of chemical reactions at a substrate surface. Our treatment is mainly restricted to discharges at low pressures, <1 Torr, which deliver activation energy, but not heat, to the surface. Plasma-based surface processes are in-dispensable for manufacturing the integrated circuits used by the electronics industry, and we use thin-film processes drawn from this field as examples. Plasma processing is also an important technology in the aerospace, automotive, steel, biomedical, and toxic waste management industries.

In our treatment of the material, we emphasize the development of a strong foundation in the fundamental physical and chemical principles that govern both dis-charges and gas- and surface-phase processes. We place little emphasis on describing state-of-the-art discharges and thin-film processes; while these change with time, the fundamentals abide. Our treatment is quantitative and emphasizes the physical insight and skills needed both to do back-of-the-envelope calculations and to do first-cut analyses or designs of discharges and thin-film processes. Practical graphs and tables are included to assist in the analysis. We give many examples throughout the book.

The book is both a graduate text, including exercises for the student, and a research monograph for practicing engineers and scientists. We assume that the reader has the usual undergraduate background in mathematics (2 years), physics ($1\frac{1}{2}$ years), and chemistry ($\frac{1}{2}$ or 1 year). Some familiarity with partial differential equations as commonly taught in courses on electromagnetics or fluid dynamics at the junior or senior undergraduate level is also assumed.

After an introductory chapter, the book is divided into four parts: low-pressure partially ionized plasmas (Chapters 2–6); gas and surface physics and chemical dynamics (Chapters 7–9); plasma discharges (Chapters 10–14); and plasma processing (Chapters 15 and 16). Atomic and molecular collision processes have been divided into two relatively self-contained chapters (Chapters 3 and 8, respectively) inserted before the corresponding chapters on kinetics in each case. This material may be read lightly or thoroughly as desired. Plasma diagnostics appear in concluding sections (Sections 4.6, 6.6, 8.6, and 11.6) of various chapters and often also serve as applications of the ideas developed in the chapters.

For the last five years, the authors have taught a one-semester course based on this material to a mixed group of mainly graduate students in electrical, chemical, and nuclear engineering, materials science, and physics. A typical syllabus follows for 30 lectures, each $1\frac{1}{2}$ hours in length:

Chapter	Lectures
1	1
2	2
3	2 (light coverage)
4	1 (sections 4.1 and 4.2 excluding waves, only)
5	2 (sections 5.1–5.3 only)
6	3 (omit section 6.4)
7	2
8	2 (light coverage, omit section 8.6)
9	3
10	1 (omit section 10.3)
11	2 (sections 11.1 and 11.2 only)
12	1 (section 12.1 only)
13	1 (section 13.1 only)
14	2
15	3
16	2 (omit section 16.4)

The core ideas of the book are developed in the sections of Chapters 2, 4–7, 9, and 10 listed in the syllabus. Atomic and molecular collisions (Chapters 3 and 8) can be emphasized more or less, but some coverage is desirable. The remaining chapters (Chapters 11–16), as well as some sections within each chapter, are relatively self-contained and topics can be chosen according to the interests of the instructor. More specialized material on guiding center motion (Section 4.3), dynamics (Section 4.4), waves (Section 4.5) and diffusion in magnetized plasmas (Sections 5.4 and 5.5) can generally be deferred until familiarity with the core material has been developed.

Our colleagues C.K. Birdsall and V. Vahedi and the plasma simulation group at Berkeley have developed user-friendly programs for PCs and workstations for computer-aided instruction and demonstrations. A number of concepts in discharge dynamics have been illustrated using various output results from these programs (see Figures 1.11, 2.2, and 6.3). We typically do four or five 20-minute simulation demonstrations in the course during the semester using this software. The software and manuals can be obtained by contacting the Software Distribution Office, Industrial Liaison Program, Department of Electrical Engineering and Computer Sciences, University of California, Berkeley, CA 94720; the electronic mail address, telephone, and fax numbers are software@eecs.berkeley.edu, (510) 643-6687, and (510) 643-6694, respectively.

This book has been three years in writing. We have received encouragement and benefited from discussions with many friends and colleagues. We acknowledge here those who contributed significantly to our enterprise. We are indebted to D.L. Flamm who was a MacKay Visiting Lecturer at Berkeley in 1988–89 and co-taught (with A.J.L.) an offering of our course in which he emphasized the chemical principles of plasma processing. One of the authors (M.A.L.) has taught abbreviated versions of the material in this book to process engineers in various short courses, along with his colleagues C.K. Birdsall, D.B. Graves, and V. Vahedi. We have benefited greatly from their insight and suggestions. Our colleagues N. Cheung, D. Graves, D. Hess, and S. Savas, our postdoctoral scholars C. Pico and R. Stewart, and our graduate students D. Carl, K. Kalpakjian, C. Lee, R. Lynch, G. Misium, R. Moroney, K. Niazi, A. Sato, P. Wainman, A. Wendt, M. Williamson, and B. Wood have taught us much, and some of their work has been incorporated into our text. Some of the material in Chapters 10, 12, and 13 is based on a review article by R.A. Gottscho and one of the authors (M.A.L.) in *Physics of Thin Films*, Vol. 18, edited by M. Francombe and J.L. Vossen, Academic Press, New York, 1994. We thank V.A. Godyak, M.B. Lieberman, and S. Brown for reviewing several chapters and suggesting clarifications of the text. W.D. Getty has used a preprint of our manuscript to teach a course similar to ours, and the final text has benefited from his comments and suggestions. Many of the ideas expressed in the book were developed by the authors while working on grants and contracts supported by the National Science Foundation, the Department of Energy, the Lawrence Livermore National Laboratory, the State of California MICRO Program, the California Competitive Technology Program, SEMATECH and the Semiconductor Research Corporation, IBM, Applied Materials, and Motorola. The authors gratefully acknowledge the hospitality of M.G. Haines at Imperial College, London (M.A.L.), and of R. Boswell at the Australian National University, Canberra (A.J.L.), where much of the manuscript was developed. We gratefully thank E. Lichtenberg and P. Park for typing portions of the manuscript.

Michael A. Lieberman
Allan J. Lichtenberg

SYMBOLS AND ABBREVIATIONS

Symbols

a	radius (m); atomic radius; a_0, Bohr radius; a_j, chemical activity of species j; a_v, etching anisotropy.
$\mathbf{a}$	acceleration (m/s^2);
A	area (m^2); a constant; A_R, reduced mass (amu).
$\not{b}$	impact parameter (m); radius (m).
B	magnetic induction (T); a constant; B_{rot}, rotational constant of molecule.
c	velocity of light in vacuum.
C	a constant; capacitance (F/m); C_V, specific heat at constant volume (J/mol-K); C_p, specific heat at constant pressure.
$\mathcal{C}$	a contour or closed loop.
d	denotes an exact differential.
đ	denotes a nonexact differential (Chapter 7).
d	distance (m); plasma size (m).
D	diffusion coefficient (m^2/s); displacement vector (C/m^2); D_a, ambipolar diffusion coefficient; D_{SiO_2}, deposition rate of silicon dioxide (m/s).
e	unsigned charge on an electron (1.602×10^{-19} C).
e	the natural base (2.718).
E	electric field (V/m); etch (or deposition) rate (Å/min).
$\mathcal{E}$	the voltage equivalent of the energy (V); i.e., energy (J) $= e\mathcal{E}$ (V).
f	frequency (Hz); distribution function (m^{-6}-s^3); f_m, Maxwellian distribution; f_{pe}, electron plasma frequency; f_{pi}, ion plasma frequency.
$\mathbf{f}_c$	collisional force per unit volume (N/m^3).
F	force (N).
g	degeneracy; $\bar{g}$, statistical weight; energy distribution function.
g	denotes a gas.
G	Gibbs free energy (J); volume ionization rate (m^{-3}-s^{-1}); G_f, Gibbs free energy of formation; G_r, Gibbs free energy of reaction; conductance (Ω^{-1}).
h	center-to-edge density ratio.
H	enthalpy (J); magnetic field (A/m); height (m); H_f, enthalpy of formation; H_r, enthalpy of reaction.
H	Boltzmann H function.
i	integer.
I	electrical current (A); differential scattering cross section (m^2/sr); I_{AB}, I_{mol}, moment of inertia of molecule (kg-m^2).

I	modified Bessel function of the first kind.
j	$\sqrt{-1}$; integer.
J	electrical current density (A/m^2); rotational quantum number.
J	Bessel function of the first kind.
$\mathcal{J}$	$\mathcal{J}_j$ denotes chemical species j.
k	Boltzmann's constant (1.381×10^{-23} J/K); wave number or wave vector (m^{-1}).
K	first-order (s^{-1}), second-order (m^3/s), or third-order (m^6/s) rate constant.
K	modified Bessel function of the second kind.
$\mathcal{K}$	equilibrium constant.
l	discharge length (m); antenna length (m); quantum number; integer.
l	denotes a liquid.
ℓ	denotes length for a line integral.
L	length (m); volume loss rate ($m^{-3}\text{-}s^{-1}$); inductance (H).
m	electron mass (9.11×10^{-31} kg); mass (kg); azimuthal mode number; m_l, m_s, and m_J, quantum numbers for axial component of orbital, spin, and total angular momentum.
M	ion mass (kg).
$\mathcal{M}$	number of chemical species.
n	particle density (m^{-3}); principal quantum number (an integer); n_i, ion density; n_e, electron density; n_g, neutral gas density.
n'	area density (m^{-2}); n'_0, area density of surface sites.
N	quantity of a substance (mol); index of refraction of a wave.
$\mathcal{N}$	number of turns.
p	pressure (N/m^2); particle momentum (kg-m/s); $p°$, standard pressure (1 bar or 1 atm); p_d, electric dipole moment (C-m); p_{ohm}, ohmic power density (W/m^3).
P	power (W); probability.
q	electric charge (C).
q	heat flow vector (W/m^2).
Q	heat (J).
$\mathcal{Q}$	resonant circuit or cavity quality factor.
r	radial position (m); r_c, gyroradius; r_{ce}, electron gyroradius.
R	gas constant (8.314 J/K-mol); cylinder radius (m); center of mass coordinate (m); nuclear separation (m); reaction rate ($m^{-3}\text{-}s^{-1}$); resistance (Ω).
s	sheath thickness (m); sticking coefficient; $\bar{s}$, thermal sticking coefficient; s_v or s_h, etching selectivity.
s	denotes a solid.
S	energy flux ($W/m^2\text{-}s$); entropy (J/K); closed surface area (m^2); S_p, pumping speed (m^3/s).

S	denotes a closed surface.
t	time (s).
T	temperature (K); T_0, standard temperature (298 K).
T	temperature in units of volts (V).
u	average velocity (m/s); u_B, Bohm velocity; u_E, $\mathbf{E} \times \mathbf{B}$ velocity; u_D, diamagnetic drift velocity.
U	energy (J); internal energy (J).
v	velocity (m/s); vibrational quantum number; $\bar{v}$, average speed; v_{th}, thermal velocity; v_R, relative velocity.
V	voltage or electric potential (V); $\tilde{V}$, rf voltage; $\bar{V}$, dc or time-average voltage.
$\mathcal{V}$	volume (m^3).
w	energy per unit volume (J/m^3); width (m).
W	kinetic energy (J); work (J).
x	rectangular coordinate (m); x_j, mole fraction of species j; x_{iz}, fractional ionization.
X	reactance (Ω).
y	rectangular coordinate (m).
Y	admittance (Ω^{-1}).
z	rectangular or axial cylindrical coordinate (m).
Z	relative charge on an ion, in units of e; impedance (Ω).
α	spatial rate of variation (m^{-1}); spatial attenuation or decay constant (m^{-1}); first Townsend coefficient (m^{-1}); ratio of negative ion to electron density; α_j, stochiometric coefficient of species j; α_p, atomic or molecular polarizability (m^3).
β	spatial rate of variation (m^{-1}); a constant.
γ	secondary electron emission coefficient; ratio of electron-to-ion temperature; ratio of specific heats.
Γ	particle flux (m^{-2}-s^{-1}).
Γ	the Gamma function.
δ	Dirac delta function; layer thickness (m); δ_p, collisionless skin depth (m); δ_c, collisional skin depth (m); δ_e, anomalous skin depth (m).
Δ	denotes the change of a quantity.
ϵ	dielectric constant (F/m); ϵ_0, vacuum permittivity (8.854×10^{-12} F/m); ϵ_p, plasma dielectric constant.
ζ	a small displacement (m); ζ_L, fractional energy loss for elastic collision.
θ	angle (rad); spherical polar angle; scattering angle in laboratory system; fractional surface coverage.
Θ	scattering angle in center of mass system (rad).
η	efficiency factor.

κ	relative dielectric constant; κ_p, relative plasma dielectric constant; κ_T, thermal conductivity.
λ	mean free path (m); λ_c, collisional mean free path; λ_e, electron mean free path; λ_i, ion mean free path; λ_{De}, electron Debye length (m).
Λ	diffusion length (m); ratio of Debye length to minimum impact parameter.
μ	mobility (m^2/V-s); chemical potential (J/mol); μ_0, vacuum permeability ($4\pi \times 10^{-7}$ H/m); μ_{mag}, magnetic moment.
ν	collision or interaction frequency (s^{-1} or Hz); ν_c, collision frequency.
ξ	a constant.
π	3.1416.
Π	stress tensor (N/m^2).
ρ	volume charge density (C/m^3); ρ_S, surface charge density (C/m^2).
σ	cross section (m^2); σ_{dc}, dc electrical conductivity (Ω^{-1}-m^{-1}); σ_{rf}, rf electrical conductivity.
τ	mean free time (s); time constant (s); τ_c, collision time.
ϕ	angle (rad); spherical azimuthal angle.
φ	magnetic flux (T-m^2).
Φ	electric potential (V); Φ_p, plasma potential; Φ_w, wall potential.
χ	angle (rad); χ_{01}, first zero of zero order Bessel function.
ψ	spherical polar angle in velocity space.
Ψ	helix pitch (rad).
ω	radian frequency (rad/s); ω_{pe}, electron plasma frequency; ω_c, gyration frequency; ω_{ce}, electron gyration frequency.
Ω	solid angle (sr).
A	scalar.
$\mathbf{A}$	vector.
$\hat{A}$	unit vector (has unit magnitude).
$\tilde{A}$	oscillating or rf part.
$\bar{A}$	average or dc part; equilibrium value.
$\dot{A}$	dA/dt.
$\ddot{A}$	d^2A/dt^2.
$\langle A \rangle$	average.

Subscript Abbreviations

a	activation.
abs	absorbed.
adet	associative detachment.
ads	adsorbed.

aff	affinity.
appl	applied.
at	atomic, atom.
att	attachment.
c	denotes collision or collisional, except ω_c and r_c denote gyration frequency and gyration radius, respectively.
chemi	chemisorption.
cond	conduction
cx	charge transfer (charge exchange).
dc	constant in time (direct current).
desor	desorption.
det	detachment.
diss	dissociation, dissipation.
diz	dissociative ionization.
e	denotes electron.
ecr	electron cyclotron resonance
eff	effective.
el	elastic.
esc	escape.
ex	excitation.
ext	external.
f	formation.
fin	final.
g	denotes gas atom.
h	denotes hot or tail electrons; denotes horizontal.
i	denotes positive ion.
in	in.
inel	inelastic.
init	initial.
inc	incident.
ind	induced.
iz	ionization.
l	left-hand circularly polarized.
L	Langevin (capture).
loss	loss.
m	denotes momentum transfer, except H_m, S_m, and G_m denote per mole.
mag	magnetic; magnetization.
mol	molecule.
ohm	ohmic.

out	out.
ox	oxide.
p	usually denotes plasma.
ph	phase.
physi	physisorption.
pol	polarization.
poly	polysilicon.
pr	photoresist.
q	quenching.
r	right-hand circularly polarized, reaction.
R	denotes reduced or relative value.
rad	radiation.
rec	recombination.
refl	reflected.
res	resonance.
rf	radio frequency.
rot	rotational.
s	denotes sheath edge.
S	denotes surface.
sc	scattering.
se	denotes secondary electron.
sput	sputtering.
stoc	stochastic.
subl	sublimation.
T	denotes total.
th	thermal.
thr	threshold.
trans	transmitted.
v	denotes vertical.
vap	vaporization.
vib	vibrational.
0	denotes initial value, uniform value, or central value.
∥	parallel.
⊥	perpendicular.

Physical Constants and Conversion Factors

Quantity	Symbol	Value
Boltzmann constant	k	1.3807×10^{-23} J/K
Elementary charge	e	1.6022×10^{-19} C
Electron mass	m	9.1095×10^{-31} kg
Proton mass	M	1.6726×10^{-27} kg
Proton/electron mass ratio	M/m	1836.2
Planck constant	h	6.6262×10^{-34} J-s
	$\hbar = h/2\pi$	1.0546×10^{-34} J-s
Speed of light in vacuum	c	2.9979×10^{8} m/s
Permittivity of free space	ϵ_0	8.8542×10^{-12} F/m
Permeability of free space	μ_0	$4\pi \times 10^{-7}$ H/m
Bohr radius	$a_0 = 4\pi\epsilon_0\hbar^2/e^2m$	5.2918×10^{-11} m
Atomic cross section	πa_0^2	8.7974×10^{-21} m^2
Temperature T associated with T $= 1$ V		11605 K
Energy associated with $\mathcal{E} = 1$ V		1.6022×10^{-19} J
Avogadro number (molecules/mol)	N_A	6.0220×10^{23}
Gas constant	$R = kN_A$	8.3144 J/K-mol
Atomic mass unit		1.6606×10^{-27} kg
Standard temperature (25 °C)	T_0	298.15 K
Standard pressure (760 Torr $= 1$ atm)	p°	1.0133×10^{5} Pa
Loschmidt's number (gas density at STP)	n°	2.6868×10^{25} m^{-3}
Pressure of 1 Torr		133.32 Pa
Energy per mole at T_0	RT_0	2.4789 kJ/mol
calorie (cal)		4.1868 J

Practical Formulae

In the following practical formulae, n_e is in units of cm^{-3}, T_e is in volts, and B is in gauss (1 tesla $= 10^4$ gauss).

Electron plasma frequency	$\omega_{pe} = (e^2 n_e / \epsilon_0 m)^{1/2}$	$f_{pe} = 9000\sqrt{n_e}$ Hz
Electron gyration frequency	$\omega_{ce} = eB/m$	$f_{ce} = 2.8B$ MHz
Electron Debye length	$\lambda_{De} = (\epsilon_0 T_e / e n_e)^{1/2}$	$\lambda_{De} = 740\sqrt{T_e/n_e}$ cm
Mean electron speed	$\bar{v}_e = (8eT_e / \pi m)^{1/2}$	$\bar{v}_e = 6.7 \times 10^7 \sqrt{T_e}$ cm/s
Bohm velocity	$u_B = (eT_e / M)^{1/2}$	$u_B = 9.8 \times 10^5 \sqrt{T_e / A_R}$ cm/s

CHAPTER 1

INTRODUCTION

1.1 MATERIALS PROCESSING

Chemically reactive plasma discharges are widely used to modify the surface properties of materials. Plasma processing technology is vitally important to several of the largest manufacturing industries in the world. Plasma-based surface processes are indispensable for manufacturing the very large scale integrated circuits (ICs) used by the electronics industry. Such processes are also critical for the aerospace, automotive, steel, biomedical, and toxic waste management industries. Materials and surface structures can be fabricated that are not attainable by any other commercial method, and the surface properties of materials can be modified in unique ways. For example, 0.2-μm-wide, 4-μm-deep trenches can be etched into silicon films or substrates (Fig. 1.1). A human hair is 50–100 μm in diameter, so hundreds of these trenches would fit endwise within a human hair. Unique materials such as diamond films and amorphous silicon for solar cells have also been produced, and plasma-based hardening of surgically implanted hip joints and machine tools have extended their working lifetimes manyfold.

It is instructive to look closer at integrated circuit fabrication, which is the key application that we describe in this book. As a very incomplete list of plasma processes, argon or oxygen discharges are used to sputter-deposit aluminum, tungsten, or high-temperature superconducting films; oxygen discharges can be used to grow SiO_2 films on silicon; SiH_2Cl_2/NH_3 and $Si(OC_2H_5)_4/O_2$ discharges are used for the

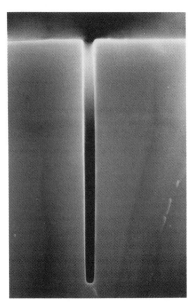

FIGURE 1.1. Trench etch (0.2 μm wide by 4 μm deep) in single-crystal silicon, showing the extraordinary capabilities of plasma processing; such trenches are used for device isolation and charge storage capacitors in integrated circuits.

plasma-enhanced chemical vapor deposition of Si_3N_4 and SiO_2 films, respectively; BF_3 discharges can be used to implant dopant (B) atoms into silicon; $CF_4/Cl_2/O_2$ discharges are used to selectively remove silicon films; and oxygen discharges are used to remove photoresist or polymer films. These types of steps (deposit or grow, dope or modify, etch or remove) are repeated again and again in the manufacture of a modern integrated circuit. They are the equivalent, on a micrometer-size scale, of centimeter-size manufacture using metal and components, bolts and solder, and drill press and lathe. For microfabrication of an IC, one-third of the tens to hundreds of fabrication steps are typically plasma-based.

Figure 1.2 shows a typical set of steps to create a metal film patterned with submicrometer features on a large area (200-mm-diameter) wafer substrate. In (a), the film is deposited; in (b), a photoresist layer is deposited over the film; in (c), the resist is selectively exposed to light through a pattern; and in (d), the resist is developed, removing the exposed resist regions and leaving behind a patterned resist mask. In (e), this pattern is transferred into the film by an etch process; the mask protects the underlying film from being etched. In (f), the remaining resist mask is removed. Of these six steps, plasma processing is generally used for film deposition (a) and etch (e), and may also be used for resist development (d) and removal (f).

The etch process in (e) is illustrated as leading to vertical sidewalls aligned with the resist mask; i.e, the mask pattern has been faithfully transferred into the metal film. This can be accomplished by an etch process that removes material in the vertical direction only. The horizontal etch rate is zero. Such *anisotropic* etches are easily produced by plasma processing. On the other hand, one might imagine

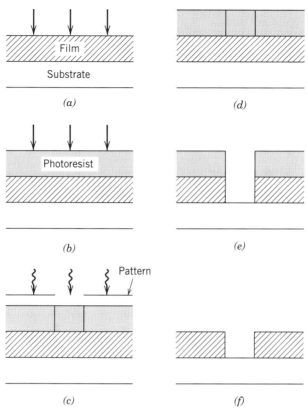

FIGURE 1.2. Deposition and pattern transfer in manufacturing an integrated circuit: (a) metal deposition; (b) photoresist deposition; (c) optical exposure through a pattern; (d) photoresist development; (e) anisotropic plasma etch; (f) remaining photoresist removal.

that exposing the masked film (d) to a liquid (or vapor phase) etchant will lead to the undercut *isotropic* profile shown in Fig. 1.3a (compare to Fig. 1.2e), which is produced by equal vertical and horizontal etch rates. Many years ago, feature spacings (e.g., between trenches) were tens of micrometers, much exceeding required film thicknesses. Undercutting was then acceptable. This is no longer true with submicrometer feature spacings. The reduction in feature sizes and spacings makes anisotropic etch processes essential. In fact, strictly vertical etches are sometimes not desired; one wants controlled sidewall angles. Plasma processing is the only commercial technology capable of such control. Anisotropy is a critical process parameter in IC manufacture and has been a major force in driving the development of plasma processing technology.

The etch process applied to remove the film in Fig. 1.2d is shown in Fig. 1.2e as not removing either the photoresist or the underlying substrate. This *selectivity* is another critical process parameter for IC manufacture. Whereas wet etches have been developed having essentially infinite selectivity, highly selective plasma etch

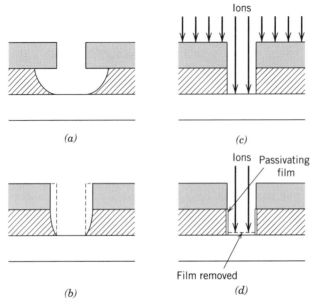

FIGURE 1.3. Plasma etching in integrated circuit manufacture: (a) example of isotropic etch; (b) sidewall etching of the resist mask leads to a loss of anisotropy in film etch; (c) illustrating the role of bombarding ions in anisotropic etch; (d) illustrating the role of sidewall passivating films in anisotropic etch.

processes are not easily designed. Selectivity and anisotropy often compete in the design of a plasma etch process, with results as shown in Fig. 1.3b. Compare this to the idealized result shown in Fig. 1.2e. Assuming that film-to-substrate selectivity is a critical issue, one might imagine simply turning off the plasma after the film has been etched through. This requires a good endpoint detection system. Even then, variations in film thickness and etch rate across the area of the wafer imply that the etch cannot be stopped at the right moment everywhere. Hence, depending on the process *uniformity*, there is a need for some selectivity. These issues are considered further in Chapter 15.

Here is a simple recipe for etching silicon using a plasma discharge. Start with an inert molecular gas, such as CF_4. Excite the discharge to sustain a plasma by electron–neutral dissociative ionization,

$$e + CF_4 \rightarrow 2e + CF_3^+ + F$$

and to create reactive species by electron–neutral dissociation,

$$e + CF_4 \rightarrow e + F + CF_3$$
$$\rightarrow e + 2F + CF_2$$

The etchant F atoms react with the silicon substrate, yielding the volatile etch product SiF_4:

$$Si(s) + 4F(g) \rightarrow SiF_4(g)$$

Here, s and g indicate solid and gaseous forms, respectively. Finally, the product is pumped away. It is important that CF_4 does not react with silicon, and that the etch product SiF_4 is volatile, so that it can be removed. This process etches silicon isotropically. For an anisotropic etch, there must be high-energy ion (CF_3^+) bombardment of the substrate. As illustrated in Figs. 1.3c and d, energetic ions leaving the discharge during the etch bombard the bottom of the trench but do not bombard the sidewalls, leading to anisotropic etching by one of two mechanisms. Either the ion bombardment increases the reaction rate at the surface (Fig. 1.3c), or it exposes the surface to the etchant by removing passivating films that cover the surface (Fig. 1.3d).

Similarly, Cl and Br atoms created by dissociation in a discharge are good etchants for silicon, F atoms and CF_2 molecules for SiO_2, O atoms for photoresist, and Cl atoms for aluminum. In all cases, a volatile etch product is formed. However, F atoms do not etch aluminum, and there is no known etchant for copper, because the etch products are not volatile at reasonable substrate temperatures.

We see the importance of the basic physics and chemistry topics treated in this book: (1) plasma physics (Chapters 2 and 4–6), to determine the electron and ion densities, temperatures, and ion bombardment energies and fluxes for a given discharge configuration; and (2) gas-phase chemistry and (3) surface physics and chemistry (Chapters 7 and 9), to determine the etchant densities and fluxes and the etch rates with and without ion bombardment. The data base for these fields of science is provided by (4) atomic and molecular physics, which we discuss in Chapters 3 and 8. We also discuss applications of equilibrium thermodynamics (Chapter 7) to plasma processing. The measurement and experimental control of plasma and chemical properties in reactive discharges is itself a vast subject. We provide brief introductions to some simple plasma diagnostic techniques throughout the text.

We have motivated the study of the fundamentals of plasma processing by examining anisotropic etches for IC manufacture. Other characteristics motivate its use for deposition, surface modification, and isotropic etch requirements. For example, a central feature of the *low-pressure* processing discharges that we consider in this book is that the plasma itself, as well as the plasma–substrate system, is not in thermal equilibrium. This enables substrate temperatures to be relatively low, compared to those required in conventional thermal processes, while maintaining adequate deposition or etch rates. Putting it another way, plasma processing rates are greatly enhanced over thermal processing rates at the same substrate temperature. For example, Si_3N_4 films can be deposited over aluminum films by plasma-enhanced chemical vapor deposition (PECVD), whereas adequate deposition rates cannot be achieved by conventional chemical vapor deposition (CVD) without melting the aluminum film. Chapter 16 gives further details.

The nonequilibrium nature of plasma processing has been known for many years, as illustrated by the laboratory data in Fig. 1.4. In time sequence, this shows first, the

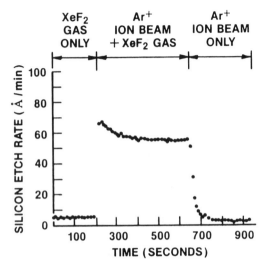

FIGURE 1.4. Experimental demonstration of ion-enhanced plasma etching. (Coburn and Winters, 1979).

equilibrium chemical etch rate of silicon in the XeF_2 etchant gas; next, the tenfold increase in etch rate with the addition of argon ion bombardment of the substrate, simulating plasma-assisted etching; and finally, the very low "etch rate" due to the physical sputtering of silicon by the ion bombardment alone.

A more recent application is the use of plasma-immersion ion implantation (PIII) to implant ions into materials at dose rates that are tens to hundreds of times larger than those achievable with conventional (beam-based) ion implantation systems. In PIII, a series of negative high-voltage pulses are applied to a substrate that is immersed directly into a discharge, thus accelerating plasma ions into the substrate. The development of PIII has opened a new implantation regime characterized by very high dose rates, even at very low energies, and by the capability to implant both large area and irregularly shaped substrates, such as flat panel displays or machine tools and dies. This is illustrated in Fig. 1.5. Further details are given in Chapter 16.

1.2 PLASMAS AND SHEATHS

Plasmas

A plasma is a collection of free charged particles moving in random directions that is, on the average, electrically neutral (see Fig. 1.6a). This book deals with weakly ionized plasma discharges, which are plasmas having the following features: (1) they are driven electrically; (2) charged particle collisions with neutral gas molecules are important; (3) there are boundaries at which surface losses are important; (4) ionization of neutrals sustains the plasma in the steady state and (5) the electrons are not in thermal equilibrium with the ions. A simple discharge is shown schematically

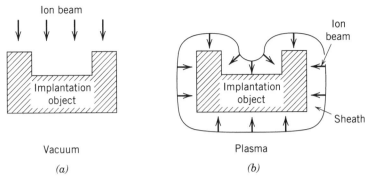

FIGURE 1.5. Illustrating ion implantation of an irregular object: (a) In a conventional ion beam implanter, the beam is electrically scanned and the target object is mechanically rotated and tilted to achieve uniform implantation; (b) in plasma-immersion ion implantation (PIII), the target is immersed in a plasma, and ions from the plasma are implanted with a relatively uniform spatial distribution.

in Fig. 1.6b. It consists of a voltage source that drives current through a low-pressure gas between two parallel conducting plates or electrodes. The gas "breaks down" to form a plasma, usually weakly ionized, i.e., the plasma density is only a small fraction of the neutral gas density. We describe some qualitative features of plasmas in this section; discharges are described in the following section.

Plasmas are often called a fourth state of matter. As we know, a solid substance in thermal equilibrium generally passes into a liquid state as the temperature is increased at a fixed pressure. The liquid passes into a gas as the temperature is further increased. At a sufficiently high temperature, the molecules in the gas decompose to form a gas of atoms that move freely in random directions, except for infrequent collisions between atoms. If the temperature is further increased, then the atoms decompose into freely moving charged particles (electrons and positive ions), and the substance enters the plasma state. This state is characterized by a common charged particle density $n_e \approx n_i \approx n$ particles/m^3 and, in equilibrium, a temperature $T_e = T_i = T$. The temperatures required to form plasmas from pure substances in thermal

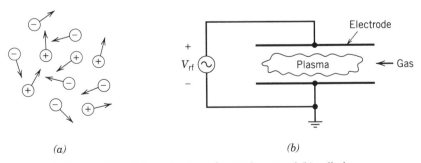

FIGURE 1.6.   Schematic view of (a) a plasma and (b) a discharge.

equilibrium range from roughly 4000 K for easy-to-ionize elements like cesium to 20000 K for hard-to-ionize elements like helium. The fractional ionization of a plasma is

$$x_{iz} = \frac{n_i}{n_g + n_i}$$

where n_g is the neutral gas density. x_{iz} is near unity for fully ionized plasmas, and $x_{iz} \ll 1$ for weakly ionized plasmas.

Much of the matter in the universe is in the plasma state. This is true because stars, as well as most interstellar matter, are plasmas. Although stars are plasmas in thermal equilibrium, the light and heavy charged particles in low-pressure processing discharges are *almost never* in thermal equilibrium, either between themselves or with their surroundings. Because these discharges are electrically driven and are weakly ionized, the applied power preferentially heats the mobile electrons, while the heavy ions efficiently exchange energy by collisions with the background gas. Hence, $T_e \gg T_i$ for these plasmas.

Figure 1.7 identifies different kinds of plasmas on a log n versus log T_e diagram. There is an enormous range of densities and temperatures for both laboratory and space plasmas. Two important types of processing discharges are indicated on the figure. Low-pressure discharges are characterized by $T_e \approx$ 1–10 V, $T_i \ll T_e$, and $n \approx 10^8–10^{13}$ cm^{-3}. These discharges are used as miniature chemical factories in which feedstock gases are broken into positive ions and chemically reactive etchants, deposition precursors, etc., which then flow to and physically or chemically react at the substrate surface. While energy is delivered to the substrate also, e.g., in the form of bombarding ions, the energy flux is there to promote the chemistry at the substrate, and not to heat the substrate. The gas pressures for these discharges are low: $p \approx$ 1 mTorr–1 Torr. *These discharges and their use for processing are the principal subject of this book.* We give the quantitative framework for their analysis in Chapter 10.

High-pressure arc discharges are also used for processing. These discharges have $T_e \approx$ 0.1–2 V and $n \approx 10^{14}–10^{19}$ cm^{-3}, and the light and heavy particles are more nearly in thermal equilibrium, with $T_i \lesssim T_e$. These discharges are used mainly to deliver heat to the substrate, e.g., to increase surface reaction rates, to melt, sinter, or evaporate materials, or to weld or cut refractory materials. Operating pressures are typically near atmospheric pressure (760 Torr). High-pressure discharges of this type are beyond the scope of this book.

Figure 1.8 shows the densities and temperatures (or average energies) for various species in a typical rf-driven capacitively coupled low-pressure discharge; e.g., for silicon etching using CF_4, as described in Section 1.1. We see that the feedstock gas, etchant atoms, etch product gas, and plasma ions have roughly the same temperature, which does not exceed a few times room temperature (0.026 V). The etchant F and product SiF_4 densities are significant fractions of the CF_4 density, but the fractional ionization is very low: $n_i \sim 10^{-5} n_g$. The electron temperature T_e is two orders of magnitude larger than the ion temperature T_i. However, we note that the energy of ions

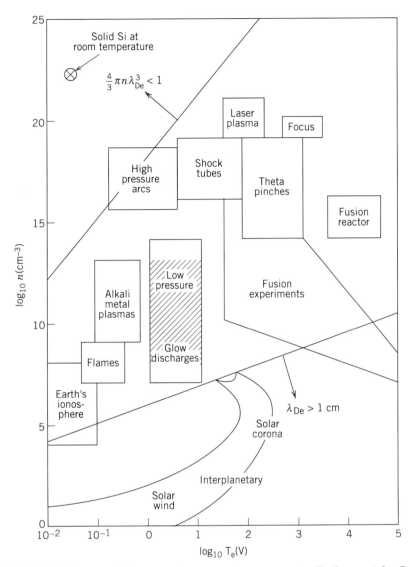

FIGURE 1.7. Space and laboratory plasmas on a $\log n$ versus $\log T_e$ diagram (after Book, 1987). λ_{De} is defined in Section 2.4.

bombarding the substrate can be 100–1000 V, much exceeding T_e. The acceleration of low-temperature ions across a thin *sheath* region where the plasma and substrate meet is central to all processing discharges. We describe this qualitatively below and quantitatively in later chapters.

Although n_i, n_e may be five orders of magnitude lower that n_g, the charged particles play central roles in sustaining the discharge and in processing. Because $T_e \gg T_i$, it is the electrons that dissociate the feedstock gas to create the free radicals,

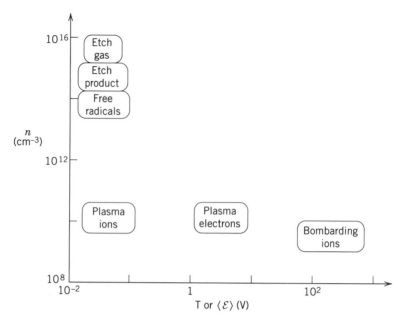

FIGURE 1.8. Densities and energies for various species in a low-pressure capacitive rf discharge.

etchant atoms, and deposition precursors, required for the chemistry at the substrate. Electrons also ionize the gas to create the positive ions that subsequently bombard the substrate. As we have seen, energetic ion bombardment can increase chemical reaction rates at the surface, clear inhibitor films from the surface, and physically sputter materials from or implant ions into the surface.

T_e is generally less than the threshold energies $\mathcal{E}_{diss}$ or $\mathcal{E}_{iz}$ for dissociation and ionization of the feedstock gas molecules. Nevertheless, dissociation and ionization occur because electrons have a distribution of energies. Letting $g_e(\mathcal{E})\,d\mathcal{E}$ be the number of electrons per unit volume with energies lying between $\mathcal{E}$ and $\mathcal{E} + d\mathcal{E}$, then the distribution function $g_e(\mathcal{E})$ is sketched in Fig. 1.9. Electrons having energies below

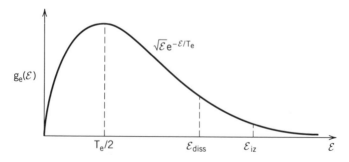

FIGURE 1.9.   Electron distribution function in a weakly ionized discharge.

$\mathcal{E}_{\text{diss}}$ or $\mathcal{E}_{\text{iz}}$ cannot dissociate or ionize the gas. We see that dissociation and ionization are produced by the high-energy tail of the distribution. Although the distribution is sketched in the figure as if it were Maxwellian at the bulk electron temperature T_e, this may not be the case. The tail distribution might be depressed below or enhanced above a Maxwellian by electron heating and electron–neutral collision processes. Two temperature distributions are sometimes observed, with T_e for the bulk electrons lower than T_h for the energetic electron tail.

Sheaths

Plasmas, which are quasineutral ($n_i \approx n_e$), are joined to wall surfaces across thin positively charged layers called *sheaths*. To see why, first note that the electron thermal velocity $(eT_e/m)^{1/2}$ is at least 100 times the ion thermal velocity $(eT_i/M)^{1/2}$ because $m/M \ll 1$ and $T_e \gtrsim T_i$. (Here, T_e and T_i are given in units of volts.) Consider a plasma of width l with $n_e = n_i$ initially confined between two grounded ($\Phi = 0$) absorbing walls (Fig. 1.10a). Because the net charge density $\rho = e(n_i - n_e)$ is zero, the electric potential Φ and the electric field E_x is zero everywhere. Hence, the fast-moving electrons are not confined and will rapidly be lost to the walls. On a very short timescale, however, some electrons near the walls are lost, leading to the situation shown in Fig. 1.10b. Thin ($s \ll l$) positive ion sheaths form near each wall in which $n_i \gg n_e$. The net positive ρ within the sheaths leads to a potential profile $\Phi(x)$ that is positive within the plasma and falls sharply to zero near both walls. This acts as a confining potential "valley" for electrons and a "hill" for ions because the

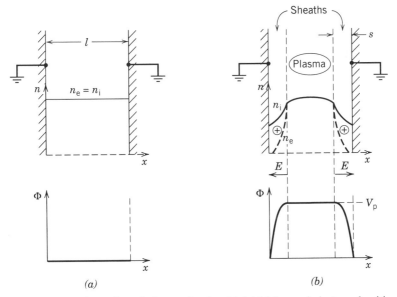

FIGURE 1.10. The formation of plasma sheaths: (a) initial ion and electron densities and potential; (b) densities, electric field, and potential after formation of the sheath.

electric fields within the sheaths point from the plasma to the wall. Thus the force $-eE_x$ acting on electrons is directed into the plasma; this reflects electrons traveling toward the walls back into the plasma. Conversely, ions from the plasma that enter the sheaths are accelerated into the walls. If the plasma potential (with respect to the walls) is V_p, then we expect that $V_p \sim$ a few T_e in order to confine most of the electrons. The energy of ions bombarding the walls is then $\mathcal{E}_i \sim$ a few T_e. Charge uncovering is treated quantitatively in Chapter 2, and sheaths in Chapter 6.

Figure 1.11 shows sheath formation as obtained from a particle-in-cell (PIC) plasma simulation. We use PIC results throughout this book to illustrate various discharge phenomena. In this simulation, the left wall is grounded, the right wall is floating (zero net current), and the positive ion density is uniform and constant in time. The electrons are modeled as $\mathcal{N}$ sheets having charge-to-mass ratio $-e/m$ that move in one dimension (along x) under the action of the time-varying fields produced by all the other sheets, the fixed ion charge density, and the charges on the walls. Electrons do not collide with other electrons, ions, or neutrals in this simulation. Four thousand sheets were used with $T_e = 1$ V and $n_i = n_e = 10^{13}$ m^{-3} at time $t = 0$. In (a), (b), (c), and (d), we see the v_x–x electron phase space, electron density, electric field, and potential after the sheath has formed, at $t = 0.77$ μs. The time history of $\mathcal{N}$ is shown in (e); 40 sheets have been lost to form the sheaths. Figures 1.11a–d show the absence of electrons near each wall over a sheath width $s \approx 6$ mm. Except for fluctuations due to the finite $\mathcal{N}$, the field in the bulk plasma is near zero, and the fields in the sheaths are large and point from the plasma to the walls. (E_x is negative

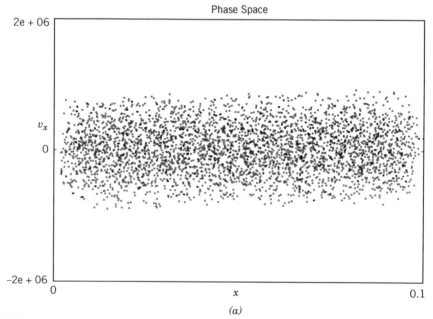

(a)

FIGURE 1.11. PIC simulation of positive ion sheath formation: (a) v_x–x electron phase space, with horizontal scale in meters.

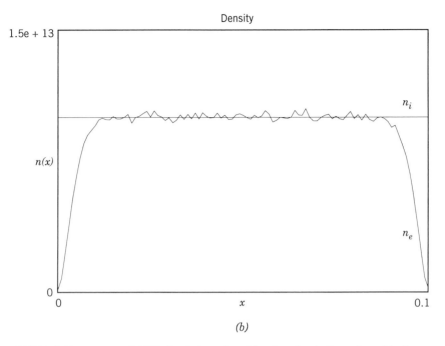

(b)

FIGURE 1.11. (*continued*) PIC simulation of positive ion sheath formation: (b) electron density n_e.

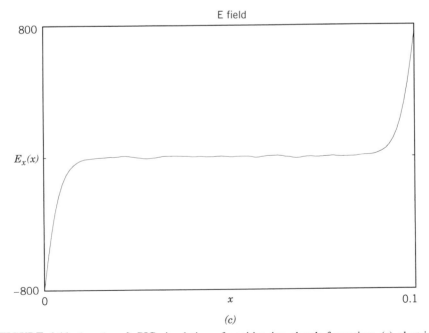

(c)

FIGURE 1.11. (*continued*) PIC simulation of positive ion sheath formation: (c) electric field E_x.

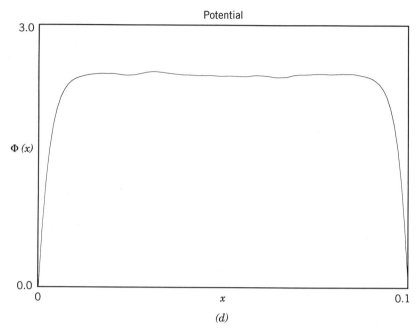

FIGURE 1.11. (*continued*) PIC simulation of positive ion sheath formation: (d) potential Φ.

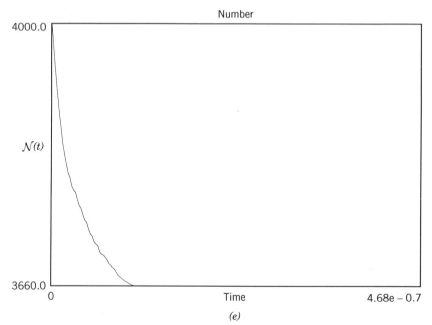

FIGURE 1.11. (*continued*) PIC simulation of positive ion sheath formation: (e) electron number $\mathcal{N}$ versus time t in seconds.

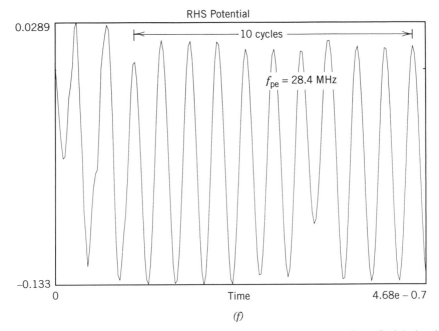

FIGURE 1.11. (*continued*) PIC simulation of positive ion sheath formation: (f) right hand potential V_r versus time t.

at the left wall and positive at the right wall to repel plasma electrons.) The potential in the center of the discharge is $V_p \approx 2.5$ V and falls to zero at the left wall (this wall is grounded by definition). The potential at the right wall is also low, but we see in (f) that it oscillates in time. We will see in Chapter 4 that these are *plasma oscillations*. We would not see them if the initial sheet positions and velocities were chosen exactly symmetrically about the midplane, or if many more sheets were used in the simulation.

If the ions were also modeled as moving sheets, then on a longer time scale we would see ion acceleration within the sheaths, and a consequent drop in ion density near the walls, as sketched in Fig. 1.10b. We return to this in Chapter 6.

The separation of discharges into bulk plasma and sheath regions is an important paradigm that applies to all discharges. The bulk region is quasineutral, and both instantaneous and time-averaged fields are low. The bulk plasma dynamics are described by diffusive ion loss at high pressures and by free-fall ion loss at low pressures. In the positive space charge sheaths, high fields exist, leading to dynamics that are described by various ion space charge sheath laws, including low-voltage sheaths and various high-voltage sheath models, such as collisionless and collisional Child laws and their modifications. The plasma and sheath dynamics must be joined at their interface. As will be seen in Chapter 6, the usual joining condition is to require that the mean ion velocity at the plasma–sheath edge be equal to the ion-sound (Bohm) velocity: $u_B = (eT_e/M)^{1/2}$, where e and M are the charge and mass of the ion and T_e is the electron temperature in volts.

1.3 DISCHARGES

Rf Diodes

Capacitively driven radio frequency (rf) discharges—so-called *rf diodes*—are commonly used for materials processing. An idealized discharge in plane parallel geometry, shown in Fig. 1.12a, consists of a vacuum chamber containing two planar electrodes separated by a spacing l and driven by an rf power source. The substrates are placed on one electrode, feedstock gases are admitted to flow through the discharge, and effluent gases are removed by the vacuum pump. Coaxial discharge geometries, such as the "hexode" shown in Fig. 1.12b, are also in widespread use. Typical parameters are shown in Table 1.1. The typical rf driving voltage is $V_{rf} = 100–1000$ V, and the plate separation is $l = 2–10$ cm. When operated at low pressure, with the wafer mounted on the powered electrode, and used to remove substrate material, such reactors are commonly called reactive ion etchers (RIEs)—a misnomer, since the etching is a chemical process enhanced by energetic ion bombardment of the substrate, rather than a removal process due to reactive ions alone.

For anisotropic etching, typically pressures are in the range 10–100 mTorr, power densities are 0.1–1 W/cm^2, the driving frequency is 13.56 MHz, and multiple wafer systems are common. Typical plasma densities are relatively low, 10^9–10^{11} cm^{-3}, and the electron temperature is of order 3 V. Ion acceleration energies (sheath voltages) are high, greater than 200 V, and fractional ionization is low. The degree of dissociation of the molecules into reactive species is seldom measured but can range widely from less than 0.1% to nearly 100% depending on gas composition and plasma conditions. For deposition and isotropic etch applications, pressures tend to be higher, ion bombarding energies are lower, and frequencies can be lower than the commonly used standard of 13.56 MHz.

The operation of capacitively driven discharges is reasonably well understood. As shown in Fig. 1.13 for a symmetrically driven discharge, the mobile plasma electrons, responding to the instantaneous electric fields produced by the rf driving voltage, oscillate back and forth within the positive space charge cloud of the ions. The massive ions respond only to the time-averaged electric fields. Oscillation of the electron cloud creates sheath regions near each electrode that contain net positive

TABLE 1.1. Range of Parameters for Rf Diode and High-Density Discharges

Parameter	Rf Diode	High-Density Source
Pressure p (m Torr)	10–1000	0.5–50
Power P (W)	50–2000	100–5000
Frequency f (MHz)	0.05–13.56	0–2450
Volume $\mathcal{V}$ (L)	1–10	2–50
Cross-sectional area A (cm^2)	300–2000	300–500
Magnetic field B (kG)	0	0–1
Plasma density n (cm^{-3})	10^9–10^{11}	10^{10}–10^{12}
Electron temperature T_e (V)	1–5	2–7
Ion acceleration energy $\mathcal{E}_i$ (V)	200–1000	20–500
Fractional ionization x_{iz}	10^{-6}–10^{-3}	10^{-4}–10^{-1}

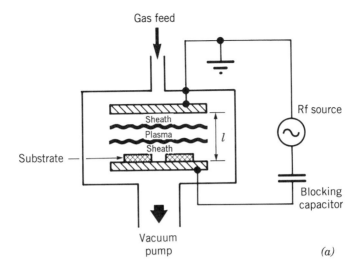

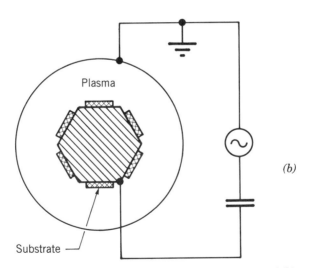

FIGURE 1.12. Capacitive rf discharges in (a) plane parallel geometry and (b) coaxial "hexode" geometry. (From "Design of High-Density Sources for Materials Processing" from the work "Physics of Thin Films," Vol. 18, by Academic Press, Inc., Publisher in Press)

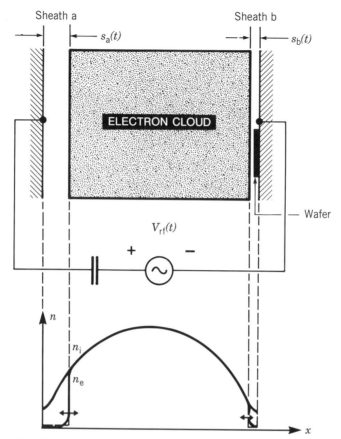

FIGURE 1.13. The physical model of an rf diode. (From "Design of High-Density Sources for Materials Processing" from the work "Physics of Thin Films," Vol. 18, by Academic Press, Inc., Publisher in Press)

charge when averaged over an oscillation period; i.e., the positive charge exceeds the negative charge in the system, with the excess appearing within the sheaths. This excess produces a strong time-averaged electric field within each sheath directed from the plasma to the electrode. Ions flowing out of the bulk plasma near the center of the discharge can be accelerated by the sheath fields to high energies as they flow to the substrate, leading to energetic-ion enhanced processes. Typical ion-bombarding energies $\mathcal{E}_i$ can be as high as $V_{rf}/2$ for symmetric systems (Fig. 1.13) and as high as V_{rf} at the powered electrode for asymmetric systems (Fig. 1.12). A quantitative description of capacitive discharges is given in Chapter 11.

We note that the positive ions continuously bombard the electrode over an rf cycle. In contrast, electrons are lost to the electrode only when the oscillating cloud closely approaches the electrode. During that time, the instantaneous sheath potential collapses to near zero, allowing sufficient electrons to escape to balance the ion charge delivered to the electrode. Except for such brief moments, the instantaneous potential of the discharge must always be positive with respect to any large electrode and wall surface; otherwise the mobile electrons would quickly leak out. Electron confinement is ensured by the presence of positive space charge sheaths near all surfaces.

We will see that a crucial limiting feature of rf diodes is that the ion-bombarding flux $\Gamma_i = nu_B$ and bombarding energy $\mathcal{E}_i$ cannot be varied independently. The situation is analogous to the lack of independent voltage and current control in diode vacuum tubes or semiconductor pn junctions. For a reasonable (but relatively low) ion flux, as well as a reasonable dissociation of the feedstock gas, sheath voltages at the driven electrode are high. For wafers placed on the driven electrode, this can result in undesirable damage, or loss of linewidth control. Furthermore, the combination of low ion flux and high ion energy leads to a relatively narrow process window for many applications. The low process rates resulting from the limited ion flux in rf diodes often mandates multiwafer or batch processing, with consequent loss of wafer-to-wafer reproducibility. Higher ion and neutral fluxes are generally required for single-wafer processing in a clustered tool environment, in which a single wafer is moved by a robot through a series of process chambers. Clustered tools are used to control interface quality and are said to have the potential for significant cost savings in fabricating integrated circuits. Finally, low fractional ionization poses a significant problem for processes where the feedstock costs and disposal of effluents are issues.

To meet the linewidth, selectivity, and damage control demands for next-generation fabrication, the mean ion bombarding energy, and its energy distribution, should be controllable independently of the ion and neutral fluxes. Some control over ion-bombarding energy can be achieved by putting the wafer on the undriven electrode and independently biasing this electrode with a second rf source. Although these so-called *rf triode* systems are in use, processing rates are still low at low pressures and sputtering contamination is an issue.

Various magnetically enhanced rf diodes and triodes have also been developed to improve performance of the rf reactor. These include, for example, magnetically enhanced reactive ion etchers (MERIEs), in which a dc magnetic field of 50–300 G is applied parallel to the powered electrode, on which the wafer sits. The magnetic field increases the efficiency of power transfer from the source to the plasma and also enhances plasma confinement. This results in a reduced sheath voltage and an increased plasma density when the magnetic field is applied. However, the plasma generated is strongly nonuniform both radially and azimuthally. To increase process uniformity (at least azimuthally), the magnetic field is slowly rotated in the plane of the wafer, e.g., at a frequency of 0.5 Hz. While this is an improvement, MERIE systems may not have good uniformity, which may limit their applicability to next-generation, submicrometer device fabrication.

High-Density Sources

The limitations of rf diodes and their magnetically enhanced variants have led to the development of a new generation of low-pressure, high-density plasma sources. A few examples are shown schematically in Fig. 1.14, and typical source and plasma parameters are given in Table 1.1. A quantitative description is given in Chapters 12

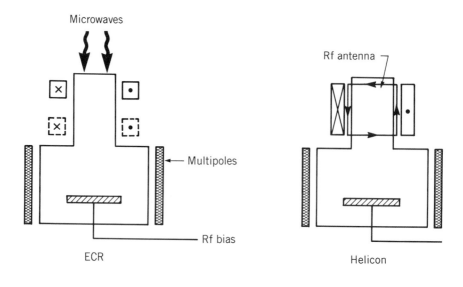

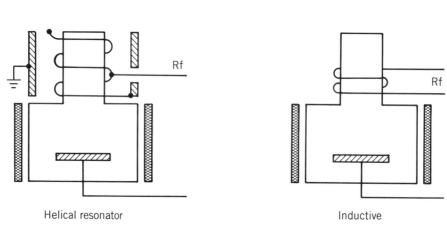

FIGURE 1.14. Some high-density "remote" sources. (From "Design of High-Density Sources for Materials Processing" from the work "Physics of Thin Films," Vol. 18, by Academic Press, Inc., Publisher in Press)

and 13. In addition to high density and low pressure, a common feature is that the rf or microwave power is coupled to the plasma across a dielectric window, rather than by direct connection to an electrode in the plasma, as for an rf diode. This noncapacitive power transfer is the key to achieving low voltages across all plasma sheaths at electrode and wall surfaces. Dc voltages, and hence ion acceleration energies, are then typically 20–30 V at all surfaces. To control the ion energy, the electrode on which the wafer is placed can be independently driven by a capacitively coupled rf source. Hence independent control of the ion/radical fluxes (through the source power) and the ion-bombarding energy (through the wafer electrode power) is possible.

The common features of power transfer across dielectric windows and separate bias supply at the wafer electrode are illustrated in Fig. 1.14. However, sources differ significantly in the means by which power is coupled to the plasma. For the electron cyclotron resonance (ECR) source shown in Fig. 1.14a, one or more electromagnet coils surrounding the cylindrical source chamber generate an axially varying dc magnetic field. Microwave power is injected axially through a dielectric window into the source plasma, where it excites a right-hand circularly polarized wave that propagates to a resonance zone, for cold electrons at $\omega = \omega_{ce}$, where the wave is absorbed. Here $\omega = 2\pi f$ is the applied radian frequency and $\omega_{ce} = eB/m$ is the electron gyration frequency at resonance. For the typical microwave frequency used, $f = 2450$ MHz, the resonant magnetic field is $B \approx 875$ G. The plasma streams out of the source into the process chamber in which the wafer is located.

A helicon source is shown in Fig. 1.14b. A weak (50–200 G) dc axial magnetic field together with an rf-driven antenna placed around the dielectric cylinder that forms the source chamber allows excitation of a helicon wave within the source plasma. Resonant wave–particle interaction is believed to transfer the wave energy to the plasma. For the helical resonator source shown in Fig. 1.14c, the external helix and conducting cylinder surrounding the dielectric discharge chamber form a slow wave structure, i.e., supporting an electromagnetic wave with phase velocity much less than the velocity of light. Efficient coupling of the rf power to the plasma is achieved by excitation of a resonant axial mode. An inductive (or transformer) coupled source is shown in Fig. 1.14d. Here the plasma acts as a single-turn, lossy conductor that is coupled to a multiturn nonresonant rf coil across the dielectric discharge chamber; rf power is inductively coupled to the plasma by transformer action. In contrast to ECR and helicon sources, a dc magnetic field is not required for efficient power coupling in helical resonator or inductive sources.

Figure 1.14 also illustrates the use of high-density sources to feed plasma into a relatively distinct, separate process chamber in which the wafer is located. As shown in the figure, the process chamber can be surrounded by dc multipole magnetic fields to enhance plasma confinement near the process chamber surfaces, while providing a magnetic near-field-free plasma environment at the wafer. Such configurations are often called "remote" sources, a misnomer since at low pressures considerable plasma and free radical production occurs within the process chamber near the wafer. Sometimes, the source and process chambers are more integral, e.g., the wafer is placed very near to the source exit, to obtain increased ion and radical fluxes, reduced spread in ion energy, and improved process uniformity. But the wafer is then exposed to higher levels of damaging radiation.

Although the need for low pressures, high fluxes, and controllable ion energies has motivated high-density source development, there are many issues that need to be resolved. A critical issue is achieving the required process uniformity over 200- to 300–mm wafer diameters. In contrast to the nearly one-dimensional geometry of typical rf diodes (two closely spaced parallel electrodes), high-density cylindrical sources can have length-to-diameter ratios of order or exceeding unity. Plasma formation and transport in such geometries are inherently radially nonuniform. Another critical issue is efficient power transfer (coupling) across dielectric windows over a wide operating range of plasma parameters. Degradation of and deposition on the window can also lead to irreproducible source behavior and the need for frequent, costly cleaning cycles. Low-pressure operation leads to severe pumping requirements for high deposition or etching rates and hence to the need for large, expensive vacuum pumps. Furthermore, plasma and radical concentrations become strongly sensitive to reactor surface conditions, leading to problems of reactor aging and process irreproducibility. Finally, dc magnetic fields are required for some source concepts. These can lead to magnetic field-induced process nonuniformities and damage, as seen, for example, in MERIE systems.

Figure 1.15 illustrates schematically the central problem of discharge analysis, using the example of an rf diode. Given the *control* parameters for the power source (frequency ω, driving voltage V_{rf} or absorbed power P_{abs}), the feedstock gas (pressure p, flow rate, and chemical composition), and the geometry (simplified here to the discharge length l), then find the *plasma* parameters, including the plasma density n_i, the etchant density n_F, the ion and etchant fluxes Γ_i and Γ_F hitting the substrate, the electron and ion temperatures T_e and T_i, the ion bombarding energy $\mathcal{E}_i$, and the sheath thickness s. The control parameters are the "knobs" that can be "turned" in order to "tune" the properties of the discharge.

The tuning range for a given discharge is generally limited. Sometimes one type of discharge will not do the job no matter how it is tuned, so another type must be selected. As suggested in Figs. 1.12 and 1.14, a bewildering variety of discharges are used for processing. Some are driven by rf, some by dc, and some by microwave power sources. Some use magnetic fields to increase the plasma confinement or the efficiency of power absorption. One purpose of this book is to guide the reader toward making wise choices when designing discharges used for processing.

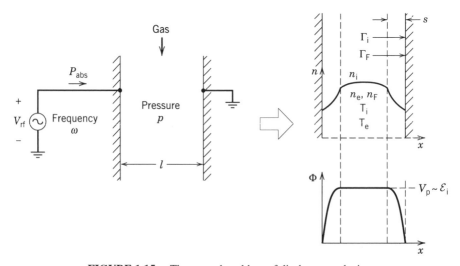

FIGURE 1.15.   The central problem of discharge analysis.

1.4 SYMBOLS AND UNITS

The choice of symbols is always vexing. While various fields each have their consistent set of symbols to represent physical quantities, these overlap between different fields, e.g., plasma physics and gas-phase chemistry. For example, H is standard for enthalpy in chemistry but is also standard for magnetic field in plasma physics. This also occurs within a given field; e.g., k is standard for Boltzmann's constant but also for wave number. Then there is always the occasional symbol that must stand for many things in different contexts. We sometimes distinguish these by using different lettering (Roman, italic, script, boldface); e.g., I is a current and I is a modified Bessel function; M is an ion mass and $\mathcal{M}$ is the number of chemical species. We can often distinguish commonly used symbols by the use of subscripts; e.g., σ denotes a cross section, but σ_{rf} and σ_{dc} denote electrical conductivities; we have done this whenever the notation is not too cumbersome. The meaning should be clear from the context, in most cases. To help avoid confusion, we have provided a table of symbols and abbreviations in the front matter of this book. These give the normal usage of symbols and their units.

As far as possible, we use the SI (MKS) system of units: meters (m), kilograms (kg), seconds (s), and coulombs (C) for charge. In these units, the charge on an electron is $-e \approx -1.602 \times 10^{-19}$ C. The unit of energy is the joule (J), but we often use the symbol $\mathcal{E}$ for the voltage that is the equivalent of the energy; i.e.,

$$U(\text{joules}) = e\mathcal{E}$$

where $\mathcal{E}$ is in volts. We also occasionally use the calorie (cal): 1 cal ≈ 4.187 J. The SI unit of pressure is the pascal (Pa), but we more commonly give gas pressures in Torr:

$$1 \text{ Torr} \approx 133.3 \text{ Pa}$$

We occasionally use 1 atm $\approx 1.013 \times 10^5$ Pa ≈ 760 Torr and 1 bar $= 10^5$ Pa to refer to gas pressures. The SI unit for the magnetic induction B is tesla (T), but we more often give B in gauss (G): 1 T $= 10^4$ G. We use the symbol T to refer to the temperature in kelvins (K). The energy equivalent temperature in joules is kT, where $k \approx 1.381 \times 10^{-23}$ J/K is Boltzmann's constant. We often use the roman typeface symbol T for the voltage equivalent of the temperature, where

$$e\text{T(volts)} = kT(\text{kelvins})$$

Hence room temperature $T = 297$ K is equivalent to T ≈ 0.026 V. Even within the standard unit system, quantities are often designated by subunits. For example, cross sections are often given in cm^2 rather than m^2 in tables, and wavelengths at microwave frequencies are commonly given in cm rather than in meters.

To assist our readers in making calculations, we give the commonly used constants in the SI system of units and the most common conversions between units in the front matter of the book. It is sometimes tempting to make a calculation in nonstandard units. For example, the collision frequency $\nu = n\sigma v$, which has units ($m^{-3} \cdot m^2 \cdot m\,s^{-1}$), could equally well be calculated in the commonly used units ($cm^{-3} \cdot cm^2 \cdot cm\,s^{-1}$), since the length units cancel. However, we urge the student not to take such shortcuts, but to systematically convert to standard units, before making a calculation.

CHAPTER 2

BASIC PLASMA EQUATIONS
AND EQUILIBRIUM

2.1 INTRODUCTION

The plasma medium is complicated in that the charged particles are both affected by external electric and magnetic fields and contribute to them. The resulting self-consistent system is nonlinear and very difficult to analyze. Furthermore, the interparticle collisions, although also electromagnetic in character, occur on space and time scales that are usually much shorter than those of the applied fields or the fields due to the average motion of the particles.

To make progress with such a complicated system, various simplifying approximations are needed. The interparticle collisions are considered independently of the larger scale fields to determine an *equilibrium distribution* of the charged-particle velocities. The velocity distribution is averaged over velocities to obtain the *macroscopic motion*. The macroscopic motion takes place in external applied fields and in the macroscopic fields generated by the average particle motion. These self-consistent fields are nonlinear, but may be linearized in some situations, particularly when dealing with waves in plasmas. The effect of spatial variation of the distribution function leads to pressure forces in the macroscopic equations. The collisions manifest themselves in particle generation and loss processes, as an average friction force between different particle species, and in energy exchanges among species. In this chapter we consider the basic equations that govern the plasma medium, concentrating attention on the macroscopic system. The complete derivation of these equations, from

25

fundamental principles, is beyond the scope of the text. We shall make the equations plausible and, in the easier instances, supply some derivations in appendices. For the reader interested in more rigorous treatment, references to the literature will be given.

In Section 2.2 we introduce the macroscopic field equations and the current and voltage. In Section 2.3 we introduce the fundamental equation of plasma physics, for the evolution of the particle distribution function, in a form most applicable for weakly ionized plasmas. We then define the macroscopic quantities and indicate how the macroscopic equations are obtained by taking moments of the fundamental equation. References given in the text can be consulted for more details of the averaging procedure. Although the macroscopic equations depend on the equilibrium distribution, their form is independent of the equilibrium. To solve the equations for particular problems the equilibrium must be known. In Section 2.4 we introduce the equilibrium distribution and obtain some consequences arising from it and from the field equations. The form of the equilibrium distribution will be shown to be a consequence of the interparticle collisions, in Appendix B.

2.2 FIELD EQUATIONS, CURRENT, AND VOLTAGE

Maxwell's Equations

The usual macroscopic form of Maxwell's equations are

$$\nabla \times \mathbf{E} = -\mu_0 \frac{\partial \mathbf{H}}{\partial t} \tag{2.2.1}$$

$$\nabla \times \mathbf{H} = \epsilon_0 \frac{\partial \mathbf{E}}{\partial t} + \mathbf{J} \tag{2.2.2}$$

$$\epsilon_0 \nabla \cdot \mathbf{E} = \rho \tag{2.2.3}$$

and

$$\mu_0 \nabla \cdot \mathbf{H} = 0 \tag{2.2.4}$$

where $\mathbf{E}(\mathbf{r}, t)$ and $\mathbf{H}(\mathbf{r}, t)$ are the electric and magnetic field vectors and where $\mu_0 = 4\pi \times 10^{-7}$ H/m and $\epsilon_0 \approx 8.854 \times 10^{-12}$ F/m are the permeability and permittivity of free space. The sources of the fields, the charge density $\rho(\mathbf{r}, t)$ and the current density $\mathbf{J}(\mathbf{r}, t)$, are related by the charge continuity equation (Problem 2.1):

$$\frac{\partial \rho}{\partial t} + \nabla \cdot \mathbf{J} = 0 \tag{2.2.5}$$

In general,

$$\mathbf{J} = \mathbf{J}_{\text{cond}} + \mathbf{J}_{\text{pol}} + \mathbf{J}_{\text{mag}}$$

where the conduction current density $\mathbf{J}_{cond}$ is due to the motion of the free charges, the polarization current density $\mathbf{J}_{pol}$ is due to the motion of bound charges in a dielectric material, and the magnetization current density $\mathbf{J}_{mag}$ is due to the magnetic moments in a magnetic material. In a plasma in vacuum, $\mathbf{J}_{pol}$ and $\mathbf{J}_{mag}$ are zero and $\mathbf{J} = \mathbf{J}_{cond}$.

If (2.2.3) is integrated over a volume $\mathcal{V}$, enclosed by a surface $\mathcal{S}$, then we obtain its integral form, Gauss' law:

$$\epsilon_0 \oint_{\mathcal{S}} \mathbf{E} \cdot d\mathbf{A} = q \tag{2.2.6}$$

where q is the total charge inside the volume. Similarly, integrating (2.2.5), we obtain

$$\frac{dq}{dt} + \oint_{\mathcal{S}} \mathbf{J} \cdot d\mathbf{A} = 0$$

which states that the rate of increase of charge inside $\mathcal{V}$ is supplied by the total current flowing across $\mathcal{S}$ into $\mathcal{V}$, i.e., that charge is conserved.

In (2.2.2), the first term on the RHS is the displacement current density flowing in the vacuum, and the second term is the conduction current density due to the free charges. We can introduce the total current density

$$\mathbf{J}_T = \epsilon_0 \frac{\partial \mathbf{E}}{\partial t} + \mathbf{J} \tag{2.2.7}$$

and taking the divergence of (2.2.2), we see that

$$\nabla \cdot \mathbf{J}_T = 0 \tag{2.2.8}$$

In one dimension, this reduces to $dJ_{Tx}/dx = 0$, such that $J_{Tx} = J_{Tx}(t)$, independent of x. Hence, for example, the total current flowing across a spatially nonuniform one-dimensional discharge is independent of x, as illustrated in Fig. 2.1. A generalization

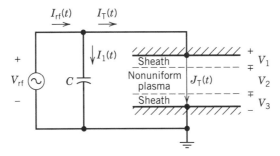

FIGURE 2.1. Kirchhoff's circuit laws: The total current J_T flowing across a nonuniform one-dimensional discharge is independent of x; the sum of the currents entering a node is zero ($I_{rf} = I_T + I_1$); the sum of voltages around a loop is zero ($V_{rf} = V_1 + V_2 + V_3$).

of this result is *Kirchhoff's current law*, which states that the sum of the currents entering a *node*, where many current-carrying conductors meet, is zero. This is also shown in Fig. 2.1, where $I_{rf} = I_T + I_1$.

If the time variation of the magnetic field is negligible, as is often the case in plasmas, then from Maxwell's equations $\nabla \times \mathbf{E} \approx 0$. Since the curl of a gradient is zero, this implies that the electric field can be derived from the gradient of a scalar potential,

$$\mathbf{E} = -\nabla\Phi \tag{2.2.9}$$

Integrating (2.2.9) around any closed loop C gives

$$\oint_C \mathbf{E} \cdot d\ell = -\oint_C \nabla\Phi \cdot d\ell = -\oint_C d\Phi = 0 \tag{2.2.10}$$

Hence, we obtain *Kirchhoff's voltage law*, which states that the sum of the voltages around any loop is zero. This is illustrated in Fig. 2.1, for which we obtain

$$V_{rf} = V_1 + V_2 + V_3$$

i.e., the source voltage V_{rf} is equal to the sum of the voltages V_1 and V_3 across the two sheaths and the voltage V_2 across the bulk plasma. Note that currents and voltages can have positive or negative values; the directions for which their values are designated as positive must be specified, as shown in the figure.

If (2.2.9) is substituted in (2.2.3), we obtain

$$\nabla^2\Phi = -\frac{\rho}{\epsilon_0} \tag{2.2.11}$$

Equation (2.2.11), *Poisson's equation,* is one of the fundamental equations that we shall use. As an example of its application, consider the potential in the center ($x = 0$) of two grounded ($\Phi = 0$) plates separated by a distance $l = 10$ cm and containing a uniform ion density $n_i = 10^{10}$ cm^{-3}, without the presence of neutralizing electrons. Integrating Poisson's equation

$$\frac{d^2\Phi}{dx^2} = -\frac{en_i}{\epsilon_0}$$

using the boundary conditions that $\Phi = 0$ at $x = \pm l/2$ and that $d\Phi/dx = 0$ at $x = 0$ (by symmetry), we obtain

$$\Phi = \frac{1}{2}\frac{en_i}{\epsilon_0}\left[\left(\frac{l}{2}\right)^2 - x^2\right]$$

The maximum potential in the center is 2.3×10^5 V, which is impossibly large for a real discharge. Hence, the ions must be mostly neutralized by electrons, leading to a quasineutral plasma.

Figure 2.2 shows a PIC simulation time history over 10^{-10} s of (a) the v_x–x phase space, (b) the number $\mathcal{N}$ of sheets versus time, and (c) the potential Φ versus x for 100 unneutralized ion sheets (with e/M for argon ions). We see the ion acceleration in (a), the loss of ions in (b), and the parabolic potential profile in (c); the maximum potential decreases as ions are lost from the system. We consider quasineutrality further in Section 2.4.

Electric and magnetic fields exert forces on charged particles given by the *Lorentz force law*:

$$\mathbf{F} = q(\mathbf{E} + \mathbf{v} \times \mathbf{B}) \tag{2.2.12}$$

where $\mathbf{v}$ is the particle velocity and $\mathbf{B} = \mu_0\mathbf{H}$ is the *magnetic induction vector*. The charged particles move under the action of the Lorentz force. The moving charges in turn contribute to both ρ and $\mathbf{J}$ in the plasma. If ρ and $\mathbf{J}$ are linearly related to $\mathbf{E}$ and $\mathbf{B}$, then the field equations are linear. As we shall see, this is not generally the case for a plasma. Nevertheless, linearization may be possible in some cases for which the plasma may be considered to have an *effective dielectric constant;* that is, the "free charges" play the same role as "bound charges" in a dielectric. We consider this further in Chapter 4.

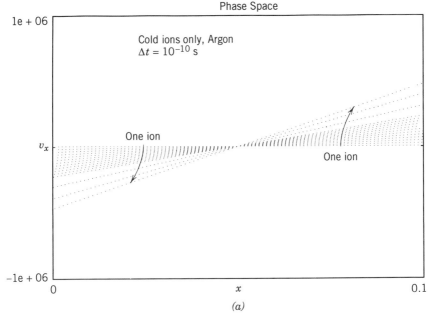

FIGURE 2.2. PIC simulation of ion loss in a plasma containing ions only: (a) v_x–x ion phase space, showing the ion acceleration trajectories.

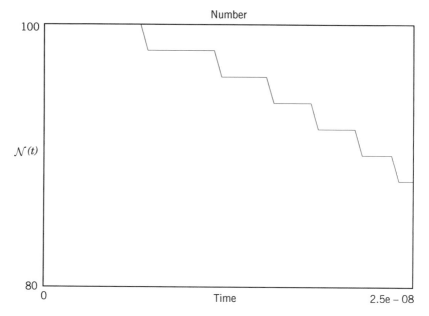

FIGURE 2.2. (*continued*) PIC simulation of ion loss in a plasma containing ions only: (b) number $\mathcal{N}$ of ion sheets versus t, with the steps indicating the loss of a single sheet.

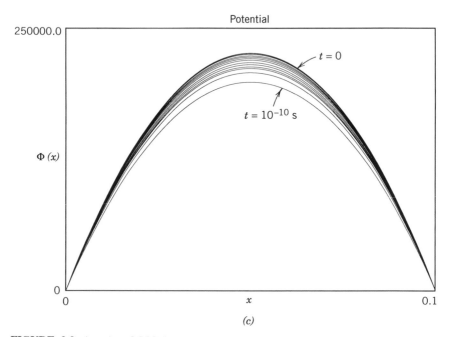

FIGURE 2.2. (*continued*) PIC simulation of ion loss in a plasma containing ions only: (c) the potential Φ versus x during the first 10^{-10} s of ion loss.

2.3 THE CONSERVATION EQUATIONS

Boltzmann's Equation

For a given species, we introduce a *distribution function* $f(\mathbf{r}, \mathbf{v}, t)$ in the six-dimensional *phase space* $(\mathbf{r}, \mathbf{v})$ of particle positions and velocities, with the interpretation that

$$f(\mathbf{r}, \mathbf{v}, t)\, d^3 r \, d^3 v = \text{number of particles inside a six-dimensional phase}$$
$$\text{space volume } d^3 r \, d^3 v \text{ at } (\mathbf{r}, \mathbf{v}) \text{ at time } t$$

The six coordinates $(\mathbf{r}, \mathbf{v})$ are considered to be independent variables. We illustrate the definition of f and its phase space in one dimension in Fig. 2.3. As particles drift in phase space or move under the action of macroscopic forces, they flow into or out of the fixed volume $dx\, dv_x$. Hence the distribution function f should obey a continuity equation which can be derived as follows. In a time dt,

$f(x, v_x, t)\, dx\, a_x(x, v_x, t)\, dt$ particles flow into $dx\, dv_x$ across face 1

$f(x, v_x + dv_x, t)\, dx\, a_x(x, v_x + dv_x, t)\, dt$ particles flow out of $dx\, dv_x$ across face 2

$f(x, v_x, t)\, dv_x\, v_x\, dt$ particles flow into $dx\, dv_x$ across face 3

$f(x + dx, v_x, t)\, dv_x\, v_x\, dt$ particles flow out of $dx\, dv_x$ across face 4

where $a_x \equiv dv_x/dt$ and $v_x \equiv dx/dt$ are the flow velocities in the v_x and x directions, respectively. Hence

$$f(x, v_x, t + dt)\, dx\, dv_x - f(x, v_x, t)\, dx\, dv_x$$
$$= \left[f(x, v_x, t)a_x(x, v_x, t) - f(x, v_x + dv_x, t)a_x(x, v_x + dv_x, t) \right] dx\, dt$$
$$+ \left[f(x, v_x, t)v_x - f(x + dx, v_x, t)v_x \right] dv_x\, dt$$

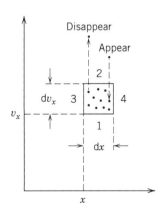

FIGURE 2.3. One-dimensional v_x–x phase space, illustrating the derivation of the Boltzmann equation and the change in f due to collisions.

Dividing by $dx\, dv_x\, dt$, we obtain

$$\frac{\partial f}{\partial t} = -\frac{\partial}{\partial x}(fv_x) - \frac{\partial}{\partial v_x}(fa_x) \tag{2.3.1}$$

Noting that v_x is independent of x and assuming that the acceleration $a_x = F_x/m$ of the particles does not depend on v_x, then (2.3.1) can be rewritten:

$$\frac{\partial f}{\partial t} + v_x\frac{\partial f}{\partial x} + a_x\frac{\partial f}{\partial v_x} = 0$$

The three-dimensional generalization,

$$\frac{\partial f}{\partial t} + \mathbf{v}\cdot\nabla_r f + \mathbf{a}\cdot\nabla_v f = 0 \tag{2.3.2}$$

with $\nabla_r = (\hat{x}\,\partial/\partial x + \hat{y}\,\partial/\partial y + \hat{z}\,\partial/\partial z)$ and $\nabla_v = (\hat{x}\,\partial/\partial v_x + \hat{y}\,\partial/\partial v_y + \hat{z}\,\partial/\partial v_z)$ is called the *collisionless Boltzmann equation* or *Vlasov equation*.

In addition to flows into or out of the volume across the faces, particles can "suddenly" appear in or disappear from the volume due to very short time scale interparticle collisions, which are assumed to occur on a time scale shorter than the evolution time of f in (2.3.2). Such collisions can practically instantaneously change the velocity (but not the position) of a particle. Examples of particles suddenly appearing or disappearing are shown in Fig. 2.3. We account for this effect, which changes f, by adding a "collision term" to the right-hand side of (2.3.2), thus obtaining the *Boltzmann equation*:

$$\frac{\partial f}{\partial t} + \mathbf{v}\cdot\nabla_r f + \frac{\mathbf{F}}{m}\cdot\nabla_v f = \frac{\partial f}{\partial t}\bigg|_c \tag{2.3.3}$$

The collision term in integral form will be derived in Appendix B. The preceding heuristic derivation of the Boltzmann equation can be made rigorous from various points of view, and the interested reader is referred to texts on plasma theory, such as Holt and Haskel (1965).

Macroscopic Quantities

The complexity of the dynamical equations is greatly reduced by averaging over the velocity coordinates of the distribution function to obtain equations depending on the spatial coordinates and the time only. The averaged quantities, such as species density, mean velocity, and energy density are called macroscopic quantities, and the equations describing them are the macroscopic conservation equations. To obtain these averaged quantities we take *velocity moments* of the distribution function, and the equations are obtained from the moments of the Boltzmann equation.

The average quantities that we are concerned with are the particle density,

$$n(\mathbf{r}, t) = \int f \, d^3 v \qquad (2.3.4)$$

the particle flux

$$\Gamma(\mathbf{r}, t) = n\mathbf{u} = \int \mathbf{v} f \, d^3 v \qquad (2.3.5)$$

where $\mathbf{u}(\mathbf{r}, t)$ is the mean velocity, and the particle kinetic energy per unit volume

$$w = \frac{3}{2}p + \frac{1}{2}mu^2 n = \frac{1}{2}m \int v^2 f \, d^3 v \qquad (2.3.6)$$

where $p(\mathbf{r}, t)$ is the isotropic pressure, which we define below. In this form, w is sum of the *internal* energy density $\frac{3}{2}p$ and the *flow* energy density $\frac{1}{2}mu^2 n$.

Particle Conservation

The lowest moment of the Boltzmann equation is obtained by integrating all terms of (2.3.3) over velocity space. The integration yields the macroscopic *continuity equation*:

$$\frac{\partial n}{\partial t} + \nabla \cdot (n\mathbf{u}) = G - L \qquad (2.3.7)$$

The collision term in (2.3.3), which yields the right-hand side of (2.3.7), is equal to zero when integrated over velocities, except for collisions that create or destroy particles, designated as G and L, respectively (e.g., ionization, recombination). In fact, (2.3.7) is transparent since it physically describes the conservation of particles. If (2.3.7) is integrated over a volume $\mathcal{V}$ bounded by a closed surface $\mathcal{S}$, then (2.3.7) states that the net number of particles per second generated within $\mathcal{V}$ either flows across the surface $\mathcal{S}$ or increases the number of particles within $\mathcal{V}$. For common low-pressure discharges in the steady state, G is usually due to ionization by electron–neutral collisions:

$$G = \nu_{iz} n_e$$

where ν_{iz} is the ionization frequency. The volume loss rate L, usually due to recombination, is often negligible. Hence

$$\nabla \cdot (n\mathbf{u}) = \nu_{iz} n_e \qquad (2.3.8)$$

in a typical discharge. However, note that the continuity equation is clearly not sufficient to give the evolution of the density n, since it involves another quantity, the mean particle velocity $\mathbf{u}$.

Momentum Conservation

To obtain an equation for $\mathbf{u}$, a first moment is formed by multiplying the Boltzmann equation by $\mathbf{v}$ and integrating over velocity. The details are complicated and involve evaluation of tensor elements. The calculation can be found in most plasma theory texts, e.g., Krall and Trivelpiece (1973). The result is

$$mn\left[\frac{\partial \mathbf{u}}{\partial t} + (\mathbf{u} \cdot \nabla)\mathbf{u}\right] = qn(\mathbf{E} + \mathbf{u} \times \mathbf{B}) - \nabla \cdot \mathbf{\Pi} + \mathbf{f}\Big|_c \qquad (2.3.9)$$

The left-hand side is the species mass density times the convective derivative of the mean velocity, representing the mass density times the acceleration. The convective derivative has two terms: the first term $\partial \mathbf{u}/\partial t$ represents an acceleration due to an explicitly time-varying $\mathbf{u}$; the second "inertial" term $(\mathbf{u} \cdot \nabla)\mathbf{u}$ represents an acceleration even for a steady fluid flow $(\partial/\partial t \equiv 0)$ having a spatially varying $\mathbf{u}$. For example, if $\mathbf{u} = \hat{x}\, u_x(x)$ increases along x, then the fluid is accelerating along x (Problem 2.3). This second term is nonlinear in $\mathbf{u}$ and can often be neglected in discharge analysis.

The mass times acceleration is acted upon, on the right-hand side, by the body forces, with the first term being the electric and magnetic force densities. The second term is the force density due to the divergence of the pressure tensor, which arises due to the integration over velocities

$$\Pi_{ij} = mn\langle(v_i - u)(v_j - u)\rangle_v \qquad (2.3.10)$$

where the subscripts i, j give the component directions and $\langle \cdot \rangle_v$ denotes the velocity average of the bracketed quantity over f^*. For weakly ionized plasmas it is almost never used in this form, but rather an isotropic version is employed:

$$\mathbf{\Pi} = \begin{pmatrix} p & 0 & 0 \\ 0 & p & 0 \\ 0 & 0 & p \end{pmatrix} \qquad (2.3.11)$$

such that

$$\nabla \cdot \mathbf{\Pi} = \nabla p \qquad (2.3.12)$$

a pressure gradient, with

$$p = \frac{1}{3}mn\langle(v - u)^2\rangle_v \qquad (2.3.13)$$

being the scalar pressure. Physically, the pressure gradient force density arises as illustrated in Fig. 2.4, which shows a small volume acted upon by a pressure that is an increasing function of x. The net force on this volume is $p(x)\,dA - p(x + dx)\,dA$ and the volume is $dA\,dx$. Hence the force per unit volume is $-\partial p/\partial x$.

---

*We assume f is normalized so that $\langle f \rangle_v = 1$.

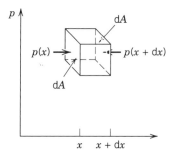

FIGURE 2.4. The force density due to the pressure gradient.

The third term on the right in (2.3.9) represents the time rate of momentum transfer per unit volume due to collisions with other species. For electrons or positive ions the most important transfer is often due to collisions with neutrals. The transfer is usually approximated by a Krook collision operator

$$\mathbf{f}\Big|_c = -\sum_\beta mn\nu_{m\beta}(\mathbf{u} - \mathbf{u}_\beta) - m\mathbf{u}(G - L) \qquad (2.3.14)$$

where the summation is over all other species, with $\mathbf{u}_\beta$ the mean velocity of species β and $\nu_{m\beta}$, the momentum transfer frequency for collisions with species β. We discuss the Krook form of the collision operator in Chapter 3 and Appendix B. The last term in (2.3.14) is generally small and gives the momentum transfer due to the creation or destruction of particles. For example, if ions are created at rest, then they exert a drag force on the moving ion fluid because they act to lower the average fluid velocity.

A common form of the average force (momentum conservation) equation is obtained from (2.3.9) for slow time variation, neglecting the inertial ($\mathbf{u} \cdot \nabla\mathbf{u}$) and magnetic forces, and taking $\mathbf{u}_\beta = 0$ in the Krook collision term for collisions with one neutral species. The result is

$$0 = qn\mathbf{E} - \nabla p - mn\nu_m\mathbf{u} \qquad (2.3.15)$$

where only the electric field, pressure gradient, and frictional forces appear. However, for fast time variation, we will also consider the acceleration term $mn\,\partial\mathbf{u}/\partial t$ on the left-hand side of (2.3.9).

Equations (2.3.7) and (2.3.9) together still do not form a closed set, since the pressure tensor Π (or scalar pressure p) is not determined. The usual procedure to close the equations is to use a thermodynamic *equation of state* to relate p to n. The *isothermal* relation for an equilibrium Maxwellian distribution is

$$p = nkT \qquad (2.3.16)$$

so that

$$\nabla p = kT\nabla n \qquad (2.3.17)$$

where T is the temperature in kelvins and k is Boltzmann's constant ($k = 1.381 \times 10^{-23}$ J/K). This holds for slow time variations, where temperatures are allowed to equilibrate. In this case, the fluid can exchange energy with its surroundings, and we also require an energy conservation equation (see below) to determine p and T. For a room temperature (297 K) neutral gas having density n_g and pressure p, (2.3.16) yields

$$n_g(\text{cm}^{-3}) \approx 3.250 \times 10^{16} \, p \, (\text{Torr}) \tag{2.3.18}$$

Alternatively the *adiabatic* equation of state is

$$p = Cn^\gamma \tag{2.3.19}$$

such that

$$\frac{\nabla p}{p} = \gamma \frac{\nabla n}{n} \tag{2.3.20}$$

where γ is the ratio of specific heat at constant pressure to that at constant volume. The specific heats are defined in Section 7.2; $\gamma = 5/3$ for a perfect gas; for one-dimensional adiabatic motion, $\gamma = 3$. The adiabatic relation holds for fast time variations, such as in waves, when the fluid does not exchange energy with its surroundings; hence an energy conservation equation is not required. For almost all applications to discharge analysis, we use the isothermal equation of state.

Energy Conservation

The energy conservation equation is obtained by multiplying the Boltzmann equation by $\frac{1}{2}mv^2$ and integrating over velocity. The integration and some other manipulation yield

$$\frac{\partial}{\partial t}\left(\frac{3}{2}p\right) + \nabla \cdot \frac{3}{2}(p\mathbf{u}) + p\nabla \cdot \mathbf{u} + \nabla \cdot \mathbf{q} = \frac{\partial}{\partial t}\left(\frac{3}{2}p\right)\bigg|_c \tag{2.3.21}$$

Here $\frac{3}{2}p$ is the energy density (J/m^3), $\frac{3}{2}p\mathbf{u}$ is the macroscopic energy flux (W/m^2), representing the flow of the internal energy density at the fluid velocity $\mathbf{u}$, $p\nabla \cdot \mathbf{u}$ (W/m^3) gives the heating or cooling of the fluid due to compression or expansion of its volume (Problem 2.4), $\mathbf{q}$ is the heat flow vector (W/m^2), which gives the microscopic energy flux, and the collisional term includes all collisional processes that change the energy density. These include ionization, excitation, elastic scattering, and frictional (ohmic) heating. The equation is usually closed by setting $\nabla \cdot \mathbf{q} = 0$ or by letting $\mathbf{q} = -\kappa_T \nabla T$, where κ_T is the thermal conductivity. For most steady-state discharges the macroscopic energy flux is balanced against the collisional processes, giving the simpler equation

$$\nabla \cdot \left(\frac{3}{2}p\mathbf{u}\right) = \frac{\partial}{\partial t}\left(\frac{3}{2}p\right)\Big|_{c} \qquad (2.3.22)$$

Equation (2.3.22), together with the continuity equation (2.3.8), will often prove sufficient for our analysis.

Summary

Summarizing our results for the macroscopic equations describing the electron and ion fluids, we have in their most usually used forms the continuity equation

$$\nabla \cdot (n\mathbf{u}) = \nu_{iz}n_e \qquad (2.3.8)$$

the force equation, with the acceleration term on the left-hand side,

$$mn\frac{d\mathbf{u}}{dt} = qn\mathbf{E} - \nabla p - mn\nu_m\mathbf{u} \qquad (2.3.15)$$

the isothermal equation of state

$$p = nkT \qquad (2.3.16)$$

and the energy-conservation equation

$$\nabla \cdot \left(\frac{3}{2}p\mathbf{u}\right) = \frac{\partial}{\partial t}\left(\frac{3}{2}p\right)\Big|_{c} \qquad (2.3.22)$$

These equations hold for each charged species, with the total charges and currents summed in Maxwell's equations. For example, with electrons and one positive ion species with charge Ze, we have

$$\rho = e\,(Zn_i - n_e) \qquad (2.3.23)$$
$$\mathbf{J} = e\,(Zn_i\mathbf{u}_i - n_e\mathbf{u}_e) \qquad (2.3.24)$$

These equations are still very difficult to solve without simplifications. They consist of 18 unknown quantities n_i, n_e, p_i, p_e, T_i, T_e, $\mathbf{u}_i$, $\mathbf{u}_e$, $\mathbf{E}$, and $\mathbf{B}$, with the vectors each counting for three. Various simplifications used to make the solutions to the equations tractable will be employed as the individual problems allow.

2.4 EQUILIBRIUM PROPERTIES

Electrons are generally in near-thermal equilibrium at temperature T_e in discharges, whereas positive ions are *almost never* in thermal equilibrium. Neutral gas molecules may or may not be in thermal equilibrium, depending on the generation and loss processes. For a single species in thermal equilibrium with itself (e.g., electrons), in the absence of time variation, spatial gradients, and accelerations, the Boltzmann equation (2.3.3) reduces to

$$\frac{\partial f}{\partial t}\bigg|_c = 0 \tag{2.4.1}$$

where the subscript c here represents the collisions of a particle species with itself. We show in Appendix B that the solution of (2.4.1) has a Gaussian speed distribution of the form

$$f(v) = Ce^{-\xi^2 m v^2} \tag{2.4.2}$$

The two constants C and ξ can be obtained by using the thermodynamic relation

$$w = \frac{1}{2}mn\langle v^2 \rangle_v = \frac{3}{2}nkT \tag{2.4.3}$$

i.e., that the average energy of a particle is $\frac{1}{2}kT$ per translational degree of freedom, and by using a suitable normalization of the distribution. Normalizing $f(v)$ to n, we obtain

$$C \int_0^{2\pi} d\phi \int_0^{\pi} \sin\theta \, d\theta \int_0^{\infty} \exp\left(-\xi^2 m v^2\right) v^2 \, dv = n \tag{2.4.4}$$

and using (2.4.3), we obtain

$$\frac{1}{2}mC \int_0^{2\pi} d\phi \int_0^{\pi} \sin\theta \, d\theta \int_0^{\infty} \exp\left(-\xi^2 m v^2\right) v^4 dv = \frac{3}{2}nkT \tag{2.4.5}$$

where we have written the integrals over velocity space in spherical coordinates. The angle integrals yield the factor 4π. The v integrals are evaluated using the relation[*]

$$\int_0^{\infty} e^{-u^2} u^{2i} du = \frac{(2i-1)!!}{2^{i+1}}\sqrt{\pi},$$

where i is an integer ≥ 1.   (2.4.6)

Solving for C and ξ we have

$$f(v) = n\left(\frac{m}{2\pi kT}\right)^{3/2} \exp\left(-\frac{mv^2}{2kT}\right) \tag{2.4.7}$$

which is the *Maxwellian distribution*.

---

[*]!! denotes the double factorial function; e.g., $7!! = 7 \times 5 \times 3 \times 1$.

Similarly, other averages can be performed. The average speed $\bar{v}$ is given by

$$\bar{v} = \left(m/2\pi kT\right)^{3/2} \int_0^\infty v \left[\exp\left(-\frac{v^2}{2v_{\text{th}}^2}\right)\right] 4\pi v^2 \, dv \tag{2.4.8}$$

where $v_{\text{th}} = (kT/m)^{1/2}$ is the thermal velocity. We obtain

$$\bar{v} = \left(\frac{8kT}{\pi m}\right)^{1/2} \tag{2.4.9}$$

The directed flux Γ_z in (say) the $+z$ direction is given by $n\langle v_z\rangle_{\text{v}}$, where the average is taken over $v_z > 0$ only. Writing $v_z = v\cos\theta$ we have in spherical coordinates

$$\Gamma_z = n\left(\frac{m}{2\pi kT}\right)^{3/2} \int_0^{2\pi} d\phi \int_0^{\pi/2} \sin\theta \, d\theta \int_0^\infty v\cos\theta \exp\left(-\frac{v^2}{2v_{\text{th}}^2}\right) v^2 \, dv$$

Evaluating the integrals, we find

$$\Gamma_z = \frac{1}{4}n\bar{v} \tag{2.4.10}$$

Γ_z is the number of particles per square meter per second crossing the $z = 0$ surface in the positive direction. Similarly, the average energy flux $S_z = n\langle\frac{1}{2}mv^2 v_z\rangle_{\text{v}}$ in the $+z$ direction can be found: $S_z = 2kT\Gamma_z$. We see that the average kinetic energy W per particle crossing $z = 0$ in the positive direction is

$$W = 2kT \tag{2.4.11}$$

It is sometimes convenient to define the distribution in terms of other variables. For example, we can define a distribution of energies $W = \frac{1}{2}mv^2$ by

$$4\pi g\left(W\right) dW = 4\pi f\left(v\right) v^2 dv$$

Evaluating dv/dW, we see that g and f are related by

$$g\left(W\right) = \frac{v\left(W\right) f\left[v\left(W\right)\right]}{m} \tag{2.4.12}$$

where $v(W) = (2W/m)^{1/2}$.

Boltzmann's Relation

A very important relation can be obtained for the density of electrons in thermal equilibrium at varying positions in a plasma under the action of a spatially varying potential. In the absence of electron drifts ($\mathbf{u}_e \equiv 0$), the inertial, magnetic, and frictional forces are zero, and the electron force balance is, from (2.3.15),

$$en_e\mathbf{E} + \nabla p_e = 0 \tag{2.4.13}$$

Setting $\mathbf{E} = -\nabla\Phi$ and assuming $p_e = n_e kT_e$, (2.4.13) becomes

$$-en_e\nabla\Phi + kT_e\nabla n_e = 0$$

or, rearranging,

$$\nabla(e\Phi - kT_e \ln n_e) = 0 \tag{2.4.14}$$

Integrating, we have

$$e\Phi - kT_e \ln n_e = \text{const}$$

or

$$n_e(\mathbf{r}) = n_0 e^{e\Phi(\mathbf{r})/kT_e} \tag{2.4.15}$$

which is *Boltzmann's relation* for electrons. We see that electrons are "attracted" to regions of positive potential. We shall generally write Boltzmann's relation in more convenient units

$$n_e = n_0 e^{\Phi/T_e} \tag{2.4.16}$$

where T_e is now expressed in volts, as is Φ.

For positive ions in thermal equilibrium at temperature T_i, a similar analysis shows that

$$n_i = n_0 e^{-\Phi/T_i} \tag{2.4.17}$$

Hence positive ions in thermal equilibrium are "repelled" from regions of positive potential. However, positive ions are almost never in thermal equilibrium in low-pressure discharges because the ion drift velocity $\mathbf{u}_i$ is large, leading to inertial or frictional forces in (2.3.15) that are comparable to the electric field or pressure gradient forces.

Debye Length

The characteristic length scale in a plasma is the electron Debye length λ_{De}. As we will show, the Debye length is the distance scale over which significant charge densities can spontaneously exist. For example, low-voltage (undriven) sheaths are typically a few Debye lengths wide. To determine the Debye length, let us introduce a sheet of negative charge having surface charge density $\rho_S < 0$ C/m^2 into an infinitely extended plasma having equilibrium densities $n_e = n_i = n_0$. For simplicity we assume immobile ions, such that $n_i = n_0$ after the sheet is introduced. However, the negative sheet "repels" nearby electrons, leading to a reduced electron density near the sheet. The situation after introduction of the sheet is shown in Fig. 2.5. To determine the potential and density variation, we use Poisson's equation, which in one dimension can be written

$$\frac{d^2\Phi}{dx^2} = -\frac{e}{\epsilon_0}(n_i - n_e) \tag{2.4.18}$$

Setting $n_e = n_0 \exp(\Phi/T_e)$, from the Boltzmann relation (2.4.16), and taking $n_i = n_0$, Poisson's equation becomes

$$\frac{d^2\Phi}{dx^2} = \frac{e n_0}{\epsilon_0}\left(e^{\Phi/T_e} - 1\right) \tag{2.4.19}$$

Expanding $\exp(\Phi/T_e)$ in a Taylor series for $\Phi \ll T_e$, (2.4.19) becomes, to lowest order in Φ/T_e,

$$\frac{d^2\Phi}{dx^2} = \frac{e n_0}{\epsilon_0}\frac{\Phi}{T_e} \tag{2.4.20}$$

The symmetric solution of (2.4.20) that vanishes at $x = \pm\infty$ is

$$\Phi = \Phi_0 e^{-|x|/\lambda_{De}} \tag{2.4.21}$$

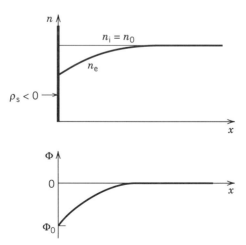

FIGURE 2.5. Calculation of the electron Debye length λ_{De}. A negatively charged sheet is introduced into a plasma containing electrons in thermal equilibrium.

where

$$\lambda_{De} = \left(\frac{\epsilon_0 T_e}{e n_0}\right)^{1/2} \tag{2.4.22}$$

In practical units, we find

$$\lambda_{De} \text{ (cm)} \approx 743\sqrt{T_e/n_e} \tag{2.4.23}$$

with T_e in volts and n_e in cm^{-3}. We find for $T_e = 4$ V and $n_e = 10^{10}$ cm^{-3} that $\lambda_{De} = 0.14$ mm. It is on space scales larger than a Debye length that the plasma will tend to remain neutral.

The Debye length is useful in many contexts. In the next chapter we shall see that it serves as a characteristic scale length to shield the Coulomb potentials of individual charged particles when they collide. Although we have calculated the above effect for electron shielding, it is also possible on slower time scales for the ions to contribute. We leave the calculation for a problem.

Quasineutrality

The potential variation across a plasma of length $l \gg \lambda_{De}$ can be estimated from Poisson's equation (2.2.11):

$$\nabla^2 \Phi \sim \frac{\Phi}{l^2} \sim \left| \frac{e}{\epsilon_0} (Zn_i - n_e) \right| \tag{2.4.24}$$

We generally expect that

$$\Phi \lesssim T_e = \frac{e}{\epsilon_0} n_e \lambda_{De}^2 \tag{2.4.25}$$

where the equality on the right follows from the definition of λ_{De}. Combining (2.4.24) and (2.4.25) we have

$$\frac{|Zn_i - n_e|}{n_e} \lesssim \frac{\lambda_{De}^2}{l^2} \tag{2.4.26}$$

For $\lambda_{De}^2/l^2 \ll 1$, (2.4.26) implies that

$$|Zn_i - n_e| \ll n_e \tag{2.4.27}$$

such that we can set

$$Zn_i = n_e \tag{2.4.28}$$

except when used in Poisson's equation. Relation (2.4.27) is the basic statement of *quasineutrality* of a plasma, and is often called the *plasma approximation*. We shall see in Chapter 6 that the plasma approximation is violated within a plasma sheath, in proximity to a material wall, either because the sheath thickness $s \approx \lambda_{De}$, or because $\Phi \gg T_e$.

PROBLEMS

2.1. Charge Conservation   Derive the conservation of charge law (2.2.5) from Maxwell's equations.

2.2. Homogeneous Discharge Model   A plasma is confined between two grounded ($\Phi = 0$) parallel plates located at $x = 0$ and $x = l$. The ion density is $n_i(x) = n_0$ for $0 < x < l$. The electron density is $n_e(x) = n_0$ for $s < x < l - s$ and is $n_e(x) = 0$ in the "sheath" regions $0 < x < s$ and $l - s < x < l$.

  (a) Solve Poisson's equation to determine the potential $\Phi(x)$ everywhere within the discharge $0 < x < l$. Find $\Phi_0 = \Phi(l/2)$ in the center of the discharge. Plot $\Phi(x)$ versus x for $0 < x < l$ for $s = l/8$.

  (b) Plot the electric field E_x versus x and show that it acts to confine electrons within the bulk plasma at both sheaths.

  (c) Choosing $\Phi_0 = 4T_e$, find an expression for s and show that s is of the order of an electron Debye length.

2.3. Bernoulli's Law   Starting from the force equation (2.3.9), derive Bernoulli's law for an incompressible fluid in steady one-dimensional flow:

$$\frac{1}{2}mnu^2(x) + p(x) = \text{const}$$

How would you use this effect to measure the change in the velocity of a fluid as it flows through a constriction in a pipe?

2.4. Compressional Heating of a Fluid   Show using a one-dimensional calculation that the relative rate of change with time of a small volume ΔV moving with the fluid velocity $\mathbf{u}$ can be written as

$$\frac{1}{\Delta V}\frac{d(\Delta V)}{dt} = \nabla \cdot \mathbf{u}$$

Hence, show from (2.3.21) that if the fluid expands, its internal energy decreases.

2.5. Averages Over a Maxwellian Distribution

  (a) Show by integrating (2.4.8) that the average speed of electrons in a Maxwellian distribution is $\bar{v}_e = (8eT_e/\pi m)^{1/2}$.

(b) Show by integrating the equation above (2.4.10) that the average one-way particle flux is $\Gamma_e = n_e \bar{v}_e/4$.

(c) Find the average one-way energy flux S_e by integrating the energy flux over a Maxwellian distribution. Comparing S_e to Γ_e, show that (2.4.11) holds, i.e., the average kinetic energy per particle crossing a surface is $W_e = 2kT_e$.

2.6. Debye Length   A conducting sphere of radius a is immersed in an infinite uniform plasma having density n_0, electrons in thermal equilibrium at temperature T_e, and ions in thermal equilibrium at temperature T_i. A small dc voltage $V_0 \ll T_e, T_i$ is applied to the sphere.

(a) Starting from Poisson's equation in spherical coordinates and using Boltzmann's relation for both the electrons at temperature T_e and the positively charged ions at temperature T_i, derive an expression for the potential $\Phi(r)$ everywhere in the plasma.

(b) Find an expression for the Debye length, and show that the Debye length is determined mainly by the temperature of the colder species.

Hints: Note that for spherical symmetry, $\nabla^2 \Phi = (1/r)d^2(r\Phi)/dr^2$. Also note that in a typical discharge, the ions are not in thermal equilibrium. Thus, even though $T_i \ll T_e$ in a discharge, the effective Debye length is usually determined by the electrons alone.

CHAPTER 3

ATOMIC COLLISIONS

3.1 BASIC CONCEPTS

When two particles collide, various phenomena may occur. As examples, one or both particles may change their momentum or their energy, neutral particles can become ionized, and ionized particles can become neutral. We introduce the fundamentals of collisions between electrons, positive ions, and gas atoms in this chapter, concentrating on simple classical estimates of the important processes in noble gas discharges such as argon. For electrons colliding with atoms, the main processes are elastic scattering in which primarily the electron momentum is changed, and inelastic processes such as excitation and ionization. For ions colliding with atoms, the main processes are elastic scattering in which momentum and energy are exchanged, and resonant charge transfer. Other important processes occur in molecular gases. These include dissociation, dissociative recombination, processes involving negative ions, such as attachment, detachment, and positive–negative ion charge transfer, and processes involving excitation of molecular vibrations and rotations. We defer consideration of collisions in molecular gases to Chapter 8.

Elastic and Inelastic Collisions

Collisions conserve momentum and energy: the total momentum and energy of the colliding particles after collision are equal to that before collision. Electrons

45

and fully stripped ions possess only kinetic energy. Atoms and partially stripped ions have internal energy level structures and can be excited, deexcited, or ionized, corresponding to changes in potential energy. It is the total energy, which is the sum of the kinetic and potential energy, that is conserved in a collision.

If the internal energies of the collision partners do not change, then the sum of kinetic energies is conserved and the collision is said to be *elastic*. Although the total kinetic energy is conserved, kinetic energy is generally exchanged between particles. If the sum of kinetic energies is not conserved, then the collision is *inelastic*. Most inelastic collisions involve excitation or ionization, such that the sum of kinetic energies after collision is less than that before collision. However, *superelastic* collisions can occur in which an excited atom can be deexcited by a collision, increasing the sum of kinetic energies.

Collision Parameters

The fundamental quantity that characterizes a collision is its *cross section* $\sigma(v_R)$, where v_R is the relative velocity between the particles before collision. To define this, we consider first the simplest situation shown in Fig. 3.1, in which a flux $\Gamma = nv$ of particles having mass m, density n, and fixed velocity v is incident on a half-space $x > 0$ of stationary, infinitely massive "target" particles having density n_g. In this case, $v_R = v$. Let dn be the number of incident particles per unit volume at x that undergo an "interaction" with the target particles within a differential distance dx, removing them from the incident beam. Clearly, dn is proportional to n, n_g, and dx for infrequent collisions within dx. Hence we can write

$$dn = -\sigma n n_g \, dx \qquad (3.1.1)$$

where the constant of proportionality σ that has been introduced has units of area and is called the cross section for the interaction. The minus sign denotes *removal* from the beam. To define a cross section, the "interaction" must be specified, e.g., ionization of the target particle, excitation of the incident particle to a given energy state, or scattering of the incident particle by an angle exceeding $\pi/2$. Multiplying

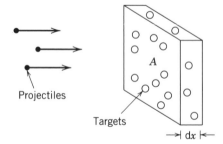

FIGURE 3.1. A flux of incident particles collides with a population of target particles in the half-space $x > 0$.

(3.1.1) by v, we find a similar equation for the flux:

$$d\Gamma = -\sigma \Gamma n_g \, dx \tag{3.1.2}$$

For a simple interpretation of σ, let the incident and target particles be hard elastic spheres of radii a_1 and a_2, and let the "interaction" be a collision between the spheres. In a distance dx there are $n_g \, dx$ targets within a unit area perpendicular to x. Draw a circle of radius $a_{12} = a_1 + a_2$ in the $x = \text{const}$ plane about each target. A collision occurs if the centers of the incident and target particles fall within this radius. Hence the fraction of the unit area for which a collision occurs is $n_g \, dx \, \pi a_{12}^2$. The fraction of incident particles that collide within dx is then

$$\frac{d\Gamma}{\Gamma} = \frac{dn}{n} = -n_g \sigma \, dx \tag{3.1.3}$$

where

$$\sigma = \pi a_{12}^2 \tag{3.1.4}$$

is the hard sphere cross section. In this particular case, σ is independent of v.

Equation (3.1.2) is readily integrated to give the uncollided flux

$$\Gamma(x) = \Gamma_0 e^{-n_g \sigma x} \equiv \Gamma_0 e^{-x/\lambda} \tag{3.1.5}$$

The quantity

$$\lambda = \frac{1}{n_g \sigma} \tag{3.1.6}$$

is the *mean free path* for the decay of the beam, that is, the distance over which the uncollided flux decreases by $1/e$ from its initial value Γ_0 at $x = 0$. If the velocity of the beam is v, then the mean time between interactions is

$$\tau = \frac{\lambda}{v} \tag{3.1.7}$$

Its inverse is the *interaction* or *collision frequency*

$$\nu \equiv \tau^{-1} = n_g \sigma v \tag{3.1.8}$$

and is the number of interactions per second that an incident particle has with the target particle population. We can also define the collision frequency per unit density, which is called the *rate constant*

$$K = \sigma v \tag{3.1.9}$$

and, trivially, from (3.1.8) and (3.1.9)

$$\nu = Kn_g \tag{3.1.10}$$

Differential Scattering Cross Section

Let us consider only those interactions that scatter the particles by $\theta = 90°$ or more. For hard spheres, taking the angle of incidence equal to the angle of reflection, the $90°$ collision occurs on the $\chi = 45°$ diagonal (see Fig. 3.2), therefore having a cross section

$$\sigma_{90} = \frac{\pi a_{12}^2}{2}, \tag{3.1.11}$$

which is a factor of two smaller than (3.1.4). Of course, multiple collisions at smaller angles (radii larger than $a_{12}/\sqrt{2}$) also eventually scatter incident particles through $90°$. This indeterminacy indicates that a more precise way of determining the scattering cross section is required. For this purpose we introduce a *differential scattering cross section* $I(v, \theta)$. Consider a beam of particles incident on a scattering center (again assumed fixed), as shown in Fig. 3.3. We assume that the scattering force is symmetric about the line joining the centers of the two particles. A particle incident at a distance b off-center from the target particle is scattered through an angle θ, as shown in Fig. 3.3. The quantity b is the *impact parameter* and θ is the *scattering angle* (see also Fig. 3.2). Now, flux conservation requires that for incoming flux Γ,

$$\Gamma\, 2\pi b\, db = -\Gamma I(v, \theta) 2\pi \sin\theta\, d\theta \tag{3.1.12}$$

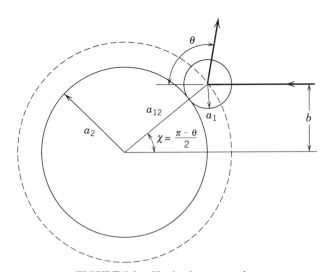

FIGURE 3.2.   Hard-sphere scattering.

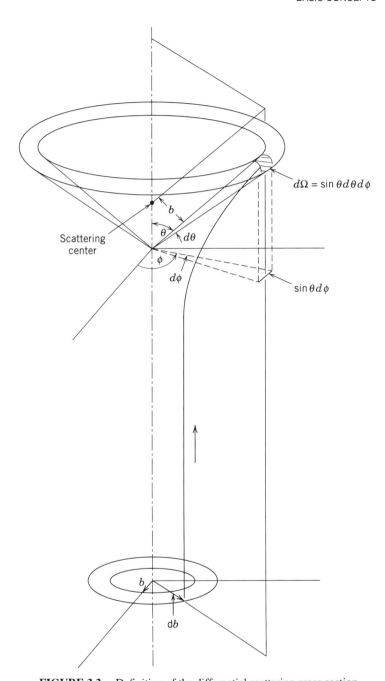

FIGURE 3.3.   Definition of the differential scattering cross section.

i.e., that all particles entering through the differential annulus $2\pi b \, db$ leave through a differential solid angle $d\Omega = 2\pi \sin \theta \, d\theta$. The minus sign is because an increase in b leads to a decrease in θ. The proportionality constant is just $I(v, \theta)$, which has the dimensions of area per steradian. From (3.1.12) we obtain

$$I(v, \theta) = \frac{b}{\sin \theta} \left| \frac{db}{d\theta} \right| \tag{3.1.13}$$

The quantity $db/d\theta$ is determined from the scattering force, and the absolute value is used since $db/d\theta$ is negative. We will calculate $I(v, \theta)$ for various potentials in Section 3.2.

We can calculate the *total scattering cross section* σ_{sc} by integrating I over the solid angle

$$\sigma_{sc} = 2\pi \int_0^\pi I(v, \theta) \sin \theta \, d\theta \tag{3.1.14}$$

It is clear that $\sigma_{sc} = \sigma$ for scattering through any angle, as defined in (3.1.2). It is often useful to define a different cross section

$$\sigma_m = 2\pi \int_0^\pi (1 - \cos \theta) I(v, \theta) \sin \theta \, d\theta \tag{3.1.15}$$

The factor $(1 - \cos \theta)$ is the fraction of the initial momentum mv lost by the incident particle, and thus (3.1.15) is the *momentum transfer cross section*. It is σ_m that is appropriate for calculating the frictional drag in the force equation (2.3.9). For a single velocity, we would just have $\nu_m = \sigma_m v$, where σ_m is generally a function of velocity. In the macroscopic force equation (2.3.15), ν_m must be obtained by averaging over the particle velocity distributions, which we do in Section 3.5.

We illustrate the use of the differential scattering cross section to calculate the total scattering and momentum transfer cross sections for the hard-sphere model shown in Fig. 3.2. The impact parameter is $b = a_{12} \sin \chi$, and differentiating, $db = a_{12} \cos \chi \, d\chi$, so that

$$b \, db = a_{12}^2 \sin \chi \cos \chi \, d\chi = \frac{1}{2} a_{12}^2 \sin 2\chi \, d\chi \tag{3.1.16}$$

From Fig. 3.2 the scattering angle $\theta = \pi - 2\chi$, such that (3.1.16) can be written

$$b \, db = -\frac{1}{4} a_{12}^2 \sin \theta \, d\theta \tag{3.1.17}$$

Substituting (3.1.17) into (3.1.13), we have

$$I(v, \theta) = \frac{1}{4} a_{12}^2 \tag{3.1.18}$$

Using the definitions of σ_{sc} and σ_m in (3.1.14) and (3.1.15), respectively, we find

$$\sigma_{sc} = \sigma_m = \pi a_{12}^2 \qquad (3.1.19)$$

for hard-sphere collisions. In general, $\sigma_{sc} \neq \sigma_m$ for other scattering forces. For electron collisions with atoms the electron radius is negligible compared to the atomic radius so that $a_{12} \approx a$, the atomic radius. Although the value of $a \approx 10^{-8}$ cm gives $\sigma_{sc} = \sigma_m \approx 3 \times 10^{-16}$ cm^2, which is reasonable, it does not capture the scaling of the cross section with speed.

In the following sections of this chapter we consider collisional processes in more detail. Except for Coulomb collisions, we confine our attention to electron–atom and ion–atom processes. After a discussion of collision dynamics in Section 3.2, we describe elastic collisions in Section 3.3 and inelastic collisions in Section 3.4. We reserve a discussion of some aspects of inelastic collisions until Chapter 8, in which a more complete range of atomic and molecular processes is considered. In Section 3.5, we describe the averaging over particle velocity distributions that must be done to obtain the collisional rate constants. Experimental values for argon are also given in Section 3.5; these are needed for discussing energy transfer and diffusive processes in the succeeding chapters. A more detailed account of collisional processes, together with many results of experimental measurements, can be found in McDaniel (1989), McDaniel et al. (1993), Massey et al. (1969–74), Smirnov (1981), and Raizer (1991).

3.2 COLLISION DYNAMICS

Center-of-Mass Coordinates

In a collision between projectile and target particles there is recoil of the target as well as deflection of the projectile. In fact, both may be moving, and, in the case of like-particle collisions, not distinguishable. To describe this more complicated state, a center of mass (CM) coordinate system can be introduced in which projectiles and targets are treated equally. Without loss of generality, we can transform to a coordinate system in which one of the particles is stationary before the collision. Hence, we consider a general collision in the laboratory frame between two particles having mass m_1 and m_2, position $\mathbf{r}_1$ and $\mathbf{r}_2$, velocity $\mathbf{v}_1$ and $\mathbf{v}_2 \equiv 0$, and scattering angle θ_1 and θ_2, as shown in Fig. 3.4a. We assume that the force $\mathbf{F}$ acts along the line joining the centers of the particles, with $\mathbf{F}_{12} = -\mathbf{F}_{21}$.

The center-of-mass coordinates may be defined by the linear transformation

$$\mathbf{R} = \frac{m_1 \mathbf{r}_1 + m_2 \mathbf{r}_2}{m_1 + m_2} \qquad (3.2.1)$$

and

$$\mathbf{r} = \mathbf{r}_1 - \mathbf{r}_2 \qquad (3.2.2)$$

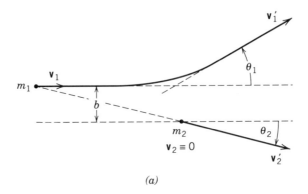

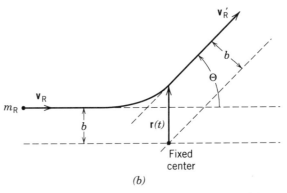

FIGURE 3.4. The relation between the scattering angles in (a) the laboratory system and (b) the center of mass (CM) system.

with the accompanying CM velocity

$$\mathbf{V} = \frac{m_1 \mathbf{v}_1 + m_2 \mathbf{v}_2}{m_1 + m_2} \tag{3.2.3}$$

and the relative velocity

$$\mathbf{v}_R = \mathbf{v}_1 - \mathbf{v}_2 \tag{3.2.4}$$

The force equations for the two particles are:

$$m_1 \dot{\mathbf{v}}_1 = \mathbf{F}_{12}(r), \qquad m_2 \dot{\mathbf{v}}_2 = \mathbf{F}_{21}(r) = -\mathbf{F}_{12}(r) \tag{3.2.5}$$

Adding these equations we get the result for the CM motion that $\dot{\mathbf{V}} = 0$, such that the CM moves with constant velocity throughout the collision. Now dividing the first of

(3.2.5) by m_1 and the second by m_2, and using the definition in (3.2.4) we have

$$m_R \dot{\mathbf{v}}_R = \mathbf{F}_{12}(r) \tag{3.2.6}$$

which is the equation of motion of a "fictitious" particle with a reduced mass

$$m_R = \frac{m_1 m_2}{m_1 + m_2} \tag{3.2.7}$$

in a fixed central force $\mathbf{F}_{12}(r)$. The fictitious particle has mass m_R, position $\mathbf{r}(t)$, velocity $\mathbf{v}_R(t)$, and scattering angle Θ, as shown in Fig. 3.4b. This result holds for any central force, including the hard-sphere, Coulomb, and polarization forces that we subsequently consider. If (3.2.6) can be solved to obtain the motion, including Θ, then we can transform back to the laboratory frame to get the actual scattering angles θ_1 and θ_2. It is easy to show from momentum conservation (Problem 3.1) that

$$\tan \theta_1 = \frac{\sin \Theta}{(m_1/m_2)(v_R/v_R') + \cos \Theta} \tag{3.2.8a}$$

and

$$\tan \theta_2 = \frac{\sin \Theta}{v_R/v_R' - \cos \Theta} \tag{3.2.8b}$$

where v_R and v_R' are the speeds in the CM system before and after the collision, respectively.

For an elastic collision, the scattering force can be written as the gradient of a potential that vanishes as $r = |\mathbf{r}| \to \infty$:

$$\mathbf{F}_{12} = -\nabla U(r) \tag{3.2.9}$$

It follows that the kinetic energy of the particle is conserved for the collision in the CM system. Hence $v_R' = v_R$, and we obtain from (3.2.8) that

$$\tan \theta_1 = \frac{\sin \Theta}{m_1/m_2 + \cos \Theta} \tag{3.2.10}$$

and, using the double-angle formula for the tangent,

$$\theta_2 = \frac{1}{2}(\pi - \Theta) \tag{3.2.11}$$

For electron collisions with ions or neutrals, $m_1/m_2 \ll 1$ and we obtain $m_R \approx m_1$ and $\theta_1 \approx \Theta$. For collision of a particle with an equal mass target, $m_1 = m_2$, we obtain $m_R = m_1/2$ and $\theta_1 = \Theta/2$. Hence for hard-sphere elastic collisions against an initially stationary equal mass target, the maximum scattering angle is $90°$.

Since the same particles are scattered into the differential solid angle $2\pi \sin \Theta \, d\Theta$ in the CM system as are scattered into the corresponding solid angle $2\pi \sin \theta_1 \, d\theta_1$ in the laboratory system, the differential scattering cross sections are related by

$$I(v_R, \Theta) \, 2\pi \sin \Theta \, d\Theta = I(v_R, \theta_1) \, 2\pi \sin \theta_1 \, d\theta_1 \qquad (3.2.12)$$

where $d\Theta/d\theta_1$ can be found by differentiating (3.2.10).

Energy Transfer

Elastic collisions can be an important energy transfer process in gas discharges, and can also be important for understanding inelastic collision processes such as ionization, as we will see in Section 3.4. For the elastic collision of a projectile of mass m_1 and velocity $\mathbf{v}_1$ with a stationary target of mass m_2, the conservation of momentum along and perpendicular to $\mathbf{v}_1$ and the conservation of energy can be written in the laboratory system as

$$m_1 v_1 = m_1 v_1' \cos \theta_1 + m_2 v_2' \cos \theta_2 \qquad (3.2.13)$$

$$0 = m_1 v_1' \sin \theta_1 - m_2 v_2' \sin \theta_2 \qquad (3.2.14)$$

$$\frac{1}{2} m_1 v_1^2 = \frac{1}{2} m_1 v_1'^2 + \frac{1}{2} m_2 v_2'^2 \qquad (3.2.15)$$

where the primes denote the values after the collision. We can eliminate v_1' and θ_1 and solve (3.2.13)–(3.2.15) to obtain

$$\frac{1}{2} m_2 v_2'^2 = \frac{1}{2} m_1 v_1^2 \frac{4 m_1 m_2}{(m_1 + m_2)^2} \cos^2 \theta_2 \qquad (3.2.16)$$

Since the initial energy of the projectile is $\frac{1}{2} m_1 v_1^2$ and the energy gained by the target is $\frac{1}{2} m_2 v_2'^2$, the fraction of energy lost by the projectile *in the laboratory system* is

$$\zeta_L = \frac{4 m_1 m_2}{(m_1 + m_2)^2} \cos^2 \theta_2 \qquad (3.2.17)$$

Using (3.2.11) in (3.2.17), we obtain

$$\zeta_L = \frac{2 m_1 m_2}{(m_1 + m_2)^2} (1 - \cos \Theta) \qquad (3.2.18)$$

where Θ is the scattering angle *in the CM system*. We average over the differential scattering cross section to obtain the average loss:

$$\langle \zeta_L \rangle_\Theta = \frac{2m_1 m_2}{(m_1 + m_2)^2} \frac{\int (1 - \cos\Theta) I(v_R, \Theta) 2\pi \sin\Theta \, d\Theta}{\int I(v_R, \Theta) 2\pi \sin\Theta \, d\Theta}$$

$$= \frac{2m_1 m_2}{(m_1 + m_2)^2} \frac{\sigma_m}{\sigma_{sc}} \tag{3.2.19}$$

where σ_{sc} and σ_m are defined in (3.1.14) and (3.1.15).

For hard-sphere scattering of electrons against atoms, we have $m_1 = m$ (electron mass) and $m_2 = M$ (atom mass), and $\sigma_{sc} = \sigma_m$ by (3.1.9), such that $\langle \zeta_L \rangle_\Theta = 2m/M \sim 10^{-4}$. Hence electrons transfer little energy due to elastic collisions with heavy particles, allowing $T_e \gg T_i$ in a typical discharge. On the other hand, for $m_1 = m_2$, we obtain $\langle \zeta_L \rangle_\Theta = \frac{1}{2}$, leading to strong elastic energy exchange among heavy particles and hence to a common temperature.

Small Angle Scattering

In the general case, (3.2.6) must be solved to determine the CM trajectory and the scattering angle Θ. We outline this approach and give some results in Appendix A. Here we restrict attention to small-angle scattering ($\Theta \ll 1$) for which the fictitious particle moves with uniform velocity v_R along a trajectory that is practically unaltered from a straight line. In this case, we can calculate the transverse momentum impulse $\Delta p_\perp$ delivered to the particle as it passes the center of force at $r = 0$ and use this to determine Θ. For a straight-line trajectory, as shown in Fig. 3.5, the particle distance from the center of force is

$$r = (b^2 + v_R^2 t^2)^{1/2} \tag{3.2.20}$$

where b is the impact parameter and t is the time. We assume a central force of the form (3.2.9) with

$$U(r) = \frac{C}{r^i} \tag{3.2.21}$$

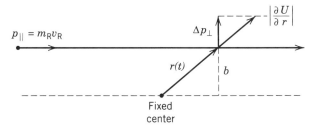

FIGURE 3.5. Calculation of the differential scattering cross section for small-angle scattering. The center of mass trajectory is practically a straight line.

where i is an integer. The component of the force acting on the particle perpendicular to the trajectory is $(b/r) \left| dU/dr \right|$. Hence the momentum impulse is

$$\Delta p_\perp = \int_{-\infty}^{\infty} \frac{b}{r} \left| \frac{dU}{dr} \right| dt \qquad (3.2.22)$$

Differentiating (3.2.20) to obtain

$$dt = \frac{r}{v_R} \frac{dr}{(r^2 - b^2)^{1/2}}$$

substituting into (3.2.22), and dividing by the incident momentum $p_\parallel = m_R v_R$, we obtain

$$\Theta = \frac{\Delta p_\perp}{p_\parallel} = \frac{2b}{m_R v_R^2} \int_b^{\infty} \left| \frac{dU}{dr} \right| \frac{dr}{(r^2 - b^2)^{1/2}} \qquad (3.2.23)$$

The integral in (3.2.23) can be evaluated in closed form (Smirnov, 1981, p. 384) to obtain

$$\Theta = \frac{A}{W_R b^i} \qquad (3.2.24)$$

where $W_R = \frac{1}{2} m_R v_R^2$ is the CM energy and

$$A = \frac{C\sqrt{\pi} \Gamma\left[(i+1)/2\right]}{2\Gamma\left[(i+2)/2\right]} \qquad (3.2.25)$$

with Γ, the Gamma function.* Inverting (3.2.24), we obtain

$$b = \left(\frac{A}{W_R \Theta} \right)^{1/i} \qquad (3.2.26)$$

and differentiating, we obtain

$$db = -\frac{1}{i} \left(\frac{A}{W_R} \right)^{1/i} \frac{d\Theta}{\Theta^{1+1/i}} \qquad (3.2.27)$$

Substituting (3.2.26) and (3.2.27) into (3.1.13), with $\sin \Theta \approx \Theta$, we obtain the differential scattering cross section for small angles:

$$I(v_R, \Theta) = \frac{1}{i} \left(\frac{A}{W_R} \right)^{2/i} \frac{1}{\Theta^{2+2/i}} \qquad (3.2.28)$$

*$\Gamma(l) = (l-1)! = l\Gamma(l-1)$ with $\Gamma(1/2) = \sqrt{\pi}$.

TABLE 3.1. Scaling of Cross Section σ, Interaction Frequency ν, and Rate Constant K, With Relative Velocity v_R, for Various Scattering Potentials U

Process	$U(r)$	σ	ν or K
Coulomb	$1/r$	$1/v_R^4$	$1/v_R^3$
Permanent dipole	$1/r^2$	$1/v_R^2$	$1/v_R$
Induced dipole	$1/r^4$	$1/v_R$	const
Hard sphere	$1/r^i, i \to \infty$	const	v_R

The variation of σ, ν, and K with v_R are determined from (3.2.28) and the basic definitions in Section 3.1. If (3.2.28) is substituted into (3.1.14) or (3.1.15), then we see that a scattering potential $U \propto r^{-i}$ leads to $\sigma \propto v_R^{-4/i}$ and $\nu \propto K \propto v_R^{-(4/i)+1}$. These scalings are summarized in Table 3.1 for the important scattering processes, which we describe in the next section.

3.3 ELASTIC SCATTERING

Coulomb Collisions

The most straightforward elastic scattering process is a Coulomb collision between two charged particles q_1 and q_2, representing an electron–electron, electron–ion, or ion–ion collision. The Coulomb potential is $U(r) = q_1 q_2 / 4\pi\epsilon_0 r$ such that $i = 1$ and we obtain

$$A = C = \frac{q_1 q_2}{4\pi\epsilon_0}$$

from (3.2.25). Using this in (3.2.27), we find

$$I = \left(\frac{b_0}{\Theta^2} \right)^2 \tag{3.3.1}$$

where

$$b_0 = \frac{q_1 q_2}{4\pi\epsilon_0 W_R} \tag{3.3.2}$$

is called the *classical distance of closest approach*. The differential scattering cross section can also be calculated exactly, which we do in Appendix A, obtaining the result

$$I = \left[\frac{b_0}{4 \sin^2 (\Theta/2)} \right]^2 \tag{3.3.3}$$

However, due to the long range of the Coulomb forces, the integration of I over small Θ (large b) leads to an infinite scattering cross section and to an infinite momentum transfer cross section, such that an upper bound to b, b_{max}, must be assigned. This is done by setting $b_{max} = \lambda_{De}$, the Debye shielding distance for a charge immersed in a plasma, which we calculated in Section 2.4. For momentum transfer, the dependence of σ_m on λ_{De} is logarithmic (Problem 3.4), and the exact choice of b_{max} (or Θ_{min}) makes little difference. For scattering, $\sigma_{sc} \sim \pi\lambda_{De}^2$, which is a very large cross section that depends sensitively on the choice of b_{max}. However, we are generally not interested in scattering through very small angles, which do not appreciably affect the discharge properties. The cross section for scattering through a large angle, say $\Theta \geq \pi/2$, is of more interest.

There are two processes that lead to a large scattering angle Θ for a Coulomb collision: (1) a single collision scatters the particle by a large angle; (2) the cumulative effect of many small-angle collisions scatters the particle by a large angle. The two processes are illustrated in Fig. 3.6; the latter process is diffusive and, as we will see, dominates the former.

To estimate the cross section $\sigma_{90}(\text{sgl})$ for a single large-angle collision we integrate (3.3.3) over solid angles from $\pi/2$ to π to obtain (Problem 3.5)

$$\sigma_{90}(\text{sgl}) = \frac{1}{4}\pi b_0^2 \tag{3.3.4}$$

To estimate $\sigma_{90}(\text{cum})$ for the cumulative effect of many collisions to produce a $\pi/2$ deflection, we first determine the mean square scattering angle $\langle\Theta^2\rangle_1$ for a single collision by averaging Θ^2 over all permitted impact parameters. Since the collisions are predominantly small angle for Coulomb collisions, we can use (3.2.24), which is

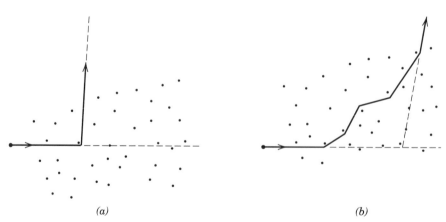

(a)   *(b)*

FIGURE 3.6. The processes that lead to large-angle Coulomb scattering: (a) single large-angle event; (b) cumulative effect of many small-angle events.

$\Theta = b_0/b$. Hence

$$\langle\Theta^2\rangle_1 = \frac{1}{\pi b_{max}^2} \int_{b_{min}}^{b_{max}} \left(\frac{q_1 q_2}{4\pi\epsilon_0 W_R}\right)^2 \frac{2\pi b\, db}{b^2} \tag{3.3.5}$$

The integration has a logarithmic singularity at both $b = 0$ and $b = \infty$ which is cut off by the finite limits. The singularity at the lower limit is due to the small-angle approximation. Setting $b_{min} = b_0$ is found to approximate a more accurate calculation. The upper limit, as already mentioned, is $b_{max} = \lambda_{De}$. Using these values and integrating, we obtain

$$\langle\Theta^2\rangle_1 = \frac{2\pi b_0^2}{\pi b_{max}^2} \ln\Lambda \tag{3.3.6}$$

where $\Lambda = \lambda_{De}/b_0 \gg 1$.

The number of collisions per second, each having a cross section of πb_{max}^2 or smaller, is $n_g \pi b_{max}^2 v_R$, where n_g is the target particle density. Since the spreading of the angle is diffusive, we can then write

$$\langle\Theta^2\rangle(t) = \langle\Theta^2\rangle_1\, n_g \pi b_{max}^2 v_R t$$

Setting $t = \tau_{90}$ at $\langle\Theta^2\rangle = (\pi/2)^2$ and using (3.3.6), we obtain (see also Spitzer, 1956, Chapter 5)

$$\nu_{90} = \tau_{90}^{-1} = n_g v_R \frac{8}{\pi} b_0^2 \ln\Lambda$$

Writing $\nu_{90} = n_g \sigma_{90} v_R$, we see that

$$\sigma_{90} = \frac{8}{\pi} b_0^2 \ln\Lambda \tag{3.3.7}$$

Although Λ is a large number, typically $\ln\Lambda \approx 10$ for the types of plasmas we are considering.

Comparing $\sigma_{90}(sgl)$ to σ_{90}, we see that due to the large range of the Coulomb fields, the effective cross section for many small-angle collisions to produce an rms deflection of $\pi/2$ is larger by a factor $(32/\pi^2) \ln\Lambda$. Because of this enhancement, it is possible for electron–ion or ion–ion particle collisions to play a role in weakly ionized plasmas (say one percent ionized). Another important characteristic of Coulomb collisions is the strong velocity dependence. From (3.3.2) we see that $b_0 \propto 1/v_R^2$. Thus, from (3.3.4) or (3.3.7)

$$\sigma_{90} \propto \frac{1}{v_R^4} \tag{3.3.8}$$

such that low-velocity particles are preferentially scattered. The temperature of the species is therefore important in determining the relative importance of the various species in the collisional processes, as we shall see in subsequent sections.

Polarization Scattering

The main collisional processes in a weakly ionized plasma are between charged and neutral particles. For electrons at low energy and for ions scattering against neutrals, the dominant process is relatively short-range polarization scattering. At higher energies for electrons, the collision time is shorter and the atoms do not have time to polarize. In this case the scattering becomes more Coulomb-like, but with b_{max} at an atomic radius, and inelastic processes such as ionization become important also. The condition for polarization scattering is $v_R \lesssim v_{at}$, where v_{at} is the characteristic electron velocity in the atom, which we obtain in the next section. Because of the short range of the polarization potential, we need not be concerned with an upper limit for the integration over b, but the potential is more complicated. We determine the potential from a simple model of the atom as a point charge of value $+q_0$, surrounded by a uniform negative charge sphere (valence electrons) of total charge $-q_0$, such that the charge density is $\rho = -q_0/\frac{4}{3}\pi a^3$, where a is the atomic radius. An incoming electron (or ion) can polarize the atom by repelling (or attracting) the charge cloud quasistatically. The balance of forces on the central point charge due to the displaced charge cloud and the incoming charged particle, taken to have charge q, is shown in Fig. 3.7, where the center of the charge cloud and the point charge are displaced by a distance d. Applying Gauss' law to a sphere of radius d around the center of the cloud,

$$4\pi\epsilon_0 d^2 E_{ind} = -q_0 \frac{d^3}{a^3}$$

we obtain the induced electric field acting on the point charge due to the displaced cloud

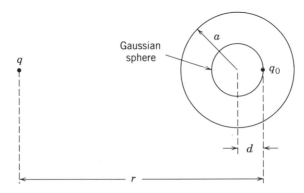

FIGURE 3.7. Polarization of an atom by a point charge q.

$$E_{\text{ind}} = -\frac{q_0 d}{4\pi\epsilon_0 a^3}$$

The electric field acting on the point charge due to the incoming charge is

$$E_{\text{appl}} = \frac{q}{4\pi\epsilon_0 r^2}$$

For force balance on the point charge, the sum of the fields must vanish, yielding an induced dipole moment for the atom:

$$p_{\text{d}} = q_0 d = \frac{qa^3}{r^2} \qquad (3.3.9)$$

The induced dipole, in turn, exerts a force on the incoming charged particle:

$$\mathbf{F} = \frac{2p_{\text{d}}q}{4\pi\epsilon_0 r^3}\hat{\mathbf{r}} = \frac{2q^2 a^3}{4\pi\epsilon_0 r^5}\hat{\mathbf{r}} \qquad (3.3.10)$$

Integrating $\mathbf{F}$ with respect to r, we obtain the attractive potential energy:

$$U(r) = -\frac{q^2 a^3}{8\pi\epsilon_0 r^4} \qquad (3.3.11)$$

The *polarizability* for this simple atomic model is defined as $\alpha_{\text{p}} = a^3$. The relative polarizabilities $\alpha_{\text{R}} = \alpha_{\text{p}}/a_0^3$, where a_0 is the Bohr radius, for some simple atoms and molecules are given in Table 3.2.

The orbits for scattering in the polarization potential are complicated (McDaniel, 1989). As shown in Fig. 3.8, there are two types of orbits. For impact parameter $b > b_{\text{L}}$, the orbit has a hyperbolic character, and for $b \gg b_{\text{L}}$, the straight-line trajectory analysis in Section 3.2 can be applied (Problem 3.6). For $b < b_{\text{L}}$, the incoming particle is "captured" and the orbit spirals into the core, leading to a large scattering angle. Either the incoming particle is "reflected" by the core and spirals out again, or the two particles strongly interact, leading to inelastic changes of state.

The critical impact parameter b_{L} can be determined from the conservation of energy and angular momentum for the incoming particle having mass m and speed v_0, with the mass of the scatterer taken to be infinite for ease of analysis. In cylindrical coordinates (see Fig. 3.8a), we obtain

$$\frac{1}{2}mv_0^2 = \frac{1}{2}m(\dot{r}^2 + r^2\dot{\phi}^2) + U(r) \qquad (3.3.12a)$$

$$mv_0 b = mr^2\dot{\phi} \qquad (3.3.12b)$$

TABLE 3.2. Relative Polarizabilities $\alpha_R = \alpha_p/a_0^3$ of Some Atoms and Molecules, Where a_0 is the Bohr Radius

Atom or Molecule	α_R
H	4.5
C	12.
N	7.5
O	5.4
Ar	11.08
CCl_4	69.
CF_4	19.
CO	13.2
CO_2	17.5
Cl_2	31.
H_2O	9.8
NH_3	14.8
O_2	10.6
SF_6	30.

Source. Smirnov (1981).

At closest approach, $\dot{r} = 0$ and $r = r_{min}$. Substituting these into (3.3.12) and eliminating $\dot{\phi}$, we obtain a quadratic equation for r_{min}^2:

$$v_0^2 r_{min}^4 - v_0^2 b^2 r_{min}^2 + \frac{2\alpha_p q^2}{4\pi\epsilon_0 m} = 0$$

Using the quadratic formula to obtain the solution for r_{min}^2, we see that there is no real solution for r_{min}^2 when

$$(v_0^2 b^2)^2 - 4v_0^2 \frac{2\alpha_p q^2}{4\pi\epsilon_0 m} \leq 0$$

Choosing the equality at $b = b_L$, we solve for b_L to obtain

$$\sigma_L = \pi b_L^2 = \left(\frac{\pi\alpha_p q^2}{\epsilon_0 m} \right)^{1/2} \frac{1}{v_0} \tag{3.3.13}$$

which is known as the *Langevin* or *capture* cross section. If the target particle has a finite mass m_2 and velocity $\mathbf{v}_2$ and the incoming particle has a mass m_1 and velocity $\mathbf{v}_1$, then (3.3.13) holds provided m is replaced by the reduced mass $m_R = m_1 m_2/(m_1 + m_2)$ and v_0 is replaced by the relative velocity $v_R = |\mathbf{v}_1 - \mathbf{v}_2|$. We note that the cross section $\sigma_L \propto 1/v_R$. Hence the collision frequency for capture is

$$\nu_L = n_g \sigma_L v_R = n_g K_L \tag{3.3.14}$$

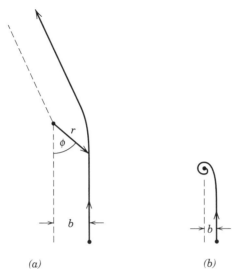

(a)   *(b)*

FIGURE 3.8. Scattering in the polarization potential, showing (a) hyperbolic and (b) captured orbits.

where

$$K_{\mathrm{L}} = \left(\frac{\pi \alpha_p q^2}{\epsilon_0 m_{\mathrm{R}}} \right)^{1/2} \tag{3.3.15}$$

is the rate constant for capture and n_{g} is the target particle density. Both v_{L} and K_{L} are independent of velocity. In practical units, the rate constants for electrons and ions are (with $q = \pm e$)

$$K_{\mathrm{Le}} = 3.85 \times 10^{-8} \alpha_{\mathrm{R}}^{1/2} \ \mathrm{cm}^3/\mathrm{s} \tag{3.3.16}$$

$$K_{\mathrm{Li}} = 8.99 \times 10^{-10} \left(\frac{\alpha_{\mathrm{R}}}{A_{\mathrm{R}}} \right)^{1/2} \ \mathrm{cm}^3/\mathrm{s} \tag{3.3.17}$$

where A_{R} is the reduced mass in atomic mass units (hydrogen ≈ 1 amu) and α_{R} is the relative polarizability. Because $\sigma_{\mathrm{L}} \propto 1/\sqrt{\mathcal{E}}$, where $\mathcal{E}$ is the collision energy in the center of mass system, the Langevin cross section dominates the elastic and inelastic collisional behavior at thermal energies ($\mathcal{E} \sim 0.026$ V), especially for ion–neutral collisions. Some molecules (but not atoms) have permanent dipole moments, leading to a scattering potential $U \propto 1/r^2$ and an enhanced Langevin cross section. We describe this briefly in Chapter 8.

What is the actual velocity dependence of elastic electron–atom collisions? At low energies we might expect quantum effects to be significant, which is indeed the case, such that some gases show low-energy resonances in their cross sections. An example

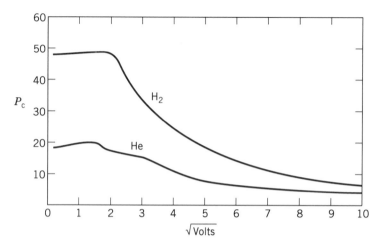

FIGURE 3.9. Probability of collision P_c for electrons in H_2 and He; the cross section is $\sigma \approx 2.87 \times 10^{-17} P_c$ cm^2 (after Brown, 1959).

of a simple velocity dependence is shown for hydrogen and helium in Fig. 3.9. Here a normalized cross section unit is used called the *probability of collision P_c*, defined as the average number of collisions in 1 cm of path through a gas at 1 Torr at 273 K. The elastic collision frequency in these units is

$$\nu_{el} = vp_0P_c$$

where $p_0 = 273p/T$. We see from the figure that at low energy the cross section is hard-sphere-like, being independent of velocity. At higher energies $\sigma \propto v^{-1}$ and thus the polarization potential governs the behavior.

The low-energy cross sections can, in fact, be quite complicated, depending on quantum mechanical effects. For example, in many gases the quantum mechanical wave diffraction of the electron around the atom at low energy leads to a "hole" in the elastic collision frequency at some low energy. This is true for some noble gases, as seen in Fig. 3.10, as well as some processing gases, such as CF_4. At higher (but still moderate) energy the approximate proportionality for polarization scattering $\sigma \propto v^{-1}$ is still found.

3.4 INELASTIC COLLISIONS

Atomic Energy Levels

The physics and spectroscopy of atoms is a vast area, and we give only a brief summary here. The reader should consult textbooks such as Bransden and Joachain (1983) and Thorne (1988) for a more thorough treatment. Atoms consist of one or more electrons bound to a heavy positive nucleus. In a classical description, electrons

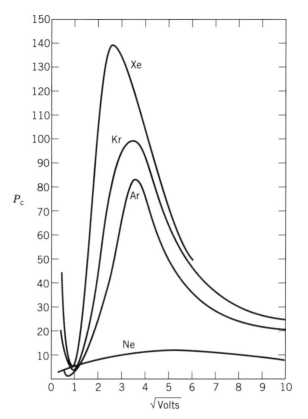

FIGURE 3.10. Probability of collision P_c for electrons in Ne, Ar, Kr, and Xe, showing the Ramsauer minima for Ar, Kr, and Xe; the cross section is $\sigma \approx 2.87 \times 10^{-17} P_c$ cm^2 (after Brown, 1959).

move in a circular orbits whose radii a are set by the balance between the inward electrostatic and the outward centrifugal forces. For the hydrogen atom, the inward force is the Coulomb force of the proton, leading to the force balance:

$$\frac{e^2}{4\pi\epsilon_0 a^2} = \frac{mv^2}{a} \qquad (3.4.1)$$

From (3.4.1), all radii (and corresponding velocities) are possible. A quantum description limits the orbits to those for which the angular momentum is an integral multiple of $\hbar$,

$$mva = n\hbar \qquad (3.4.2)$$

where here $n \geq 1$ is an integer called the *principal quantum number*, and $\hbar = h/2\pi$, with Planck's constant $h \approx 6.626 \times 10^{-34}$ J-s. Solving (3.4.1) and (3.4.2) yields the quantized radii

$$a_n = n^2 a_0 \tag{3.4.3}$$

where, for the lowest level ($n = 1$),

$$a_0 = \frac{4\pi\epsilon_0 \hbar^2}{e^2 m} \approx 5.29 \times 10^{-11} \text{ m} \tag{3.4.4}$$

is the *Bohr radius*. The velocity is

$$v_n = \frac{v_{\text{at}}}{n}$$

where

$$v_{\text{at}} = \frac{e^2}{4\pi\epsilon_0 \hbar} \approx 2.19 \times 10^6 \text{ m/s} \tag{3.4.5}$$

is the electron velocity in the first Bohr orbit. The characteristic atomic timescale is then

$$t_{\text{at}} = \frac{a_0}{v_{\text{at}}} \approx 2.42 \times 10^{-17} \text{ s} \tag{3.4.6}$$

The electron energy W_n is the sum of the kinetic and potential energy,

$$W_n = \frac{1}{2} m v_n^2 - \frac{e^2}{4\pi\epsilon_0 a_n} \tag{3.4.7}$$

Defining $W_n(\text{J}) = e\mathcal{E}_n(\text{V})$, we obtain

$$\mathcal{E}_n = -\frac{\mathcal{E}_{\text{at}}}{n^2} \tag{3.4.8a}$$

where

$$\mathcal{E}_{\text{at}} = \frac{1}{2} \frac{m}{e} \left(\frac{e^2}{4\pi\epsilon_0 \hbar} \right)^2 \approx 13.61 \text{ V} \tag{3.4.8b}$$

is the ionization potential of the hydrogen atom in its lowest energy state ($n = 1$).

For a many-electron atom, a valence electron sees some effective positive charge $Z_{\text{eff}} e$. This leads to a radius for the first Bohr orbit $a_{\text{eff}} = a_0/Z_{\text{eff}}$ and to an ionization potential $\mathcal{E}_{\text{iz}} = Z_{\text{eff}}^2 \mathcal{E}_{\text{at}}$. When we combine these expressions, the radius of an atom is found to scale as

$$a_{\text{eff}} \approx a_0 \left(\frac{\mathcal{E}_{\text{at}}}{\mathcal{E}_{\text{iz}}}\right)^{1/2} \tag{3.4.9}$$

where $\mathcal{E}_{\text{at}}$ is given by (3.4.8b).

This picture, while qualitatively correct, is incomplete. Quantum mechanics specifies the state of each electron in an atom in terms of four quantum numbers, n, l, m_l and m_s (n, l, and m_l are integers), with the restrictions $l + 1 \leq n$, $|m_l| \leq l$, and with $m_s = \pm\frac{1}{2}$. The quantum numbers l and m_l specify the total orbital angular momentum and its component in a particular direction; the quantum number m_s specifies the direction of the electron spin.

For the preceding model, the energy of each level depends only on n. By the restrictions on l, m_l, and m_s, there are $2n^2$ electron states having the same energy $\mathcal{E}_n$. The energy level $\mathcal{E}_n$ is said to have *degeneracy* $2n^2$. For an atom with more than one electron, the force balance includes not only the attractive force of the nucleus, but also the repulsive forces of the other electrons. In the *central field model*, each electron moves under the influence of a spherically symmetric potential that includes the average effects of all the other electrons. This breaks the degeneracy such that the energy is a function of both n and l. Figure 3.11 shows a typical energy level diagram with the different l values displaced to the right. For historical reasons, electrons having $l = 0, 1, 2,$ and 3 are known as s, p, d, and f electrons, respectively.

The Pauli exclusion principle states that no two electrons can have the same state. Hence stable atoms are built by placing electrons into the available states in order of increasing energy. For example, the *electronic configurations* of the lowest-energy states (ground states) of hydrogen, oxygen, and argon are $1s$, $1s^2 2s^2 2p^4$, and $1s^2 2s^2 2p^6 3s^2 3p^6$. In this notation, the values of n and l specify a given electron *subshell*, and the superscript indicates the number of electrons in each subshell, which holds a maximum of $2(2l + 1)$ electrons. The *valence electrons*, which are those in the last (usually incomplete) subshell, determine the collisional and other behavior of atoms. For example, an electron collision with an argon atom can excite the atom to a higher energy level,

$$e + \text{Ar} \rightarrow \text{Ar}^* + e$$

corresponding to a change of state

$$3p^6 \rightarrow 3p^5 4s^1$$

for the valence electrons.

For the light elements (roughly $Z \lesssim 40$), the energy levels are usually labeled by the values of the permitted orbital and spin angular momentum L and S for the sum of all the valence electrons. Levels with different L values are known as S, P, D, and F levels for $L = 0, 1, 2,$ and 3, by analogy with single-electron terminology. The integer or half-integer value of S is indicated by a superscript $2S + 1$ placed to the

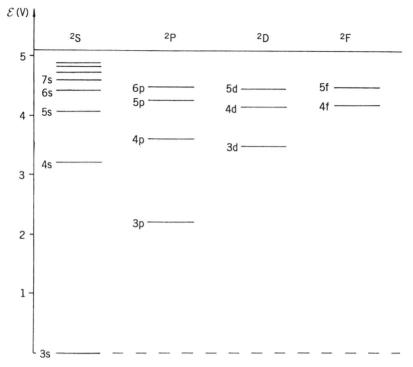

FIGURE 3.11. Atomic energy levels for the central field model of an atom, showing the dependence of the energy levels on the quantum numbers n and l; the energy levels are shown for sodium, without the fine structure (after Thorne, 1988).

left of the L value. This superscript, the *multiplicity*, gives the degeneracy of the level. The degeneracy is usually removed by the weak magnetic interactions between the spin and orbital motions, giving rise to additional small splittings of the degenerate energy levels, the so-called *fine structure*. This is specified by a quantum number J for the sum of the total orbital and spin angular momentum, which can have integer or half-integer value, and which is written as a subscript to the right of the L value. The ground states of hydrogen, oxygen, and argon in this notation are $^2S_{1/2}$, $^3P_2$ and $^1S_0$, respectively. For heavy elements, roughly $Z \gtrsim 40$, the L and S values are no longer meaningful quantum numbers, and the n and J values alone, along with the j values of the individual electrons, can be used to specify a level.

Electric Dipole Radiation and Metastable Atoms

Atoms in their ground states can be excited by collisions or radiation to higher energy bound states. In most cases, only a single-valence electron is excited. Most bound states can emit a photon by electric dipole radiation and return to some lower energy state or to the ground state:

$$e + Ar \rightarrow Ar^* + e \rightarrow Ar + e + \hbar\omega$$

Here $\hbar\omega$ is the photon energy and ω is its radian frequency. The radiation is usually in the visible or ultraviolet. Electric dipole radiation is permitted between two states only if the selection rules

$$\Delta l = \pm 1$$
$$\Delta J = 0, \pm 1 \quad \text{(but } J = 0 \rightarrow J = 0 \text{ forbidden)} \quad (3.4.10a)$$

are satisfied. For the light elements, with L and S also good quantum numbers,[*]the additional rules

$$\Delta L = 0, \pm 1 \quad \text{(but } L = 0 \rightarrow L = 0 \text{ forbidden)},$$
$$\Delta S = 0 \quad (3.4.10b)$$

must also be satisfied.

We can estimate the timescale for electric dipole radiation from the time-average energy per unit time radiated by a classical oscillating dipole $p_d(t) = p_{d0} \cos \omega t$ (Jackson, 1975, Chapter 14):

$$P_{\text{rad}} = \frac{\omega^4 p_{d0}^2}{12\pi\epsilon_0 c^3} \quad (3.4.11)$$

Dividing the energy radiated $\hbar\omega$ by P_{rad}, we obtain the radiation time

$$t_{\text{rad}} = \frac{12\pi\epsilon_0 \hbar c^3}{\omega^3 p_{d0}^2} \quad (3.4.12)$$

Taking the simple estimates $p_{d0} = ea_0$ and $\omega = e\mathcal{E}_{\text{at}}/\hbar$, with $\mathcal{E}_{\text{at}}$ given by (3.4.8b), and using (3.4.6), we obtain

$$t_{\text{rad}} = 24 \left(\frac{4\pi\epsilon_0 \hbar c}{e^2} \right)^3 t_{\text{at}} \approx 6.2 \times 10^7 t_{\text{at}}$$
$$\approx 1.5 \times 10^{-9} \text{ s} \quad (3.4.13)$$

We see that t_{rad} is long compared to the characteristic atomic timescale t_{at}. However, the characteristic time between collisions is

$$\tau \sim (n_g \pi a_0^2 \bar{v})^{-1}$$

For electrons with $T_e \sim 3$ V and $n_g \sim 3.3 \times 10^{14}$ cm^{-3} (corresponding to a gas pressure of 10 mTorr), we obtain

$$\tau_e \sim 3 \times 10^{-7} \text{ s} \quad (3.4.14)$$

For heavy particle collisions, we estimate for $T_i \sim T_e$ and $M/m \sim 10^4$ that $\tau_i \sim 100\tau_e$. Hence we have $t_{\text{rad}} \ll \tau_e, \tau_i$ in low-pressure discharges. This implies that

---

[*]If L and S are good quantum numbers, then the atomic state has both a unique orbital and spin angular momentum.

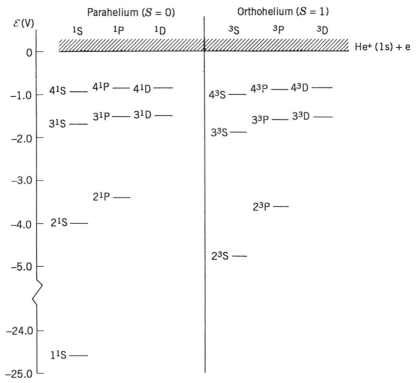

FIGURE 3.12. The energy levels of the helium atom, showing the division into singlet and triplet states and the metastable levels 2^1S and 2^3S (after Bransden and Joachain, 1983).

excited states will generally be deexcited by electric dipole radiation rather than by collisions.

Certain excited states, however, cannot satisfy the selection rules (3.4.10) for electric dipole radiation; for these states $p_{d0} \equiv 0$. While other radiative transitions may occur, such as electric quadrupole or magnetic dipole radiation, or radiationless transitions may occur, to states of nearly equal energy that subsequently do radiate, these mechanisms are generally weak, leading to transition times that can be long compared to the collision times τ_e and τ_i. The energy levels from which electric dipole radiation is forbidden are called *metastable*, and the excited atoms are called *metastable atoms*. Metastable atoms are often present at considerable densities in weakly ionized discharges, where they can be further excited, ionized, or deexcited by collisions.

An example of metastable levels is shown in Fig. 3.12 for the two-valence electron helium system. Since electric dipole transitions between $S = 0$ and $S = 1$ states are forbidden, the diagram decomposes into two nearly independent energy level systems: the *singlets* $(2S + 1 = 1)$ and the *triplets* $(2S + 1 = 3)$. Because $L = 0 \rightarrow L = 0$ is forbidden, the 2^1S and 2^3S states* are metastable.

*Here "2" denotes the value of the principal quantum number n.

These states find application in He–Ne gas lasers, where they are excited by e–He collisions and are collisionally deexcited by He*–Ne collisions to create excited Ne* atoms that subsequently radiate, leading to laser action.

Electron Ionization Cross Section

Quantum mechanics is needed to properly treat electron–atom ionization. We give here a simple classical description (Thomson, 1912) that provides a qualitative treatment. The basic idea is to determine the condition for the incident electron (having velocity v) to transfer to a valence electron (assumed to be at rest) an energy equal to the ionization energy. Using (3.3.1) with $q_1 = q_2 = -e$ and $m_1 = m_2 = m$, the electron charge and mass, we have for a small angle collision that

$$I(v, \Theta) = \left(\frac{e^2}{4\pi\epsilon_0} \right)^2 \frac{1}{W_R^2} \frac{1}{\Theta^4} \tag{3.4.15}$$

where $W_R = \frac{1}{2} m_R v^2$ is the CM energy and $m_R = m/2$ is the reduced mass. Substituting $\theta = \Theta/2$ in (3.4.15), we transform to the scattering angle in the laboratory frame, and using (3.2.12), we obtain

$$dI = I(v, \theta)\, 2\pi \sin\theta\, d\theta = 2\pi \left(\frac{e^2}{4\pi\epsilon_0} \right)^2 \frac{1}{W^2} \frac{d\theta}{\theta^3} \tag{3.4.16}$$

where $W = \frac{1}{2} m v^2$ is the energy in the laboratory system. The energy transfer to a stationary target from a moving one is

$$W_L = \zeta_L(\Theta) W \tag{3.4.17}$$

where ζ_L is given by (3.2.18). Again, making the small-angle assumption, $\cos\Theta \approx 1 - \Theta^2/2$, with equal mass electrons, we obtain

$$W_L = \frac{1}{4}\Theta^2 W = \theta^2 W \tag{3.4.18}$$

and

$$dW_L = 2\theta\, d\theta\, W \tag{3.4.19}$$

Substituting (3.4.18) and (3.4.19) in (3.4.16) we have

$$dI = \pi \left(\frac{e^2}{4\pi\epsilon_0} \right)^2 \frac{1}{W} \frac{dW_L}{W_L^2} \tag{3.4.20}$$

For ionization, we integrate W_L from the ionization energy U_{iz} (for $W > U_{iz}$) to W, obtaining

$$\sigma_{iz} = \pi \left(\frac{e^2}{4\pi\epsilon_0}\right)^2 \frac{1}{W} \left(\frac{1}{U_{iz}} - \frac{1}{W}\right) \qquad (3.4.21a)$$

or, using voltage units $W = e\mathcal{E}$, $U_{iz} = e\mathcal{E}_{iz}$,

$$\sigma_{iz} = \pi \left(\frac{e}{4\pi\epsilon_0}\right)^2 \frac{1}{\mathcal{E}} \left(\frac{1}{\mathcal{E}_{iz}} - \frac{1}{\mathcal{E}}\right) \qquad \mathcal{E} > \mathcal{E}_{iz} \qquad (3.4.21b)$$

which is the *Thompson cross section*. For $\mathcal{E} < \mathcal{E}_{iz}$, $\sigma_{iz} = 0$. The ionization cross section reaches its maximum value for $\mathcal{E} = 2\mathcal{E}_{iz}$,

$$\sigma_{iz}(\text{max}) = \frac{\pi}{4} \left(\frac{e}{4\pi\epsilon_0}\right)^2 \frac{1}{\mathcal{E}_{iz}^2}$$

and falls proportional to $\mathcal{E}^{-1}$ for $\mathcal{E} \gg \mathcal{E}_{iz}$. The cross section in (3.4.21) should be multiplied by the number of valence electrons if there is more than one.

Another classical estimate for σ_{iz} is found if the orbital electron motion and its radial distribution is taken into account. Smirnov (1981, p. 253) gives the result:

$$\sigma_{iz} = \frac{\pi}{4} \left(\frac{e}{4\pi\epsilon_0}\right)^2 \frac{1}{\mathcal{E}} \left(\frac{5}{3\mathcal{E}_{iz}} - \frac{1}{\mathcal{E}} - \frac{2\mathcal{E}_{iz}}{3\mathcal{E}^2}\right), \qquad \mathcal{E} > \mathcal{E}_{iz} \qquad (3.4.22)$$

which has twice the maximum value of the Thomson cross section at $\mathcal{E} \approx 1.85\,\mathcal{E}_{iz}$. Practical formulae for cross sections can be found in Barnett (1989). A quantum mechanical calculation shows that $\sigma_{iz} \propto \ln \mathcal{E}/\mathcal{E}$ at high energies.

The ionization rate, at a given energy, is obtained from the cross section as

$$\nu_{iz} = n_g \sigma_{iz} v$$

which falls as v^{-1} for $\mathcal{E} \gg \mathcal{E}_{iz}$. As with the collision frequency, the ionizations are usually caused by a distribution of electron energies, and particularly for a low-temperature Maxwellian (say $T_e = 4$ V) ν_{iz} is very sensitive to the exponential tail of the distribution. This also implies great sensitivity to the form of the distribution function. We shall encounter this effect, and the problems of analysis arising from it, in calculating the particle balance in discharges. In the next section we consider the effective collision parameters when integrated over the particle distributions.

Electron Excitation Cross Section

A simple classical estimate for excitation to a given energy level $\mathcal{E}_n$ can be obtained by following the Thomson procedure but integrating dI over the energy W_L transferred from $e\mathcal{E}_n$ (for $W > e\mathcal{E}_n$) to $\min(W, e\mathcal{E}_{n+1})$. For the total excitation cross section σ_{ex},

dI can be integrated from $e\mathcal{E}_2$ (for $W > e\mathcal{E}_2$) to $\min(W, U_{iz})$. We leave this as an exercise for the reader. Quantum mechanics shows that the cross sections to levels that are optically forbidden (electric dipole radiation to the ground state is forbidden) are smaller and fall off faster with energy above the peak than for electron impact excitation to optically allowed levels.

For real gases, the atomic cross sections are only approximated by the analytic expressions found here. More accurate determinations are made experimentally using crossed beam techniques. As an example, for argon, which is a commonly used gas in discharges, the electron elastic, ionization, and excitation cross sections are shown in Fig. 3.13. The ionization cross section reasonably follows the analytic estimates with $\mathcal{E}_{iz} = 15.76$ V. The analytic form (3.4.21b) with six valence electrons has $\sigma_{iz}(\max) \approx 3.9 \times 10^{-16}$ cm^2 at $\mathcal{E} \approx 31.6$ V, while the experimental values, from Fig. 3.13, are $\sigma_{iz}(\max) \approx 3.9 \times 10^{-16}$ cm^2 at $\mathcal{E} \approx 60$ V. The total excitation cross section roughly follows the ionization cross section, except that it extends to lower energies as the excitation threshold energy is roughly $\mathcal{E}_{ex} \approx \frac{3}{4}\mathcal{E}_{iz}$; for argon $\mathcal{E}_{ex} \approx 11.55$ V. The elastic scattering cross section, on the other hand, has a low energy dependence due to a quantum mechanical resonance, the Ramsauer minimum, and therefore follows neither the hard-sphere nor the polarization models. At the higher energies the electrons can penetrate into the atomic cloud and a cross section $\sigma_{el} \propto v^{-2}$ is found, which implies an admixture of polarization and Coulomb scattering.

Ion–Atom Charge Transfer

A positive ion can collide with an atom so as to capture a valence electron, resulting in a transfer of the electron from the atom to the ion. In general, the energy of the level from which the electron is released is not equal to the energy of the level into which

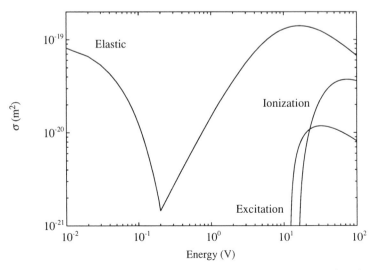

FIGURE 3.13. Ionization, excitation and elastic scattering cross sections for electrons in argon gas (compiled by Vahedi, 1993).

the electron is captured, leading to an *energy defect* ΔW, which may be positive or negative. For $\Delta W \neq 0$, the kinetic energy of the colliding particles is not conserved in the collision. If, however, the atom and ion are parent and child, then the transfer can occur with zero defect; e.g.,

$$Ar^+ \text{ (fast)} + Ar \text{ (slow)} \rightarrow Ar \text{ (fast)} + Ar^+ \text{ (slow)} \tag{3.4.23}$$

and the process is said to be *resonant*. Although the ion and atom change their internal states, their kinetic energy is conserved. The cross section for resonant charge transfer is large at low collision energies, making this an important process in weakly ionized plasmas. Here we give a simple classical estimate of charge transfer that provides a qualitative picture of the process. A more complete understanding depends on molecular phenomena that will be considered further in Chapter 8. For a more thorough treatment of the phenomena, the reader should consult the monograph by Bransden and McDowell (1992).

For the reaction

$$A^+ + B \text{ (at rest)} \rightarrow A + B^+ \text{ (at rest)} \tag{3.4.24}$$

the transfer from level n of B requires two steps: release from B and capture by A^+. For a center-to-center separation a_{12} of A^+ and B, the potential energy of the electron in level n of B is

$$W = -\frac{U_{izB}}{n^2} - \frac{e^2}{4\pi\epsilon_0 a_{12}} \tag{3.4.25}$$

where the first term, from (3.4.8), is the energy when A^+ is not present, and the second term is the additional electrostatic energy due to the nearby positive charge A^+. The potential energy $U(z)$ of an electron in the Coulomb fields of the A^+ and B^+ ions is

$$U(z) = -\frac{e^2}{4\pi\epsilon_0 z} - \frac{e^2}{4\pi\epsilon_0 |a_{12} - z|} \tag{3.4.26}$$

where z is the distance from the center of A^+ toward B. As sketched in Fig. 3.14, $U(z) \rightarrow -\infty$ at the centers of A^+ and B^+ and has its maximum value

$$U_{max} = -\frac{e^2}{\pi\epsilon_0 a_{12}} \tag{3.4.27}$$

at $z = a_{12}/2$. The condition for release from B is found by equating W to U_{max} (see figure), giving

$$a_{12} = \frac{3e^2 n^2}{4\pi\epsilon_0 U_{izB}} \tag{3.4.28}$$

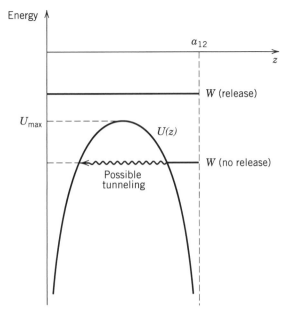

FIGURE 3.14.   Illustrating the calculation of ion–atom charge transfer.

For capture into level n' of A, the energy defect is

$$\Delta W_{AB} \approx \frac{U_{izB}}{n^2} - \frac{U_{izA}}{n'^2} \tag{3.4.29}$$

The capture is energetically possible only if

$$\frac{1}{2}m_{A+}v_{A+}^2 \gtrsim \Delta W_{AB} \tag{3.4.30}$$

At the low incident velocities of interest in weakly ionized discharges, we have $v_{A+} \ll v_{at}$, where v_{at} given by (3.4.5) is the characteristic electron velocity in the atom. In this case, capture of the released electron occurs with high probability because the collision time $t \sim a_{12}/v_{A+}$ is long compared to the atomic timescale t_{at} given in (3.4.7). Hence we estimate

$$\sigma_{cx} \approx \pi a_{12}^2 \qquad \text{for } \frac{1}{2}m_{A+}v_{A+}^2 \gtrsim \Delta W_{AB}$$
$$\approx 0 \qquad \text{otherwise} \tag{3.4.31}$$

with a_{12} given by (3.4.28). For ground-state resonant transfer ($A \equiv B$), (3.4.28) gives a cross section that is independent of energy:

$$\sigma_{cx} \approx 36\pi \left(\frac{e^2}{8\pi\epsilon_0 U_{iz}}\right)^2 \tag{3.4.32}$$

where the quantity in parentheses is approximately the atomic radius of the ground-state atom.

The cross section (3.4.32) does not show a velocity dependence. However, more detailed theoretical calculations and experiments show that σ_{cx} varies as (Rapp and Francis, 1962)

$$\sigma_{cx} \sim \frac{1}{\mathcal{E}_{iz}}(C_1 - C_2 \ln v_{A+})^2 \tag{3.4.33}$$

in the range of v_{A+} from 10^5 to 10^8 cm/s, with $C_1 \approx 1.58 \times 10^{-7}$, $C_2 \approx 7.24 \times 10^{-8}$, $\mathcal{E}_{iz}$ the ionization potential of A in volts, and σ_{cx} in cm^2. The explanation is indicated in Fig. 3.14. Even though electron release from B is not permitted classically, the electron can tunnel through the potential barrier quantum mechanically.

We can understand the form of (3.4.33) as follows (Smirnov, 1981): The ground-state valence electron in B oscillates in the Coulomb field of the nucleus with a period $\tau \approx h/e\mathcal{E}_{iz}$. The probability P that the electron tunnels across a potential barrier of height $\mathcal{E}_{iz}$ from $x = 0$ to $x = b_0$ in one oscillation is found by solving the Schrödinger equation for the electron wave function $\Psi(x)$,

$$-\frac{\hbar^2}{2m}\frac{d^2\Psi}{dx^2} = -e\mathcal{E}_{iz}\Psi \tag{3.4.34}$$

within this interval of x. We obtain $P = |\Psi(b_0)/\Psi(0)|^2 = e^{-2\alpha b_0}$, where

$$\alpha = \left(\frac{2me\mathcal{E}_{iz}}{\hbar^2}\right)^{1/2} \tag{3.4.35}$$

The time for the electron to tunnel from B to A^+ is then τP. Equating this time to the collision time b_0/v_{A+} and solving for b_0, we obtain

$$b_0 \approx \frac{1}{2\alpha}\ln\left(\frac{hv_{A+}}{e\mathcal{E}_{iz}b_0}\right) \tag{3.4.36}$$

Estimating the cross section as $\sigma_{cx} \approx \pi b_0^2$ and rearranging, we obtain the form (3.4.33).

The cross section (3.4.33) is based on the assumption of straight-line trajectories for the collision. At low collision energies, the trajectories are strongly perturbed by the polarization force and the collision partners can be "captured," as described in Section 3.3. The cross section σ_L for capture is given by (3.3.13). For such a capture, the probability of resonant charge transfer is $\frac{1}{2}$ (equal probability that the electron is found on either particle). Hence we can estimate

$$\sigma_{cx} \approx \frac{1}{2}\sigma_L \tag{3.4.37}$$

for low collision energies. The condition that the trajectories be strongly perturbed can be estimated from the dynamics in the polarization potential for typical polarizabilities to be $v_{A+} \lesssim 10^5/A_R^{1/2}$ cm/s, where A_R is the reduced mass in amu.

Experimental values for resonant charge transfer and elastic (polarization) scattering of noble gas ions in their parent gases are shown in Fig. 3.15. Because kinetic energy is conserved, resonant charge transfer acts as an elastic collision. At low energies, the cross sections are large. Because the resonant charge transfer cross section is large, the particles are practically undeflected in the CM system, leading after the charge transfer to an effective scattering angle for the ion, in the CM system, of 180° and a momentum transfer of $2m_R v_R$ for every collision. Hence the momentum transfer cross section for resonant charge transfer is

$$\sigma_{mi} = 2\sigma_{cx} \tag{3.4.38}$$

Ion–Atom Ionization

An ion colliding with an atom would be expected to transfer only a small fraction $\sim 2m/M$ of its kinetic energy $\mathcal{E}$ to a valence electron. Hence one might expect significant ionization only for $\mathcal{E} \gtrsim (M/2m)\mathcal{E}_{iz} \sim 10^4$–$10^5$ V. Experimentally, however, significant ionization is seen for $\mathcal{E} \gtrsim 100$ V; e.g., for argon ions in argon gas, $\sigma_{iz,i} \approx 10^{-16}$ cm^2 at $\mathcal{E} \sim 200$ V (Haugsjaa and Amme, 1970). This phenomenon may be due to the formation of an unstable Ar_2^+ molecular complex. We consider processes such as this in Chapter 8. Such a process may be important in the high-voltage sheaths of capacitive rf discharges.

3.5 AVERAGING OVER DISTRIBUTIONS AND SURFACE EFFECTS

Averaging over a Maxwellian Distribution

To obtain the collision quantities in a plasma we integrate over the velocity distribution functions of the particles. The collision frequency and rate constant are then

$$\nu = n_g K = n_g \langle \sigma(v_R) v_R \rangle_{\mathbf{v}_1, \mathbf{v}_2}$$
$$= n_g \int d^3 v_1 \, d^3 v_2 \, f_1(\mathbf{v}_1) f_2(\mathbf{v}_2) \, \sigma(v_R) v_R \tag{3.5.1}$$

where the distributions f_1 and f_2 have been normalized to unity and $v_R = |\mathbf{v}_1 - \mathbf{v}_2|$. If the characteristic velocities of the target particles are much less than those of the incident particles, which is often the case, then $v_R \approx |\mathbf{v}_1|$, and the $\mathbf{v}_2$ integration is trivially done. We usually take the incident distribution to be an isotropic Maxwellian, since this is the natural outcome of collisional processes, as derived in Appendix B.

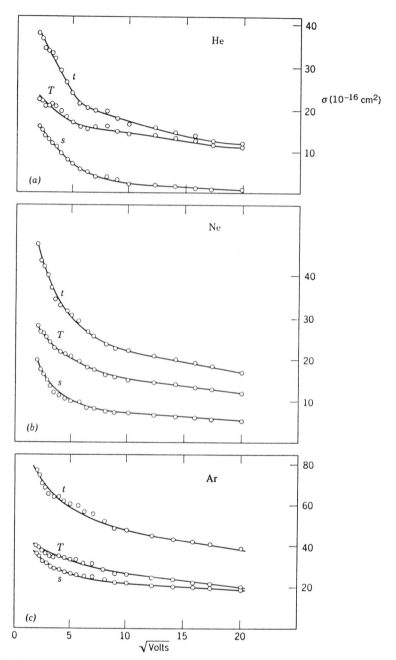

FIGURE 3.15. Experimental values for elastic scattering (s), charge transfer (T), and the sum of the two mechanisms (t) for helium, neon, and argon ions in their parent gases (McDaniel et al., 1993).

The rate constant is then (writing v for v_1)

$$K(T) = \langle \sigma(v)v \rangle_v$$

$$= \left(\frac{m}{2\pi kT} \right)^{3/2} \int_0^\infty \sigma(v)v \exp\left(-\frac{mv^2}{2kT} \right) 4\pi v^2 \, dv \qquad (3.5.2)$$

where m and T are the incident particle mass and temperature.

For a hard-sphere collision, for which $\sigma = \pi a_{12}^2$ independent of v, the integration is easily performed, yielding

$$K(T) = \pi a_{12}^2 \bar{v} \qquad (3.5.3)$$

where $\bar{v} \propto T^{1/2}$ is the mean speed from (2.4.9). For polarization scattering with $\sigma \propto 1/v$, we find $K(T) = \text{const}$, independent of T. For Coulomb scattering that has a velocity dependence $\sigma_{el} \propto 1/v^4$, from (3.3.7) (if we consider $\ln \Lambda$ as a constant for purposes of integration), calculating K_{el} as in (3.5.2) leads to a logarithmic infinity at $v = 0$. This is apparent, rather than real, as the momentum transfer rate constant K_m obtained from (3.1.15), which we use in the force equation, remains finite (see, for example, Holt and Haskell, 1965, Chapter 10). For electron–atom ionization and excitation, with $T_e \sim 4 \text{ V} \ll \mathcal{E}_{iz}, \mathcal{E}_{ex}$, the threshold energies, only the tail of the Maxwellian and the behavior of $\sigma(v)$ near threshold contribute to the rate constant, as shown in Fig. 1.9. For ionization, we can expand the Thomson cross section (3.4.21b) near $\mathcal{E} = \mathcal{E}_{iz}$ to obtain

$$\sigma_{iz}(\mathcal{E}) = \sigma_0 \frac{\mathcal{E} - \mathcal{E}_{iz}}{\mathcal{E}_{iz}} \qquad \mathcal{E} > \mathcal{E}_{iz}$$

$$= 0 \qquad \mathcal{E} \le \mathcal{E}_{iz}$$

where $\sigma_0 = \pi(e/4\pi\epsilon_0\mathcal{E}_{iz})^2$ and where $\mathcal{E} = \frac{1}{2}mv^2/e$. Inserting this into (3.5.2) and integrating, we obtain

$$K_{iz}(T_e) = \sigma_0 \bar{v}_e \left(1 + \frac{2T_e}{\mathcal{E}_{iz}} \right) e^{-\mathcal{E}_{iz}/T_e} \qquad (3.5.4)$$

where $\bar{v}_e = (8eT_e/\pi m)^{1/2}$. We leave the details to a problem.

In general, for electron collisions with atoms, the experimentally determined cross sections can be weighted by the electron distribution function and numerically integrated. Using the measured ionization, excitation, and elastic scattering cross sections for argon, given in Fig. 3.13, we obtain the rate constants $K(T_e)$ shown in Fig. 3.16. The rate constants are smoothed by the integration. Below the ionization and excitation threshold energies, there is an exponential decrease of the ionization and excitation rate constants with T_e, describing the exponentially decreasing number of electrons that are able to ionize or excite the atom.

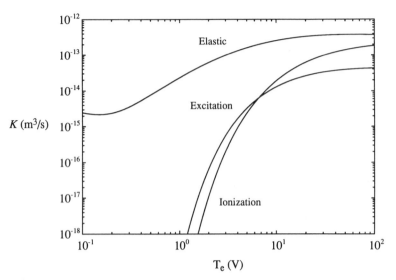

FIGURE 3.16. Electron collision rate constants K_{iz}, K_{ex} and K_m versus T_e in argon gas (compiled by Vahedi, 1993).

As crude analytical approximations to K_{iz} and K_{ex}, over a limited range of T_e, we can fit the data to an Arrhenius form, obtaining, e.g.,

$$K_{iz} \approx K_{iz0}\, e^{-\mathcal{E}_{iz}/T_e} \tag{3.5.5}$$

where $\mathcal{E}_{iz}$ is the ionization energy and where the preexponential factor for argon is $K_{iz0} \approx 5 \times 10^{-14}$ m^3/s. For elastic scattering we can do a similar fit, but we most often approximate

$$K_{el} \approx K_{el0} \approx 10^{-13} \text{ m}^3/\text{s} \tag{3.5.6}$$

For ion–atom collisions, we most often require the total ion–atom scattering cross section for low-energy ions ($T_i \sim 0.05$ V), which we estimate from the data in Fig. 3.15 to be

$$\sigma_i \approx 10^{-14} \text{ cm}^2$$

Using (3.1.6), we obtain

$$\lambda_i = \frac{1}{n_g \sigma_i} \approx \frac{1}{330 p} \text{ cm}, \qquad (p \text{ in Torr}) \tag{3.5.7}$$

Energy Loss per Electron-Ion Pair Created

A very important quantity that we use in subsequent chapters is the collisional energy loss per electron–ion pair created, $\mathcal{E}_c(T_e)$, which is defined as

$$K_{iz}\mathcal{E}_c = K_{iz}\mathcal{E}_{iz} + K_{ex}\mathcal{E}_{ex} + K_{el}\frac{3m}{M}T_e \tag{3.5.8}$$

The terms on the RHS of (3.5.8) account for the loss of electron energy due to ionization, excitation, and elastic (polarization) scattering against neutral atoms. These are usually the dominant energy losses in weakly ionized electropositive discharges. The quantity $(3m/M)T_e$ is the mean energy lost per electron for a polarization scattering, as determined using (3.2.19). The resultant values of $\mathcal{E}_c$ for argon and oxygen shown in Fig. 3.17 are obtained using data such as that given in Fig. 3.16 for argon. At high temperatures, $\mathcal{E}_c$ asymptotes to somewhat less than $2\mathcal{E}_{iz}$. At temperatures below $\mathcal{E}_{iz}$ the energy loss per ionizing collision rises as the excitation energy loss exceeds that due to ionizations, and at low temperatures the elastic energy transfer becomes important. For a typical discharge with a temperature $T_e = 4$ V, approximately 40 V of energy is lost per ionizing collision in argon.

For molecular gases, additional collisional energy losses include excitation of vibrational and rotational energy levels, molecular dissociation, and, for electronegative gases, negative ion formation. We discuss these processes in Chapter 8. In a molecular gas, $\mathcal{E}_c$ can be a factor of 2–10 times higher than for a noble gas at the same electron temperature.

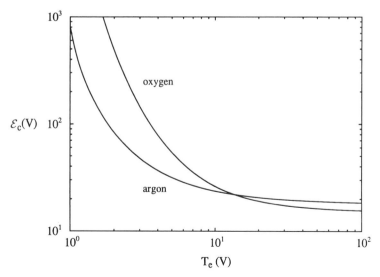

FIGURE 3.17. Collisional energy loss per electron–ion pair created, $\mathcal{E}_c$, versus T_e in argon and oxygen (compiled by Vahedi, 1993).

Surface Effects

A few facts must be described about collisions of particles with surfaces. Averaged over short time scales, electrons and positive ions arrive at surfaces in equal numbers, and almost all electron–ion pairs recombine on surfaces, leading to the reinjection of neutral atoms back into the discharge. Hence we will treat surfaces as "black holes" for charged particles. High-energy ions can also sputter neutral atoms from surfaces or can cause secondary electrons to be emitted from surfaces. If Γ_i is the incident ion flux, then, with secondary emission coefficients γ,

$$\Gamma_{\text{sput}} = \gamma_{\text{sput}}\Gamma_i$$
$$\Gamma_{\text{se}} = \gamma_{\text{se}}\Gamma_i$$

For incident ion energies of order 1 kV, we find $\gamma_{\text{sput}} \sim 1$, $\gamma_{\text{se}} \sim 0.1$–$0.2$ for metals, and $\gamma_{\text{se}} \sim 1$ for some insulators. Sputtering is an important process by which films are deposited on substrates, and secondary emission is a critical process for maintaining dc glow discharges. We describe surface processes in detail in Chapter 9. Applications of secondary electron emission and sputtering are described in Chapters 14 and 16.

PROBLEMS

3.1. Scattering Angle Transformations   Show using momentum conservation that for collision of a projectile with an initially stationary target (3.2.8) holds for the transformation of scattering angles between the laboratory and CM systems.

3.2. Hard Sphere Scattering   Using (3.2.10) and (3.2.12), find the differential scattering cross section in the laboratory system for a hard-sphere elastic collision of a projectile of mass m_1 with an initially stationary target of mass m_2.

3.3. Differential Scattering Cross Section   Using (A.13), (3.2.10) and (3.2.12), find the differential scattering cross section in the laboratory system for Coulomb scattering of an electron with an initially stationary electron.

3.4. Momentum Transfer for Coulomb Collisions   Use (3.3.1) in (3.1.15) and integrate from Θ_{min} to Θ_{max} to estimate the momentum transfer cross section $\sigma_m(v_R)$ for Coulomb collisions, where Θ_{min} and Θ_{max} are determined by setting $b = \lambda_{\text{De}}$ and $b = b_0$ in (3.2.24), respectively.

3.5. Large-Angle Coulomb Scattering   Integrate (3.3.3) over the appropriate solid angles to obtain (3.3.4).

3.6. Small-Angle Polarization Scattering   For small-angle polarization scattering, determine the differential scattering cross section (3.2.28) in the CM system using the potential (3.3.11).

3.7. Cross Sections   A point mass m having incoming speed v is scattered by a fixed (infinite mass) elastic hard sphere of radius a.

(a) Show that the differential elastic scattering cross section is $I(v, \theta) = a^2/4$.

(b) Find the elastic scattering cross section σ_{el} and the momentum transfer cross section σ_m and compare.

(c) Modeling electron–neutral elastic scattering in 20 mTorr argon gas at $25°$C as hard-sphere scattering with $a = \alpha_p^{1/3}$, where $\alpha_p = 11.08\, a_0^3$ is the polarizability of argon atoms ($a_0 \approx 0.53 \times 10^{-8}$ cm is the Bohr radius), and with v corresponding to a 5-V electron, find the mean free path λ_{el} and the collision frequency ν_{el} for scattering.

(d) For a Maxwellian distribution of electrons at temperature T_e, find the average power p_{el} per unit volume lost by the electron fluid due to hard-sphere elastic scattering (cross section σ_0) against neutrals of mass $M \gg m$ and density n_g. Show that the elastic-scattering energy loss term in the expression for $\mathcal{E}_c$ in (3.5.8) is not correct for hard-sphere scattering, and give the correct coefficient for this term. (The coefficient is correct for polarization scattering.)

3.8. Excitation Cross Section   Estimate the total cross section σ_{ex} for electron impact excitation of an atom having one valence electron in the $n = 1$ ground state to the $n > 1$ bound states. As a simple model (the Bohr atom), if $\mathcal{E}_{iz}$ is the ionization potential of the ground state, then the $n > 1$ states have energies lying between $3\mathcal{E}_{iz}/4$ and $\mathcal{E}_{iz}$.

(a) To do this, integrate the differential cross section $I(v, \theta)$ for small-angle Coulomb scattering of the incoming electron (energy $\mathcal{E}$ in volts) by the (initially stationary) valence electron over all scattering angles θ for which the energy transfer $\mathcal{E}_L$ to the valence electron lies in the energy range from $3\mathcal{E}_{iz}/4$ to $\min(\mathcal{E}_{iz}, \mathcal{E})$. The required procedure is similar to that used to obtain the Thomson ionization cross section σ_{iz}.

(b) Plot (linear scales) $\sigma_{ex}(\mathcal{E})$ and the Thomson cross section $\sigma_{iz}(\mathcal{E})$ versus $\mathcal{E}/\mathcal{E}_{iz}$ on the same graph and compare.

3.9. Ionization Rate Constant   For most gas discharges, the electron temperature $T_e \ll \mathcal{E}_{iz}$, the ionization energy of the gas atoms. Thus, electrons in the tail of the Maxwellian distribution are responsible for the ionization of the gas.

(a) Using the Thomson formula for the ionization cross section near the threshold energy $\mathcal{E} = \mathcal{E}_{iz}$, obtain the ionization rate constant K_{iz} given in (3.5.4).

(b) Plot K_{iz} (log scale) versus T_e (linear scale, in V) for $\mathcal{E}_{iz} = 15.8$ V (argon gas) and T_e in the range 1 to 6 V.

3.10. Ionization from Metastable State

(a) Following the Thomson procedure, estimate the ionization cross section per valence electron from the metastable $n = 2$ level, having energy $\mathcal{E}_2 = \mathcal{E}_{iz}/4$, and find the ratio of the maximum metastable-to-ground-state Thomson ionization cross sections.

(b) Using your results in (a) and the expression (3.5.4) for the ionization rate constant, find the ratio of the metastable to the ground-state ionization rate constants for argon with $\mathcal{E}_{iz} = 15.8$ V and $T_e = 3$ V.

3.11. Charge Transfer to a Multiply Ionized Ion   Following the approach used in Section 3.4, determine the maximum charge transfer cross section from the ground state of an atom to an ion having a positive charge of $+Ze$, where $Z > 1$.

3.12. Energy Transfer   Consider the inelastic collision of two bodies A and B to form a single body AB*, where AB* is an excited state of AB having excitation energy $\mathcal{E}_{ex}$. Let A and B have masses m_A and m_B and initial speed v_A and $v_B \equiv 0$. Using momentum and energy conservation, find the speed v_{AB*} and the excitation energy $\mathcal{E}_{ex}$ after the collision. Hence show that $\mathcal{E}_{ex}$ can never be zero; i.e., two bodies cannot collide elastically to form one body.

CHAPTER 4

PLASMA DYNAMICS

4.1 BASIC MOTIONS

The equations of motion for a particle acted on by electric and magnetic fields are

$$m\frac{d\mathbf{v}}{dt} = q\left[\mathbf{E}(\mathbf{r}, t) + \mathbf{v} \times \mathbf{B}(\mathbf{r}, t)\right] \tag{4.1.1a}$$

$$\frac{d\mathbf{r}}{dt} = \mathbf{v}(t) \tag{4.1.1b}$$

where the RHS of (4.1.1a) is the Lorentz force (2.2.12) and $\mathbf{v}(t)$ is the Lagrangian velocity. These equations cannot be solved for the general case where the force is a nonlinear function of $\mathbf{r}$, but solutions for various special cases can be found.

Motion in Constant Fields

For a constant electric field $\mathbf{E} = \mathbf{E}_0$ with $\mathbf{B} \equiv 0$, the particle moves with a constant acceleration along $\mathbf{E}_0$:

$$\mathbf{r}(t) = \mathbf{r}_0 + \mathbf{v}_0 t + \frac{1}{2}\mathbf{a}_0 t^2 \tag{4.1.2}$$

where $\mathbf{r}_0$ and $\mathbf{v}_0$ are the particle position and velocity at $t = 0$ and $\mathbf{a}_0 = q\mathbf{E}_0/m$.

85

For a constant magnetic field $\mathbf{B} = \hat{z}B_0$ which we take to lie along z, with $\mathbf{E} \equiv 0$, the components of (4.1.1a) are

$$m\frac{dv_x}{dt} = qv_yB_0 \tag{4.1.3a}$$

$$m\frac{dv_y}{dt} = -qv_xB_0 \tag{4.1.3b}$$

$$m\frac{dv_z}{dt} = 0 \tag{4.1.3c}$$

The trivial z motion is decoupled from the x and y motions. Differentiating (4.1.3a) and eliminating v_y using (4.1.3b), we obtain

$$\frac{d^2v_x}{dt^2} = -\omega_c^2 v_x \tag{4.1.4}$$

where

$$\omega_c = \frac{qB_0}{m} \tag{4.1.5}$$

is the *gyration* or *cyclotron frequency*. Solving (4.1.4) and using (4.1.3a) to obtain v_y, we find

$$v_x = v_{\perp 0}\cos(\omega_c t + \phi_0) \tag{4.1.6a}$$

$$v_y = -v_{\perp 0}\sin(\omega_c t + \phi_0) \tag{4.1.6b}$$

$$v_z = v_{z0} \tag{4.1.6c}$$

where $v_{\perp 0}$ is the speed perpendicular to $\mathbf{B}_0$, and ϕ_0 is an arbitrary phase. Integrating (4.1.1b) yields the particle position

$$x = r_c\sin(\omega_c t + \phi_0) + (x_0 - r_c\sin\phi_0) \tag{4.1.7a}$$

$$y = r_c\cos(\omega_c t + \phi_0) + (y_0 - r_c\cos\phi_0) \tag{4.1.7b}$$

$$z = z_0 + v_{z0}t \tag{4.1.7c}$$

where

$$r_c = \frac{v_{\perp 0}}{|\omega_c|} \tag{4.1.8}$$

is the *gyration radius*.

Equations (4.1.6) and (4.1.7) show that the particle moves in a circular orbit perpendicular to **B** having frequency ω_c and radius r_c about a *guiding center, x =* x_0, $y = y_0$, $z = z_0 + v_{z0}t$, that moves uniformly along z. Positive charges gyrate around the magnetic field according to the *left-hand rule,* and negative charges gyrate according to the *right-hand rule.* We can understand the motion by equating the inward Lorentz force to the outward centrifugal force:

$$|qv_{\perp 0}B_0| = \frac{mv_{\perp 0}^2}{r_c}$$

as shown in Fig. 4.1, which yields circular motion with a radius given by (4.1.8).

The gyrofrequency and radius are important frequency and length scales for mag-netized plasmas. In practical units, for electrons,

$$f_{ce} = \frac{\omega_{ce}}{2\pi} \approx 2.80 \times 10^6 B_0 \text{ Hz} \qquad (B_0 \text{ in gauss}) \qquad (4.1.9)$$

$$r_{ce} \approx \frac{3.37\sqrt{\mathcal{E}}}{B_0} \text{ cm} \qquad (\mathcal{E} \text{ in volts}) \qquad (4.1.10)$$

and for singly charged ions,

$$f_{ci} = \frac{\omega_{ci}}{2\pi} \approx \frac{1.52 \times 10^3 B_0}{A_R} \text{ Hz} \qquad (B_0 \text{ in gauss}) \qquad (4.1.11)$$

$$r_{ci} \approx \frac{1.44 \times 10^2 \sqrt{\mathcal{E}A_R}}{B_0} \text{ cm} \qquad (\mathcal{E} \text{ in volts}) \qquad (4.1.12)$$

where A_R is the ion mass in atomic mass units (amu). At $B_0 = 100$ G (0.01 T) and for a 15-V (ionizing) electron, we find $f_{ce} \approx 280$ MHz and $r_{ce} \approx 1.3$ mm, showing that electrons are well confined perpendicular to **B**.

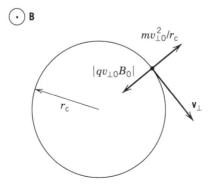

FIGURE 4.1. Charged particle gyration in a uniform magnetic field; **B** is directed out of the page.

An argon ion ($A_R = 40$) in thermal equilibrium with neutrals ($\mathcal{E} = 0.026$ V) has $f_{ci} \approx 3.8$ kHz and $r_{ci} \approx 1.4$ cm and is more weakly confined. With ambipolar acceleration (see Chapter 5), the ion can take on the electron temperature, which at 5 V would give $r_{ci} \approx 20$ cm, which is larger than a typical discharge. Hence ions are not well confined by the magnetic field. We will often model electrons as confined and ions as not confined in weakly magnetized discharges.

E × B Drifts

A simple solution is obtained for a particle moving in uniform $\mathbf{E}$ and $\mathbf{B}$ fields. Without loss of generality, we take $\mathbf{B} = \hat{z}B_0$ and $\mathbf{E} = \mathbf{E}_\perp + \hat{z}E_{z0} = \hat{x}E_{\perp 0} + \hat{z}E_{z0}$. Letting $\mathbf{v} = \hat{z}v_z(t) + \mathbf{v}_\perp(t)$ in the Lorentz force equation (4.1.1a), we obtain a uniform acceleration along z, as in (4.1.2), and the equation for the transverse motion:

$$m\frac{d\mathbf{v}_\perp}{dt} = q(\hat{x}E_{\perp 0} + \mathbf{v}_\perp \times \hat{z}B_0) \tag{4.1.13}$$

We let

$$\mathbf{v}_\perp(t) = \mathbf{v}_E + \mathbf{v}_c(t) \tag{4.1.14}$$

where $\mathbf{v}_E$ is a constant velocity. Using this in (4.1.13), we find

$$m\frac{d\mathbf{v}_c}{dt} = q(\hat{x}E_{\perp 0} + \mathbf{v}_E \times \hat{z}B_0 + \mathbf{v}_c \times \hat{z}B_0)$$

Choosing the first two terms on the RHS to cancel, we obtain

$$\mathbf{v}_E = \frac{\mathbf{E} \times \mathbf{B}}{B_0^2} \tag{4.1.15}$$

and

$$m\frac{d\mathbf{v}_c}{dt} = q\mathbf{v}_c \times \hat{z}B_0 \tag{4.1.16}$$

We can write $\mathbf{E}$ rather than $\mathbf{E}_\perp$ in (4.1.15) because $\hat{z}E_{z0} \times \mathbf{B} \equiv 0$. We have seen that the solution to (4.1.16) is gyration at frequency ω_c with gyration radius r_c. Hence the transverse motion is the sum of a guiding center drift $\mathbf{v}_E$ and a gyration:

$$\mathbf{v}_\perp(t) = \mathbf{v}_E + \mathrm{Re}\left(\mathbf{v}_{c0}e^{j\omega_c t}\right) \tag{4.1.17}$$

We note from (4.1.15) that $\mathbf{v}_E$ is perpendicular to both $\mathbf{E}$ and $\mathbf{B}$ and is independent of the mass and charge of the particles; hence electrons and ions drift with the same speed in the same direction. If $n_i = n_e$, there is no net current. Integrating (4.1.1b) using (4.1.17), we obtain

$$\mathbf{r}_\perp(t) = \mathbf{r}_{\perp 0} + \mathbf{v}_E t + \text{Re} \left(\frac{1}{j\omega_c} \mathbf{v}_{c0} e^{j\omega_c t} \right) \tag{4.1.18}$$

for the particle position. The orbits for electrons and ions are shown in Fig. 4.2 for the case where the particles are initially at rest. In this case, $|\mathbf{v}_{c0}| = |\mathbf{v}_E|$ and the kinetic energies of the drift and gyration motions are equal. The orbits are cycloids with maximum displacement $2|\mathbf{v}_E/\omega_c|$ along y as shown. Physically, $\mathbf{E}_\perp$ initially accelerates the particles along y; as they gain speed, the $\mathbf{v} \times \mathbf{B}$ force turns them back toward their initial y positions.

It is clear from the procedure used to solve (4.1.13) that *any* constant transverse force $\mathbf{F}_\perp$ acting on a gyrating particle in a constant magnetic field will give rise to a drift perpendicular to both $\mathbf{F}_\perp$ and $\mathbf{B}$:

$$\mathbf{v}_F = \frac{(\mathbf{F}_\perp/q) \times \mathbf{B}}{B_0^2} \tag{4.1.19}$$

Nonuniform magnetic fields can give rise to additional forces both along (F_z) and perpendicular ($\mathbf{F}_\perp$) to $\mathbf{B}$. We consider these forces and the resulting particle motion in Section 4.3.

Energy Conservation

Dot multiplying (4.1.1a) by $\mathbf{v}$, we obtain

$$\frac{d}{dt} \left(\frac{1}{2} mv^2 \right) = q\mathbf{v} \cdot \mathbf{E}[\mathbf{r}(t), t] \tag{4.1.20}$$

which shows that the magnetic field does no work on the particle. The rate of change of kinetic energy is equal to the power $q\mathbf{v} \cdot \mathbf{E}$ transferred from the electric field to the

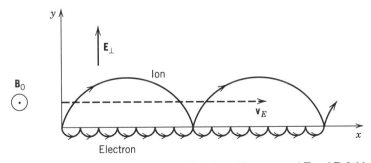

FIGURE 4.2. Motion of electrons and ions in uniform crossed $\mathbf{E}$ and $\mathbf{B}$ fields.

particle. For a static field, $\mathbf{E}(\mathbf{r}) = -\nabla\Phi$, (4.1.20) can be written as

$$\frac{d}{dt}\left(\frac{1}{2}mv^2\right) = -q\frac{d\mathbf{r}}{dt} \cdot \nabla\Phi[\mathbf{r}(t)] = -q\frac{d}{dt}\Phi[\mathbf{r}(t)]$$

which can be integrated to yield

$$\frac{1}{2}mv^2(t) + q\Phi[\mathbf{r}(t)] = \text{const} \qquad (4.1.21)$$

This expresses the energy conservation for a particle in a static electric field.

For a collection of particles (a fluid consisting of one species), the force equation

$$mn\frac{d\mathbf{u}}{dt} = qn(\mathbf{E} + \mathbf{u} \times \mathbf{B}) - \nabla p - mn\nu_m\mathbf{u} \qquad (4.1.22)$$

repeated here from (2.3.15), is more complicated, with additional terms due to pressure gradients and collisions with particles of other species. Recall that $d/dt \equiv \partial/\partial t + \mathbf{u} \cdot \nabla$ is the convective derivative and that $\mathbf{u}(\mathbf{r}, t)$ is the Eulerian fluid velocity, which is related to the Lagrangian particle velocity by $\mathbf{v}(t) = \mathbf{u}[\mathbf{r}(t), t]$. Equation (4.1.22) cannot generally be solved, even when the fields are known. Furthermore, in most cases the fields themselves are functions of the particle motions, which act as charge and current sources in the Maxwell or Poisson equations. These must be determined self-consistently with the particle motions. This coupling of particles and fields lies at the root of all plasma phenomena.

In this chapter, we describe various solutions to (4.1.1) or (4.1.22), coupling the particle motions to the fields as needed. In Section 4.2, we consider a uniform unmagnetized plasma and introduce the coupling to describe such collective phenomena as plasma oscillations, the plasma dielectric constant, and, equivalently, the plasma conductivity. The conductivity determines the ohmic power dissipation, which is an important mechanism for electron heating in discharges. We also introduce wave phenomena, which can be important for plasma heating. The remainder of the chapter is devoted to magnetized plasmas, which are finding increasing application in materials processing. Guiding center motion in nonuniform magnetic fields is described in Section 4.3. Guiding center concepts play an important role in hot electron confinement in several low-pressure, high-density source concepts, which we describe in Chapters 11, 13, and 14. The dielectric tensor for magnetized plasmas is introduced in Section 4.4, and is used in Section 4.5 to describe waves in uniform magnetized plasmas. These waves play a critical role in energy deposition in several high-density sources, such as ECRs and helicons, which we discuss in Chapter 13, and are also important for plasma diagnostics, which we introduce in Section 4.6. Wave phenomena in nonuniform or bounded plasmas will be dealt with in the application chapters that follow, when the need for the material naturally arises. The subject of waves in plasmas is vast, and the reader should consult more specialized monographs (Allis

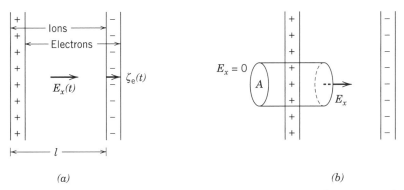

FIGURE 4.3. Plasma oscillations in a slab geometry: (a) displacement of electron cloud with respect to ion cloud; (b) calculation of the resulting electric field.

et al., 1963; Stix, 1992; Ginzburg, 1964) for more thorough treatments. We defer the study of steady-state solutions in nonuniform plasmas, which are important for particle diffusion and transport, to Chapter 5.

4.2   NONMAGNETIZED PLASMA DYNAMICS

Plasma Oscillations

As the simplest example of the coupling of particles and fields, we consider the *undriven* motion of a plasma slab of finite width l containing a density $n_e = n_i = n_0$ of cold ($T_e = 0$) electrons and infinite mass (stationary) ions. Since $n_e = n_i$, the electric field $\mathbf{E} = 0$ in the slab. Now let the slab of electrons be displaced to the right with respect to the ions by a small distance $\zeta_e(t) \ll l$ at time t, as shown in Fig. 4.3a. This leads to a surface charge density $\rho_S = en_0\zeta_e$ at the left edge due to the uncovering of the stationary ion cloud. We similarly obtain $\rho_S = -en_0\zeta_e$ at the right edge. Using Gauss' law (2.2.6) applied to the pillbox shown in Fig. 4.3b, these equal and opposite surface charges lead to an electric field within the slab:

$$E_x = \frac{en_0\zeta_e}{\epsilon_0} \tag{4.2.1}$$

The force equation for the electrons is[*]

$$m\frac{d^2\zeta_e}{dt^2} = -eE_x \tag{4.2.2}$$

---

[*]Since ζ_e is small, the $\mathbf{u} \cdot \nabla\mathbf{u}$ term in (4.1.22) is small and there is no difference between Eulerian and Lagrangian velocities.

Substituting (4.2.1) into (4.2.2) yields

$$\frac{d^2 \zeta_e}{dt^2} = -\omega_{pe}^2 \zeta_e \tag{4.2.3}$$

where

$$\omega_{pe} = \left(\frac{e^2 n_0}{\epsilon_0 m}\right)^{1/2} \tag{4.2.4}$$

the *electron plasma frequency,* is the fundamental characteristic frequency of a plasma. The solution of (4.2.3) is

$$\zeta_e(t) = \zeta_{e0} \cos(\omega_{pe} t + \phi_0) \tag{4.2.5}$$

which represents a sinusoidal oscillation of the electron cloud with respect to the ion cloud at the natural frequency ω_{pe}. In practical units,

$$f_{pe} = \frac{\omega_{pe}}{2\pi} \approx 8980 \sqrt{n_0} \text{ Hz}, \qquad (n_0 \text{ in cm}^{-3}) \tag{4.2.6}$$

Plasma frequencies for discharges are typically in the microwave region (1–10 GHz).

If the assumption of infinite mass ions is not made, then the ions also move slightly and we obtain (Problem 4.1) the natural frequency

$$\omega_p = (\omega_{pe}^2 + \omega_{pi}^2)^{1/2} \tag{4.2.7}$$

where

$$\omega_{pi} = \left(\frac{e^2 n_0}{\epsilon_0 M}\right)^{1/2} \tag{4.2.8}$$

is the ion plasma frequency. For $M \gg m$, $\omega_p \approx \omega_{pe}$.

The existence of plasma oscillations does not depend on the assumption of a slab geometry. It can be shown that any perturbed charge density oscillates at the plasma frequency (Problem 4.2). Note that the characteristic plasma scale length, velocity, and frequency are related by

$$\lambda_{De} = \frac{v_{th}}{\omega_{pe}} \tag{4.2.9}$$

Plasma oscillations are damped in time by collisions and can also be damped collisionlessly by a mechanism known as *Landau damping,* which we describe below, when considering electrostatic waves. Collisional damping usually dominates Landau damping in discharges, and the oscillations generally fall to noise levels if there are no external drives.

Dielectric Constant and Conductivity

We now consider a uniform plasma in the presence of a background gas that is *driven* by a small amplitude time-varying electric field:

$$E_x(t) = \tilde{E}_x \cos \omega t = \operatorname{Re} \tilde{E}_x e^{j\omega t} \tag{4.2.10}$$

where $\tilde{E}_x$ is the electric field amplitude. We again let the ion mass be infinite for ease of calculation, and we assume that all quantities vary sinusoidally in time at frequency ω. The electron force equation is

$$m \frac{du_x}{dt} = -eE_x - m\nu_m u_x \tag{4.2.11}$$

where ν_m is the electron–neutral collision frequency. Letting

$$u_x(t) = \operatorname{Re} \tilde{u}_x e^{j\omega t} \tag{4.2.12}$$

and using this and (4.2.10) in (4.2.11), we obtain the complex velocity amplitude

$$\tilde{u}_x = -\frac{e}{m} \frac{1}{j\omega + \nu_m} \tilde{E}_x \tag{4.2.13}$$

From (2.2.7), the total current is

$$J_{Tx} = \epsilon_0 \frac{\partial E_x}{\partial t} + J_x \tag{4.2.14}$$

where the conduction current J_x is due to the electron motion only, which, in the cold plasma approximation, is

$$\tilde{J}_x = -en_0 \tilde{u}_x \tag{4.2.15}$$

We also have that

$$\frac{\partial E_x}{\partial t} = \operatorname{Re} j\omega \tilde{E}_x e^{j\omega t}$$

such that the total current amplitude is

$$\tilde{J}_{Tx} = j\omega\epsilon_0 \tilde{E}_x - en_0 \tilde{u}_x \tag{4.2.16}$$

Using (4.2.13) in (4.2.16), we obtain

$$\tilde{J}_{Tx} = j\omega\epsilon_0 \left[1 - \frac{\omega_{pe}^2}{\omega(\omega - j\nu_m)} \right] \tilde{E}_x \tag{4.2.17}$$

which relates the total current to the electric field in the sinusoidal steady state. Hence we can introduce an effective *plasma dielectric constant*

$$\epsilon_p = \epsilon_0 \kappa_p = \epsilon_0 \left[1 - \frac{\omega_{pe}^2}{\omega(\omega - j\nu_m)} \right] \tag{4.2.18}$$

where κ_p is the relative dielectric constant. Maxwell's equation (2.2.2) can then be written

$$\nabla \times \tilde{H} = j\omega\epsilon_p\tilde{E} \tag{4.2.19}$$

where we can introduce the displacement vector $\tilde{D} = \epsilon_p\tilde{E}$, showing the correspondence of a plasma to a dielectric material.

We can also introduce a *plasma conductivity* by writing (4.2.17) in the form $\tilde{J}_{Tx} = (\sigma_p + j\omega\epsilon_0)\tilde{E}_x$, with

$$\sigma_p = \frac{\epsilon_0\omega_{pe}^2}{j\omega + \nu_m} \tag{4.2.20}$$

such that (2.2.2) becomes

$$\nabla \times \tilde{H} = (\sigma_p + j\omega\epsilon_0)\tilde{E} \tag{4.2.21}$$

Equations (4.2.19) and (4.2.21) are equivalent. Hence we can consider a plasma to be either a dielectric ϵ_p or a conductor σ_p, as we find useful. For low frequencies $\omega \ll \nu_m, \omega_{pe}$, we find that $\sigma_p \to \sigma_{dc}$, where

$$\sigma_{dc} = \frac{\epsilon_0\omega_{pe}^2}{\nu_m} = \frac{e^2 n_0}{m\nu_m} \tag{4.2.22}$$

which is the *dc plasma conductivity* in the cold plasma approximation. For electron–ion rather than electron–neutral collisions, (4.2.22) is replaced by the parallel Spitzer conductivity

$$\sigma_{ei} \approx \frac{0.019\, T_e^{3/2}}{\ln \Lambda} \quad \Omega^{-1}\text{-m}^{-1} \quad (T_e \text{ in volts}) \tag{4.2.23}$$

where $\ln \Lambda$ is defined in (3.3.6).

For high frequencies it is more useful to consider ϵ_p rather than σ_p. For $\omega \gg \nu_m$, (4.2.18) reduces to the collisionless plasma dielectric constant

$$\epsilon_p = \epsilon_0 \kappa_p = \epsilon_0 \left(1 - \frac{\omega_{pe}^2}{\omega^2} \right) \qquad (4.2.24)$$

At very high driving frequencies (in the high microwave regime) where $\omega > \omega_{pe}$, ϵ_p is positive but less than ϵ_0; hence the plasma acts as a dielectric with a relative dielectric constant less than unity. At lower frequencies, $\omega < \omega_{pe}$, which is true for most discharges driven at rf frequencies, we see that $\epsilon_p < 0$. A slab of such a plasma of width l and cross-sectional area A then has a capacitance $C = \epsilon_p A/l$ that is negative, corresponding to an impedance $Z = 1/(j\omega C)$ that is inductive (positive imaginary). Hence the plasma behaves like an inductor in this frequency regime.

Figure 4.4 illustrates the rf current and electric field amplitudes and phases in the sheath and plasma regions in the regime $\nu_m \ll \omega \ll \omega_{pe}$, which is typical for low-pressure rf discharges. From (2.2.8), $\tilde{J}_{Tx}$ is the same in the sheath and plasma regions. In the sheath regions, there is only displacement current and

$$\tilde{E}_x(\text{sheath}) = \frac{\tilde{J}_{Tx}}{j\omega\epsilon_0} \qquad (4.2.25a)$$

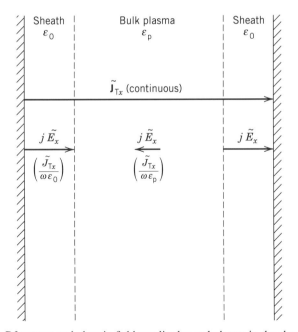

FIGURE 4.4. Rf current and electric field amplitudes and phases in the sheath and plasma regions of an rf discharge.

In the bulk plasma region,

$$\tilde{E}_x(\text{plasma}) = \frac{\tilde{J}_{Tx}}{j\omega\epsilon_p} \tag{4.2.25b}$$

Since $\epsilon_p < 0$ and $|\epsilon_p| \gg \epsilon_0$, the field in the bulk plasma is much smaller than, and 180° out of phase with, the fields in the sheaths, as shown in the figure. Hence almost all of the rf voltage is dropped across the sheath regions, and comparatively little voltage appears across the bulk plasma.

Ohmic Heating

Although the electric field within the bulk plasma is small, it gives rise to a significant electron heating due to electron–neutral collisions. The time-average power per unit volume absorbed by the plasma, p_{abs}, is given by

$$p_{\text{abs}} = \frac{1}{T} \int_0^T \mathbf{J}_T(t) \cdot \mathbf{E}(t)\, dt = \frac{1}{2}\text{Re}\left(\tilde{\mathbf{J}}_T \cdot \tilde{\mathbf{E}}^*\right) = \frac{1}{2}\text{Re}\left(\tilde{\mathbf{J}}_T^* \cdot \tilde{\mathbf{E}}\right) \tag{4.2.26}$$

where $T = 2\pi/\omega$ is the period, the asterisk denotes complex conjugation, and the latter forms follow from (4.2.10) and the equivalent expression for $\mathbf{J}_T(t)$ (Problem 4.4). If we substitute $\tilde{\mathbf{J}}_T = (\sigma_p + j\omega\epsilon_0)\tilde{\mathbf{E}}$ into (4.2.26), then we obtain the collisional (ohmic) power absorbed by the electrons in terms of the electric field amplitude $\tilde{\mathbf{E}}$:

$$p_{\text{ohm}} = \frac{1}{2}|\tilde{\mathbf{E}}|^2\, \sigma_{\text{dc}}\, \frac{\nu_m^2}{\omega^2 + \nu_m^2} \tag{4.2.27}$$

In many cases, the current density is known rather than the electric field. Letting $\tilde{\mathbf{E}} = \tilde{\mathbf{J}}_T/(\sigma_p + j\omega\epsilon_0)$ in (4.2.26), we obtain

$$p_{\text{ohm}} = \frac{1}{2}|\tilde{\mathbf{J}}_T|^2\, \text{Re}\left(\frac{1}{\sigma_p + j\omega\epsilon_0}\right) \tag{4.2.28}$$

Taking the real part of $1/(\sigma_p + j\omega\epsilon_0)$, we obtain

$$\text{Re}\left(\frac{1}{\sigma_p + j\omega\epsilon_0}\right) = \frac{1}{\sigma_{\text{dc}}}\left(\frac{\omega_{\text{pe}}^4}{(\omega_{\text{pe}}^2 - \omega^2)^2 + \omega^2\nu_m^2}\right) \tag{4.2.29}$$

For $\omega \ll \omega_{\text{pe}}$, the term in parentheses is unity and we obtain the simple result

$$p_{\text{ohm}} = \frac{1}{2}|\tilde{\mathbf{J}}_T|^2\, \frac{1}{\sigma_{\text{dc}}} \tag{4.2.30}$$

We shall apply (4.2.27) or (4.2.30) to find ohmic power absorption from waves as well as from oscillating fields. We will return to the calculation in Chapter 11 on rf discharges, where we determine $\tilde{\mathbf{J}}_T$, given the external driving source. However, we shall also find that, for low-pressure discharges, the ohmic power may not be the main source of power absorption by the plasma electrons. Rather, a mechanism of electron collisions with the oscillating sheaths can provide the principal electron heating.

Electromagnetic Waves

Waves can be important to carry energy from the surface of a plasma, where the wave is excited, into the bulk plasma, where the wave energy can be absorbed. Plasmas support both electromagnetic and electrostatic waves. Electromagnetic waves in plasmas are similar to those in dielectric materials, and propagate due to the exchange of energy between electric and magnetic forms. Letting the electric and magnetic fields of the wave vary as

$$\mathbf{E}, \mathbf{H} \sim \exp j(\omega t - \mathbf{k} \cdot \mathbf{r}) \qquad (4.2.31)$$

where $\mathbf{k}$ is the propagation vector, then for a uniform, isotropic (no applied dc magnetic field) plasma, the waves are transverse, with $\mathbf{E}$, $\mathbf{H}$, and $\mathbf{k}$ mutually perpendicular. To obtain the dispersion relation, we use (4.2.31) in (2.2.1) and (4.2.19) to obtain

$$\mathbf{k} \times \tilde{\mathbf{E}} = \omega \mu_0 \tilde{\mathbf{H}} \qquad (4.2.32)$$

and

$$\mathbf{k} \times \tilde{\mathbf{H}} = -\omega \epsilon_p \tilde{\mathbf{E}} \qquad (4.2.33)$$

Cross multiplying (4.2.32) by $\mathbf{k}$ and using (4.2.33), we obtain

$$\mathbf{k} \times (\mathbf{k} \times \tilde{\mathbf{E}}) = -\omega^2 \epsilon_p \mu_0 \qquad (4.2.34)$$

Expanding the triple cross product* and noting that $\mathbf{k} \cdot \tilde{\mathbf{E}} = 0$ for transverse waves, we obtain

$$k^2 \tilde{\mathbf{E}} = \kappa_p \frac{\omega^2}{c^2} \tilde{\mathbf{E}}$$

where k is the wave-vector magnitude, where we have written $\epsilon_p = \epsilon_0 \kappa_p$ from (4.2.18), and where we have used $c = 1/\sqrt{\mu_0 \epsilon_0}$ for the speed of light in vacuum. A nonzero $\tilde{\mathbf{E}}$ exists only if

$$k = \pm \frac{\sqrt{\kappa_p} \, \omega}{c} \qquad (4.2.35)$$

---

*$\mathbf{k} \times (\mathbf{k} \times \mathbf{E}) \equiv (\mathbf{k} \cdot \mathbf{E})\mathbf{k} - k^2 \mathbf{E}$.

which is the dispersion relation for transverse waves. Using κ_p for a cold collisionless plasma with infinite mass ions from (4.2.24), we see that the waves propagate (k is real) for $\kappa_p > 0$; i.e., for $\omega > \omega_{pe}$, and are cut off for $\omega < \omega_{pe}$. We plot ω versus k in Fig. 4.5. Because ω is generally less than ω_{pe} in a discharge, electromagnetic waves excited at the plasma surface are not able to propagate into the plasma. In this case, the fields decay exponentially into the plasma. We will consider this further in Chapter 12 to determine the power transfer in inductive discharges. On the other hand, we show in Section 4.5 that electromagnetic waves *can* propagate into a *magnetized* plasma.

The two independent polarizations have the same propagation constant k. Letting $\mathbf{k} = \hat{x}k_x$, the most general transverse wave propagates along x with a polarization that is the superposition,

$$\tilde{\mathbf{E}} = \hat{y}\tilde{E}_y + \hat{z}\tilde{E}_z$$

which specifies a general elliptical polarization. As will be seen in Section 4.5, this is *not* true for waves in a magnetized plasma.

Electrostatic Waves

In a warm plasma, waves can propagate having $\mathbf{k} \parallel \mathbf{E}$. Such waves, which are not possible in a vacuum (or dielectric), are similar to sound waves in a gas. The waves propagate due to an exchange of energy between thermal and electric forms. Thermal electron motion, not considered in deriving the dielectric constant (4.2.23), leads to an additional term in the force equation due to ∇p_e, the gradient of the electron pressure. As a result of this, the electron plasma oscillations described by (4.2.3), for which $\mathbf{k} \parallel \mathbf{E}$, are converted into electron plasma waves.

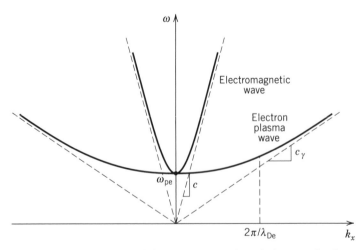

FIGURE 4.5. Dispersion ω versus k for electromagnetic and electrostatic electron plasma waves in an unmagnetized plasma.

To derive the dispersion relation, we use the property of an adiabatic equation of state (2.3.20), with $\nabla p_e/p_e = \gamma \nabla n_e/n_e$, to describe the variation of p_e, together with the usual Maxwellian relation $p_e = n_e k T_e$, with T_e constant. Substituting these quantities in (4.1.22), in the absence of a magnetic field and assuming that collisions are unimportant, we have

$$mn_e \left[\frac{\partial \mathbf{u}_e}{\partial t} + (\mathbf{u}_e \cdot \nabla)\mathbf{u}_e \right] = -en_e\mathbf{E} - \gamma k T_e \nabla n_e \tag{4.2.36}$$

where, here, k is Boltzmann's constant. We now make the usual assumptions of small signal quantities n_1, E_1, and u_1,

$$n_e = n_0 + n_1, \qquad \mathbf{E} = \hat{x}E_1, \qquad \mathbf{u}_e = \hat{x}u_1 \tag{4.2.37}$$

with no steady fields or drifts. We also assume sinusoidal wave motion, with all quantities varying as

$$n_1, E_1, u_1 \sim \exp j(\omega t - k_x x) \tag{4.2.38}$$

where k_x is the propagation constant. Unlike electromagnetic waves, the electric field is parallel to $\mathbf{k}$ so that of the field equations only the divergence equation (2.2.3) is required. We further consider that the ions are essentially stationary on the time scale of the wave frequency. Assuming that all quantities vary as in (4.2.37) and (4.2.38), substituting into the continuity equation (2.3.7) (but without sources or sinks), the force equation (4.2.36), and the divergence equation (2.2.3), we obtain the first-order equations:

$$\omega n_1 - k_x n_0 u_1 = 0 \tag{4.2.39}$$

$$j\omega mn_0 u_1 = -en_0 E_1 + jk_x \gamma k T_e n_1 \tag{4.2.40}$$

$$jk_x \epsilon_0 E_1 = en_1 \tag{4.2.41}$$

Combining (4.2.39) through (4.2.41), we can factor out the first-order quantities to find the dispersion equation:

$$\omega^2 = \omega_{pe}^2 + k_x^2 c_\gamma^2 \tag{4.2.42}$$

where

$$c_\gamma = \left(\frac{\gamma k T_e}{m} \right)^{1/2} \tag{4.2.43}$$

is the adiabatic electron sound speed. For the one-dimensional motion considered here, $\gamma = 3$. The dispersion (4.2.42) is plotted in Fig. 4.5, with the value of $k_x = 2\pi/\lambda_{De}$ indicated on the figure. As one might expect, for $k_x \gtrsim 2\pi/\lambda_{De}$, thermal disruption of the collective process would be expected to be very important, and the waves are strongly damped. This collisionless damping, called Landau damping, is discussed in most books on fully ionized plasmas (e.g., Chen, 1984, Chapter 7). For long wavelengths, $k_x \ll 2\pi/\lambda_{De}$, the waves are not strongly damped, but they may be only weakly excited.

If the ions are also considered to be mobile, under certain circumstances, new waves can appear. For cold plasmas, the electron motion dominates the behavior of the waves, such that the plasma frequency in (4.2.42) is only slightly modified, as given by (4.2.7). For equal-temperature electrons and ions, this small modification still holds. However, for $T_i \ll T_e$, as usually exists in weakly ionized discharges, the electron random motion prevents the electrons from neutralizing independent ion motion, and short wavelength ion sound waves can exist. These are usually heavily damped and therefore not of great significance. We leave details of a calculation to a problem. The calculation can be found in most texts on fully ionized plasmas, e.g., Chen (1984, Chapter 4).

4.3 GUIDING CENTER MOTION

If the electric or magnetic field varies in space, the charged particle motion becomes much more complicated, and generally analytic solutions cannot be found. One very important configuration is that of a spatially varying magnetic field in which the gyration radius is much smaller than the scale length of the field variation. In that situation, an expansion in the gyroradius can be performed that allows separation into the fast gyromotion and slow drifts of the guiding center across field lines. We have already seen this separation in Section 4.1 for the trivial case of uniform **B**, where the guiding center moves uniformly along **B**. The separation of the motion is particularly useful for calculating particle confinement in fully ionized plasmas (see, for example, Chen (1984, Chapter 2) or Schmidt (1979, Chapter 2)), but can also be applied to a number of high-density source concepts for materials processing. Here we introduce the subject and point out a few implications for weakly ionized plasmas. A more complete derivation can be found in Schmidt (1979, Chapter 2).

The basic procedure is to expand the instantaneous position into a guiding center and a gyroradius about that center,

$$\mathbf{r} = \mathbf{r}_g(t) + \mathbf{r}_c(t) \tag{4.3.1}$$

with an accompanying velocity,

$$\mathbf{v} = \mathbf{v}_g + \mathbf{v}_c \tag{4.3.2}$$

where $\mathbf{v}_g = d\mathbf{r}_g/dt$ and $\mathbf{v}_c = d\mathbf{r}_c/dt$. The magnetic field in the neighborhood of the guiding center is expanded as

$$\mathbf{B}(\mathbf{r}) = \mathbf{B}_0(\mathbf{r}) + (\mathbf{r}_c \cdot \nabla)\mathbf{B}(\mathbf{r}) \tag{4.3.3}$$

with

$$|\mathbf{r}_c \nabla \mathbf{B} / \mathbf{B}_0| \ll 1 \qquad (4.3.4)$$

With this approximation, $|\mathbf{r}_c(t)|$ can be taken as a constant over a gyroperiod. Then, averaging over a gyroperiod, the rapidly rotating terms average to zero in lowest order, resulting in an equation for the drift motion:

$$m \frac{d\mathbf{v}_g}{dt} = \mathbf{F}_{ext} + q\mathbf{v}_g \times \mathbf{B} + q\langle \mathbf{v}_c \times (\mathbf{r}_c \cdot \nabla)\mathbf{B}\rangle \qquad (4.3.5)$$

where $\langle\ \rangle$ denotes an average over a gyroperiod. The third term on the right-hand side has a product of rapidly oscillating quantities and therefore a first-order average value, giving, after some algebra,

$$m \frac{d\mathbf{v}_g}{dt} = \mathbf{F}_{ext} + q\mathbf{v}_g \times \mathbf{B}_0 - \frac{\frac{1}{2}mv_\perp^2}{B_0}\nabla B \qquad (4.3.6)$$

Here $\mathbf{F}_{ext}$ includes all external forces, $B_0 = |\mathbf{B}_0|$, and $v_\perp = |\mathbf{v}_c|$, the velocity perpendicular to the field line. All quantities are calculated on the guiding center of the orbit. We indicate the effect of the various terms in (4.3.6) with some simple examples.

Parallel Force

We justify (4.3.6) for a particle gyrating in a magnetic field $\hat{z}B_z(z)$ that is increasing along z. The magnetic field lines converge as shown in Fig. 4.6, and the Lorentz force $q\mathbf{v}_\perp \times \mathbf{B}$ has a component along z given by

$$F_z = -qv_\phi B_r \qquad (4.3.7)$$

where $v_\phi = -v_{\perp 0}$, and B_r is obtained from (2.2.4), which is, in cylindrical coordinates,

$$\frac{1}{r}\frac{\partial}{\partial r}(rB_r) + \frac{\partial B_z}{\partial z} = 0$$

This yields B_r upon integrating with respect to r:

$$B_r \approx -\frac{r_c}{2}\frac{\partial B_z}{\partial z} \qquad (4.3.8)$$

Substituting (4.3.8) in (4.3.7) and taking all quantities as constant over a gyroorbit, in keeping with our expansion, we obtain the average force acting on the guiding center to be

$$F_z = -\frac{\frac{1}{2}mv_\perp^2}{B_z}\frac{\partial B_z}{\partial z} \tag{4.3.9}$$

We see that (4.3.9) corresponds to the z component of the third term on the RHS of (4.3.6). The force F_z pushes the particle into regions of smaller B and is independent of charge. From the averaging procedure it is seen to be valid only for

$$r_c \ll \left(\frac{1}{B_z}\frac{dB_z}{dz}\right)^{-1} \tag{4.3.10}$$

which is equivalent to (4.3.4).

Adiabatic Constancy of the Magnetic Moment

For the field of Fig. 4.6 we introduce the quantity

$$\mu_{\text{mag}} = \frac{\frac{1}{2}mv_\perp^2}{B_z} = \frac{W_\perp}{B_z} \tag{4.3.11}$$

which can be shown to be the *magnetic moment* of the particle (see Problem 4.8). As the particle moves, both B_z and $W_\perp$ can change; however, the total kinetic energy of the particle is conserved because the magnetic field does no work. For the above example,

$$W_\perp(z) + W_z(z) = \text{const} \tag{4.3.12}$$

where $W_z = \frac{1}{2}mv_z^2$. If the particle moves a distance dz, then

$$dW_z = F_z\, dz = -\frac{W_\perp}{B_z}\, dB_z \tag{4.3.13}$$

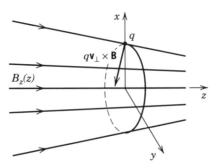

FIGURE 4.6.   Calculation of the parallel force due to a magnetic field gradient $\partial B_z/\partial z$.

Differentiating (4.3.12) yields $dW_z = -dW_\perp$; hence (4.3.13) becomes

$$\frac{dW_\perp}{W_\perp} = \frac{dB_z}{B_z} \qquad (4.3.14)$$

which can be integrated to obtain

$$\frac{W_\perp}{B_z} \equiv \mu_{mag} = \text{const} \qquad (4.3.15)$$

The magnetic moment is one example of an *adiabatic invariant,* a quantity that is approximately conserved in the motion if the scale length condition (4.3.4) is satisfied.

The constancy of μ_{mag} has an important consequence in the magnetic mirroring of charged particles in an increasing magnetic field. As B_z increases, $W_\perp$ increases to keep μ_{mag} constant, reflecting the particle when $W_\perp = W$ $(W_z = 0)$. Although this property is of primary concern in nearly collisionless plasmas where plasma confinement is of greatest interest, it can also play a significant role in confining the higher-energy electrons in cyclotron resonance or magnetron discharges, which we consider in Chapters 13 and 14.

Drift Due to Motion Along Field Lines (Curvature Drift)

Consider a curved field line in the x–z plane. As shown in Fig. 4.7, although $B_x = 0$ at the origin, $\partial B_x / \partial z$ is nonzero. The radius of curvature R of the field line is found from (see figure)

$$\frac{dz}{R} = -\frac{dB_x}{B_z}$$

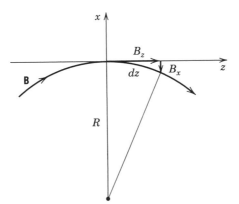

FIGURE 4.7. Calculation of the curvature drift due to a magnetic field gradient $\partial B_x / \partial z$.

which yields

$$\frac{1}{R} = -\frac{1}{B_z}\frac{\partial B_x}{\partial z} \qquad (4.3.16)$$

The centrifugal force acting on the particle is

$$\mathbf{F}_R = \frac{mv_z^2}{R}\hat{x} = \frac{2W_z}{R}\hat{x} = -\frac{2W_z}{B_z}\frac{\partial B_x}{\partial z} \qquad (4.3.17)$$

Since the force in (4.3.17) is an average force, we can substitute it into (4.1.19) to obtain the drift of the guiding center due to the field line curvature:

$$\mathbf{v}_R = \frac{2W_z}{qB_z^2}\frac{\partial B_x}{\partial z}\hat{y} \qquad (4.3.18)$$

We see that electrons and ions drift in opposite directions perpendicular to both **B** and the curvature force, giving rise to a net current. The drift given in (4.3.18) is not immediately seen in the averaged equation (4.3.6). To obtain the drifts, (4.3.6) is cross-multiplied by $\mathbf{B}/qB^2$, such that the second term on the right is $\mathbf{v}_{g\perp}$. The drift (4.3.18) is then obtained from the vector decomposition of $(m\,d\mathbf{v}_g/dt)\times\mathbf{B}/qB^2$. We leave the calculation to Problem 4.9.

Drift Due to Gyration (Gradient Drift)

Consider a magnetic field $B_z(x)$ with a gradient perpendicular to the lines of **B**, as shown in Fig. 4.8a. Viewing the motion of a gyrating particle in the x–y plane (Fig. 4.8b), we see that there is a stronger Lorentz force at the upper half of the orbit

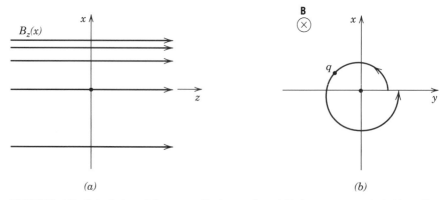

(a)   (b)

FIGURE 4.8. Calculation of the perpendicular gradient drift due to a magnetic field gradient $\partial B_z/\partial x$: (a) the magnetic field lines; (b) the motion viewed in the x–y plane.

than at the lower half, producing a smaller gyration radius at the upper half than at the lower, and leading to a net drift along y. This drift can be obtained directly from the third term on the right in (4.3.6) with $\nabla B = \nabla B_z = \partial B_z / \partial x$, which, as an average force, can be substituted into (4.1.19) to give (see also Problem 4.10)

$$\mathbf{v}_{\nabla B} = -\frac{W_\perp}{qB_z^2} \nabla_\perp B_z \times \hat{z} \tag{4.3.19}$$

Electrons and ions drift in opposite directions, giving rise to a net current, as with the curvature drift.

Note that although the two drifts found in (4.3.18) and (4.3.19) are commonly called curvature drift and gradient drift, they are really distinguished by a velocity parallel to field lines and perpendicular to field lines, respectively. Both drifts arise due to field gradients. If the zero-order magnetic fields are produced by currents external to the plasma, then from (2.2.2),

$$\nabla \times \mathbf{B} \approx 0 \tag{4.3.20}$$

inside the plasma, where we have neglected the first-order (weak) currents produced by the moving charges in the plasma. In this case,

$$\frac{\partial B_x}{\partial z} = \frac{\partial B_z}{\partial x} \tag{4.3.21}$$

and the curvature and gradient drifts can be expressed in terms of a single gradient.

Polarization Drift

Consider a uniform magnetic field $\hat{z}B_0$ and a transverse electric field $\hat{x}E(t)$ that varies slowly with time. Then the $\mathbf{E} \times \mathbf{B}$ drift velocity also varies slowly with time:

$$\mathbf{v}_E(t) = -\frac{E(t)}{B_0} \hat{y} \tag{4.3.22}$$

Hence the guiding center accelerates along $\hat{y}$. The acceleration in the lab frame is

$$\mathbf{a}(t) = -\frac{1}{B_0} \frac{\partial E}{\partial t} \hat{y}$$

In the frame of the particle, there is therefore an average inertial force transverse to $\mathbf{B}$:

$$\mathbf{F}_p = -\mathbf{F} = -m\mathbf{a} = \frac{m}{B_0} \frac{\partial E}{\partial t} \hat{y} \tag{4.3.23}$$

Using (4.1.19), this gives rise to a guiding center drift

$$\mathbf{v}_p = \frac{m}{qB_0^2} \frac{\partial \mathbf{E}}{\partial t}$$

(4.3.24)

that lies along $\mathbf{E}$ itself. Again ions and electrons drift in opposite directions, giving an additive current, which for $n_i = n_e = n_0$ is

$$\mathbf{J}_p = \frac{(M + m)n_0}{B_0^2} \frac{\partial \mathbf{E}}{\partial t}$$

(4.3.25)

We see that the electron drift component of the current is negligible due to the mass dependence in (4.3.25). Introducing a low-frequency *perpendicular dielectric constant* $\epsilon_\perp$ through the relation $\mathbf{J}_p = \epsilon_\perp \partial \mathbf{E}/\partial t$, and dropping the electron mass term, we obtain

$$\epsilon_\perp = \epsilon_0 \left(1 + \frac{Mn_0}{\epsilon_0 B_0^2} \right) = \epsilon_0 \left(1 + \frac{\omega_{pi}^2}{\omega_{ci}^2} \right)$$

(4.3.26)

For $n_0 \sim 10^{10}$ cm^{-3} and $B_0 \sim 100$ G, we obtain $\epsilon_\perp \sim 10^6 \epsilon_0$. At low frequencies, $\omega \ll \omega_{ci}$, this very large positive dielectric constant perpendicular to $\mathbf{B}$ shields a magnetized plasma from external electric fields perpendicular to $\mathbf{B}$. For electric fields along $\mathbf{B}$, we can introduce $\epsilon_\parallel = \epsilon_p$ as given in (4.2.24), which at low frequencies is large and negative, also shielding the plasma from electric fields lying along $\mathbf{B}$.

The gyration motion itself also produces currents in a nonuniform plasma. To see this, we form

$$\mathbf{M} = -\hat{z}n(\mathbf{r})\mu_{mag}$$

(4.3.27)

the magnetization of the plasma, such that $\mathbf{B}$ and $\mathbf{H}$ are related by

$$\mathbf{B} = \mu_0(\mathbf{H} + \mathbf{M})$$

(4.3.28)

Then substituting (4.3.28) into (2.2.2) yields

$$\nabla \times \mathbf{B} = \mu_0 \mathbf{J} + \mu_0 \mathbf{J}_{mag} + \epsilon_0 \frac{\partial \mathbf{E}}{\partial t}$$

(4.3.29)

where

$$\mathbf{J}_{mag} = \nabla \times \mathbf{M}$$

(4.3.30)

is the magnetization current. Equation (4.3.29) shows explicitly the three sources of $\mathbf{B}$ in a magnetized plasma: the conduction, magnetization, and polarization currents.

TABLE 4.1. Summary of Guiding Center Drifts
$(\mathbf{R_c}/R_c^2 = -\nabla B/B)$

General force drift	$\mathbf{v}_F = \dfrac{(\mathbf{F}/q) \times \mathbf{B}}{B^2}$
Electric field drift	$\mathbf{v}_E = \dfrac{\mathbf{E} \times \mathbf{B}}{B^2}$
Curvature drift	$\mathbf{v}_R = \dfrac{2W_\parallel}{q} \dfrac{\mathbf{R_c} \times \mathbf{B}}{R_c^2 B^2}$
Grad-B drift	$\mathbf{v}_{\nabla B} = \dfrac{W_\perp}{q} \dfrac{\mathbf{B} \times \nabla B}{B^3}$
Polarization drift	$\mathbf{v}_p = \dfrac{m}{qB^2} \dfrac{\partial \mathbf{E}}{\partial t}$

Since the currents of the gyrating charges act to weaken the applied field, the plasma is *diamagnetic*. As we can see from (4.3.7), the diamagnetism depends both on the plasma density and particle energies, and becomes important only in dense energetic plasmas, primarily those encountered in fusion research. In all but the highest density discharges, the weakening of an applied magnetic field due to plasma diamagnetism is small.

The guiding center motion is derived by a formal expansion of (4.1.1) in most books on fully ionized plasmas, e.g., Schmidt (1979), rather than from the more physical approach given here. A summary of the drifts is given in Table 4.1.

4.4 DYNAMICS OF MAGNETIZED PLASMAS

The response of a plasma immersed in a steady uniform magnetic field $\mathbf{B}_0$ and subject to time-varying electric and magnetic fields is very complicated. The fact that the gyromotion converts velocities being acted on by one field component to another velocity component leads to a *gyrotropic* dielectric tensor, having complex conjugate off-diagonal elements in the absence of dissipation. Furthermore, the inhibition of the electron motion perpendicular to $\mathbf{B}_0$ gives rise to an important ion response, particularly at low frequencies. Collisional dissipation further complicates the picture. Fortunately, for consideration of electromagnetic waves, the wave velocities are generally much higher than the thermal velocities, and thus the effects of the electron and ion thermal velocities can be ignored.

On the other hand, we have seen in Section 4.2, in the absence of $\mathbf{B}_0$, that electrostatic waves can resonate with thermal velocities, leading to strong temperature effects. Similarly, in magnetized plasmas there are electrostatic waves that propagate across the magnetic field, whose nature depends on thermal effects. These waves are generally of little interest for weakly ionized plasmas and will not be considered here. The interested reader is directed to the literature (e.g., Stix 1992).

Our approach in this section will first be to derive the dielectric tensor in the simplest case where only electrons participate, and the electron fluid is considered to be cold and collisionless. It is then straightforward to include the effect of collisions and the addition of a mobile ion species. Using this dielectric tensor, in any of the above approximations, we can derive the dispersion relation for waves propagating at an arbitrary angle to $\mathbf{B}_0$. Because of the complexity of the wave problem, we leave a detailed consideration of the waves to Section 4.5.

Dielectric Tensor

We begin with the force equation in rectangular coordinates as in (4.1.1), with $\mathbf{B}_0 = \hat{z}B_{z0}$. Assuming sinusoidal variation (4.2.31) of the electric field, the linearized equations for the electron motion are then

$$j\omega\tilde{v}_x = -\frac{e}{m}\tilde{E}_x - \omega_{ce}\tilde{v}_y \tag{4.4.1a}$$

$$j\omega\tilde{v}_y = -\frac{e}{m}\tilde{E}_y + \omega_{ce}\tilde{v}_x \tag{4.4.1b}$$

$$j\omega\tilde{v}_z = -\frac{e}{m}\tilde{E}_z \tag{4.4.1c}$$

where we have chosen $\omega_{ce} = eB_{z0}/m$ to be explicitly positive. Solving (4.1.1a) and (4.4.1b) simultaneously, for $\tilde{v}_x$ and $\tilde{v}_y$, we have

$$\tilde{v}_x = \frac{e}{m}\frac{j\omega\tilde{E}_x - \omega_{ce}\tilde{E}_y}{\omega^2 - \omega_{ce}^2} \tag{4.4.2a}$$

$$\tilde{v}_y = \frac{e}{m}\frac{j\omega\tilde{E}_y + \omega_{ce}\tilde{E}_x}{\omega^2 - \omega_{ce}^2} \tag{4.4.2b}$$

Using our previous assumption that $\mathbf{J} = -en_0\mathbf{v}$ and defining the dielectric properties from Maxwell's equation,

$$\nabla \times \tilde{\mathbf{H}} = j\omega\epsilon_0\tilde{\mathbf{E}} + \tilde{\mathbf{J}} \equiv j\omega\bar{\bar{\epsilon}}_p \cdot \tilde{\mathbf{E}} \tag{4.4.3}$$

we obtain

$$\bar{\bar{\epsilon}}_p = \epsilon_0\bar{\bar{\kappa}}_p = \epsilon_0 \begin{pmatrix} \kappa_\perp & -j\kappa_\times & 0 \\ j\kappa_\times & \kappa_\perp & 0 \\ 0 & 0 & \kappa_\parallel \end{pmatrix} \tag{4.4.4}$$

where

$$\kappa_\perp = 1 - \frac{\omega_{pe}^2}{\omega^2 - \omega_{ce}^2} \tag{4.4.5a}$$

$$\kappa_\times = \frac{\omega_{ce}}{\omega} \frac{\omega_{pe}^2}{\omega^2 - \omega_{ce}^2} \tag{4.4.5b}$$

$$\kappa_\parallel = 1 - \frac{\omega_{pe}^2}{\omega^2} \tag{4.4.5c}$$

The z or $\parallel$ component is the same as the dielectric constant (4.2.24) in the absence of $\mathbf{B}_0$. The other components are characteristic of a lossless gyrotropic medium, with $\epsilon_{ij} = \epsilon_{ji}^*$.

Given the collisionless electron dielectric tensor components (4.4.5), it is rather simple to include the effect of collisions, or the contribution of mobile ions. To include collisions, we recognize that each ω originating from the force equation is transformed as $\omega \rightarrow \omega - j\nu_m$. The ω's arising from Maxwell's equations, however, remain unchanged. Performing this operation, we obtain

$$\kappa_\perp = 1 - \frac{\omega - j\nu_m}{\omega} \frac{\omega_{pe}^2}{(\omega - j\nu_m)^2 - \omega_{ce}^2} \tag{4.4.6a}$$

$$\kappa_\times = \frac{\omega_{ce}}{\omega} \frac{\omega_{pe}^2}{(\omega - j\nu_m)^2 - \omega_{ce}^2} \tag{4.4.6b}$$

$$\kappa_\parallel = 1 - \frac{\omega_{pe}^2}{\omega(\omega - j\nu_m)} \tag{4.4.6c}$$

The dielectric tensor, including ion dynamics, is also easily obtained by generalizing (4.4.3). To do this, we recognize that the electron and ion currents add. Then each term in the dielectric tensor consists of a sum of electron and ion components of the same form, but with the parameters appropriate to that species. Thus, again ignoring collisions,

$$\kappa_\perp = 1 - \frac{\omega_{pe}^2}{\omega^2 - \omega_{ce}^2} - \frac{\omega_{pi}^2}{\omega^2 - \omega_{ci}^2} \tag{4.4.7a}$$

$$\kappa_\times = \frac{\omega_{ce}}{\omega} \frac{\omega_{pe}^2}{\omega^2 - \omega_{ce}^2} - \frac{\omega_{ci}}{\omega} \frac{\omega_{pi}^2}{\omega^2 - \omega_{ci}^2} \tag{4.4.7b}$$

$$\kappa_\parallel = 1 - \frac{\omega_p^2}{\omega^2} \tag{4.4.7c}$$

where $\omega_{ci} = eB_{z0}/M$ is defined to be explicitly positive, and we have combined the electron and ion plasma frequencies in (4.4.7c) using (4.2.7). Examining the size of the terms in (4.4.7a), we are often considering situations in which $\omega_{pe} \sim \omega_{ce}$. In that case we see that $\omega_{pi} \sim (M/m)^{1/2}\omega_{ci}$ such that, depending on the range of frequencies being considered, the ion motion can dominate the transverse dielectric components. We have already seen an example of this for low frequencies, $\omega \ll \omega_{ci}$, where (4.4.7a) reduces to (4.3.26). We shall return to this point in considering the wave spectrum.

The Wave Dispersion

Returning to consideration of waves of the form exp $j(\omega t - \mathbf{k} \cdot \mathbf{r})$, Maxwell's curl equations become

$$\mathbf{k} \times \tilde{\mathbf{E}} = \omega\mu_0\tilde{\mathbf{H}} \tag{4.4.8}$$

and

$$\mathbf{k} \times \tilde{\mathbf{H}} = -\omega\epsilon_0\bar{\bar{\kappa}}_p \cdot \tilde{\mathbf{E}} \tag{4.4.9}$$

where $\bar{\bar{\kappa}}_p$ is given by one of the forms in the previous subsection. Taking the cross product of $\mathbf{k}$ with (4.4.8) and substituting for $\mathbf{k} \times \tilde{\mathbf{H}}$ from (4.4.9), we obtain the equation describing electromagnetic waves in a magnetized plasma:

$$\mathbf{k} \times (\mathbf{k} \times \tilde{\mathbf{E}}) + k_0^2\bar{\bar{\kappa}}_p \cdot \tilde{\mathbf{E}} = 0 \tag{4.4.10}$$

where $k_0 = \omega/c$ is the propagation constant of a plane wave of frequency ω in free space, with c the velocity of light.

The vector equation (4.4.10) is very complicated because all of the components of $\tilde{\mathbf{E}}$ couple together. In deriving the dielectric tensor we used rectangular coordinates with $\mathbf{B}_0$ taken along the z direction for concreteness. We have one more direction to define, that of the wave vector, which we can take to lie in the x–z plane, without loss of generality. Doing this, (4.4.10) can be written as

$$\begin{bmatrix} k_z^2 & 0 & -k_xk_z \\ 0 & k_x^2 + k_z^2 & 0 \\ -k_xk_z & 0 & k_x^2 \end{bmatrix} \begin{bmatrix} \tilde{E}_x \\ \tilde{E}_y \\ \tilde{E}_z \end{bmatrix} = k_0^2 \begin{bmatrix} \kappa_\perp & -j\kappa_\times & 0 \\ j\kappa_\times & \kappa_\perp & 0 \\ 0 & 0 & \kappa_\| \end{bmatrix} \begin{bmatrix} \tilde{E}_x \\ \tilde{E}_y \\ \tilde{E}_z \end{bmatrix} \tag{4.4.11}$$

If the angle between $\mathbf{k}$ and $\mathbf{B}$ is defined as θ, then $k_z = k \cos \theta$ and $k_x = k \sin \theta$, where here $k = |\mathbf{k}|$. Furthermore, it is usual to normalize the magnitude of k as $N = k/k_0$, where N here is the *index of refraction* of the wave. Using this notation, and requiring that the determinant of the coefficients of the equation for $\tilde{\mathbf{E}}$ vanishes for a nontrivial solution, we obtain

$$\det \begin{bmatrix} N^2 \cos^2\theta - \kappa_\perp & j\kappa_\times & -N^2 \cos\theta \sin\theta \\ -j\kappa_\times & N^2 - \kappa_\perp & 0 \\ -N^2 \cos\theta \sin\theta & 0 & N^2 \sin^2\theta - \kappa_\| \end{bmatrix} = 0 \tag{4.4.12}$$

Equation (4.4.12) is the dispersion equation, which relates $k \equiv k_0 N$, ω, and θ.

4.5 WAVES IN MAGNETIZED PLASMAS

In this section, we first describe some general properties of waves in magnetized plasmas and then consider in some detail the *principal* waves, i.e., those traveling parallel to and perpendicular to $\mathbf{B}_0$. We then give a qualitative description of propagation at an arbitrary angle in the various regimes of frequency, density, and magnetic field.

Evaluating the determinant in (4.4.12), we find that the cubic terms in N^2 cancel, reducing the equation to a biquadratic form:

$$aN^4 - bN^2 + c = 0 \tag{4.5.1}$$

where

$$a = \kappa_\perp \sin^2 \theta + \kappa_\parallel \cos^2 \theta \tag{4.5.2a}$$

$$b = (\kappa_\perp^2 - \kappa_\times^2) \sin^2 \theta + \kappa_\parallel \kappa_\perp (1 + \cos^2 \theta) \tag{4.5.2b}$$

$$c = (\kappa_\perp^2 - \kappa_\times^2) \kappa_\parallel \tag{4.5.2c}$$

Hence there are in general two different solutions for N^2 for each angle θ. These solutions correspond to the two allowed polarizations for the electric field of the wave. Because the discriminant $b^2 - 4ac$ of (4.5.1) is always positive, N^2 is real, and N is either real and the wave propagates, or imaginary and the wave is cutoff. In the latter case, which may occur for one or both solutions, depending on the parameters, the wave of that polarization does not propagate but decays exponentially. The two wave polarizations are determined by the relative magnitudes of the components of the electric field. These are given by the ratios of the cofactors of any row in the matrix (4.4.12). Taking the first row, we obtain

$$\tilde{E}_x : \tilde{E}_y : \tilde{E}_z :: (\kappa_\perp - N^2)(\kappa_\parallel - N^2 \sin^2 \theta) : j\kappa_\times (N^2 \sin^2 \theta - \kappa_\parallel)$$
$$: (N^2 - \kappa_\perp) N^2 \sin \theta \cos \theta \tag{4.5.3}$$

which gives two different ratios of the field components for the two values of N^2. Since the two waves generally have different propagation constants, their electric fields do not have the same spatial variation and their polarizations cannot be summed to determine a resultant polarization that remains fixed as the waves propagate.

Although (4.5.1) can be solved for N^2 as a function of θ, the results are not particularly illuminating. It is more useful to solve for θ as a function of N^2. Before doing this, it is convenient to introduce two combinations of the dielectric components,

$$\kappa_r = \kappa_\perp - \kappa_\times \tag{4.5.4a}$$

and

$$\kappa_l = \kappa_\perp + \kappa_\times \tag{4.5.4b}$$

such that $\kappa_\perp^2 - \kappa_\times^2 = \kappa_r \kappa_l$ in (4.5.2b) and (4.5.2c). For the simplest case of no collisions and infinite mass ions, we use (4.4.5a) and (4.4.5b) to obtain

$$\kappa_r = 1 - \frac{\omega_{pe}^2}{\omega(\omega - \omega_{ce})} \tag{4.5.5a}$$

and

$$\kappa_l = 1 - \frac{\omega_{pe}^2}{\omega(\omega + \omega_{ce})} \tag{4.5.5b}$$

with the obvious extensions $\omega \pm \omega_{ce} \rightarrow \omega \pm \omega_{ce} - j\nu_m$ in (4.5.5) for adding collisions and

$$\kappa_r = 1 - \frac{\omega_{pe}^2}{\omega(\omega - \omega_{ce})} - \frac{\omega_{pi}^2}{\omega(\omega + \omega_{ci})} \tag{4.5.6a}$$

and

$$\kappa_l = 1 - \frac{\omega_{pe}^2}{\omega(\omega + \omega_{ce})} - \frac{\omega_{pi}^2}{\omega(\omega - \omega_{ci})} \tag{4.5.6b}$$

for a collisionless plasma with mobile ions. Substituting $\sin^2 \theta + \cos^2 \theta$ for 1 in (4.5.2b) and (4.5.2c), substituting a, b, and c into (4.5.1), and dividing (4.5.1) by $\cos^2 \theta$, we can solve to obtain

$$\tan^2 \theta = -\frac{\kappa_\parallel(N^2 - \kappa_r)(N^2 - \kappa_l)}{(N^2 - \kappa_\parallel)(\kappa_\perp N^2 - \kappa_r \kappa_l)} \tag{4.5.7}$$

Principal Electron Waves

(a) **k** ∥ **B₀** ($\theta = 0$)   For this case the numerator of (4.5.7) vanishes, yielding

$$\kappa_\parallel(N^2 - \kappa_r)(N^2 - \kappa_l) = 0 \tag{4.5.8}$$

The first solution $\kappa_\parallel = 0$ gives the plasma oscillations for **E** ∥ **B₀** discussed in Section 4.2. The second and third solutions give the principal waves. Using (4.5.5), these are

$$N_r^2 = 1 - \frac{\omega_{pe}^2}{\omega(\omega - \omega_{ce})} \tag{4.5.9a}$$

and

$$N_1^2 = 1 - \frac{\omega_{pe}^2}{\omega(\omega + \omega_{ce})} \tag{4.5.9b}$$

where ω_{ce} is explicitly positive. The first wave has a resonant denominator for $\omega = \omega_{ce}$, which gives the dispersion for the right-hand polarized (RHP) wave. At $\omega = \omega_{ce}$, the wave rotates in synchronism with the gyrating electrons, which then see a constant field leading to resonant energy absorption, as we will see in Chapter 13. The second wave is the left-hand polarized (LHP) wave, which is nonresonant.

To see that (4.5.9a) represents a right circularly polarized wave, we let $N^2 = \kappa_r$ in (4.5.3) to obtain

$$E_x : E_y :: \kappa_\perp - \kappa_r : -j\kappa_\times$$

and using (4.5.4a),

$$E_x : E_y :: \kappa_\times : -j\kappa_\times$$

Hence the field is given by

$$\mathbf{E} = \text{Re}\left[\tilde{E}_r(\hat{x} - j\hat{y}) \exp j(\omega t - \mathbf{k}_r \cdot \mathbf{r})\right] \tag{4.5.10a}$$

which at fixed $\mathbf{r}$ has a constant amplitude and rotates in the right-hand sense around $\mathbf{B}_0$ at frequency ω. Similarly, the LHP wave has

$$\mathbf{E} = \text{Re}\left[\tilde{E}_l(\hat{x} + j\hat{y}) \exp j(\omega t - \mathbf{k}_l \cdot \mathbf{r})\right] \tag{4.5.10b}$$

and rotates in the left-hand sense around $\mathbf{B}_0$. The most general solution propagating along z is a sum of the RHP and LHP waves given above.

The wave dispersion is easily described by first computing the resonances, $N \to \infty$, and cutoffs, $N \to 0$. Besides the resonance of the RHP wave at $\omega = \omega_{ce}$, there is a cutoff at

$$1 - \frac{\omega_{pe}^2}{\omega(\omega - \omega_{ce})} = 0$$

or, solving for ω,

$$\omega_R = \frac{\omega_{ce} + \sqrt{\omega_{ce}^2 + 4\omega_{pe}^2}}{2} \tag{4.5.11}$$

Only the + solution corresponds to positive ω, leading to an upper cutoff frequency above both ω_{pe} and ω_{ce}. For the LHP wave, a similar calculation gives a cutoff at

$$\omega_L = \frac{-\omega_{ce} + \sqrt{\omega_{ce}^2 + 4\omega_{pe}^2}}{2} \tag{4.5.12}$$

Again the + solution has been taken, which leads to a lower cutoff frequency below ω_{pe}. We should, however, have some doubts about this part of the solution, because it can occur at low frequencies where ion dynamics may be important. It is now possible to sketch an ω–k or dispersion diagram for the waves. We first obtain the other principal waves, so that we can sketch the results on a single diagram.

(b) **k** $\perp$ **B**$_0$ $(\theta = \pi/2)$ For this case the denominator of (4.5.7) vanishes, yielding

$$(N^2 - \kappa_{\parallel})(\kappa_{\perp} N^2 - \kappa_r \kappa_l) = 0 \tag{4.5.13}$$

The first solution is just the wave (4.2.35) for propagation in an unmagnetized plasma. It corresponds to a linearly polarized wave electric field lying along the dc magnetic field direction $\hat{z}$, so that the motion is unaffected by **B**$_0$, and is called the ordinary (o) wave. The second solution gives a wave having electric fields that are perpendicular to **B**$_0$, but with components both perpendicular ($\hat{y}$) and parallel ($\hat{x}$) to **k**. Solving for N, we have the extraordinary (x) wave dispersion:

$$N_x^2 = \frac{\left[1 - \dfrac{\omega_{pe}^2}{\omega(\omega - \omega_{ce})}\right]\left[1 - \dfrac{\omega_{pe}^2}{\omega(\omega + \omega_{ce})}\right]}{1 - \dfrac{\omega_{pe}^2}{\omega^2 - \omega_{ce}^2}} \tag{4.5.14}$$

We see that the numerator has the same two cutoff solutions that we found for the RHP and LHP waves. The resonance at $\omega = \omega_{ce}$ disappears, because of cancellation of the factor $\omega - \omega_{ce}$. However, a new resonance appears at the upper hybrid frequency ω_{UH} given by

$$\omega_{UH}^2 = \omega_{pe}^2 + \omega_{ce}^2 \tag{4.5.15}$$

when the numerator of $\kappa_{\perp}$ is zero.

The dispersion (ω–k) diagrams for the principal waves in an electron plasma are sketched in Fig. 4.9. All the results above the lower cutoff frequencies are reasonably representative of the dispersion when ions are also present. However, at lower frequencies, particularly near ω_{pi} and below, we expect the ion dynamics to be important. We discuss these additional wave solutions below.

First, however, we point out some important characteristics of the less cluttered dispersion of Fig. 4.9. Considering the RHP wave with $\omega_{ce} > \omega$ the wave is propagating. Now let B_0 decrease slowly in the direction of propagation until $\omega_{ce}(z) = \omega$. At this value there is a resonance at which $k_r = \infty$, and both the phase and group velocity

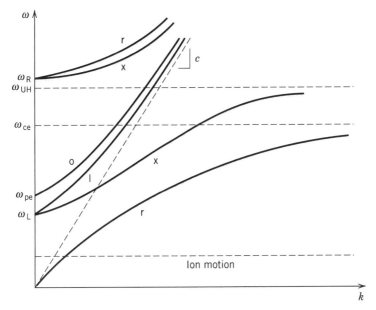

FIGURE 4.9. Dispersion ω versus k for the principal waves in a magnetized plasma with immobile ions for $\omega_{ce} > \omega_{pe}$.

go to zero. A careful analysis reveals that the wave energy is strongly absorbed at this field strength provided certain conditions on the scale length of the field variation and the density are satisfied. This phenomenon of absorption on a "magnetic beach" is an important mechanism for plasma heating and is a major subject of Chapter 13. A similar phenomenon occurs at the upper hybrid resonance for the x wave. However, this wave may not be accessible from outside the plasma, particularly at high density for which $\omega_{pe} > \omega_{ce}$ (not shown in Fig. 4.9), if the decreasing magnetic field requires the wave to pass through the upper hybrid cutoff frequency ω_R. Similarly, the left-hand wave can also be cutoff at high densities if $\omega < \omega_L$.

Principal Waves Including Ion Dynamics

(a) **k** ‖ **B₀**   Adding the ion dynamics into the dispersion equation using (4.5.6), we obtain, for the two polarizations, corresponding to (4.5.9),

$$N_r^2 = 1 - \frac{\omega_{pe}^2}{\omega(\omega - \omega_{ce})} - \frac{\omega_{pi}^2}{\omega(\omega + \omega_{ci})} \qquad (4.5.16a)$$

and

$$N_l^2 = 1 - \frac{\omega_{pe}^2}{\omega(\omega + \omega_{ce})} - \frac{\omega_{pi}^2}{\omega(\omega - \omega_{ci})} \qquad (4.5.16b)$$

where again ω_{ce} and ω_{ci} are positive. Considering first the RHP wave, we put the plasma terms under a common denominator and take $n_i = n_e$ to get

$$N_r^2 = 1 - \frac{\omega_p^2}{(\omega - \omega_{ce})(\omega + \omega_{ci})} \tag{4.5.17}$$

Similarly for the LHP wave, we have

$$N_l^2 = 1 - \frac{\omega_p^2}{(\omega + \omega_{ce})(\omega - \omega_{ci})} \tag{4.5.18}$$

(b) **k** $\perp$ **B$_0$**   In a similar manner, using the dispersion for the extraordinary (x) wave from (4.5.13), with (4.4.7*a*) and (4.5.6), we have

$$N_x^2 = \frac{\left[1 - \dfrac{\omega_{pe}^2}{\omega(\omega - \omega_{ce})} - \dfrac{\omega_{pi}^2}{\omega(\omega + \omega_{ci})}\right]\left[1 - \dfrac{\omega_{pe}^2}{\omega(\omega + \omega_{ce})} - \dfrac{\omega_{pi}^2}{\omega(\omega - \omega_{ci})}\right]}{1 - \dfrac{\omega_{pe}^2}{\omega^2 - \omega_{ce}^2} - \dfrac{\omega_{pi}^2}{\omega^2 - \omega_{ci}^2}} \tag{4.5.19}$$

The important properties of the waves are distinguished by their cutoffs and resonances. Comparing the numerator factors of (4.5.19) with (4.5.16), it is easy to see that the two cutoffs of the x wave correspond to the cutoffs of the RHP and LHP waves. In addition to the upper hybrid resonance ω_{UH}, a second resonance at the lower hybrid frequency ω_{LH} appears. For $\omega_{pi}^2 \gg \omega_{ci}^2$, (usual for materials processing discharges), we find

$$\frac{1}{\omega_{LH}^2} \approx \frac{1}{\omega_{pi}^2} + \frac{1}{\omega_{ce}\omega_{ce}} \tag{4.5.20}$$

Low-frequency wave energy can be strongly absorbed by the plasma at this resonance.

We list all of the cutoffs and resonances of these waves in Table 4.2. With these values, and noting where the propagation constant changes from real to imaginary, the dispersion diagram for the principal waves can be qualitatively sketched, as in Fig. 4.10. The high-frequency range is, of course, similar to Fig. 4.9. Near ω_{pi} and below, the waves are strongly modified by the ion dynamics. Of particular note is that for very low frequencies $\omega \ll \omega_{ci}$, the wave dispersions for RHP, LHP, and x waves all reduce to

$$k^2 = k_0^2 \left(1 + \frac{\omega_{pi}^2}{\omega_{ci}^2}\right) \tag{4.5.21}$$

which propagate down to zero frequency. The term in parentheses is just the low-frequency perpendicular dielectric constant defined in (4.3.26). For reasonably high

TABLE 4.2. Summary of Cutoffs and Resonances for the Principal Waves

Wave	Cutoffs ($k = 0$)	Resonances ($k = \infty$)
r wave	$(\omega - \omega_{ce})(\omega + \omega_{ci}) = \omega_p^2$	$\omega = \omega_{ce}$
	or	
	$\omega \approx \dfrac{\omega_{ce} + \sqrt{\omega_{ce}^2 + 4\omega_p^2}}{2}$	
l wave	$(\omega + \omega_{ce})(\omega - \omega_{ci}) = \omega_p^2$	$\omega = \omega_{ci}$
	or	
	$\omega \approx \dfrac{-\omega_{ce} + \sqrt{\omega_{ce}^2 + 4\omega_p^2}}{2}$	
x wave	Both as above	$\omega_{UH}^2 \approx \omega_p^2 + \omega_{ce}^2$
		and
		$\dfrac{1}{\omega_{LH}^2} \approx \dfrac{1}{\omega_{pi}^2} + \dfrac{1}{\omega_{ce}\omega_{ci}}$
		for $\omega_{pi} \gg \omega_{ci}$
o wave	$\omega = \omega_p$	None

density with $\omega_{pi} \gg \omega_{ci}$, the l can be discarded. The phase velocity of this wave is then

$$v_{ph} = \frac{\omega}{k} = \frac{\omega_{ci}}{\omega_{pi}} c \equiv v_A, \qquad (4.5.22)$$

where v_A is known as the Alfven velocity. Alfven waves were first described in connection with wave propagation in the earth's magnetosphere, and play important roles in low-frequency phenomena in magnetized plasmas.

The CMA Diagram

The preceding gives a far from complete picture of the waves that can propagate at an arbitrary angle to the magnetic field. The complete dispersion equation (4.5.1) can be solved numerically to obtain the propagation constant for each of the waves at an arbitrary angle to the magnetic field. A convenient presentation of the results can be described in the Clemmow–Mullaly–Allis (CMA) diagram given in Fig. 4.11. The relative phase velocities $v_{ph}/c = \omega/kc$ of the two waves are plotted in polar coordinates versus θ for various magnetic fields ($\omega_{ce}\omega_{ci}/\omega^2$) and densities ($\omega_p^2/\omega^2$) on the ordinate and abscissa, with the B-field direction for the polar ($v_{ph}(\theta)/c$) phase velocity surfaces being vertical. The principal propagating waves are indicated using the notation r, l, o, and x for the RHP, LHP, ordinary, and extraordinary waves,

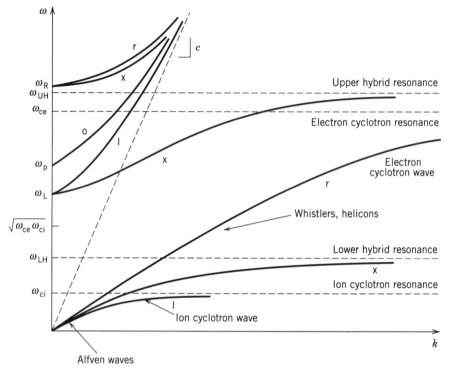

FIGURE 4.10. Dispersion ω versus k for the principal waves in a magnetized plasma with mobile ions.

respectively. The velocity of light circle is shown dashed to give the radial scale of the surfaces. The cutoffs (labeled $u = \infty$) and resonances (labeled $u = 0$) of the principal waves divide the diagram into various regions, each having its own topology for the two phase velocity surfaces. The topologies are either ellipsoids, dumbbells, or toroids, with the latter two indicating resonance ($k \to \infty$ or $v_{ph} \to 0$) at a nonzero propagation angle. In this presentation, the high-frequency region where the propagation is like that of free space is in the lower left-hand corner, while the three Alfven waves are in the upper right. Most of the information can be understood by continuation of the principal wave solutions to arbitrary angles, as the reader is invited to confirm. Since the distance from the origin to the encircling surface represents the phase velocity in that direction with respect to the velocity of light, the CMA diagram has been described as a "plasma pond" in which the shape of each surface corresponds to the outward ripple for a disturbance at its center.

Although the CMA diagram gives a reasonable picture of the electromagnetic waves in an unbounded plasma, it neither gives a complete catalog of the waves that can propagate nor accounts for boundary conditions on the wave fields or spatial variations of the plasma and magnetic field. We have already discussed electrostatic electron and ion plasma waves that can propagate in the absence of or along a magnetic

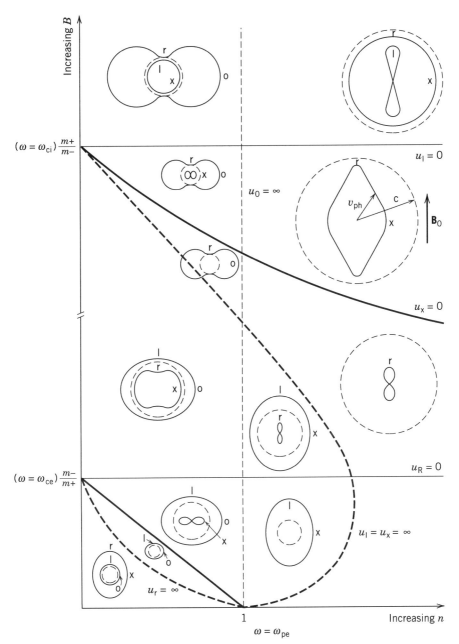

FIGURE 4.11. The CMA diagram for waves in a magnetized plasma. The cutoffs and resonances are indicated by the lines labeled $u = \infty$ and $u = 0$, respectively, where u denotes the phase velocity and the subscripts label the principal waves (after Allis et al., 1963).

field. There is also a large class of electrostatic cyclotron waves that can propagate across the B field. These latter waves are not of great interest in the context of our applications. Analysis can be found in advanced books on plasma wave theory such as Stix (1992, Chapter 9).

Variations in the plasma and B field play essential roles in plasma heated by electron cyclotron resonance interaction, as considered in Chapter 13. The boundary conditions on the electromagnetic fields can also play an important role, as discussed in that chapter. Plasma boundaries can also support additional waves. These bounded plasma waves can be of importance in various contexts, as will be described in Chapter 13. The interested reader can find a description of some of them in Krall and Trivelpiece (1973, Chapter 4), and more briefly in Chen (1984, Chapter 4).

4.6 WAVE DIAGNOSTICS

Because the propagation constant of a wave is dependent on the plasma frequency $\omega_{pe}^2 = e^2 n_e / \epsilon_0 m$, propagation measurements have been used to measure plasma density. In principle, the wave attenuation can also be used to measure the collision frequency, but this method has not been generally employed. Because the plasma frequency is often in the microwave (or submicrowave) range of frequencies, the waves used tend to have frequencies in that range, and the diagnostics are often referred to as *microwave diagnostics*. The methods of using the waves for electron density measurements vary with the plasma configuration. A few such methods are described below. A particular advantage of wave methods is that they are, in principle, noninvasive, and therefore can be used in situations where probe diagnostics (described in Section 6.6) would not be appropriate. A comprehensive account of plasma diagnostics, including wave diagnostics, can be found in Huddlestone and Leonard (1965).

Interferometer

The most commonly used wave diagnostic is the microwave interferometer. The principle of its use is that the change in phase shift across a region with and without a plasma can be measured. This in turn can be related to the change in propagation constant and hence to the plasma frequency. Starting from a wave propagating in a uniform plasma without an applied dc magnetic field, or with a linear polarization such that the electric field is directed along the dc magnetic field, the propagation constant is given by (4.2.35) as

$$k = \left(1 - \frac{\omega_p^2}{\omega^2} \right)^{1/2} k_0 \tag{4.6.1}$$

where $k_0 = \omega/c$ is the free space propagation constant. We ignore collisions in this approximation. Now consider that the wave propagates across a region of length l

in which the density may be changing slowly compared to a wavelength. The WKB solution (see Section 13.1) is that k also changes slowly such that the phase shift can be written in the form

$$\phi = \int_0^l k(x) \, dx \tag{4.6.2}$$

Substituting (4.6.1) in (4.6.2), and subtracting the free space phase shift $k_0 l$, the change in phase shift is

$$\Delta\phi = k_0 \left\{ \int_0^l \left[1 - \frac{\omega_p^2(x)}{\omega^2} \right]^{1/2} dx - l \right\} \tag{4.6.3}$$

It is often possible to choose the diagnostic frequency sufficiently high compared to the plasma frequency that the square root can be expanded. The free space part of the phase shift then conveniently cancels from (4.6.3) leaving

$$\Delta\phi \approx k_0 \int_0^l \frac{\omega_p^2(x)}{2\omega^2} dx = \frac{k_0 e^2}{2\epsilon_0 m \omega^2} \int_0^l n(x) \, dx \tag{4.6.4}$$

In this approximation we see that the line integral of the density can be directly measured in terms of a phase shift. In many configurations the density can be measured quite accurately by this method, serving as a check on the less accurate but local probe method, described in Section 6.6. If the approximation in (4.6.4) cannot be made, it is still possible to determine the same information from (4.6.3), but the calculation is not straightforward.

The actual measurement technique uses an interferometer that compares signals going through the plasma region and around it. A schematic of such an interferometer is shown in Fig. 4.12. In the absence of the plasma, the reference leg is adjusted to have a 180° phase shift at the same amplitude as the plasma leg, giving a null

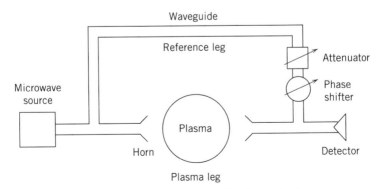

FIGURE 4.12.   A microwave interferometer for plasma density measurement.

output. With the plasma present, the phase shift across the plasma leg changes and a signal is observed. The most convenient way of using the interferometer is to have $l \gg \lambda$, such that $\Delta\phi$ can change through more than 360° (a fringe shift) for $\omega_p^2/\omega^2 \ll 1$ (see (4.6.4)). For $\Delta\phi = 180°$, the signals through the two legs are in phase and the signal is a maximum, returning to a near null signal at $\Delta\phi = 360°$. Very accurate measurements can be made in this regime in which the plasma is turned on sufficiently slowly that the number of fringe shifts and fractions thereof can be measured. Often, however, the plasma size and available detection frequencies make $l \lesssim \lambda$, and fractional fringe shifts must be measured. This can be relatively straightforward if (4.6.4) holds such that $\Delta\phi \propto n$. However, the signal amplitude must be known, and this is complicated by reflection and refraction of the wave at the plasma–dielectric interfaces.

The finite size of the plasma, compared to the wavelength of the interferometer, has other consequences that can be more serious than the limited phase shift. If the transverse dimension of the plasma is also comparable to a wavelength, then diffraction around the plasma becomes a serious problem. This is often significant when diagnosing plasma cylinders. Small transverse plasma dimensions have tended to push the interferometer frequency up, such that $\omega_p^2/\omega^2 \ll 1$. In this case the phase shift, which is proportional to this ratio, becomes small. This has led to more complicated methods of detection. For dense plasmas, laser interferometers have been used to obtain small but measurable phase shifts. The microwave interferometer has been a mainstay of fusion plasma diagnostics from their inception, since noninvasive measurement techniques are required on such plasmas. An early monograph (Heald and Wharton, 1965) recounted these techniques in detail.

An example of a 35-GHz microwave interferometer measurement of density and its comparison to density measurements using Langmuir probes (see Section 6.6) is shown in Fig. 4.13 for a planar coil, rf-driven inductive discharge. The transmitting and receiving horn antennae were placed externally to the chamber, with the microwaves transmitted through the rectangular chamber parallel to the surface of the planar coil (see Section 12.3 for further description of the discharge configuration).

The ordinary wave is not suitable for an interferometer if $\omega_p > \omega$, because the wave will not propagate. In time-varying plasmas, the cutoff itself can be used as a benchmark of qualitative plasma behavior. In a magnetic field, it is still possible to have a propagating wave along the field, provided $\omega_{ce} > \omega$, as given by (4.5.17). Although this wave is very important for plasma heating, as described in Section 13.1, it has only occasionally been used for plasma diagnostics. With $\omega_p > \omega$, methods described in the following subsections have sometimes been employed.

Cavity Perturbation

Another relatively straightforward technique for diagnosing a plasma is by the shift in frequency of a microwave cavity when a plasma fills part of the cavity. Slater's perturbation formula (Harrington, 1961, Chapter 7) can be applied to an unmagnetized plasma in the frequency range where the plasma frequency $\omega_p \ll \omega_0$, the resonant

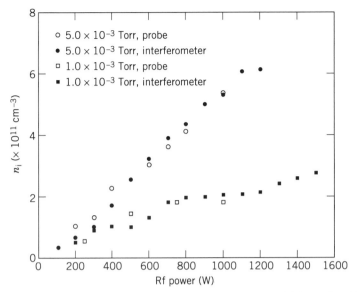

FIGURE 4.13. Mean electron density versus incident power at the midplane of an rf inductive discharge as measured by a microwave interferometer, compared with ion density as measured by a Langmuir probe (Hopwood et al., 1993b).

frequency, (and $\nu_m \ll \omega_0$) giving the relative shift in resonance frequency:

$$\frac{\Delta\omega}{\omega_0} = \frac{1}{2\omega_0^2} \frac{\int \omega_p^2 |\mathbf{E}|^2 d\mathcal{V}}{\int |\mathbf{E}|^2 d\mathcal{V}} \qquad (4.6.5)$$

where $\mathbf{E}$ is the unperturbed resonance electric field, and the integrals are over the total cavity volume. The formula can also be modified to include higher-density plasmas, provided the plasma dimensions are small compared to λ. For evaluating the integrals, most measurements have used cylindrical cavity modes such as the TM_{010} mode (see Ramo et al., 1984, Chapter 10), for which $\mathbf{E} = \hat{z}\tilde{E}_z$, where

$$\tilde{E}_z = E_0 J_0\left(\frac{\chi_{01} r}{R}\right) \qquad (4.6.6)$$

where $\chi_{01} \approx 2.405$ and R is the cavity radius. Processing chambers usually have more complicated geometry. In this case one can experimentally determine the electric field profiles of several modes. The spatial density profile can also be measured to evaluate the integral in the numerator of (4.6.5), although different estimates of the profile only slightly modify the results. Reasonable consistency in density measurements can be obtained by using more than one mode. An example of results and their comparison to Langmuir probe measurements is shown in Fig. 4.14 for a particular process chamber.

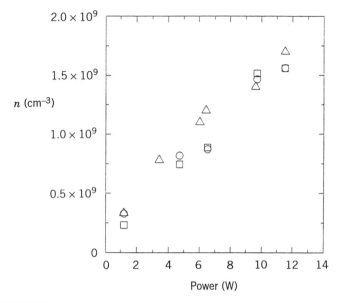

FIGURE 4.14. Electron density versus absorbed power in a 10-mTorr argon discharge. Data from 443-MHz cavity resonance (circles), 506-MHz cavity resonance (squares), and Langmuir probe (triangles) (Moroney et al., 1989).

Another easily measured quantity in a cavity is the Q defined by

$$Q \equiv \omega \frac{\text{Energy stored}}{\text{Power dissipated}} = \frac{\omega_0}{\Delta \omega} \qquad (4.6.7)$$

where $\Delta \omega$ is the frequency shift between the half power points on each side of the resonance. The second equality, given in all circuit texts, follows directly from the definition. The cavity Q with plasma is lower than that without plasma due to dissipation within the plasma. Provided $\nu_m \ll \omega_0$ the microwave cavity measurement of density is not significantly modified. However, the change in Q can be used to directly determine the collision frequency of the plasma electrons if ohmic heating is the main source of energy absorption. Experiments of this nature have been successfully performed, but have not come into general use as a plasma diagnostic.

Wave Propagation

An interesting type of diagnostic is one that uses intrinsic properties of wave propagation in bounded plasmas. For example, one method of plasma heating, described in Section 13.3, is by surface waves. The propagation properties of these waves can be measured and related to the average plasma density over which the fields are important. For waves whose fields are confined close to the plasma–dielectric interface, the propagation can give information about the edge density, in contrast to the average density obtained from the methods described above. The equations governing surface-wave propagation are given in Sec 13.3, and should be consulted if the method is to be applied.

Although more difficult to measure, it is also possible to obtain information on the electron collisionality from the wave decay. This is also considered in Section 13.3. We note, however, that the decay constant involves collisionless (Landau) damping as well as collisional damping, so that the results must be interpreted with care.

A particularly simple situation for obtaining the plasma frequency is that for which the plasma is transversely resonant. A simple calculation then yields the plasma frequency. For example, for a parallel plane geometry, let d be the length of the plasma and $2s_m$ be the total length of both sheaths. The discharge can be modeled as two capacitors in series, where the capacitances per unit area are

$$C_s \approx \frac{\epsilon_0}{2s_m} \tag{4.6.8}$$

and

$$C_p \approx \frac{\epsilon_p}{d} \approx \frac{\epsilon_0(1 - \omega_p^2/\omega^2)}{d} \tag{4.6.9}$$

Note that C_p is inductive ($C_p < 0$) for $\omega < \omega_p$. The total capacitance is then

$$C_T = \left(\frac{1}{C_p} + \frac{1}{C_s} \right)^{-1} \tag{4.6.10}$$

Substituting the expressions for C_s and C_p into (4.6.10), we obtain

$$C_T = \frac{\epsilon_0 \left(\omega^2 - \omega_p^2 \right)}{2s_m \left(\omega^2 - \omega_p^2 \right) + d\omega^2} \tag{4.6.11}$$

This expression will have a resonance when the denominator vanishes, or

$$\omega = \omega_p \left(\frac{2s_m}{2s_m + d} \right)^{1/2} \tag{4.6.12}$$

The resonance has been observed in both capacitive and inductive discharges. The densities obtained from (4.6.12) agree reasonably well with other density measurements made on the same discharge. The method can also be applied to cylindrical plasmas, and configurations in which there are dielectrics, giving somewhat more complicated expressions replacing (4.6.12). Indeed, the first application of the method was to a plasma cylinder, surrounded by a dielectric tube with split cylinder exciting electrodes. (See Parker et al. (1964) for details, including thermal effects.) The lowest-order "dipole" resonance can be approximated by the simple form

$$\omega = \frac{\omega_p}{(1 + \kappa_{eff})^{1/2}} \tag{4.6.13}$$

where κ_{eff} is the effective relative dielectric constant of the region between the plasma and the electrode.

Finally, we wish to point out that the distinction between perturbation of a cavity resonance and propagation of a plasma wave is not decisive, but only convenient. For a plasma within a cavity, a field solution for the cavity resonance predicts the shift in resonance frequency. Similarly, if an inductor is connected across the plates of a discharge, it must be included within the resonance calculation leading to (4.6.12). A device that illustrates this duality is the helical resonator, which we treat in Section 12.4. It is seen there that a plasma inside of a slow wave helical structure changes the propagation constant in a known way, such that the plasma density can be inferred. As the helix structure is operated resonantly, this manifests itself as a change in the resonant frequency. It is also possible to measure power absorption from the change in the Q of a helical resonator.

PROBLEMS

4.1. Plasma Oscillations With Mobile Ions   Show in a slab geometry that the plasma oscillation frequency is given by (4.2.7) if the ions are permitted to be mobile.

4.2. Plasma Oscillations for a Perturbed Charge Density   For a plasma with immobile uniform density ions, show that an arbitrary displacement $\zeta_e(\mathbf{r}, t)$ of the electron fluid with respect to the ions leads to a perturbed charge density $\rho = en_0 \nabla \cdot \zeta_e$. Using the divergence equation for the electric field and the equation of motion for the electron fluid, show that the charge density oscillates sinusoidally at the electron plasma frequency ω_{pe}.

4.3. A Particle-in-Cell Simulation With One Electron Sheet   A plasma having uniform density n_0 is confined between two parallel perfectly conducting planes separated by a distance l. For computer simulation, the plasma is modeled as follows: The ions are assumed to be fixed and have a uniform density n_0. The electrons are all gathered into a single sheet of charge of surface charge density $\rho_S = -en_0 l$ C/m^2, which is allowed to move in response to the electric fields seen by the sheet.

(a) Show that the equilibrium position of the electron sheet is in the center of the plasma.

(b) If the sheet is given a small displacement about its equilibrium position and then released, what happens? Find the subsequent motion of the sheet.

(c) Suppose the two parallel planes are connected together (grounded). Repeat part (b) to determine the motion.

4.4. Time-Average Power in the Sinusoidal Steady State   Show that (4.2.26) holds; i.e., if $\mathbf{J}_T(t)$ and $\mathbf{E}(t)$ are sinusoids having complex vector amplitudes $\tilde{\mathbf{J}}_T$ and $\tilde{\mathbf{E}}$, then the time average absorbed power per unit volume can be written as

$$p_{abs} = \frac{1}{2}\text{Re}\ (\tilde{\mathbf{J}}_T \cdot \tilde{\mathbf{E}}^*) = \frac{1}{2}\text{Re}\ (\tilde{\mathbf{J}}_T^* \cdot \tilde{\mathbf{E}})$$

4.5. Ohmic Heating Power in a Nonuniform Rf Discharge   An rf discharge with a nonuniform density $n(x)$ is ignited between two plane parallel electrodes located at $x = \pm l/2$. The total rf current density (conduction + displacement) is $\mathbf{J}_T(x,t) = \hat{x}J_0 \cos \omega t$. The rf electric field in the discharge is similarly given by $\mathbf{E}(x,t) = \hat{x}E_0(x) \cos(\omega t + \phi_0)$.

(a) Prove from Maxwell's equations that J_0 is a constant, independent of x.

(b) Writing $\mathbf{J}_T = \hat{x}\text{Re}(\tilde{J}e^{j\omega t})$ and $\mathbf{E} = \hat{x}\text{Re}[\tilde{E}(x)e^{j\omega t}]$, find expressions for the complex amplitudes $\tilde{J}$ and $\tilde{E}$.

(c) For a high-pressure (collisional) discharge such that $\omega \ll \nu_m \ll \omega_{pe}$, with a plasma density $n(x) = n_0 \cos(\pi x/l)$, find an expression for $E_0(x)$ and $\tilde{E}(x)$ in terms of J_0, n_0, ν_m, l, and other constants. Use the expression (4.2.22) for the dc plasma conductivity $\sigma_{dc}(x)$ with $n_0 \rightarrow n(x)$.

(d) In the limit of (c), integrate p_{ohm} over x to find the ohmic power per unit area within a discharge volume $|x| \leq d/2$, where $d < l$.

(e) Note that your result in (d) tends to infinity as $d \rightarrow l$. Comment on the correctness of this result.

4.6. Electrostatic Ion Plasma Waves   Derive the dispersion relation for electrostatic ion plasma waves in a uniform collisionless plasma containing mobile ions with $T_e \gg T_i$, and show that for long wavelengths (low frequencies), the waves propagate at the ion sound speed $(kT_e/M)^{1/2}$. Use the Boltzmann relation to relate n_e to Φ.

4.7. Guiding Center Motion   Consider a cylindrically symmetric, time-varying magnetic field that varies parabolically with axial distance z as $\mathbf{B} = \hat{z}B(t)(1 + z^2/l^2)$. Assume that $B(t)$ increases slowly from the value B_0 at time $t = 0$ to B_1 at $t = t_1$. A charged particle of mass m located at $z = 0$ has perpendicular energy $W_{\perp 0}$ and parallel energy W_{z0} at $t = 0$. Assume that the guiding center equations of motion are valid and that $\mu_{mag} = $ const.

(a) Give the final perpendicular energy $W_{\perp 1}$ at $z = 0$ (after a time t_1).

(b) Write the equation for the motion along z, assuming that the motion is fast compared to the time variation of $B(t)$. Show that the motion is a sinusoidal oscillation along z, and calculate the oscillation frequency ω_b. This shows that the particle is confined axially in the magnetic field.

(c) Assume now that $W_{z0} = 0$ and $R(t) \ll l$, where $R(t)$ is the radial distance of the guiding center of the particle from the z axis. By using Faraday's law (2.2.1) to find the induced electric field $E_\phi(t)$ and calculating the resulting $\mathbf{E} \times \mathbf{B}$ drift, show that $B(t)R^2(t) = $ const during the slow change from B_0 to B_1.

4.8. Magnetic Moment   The magnetic moment of a charged particle gyrating in a magnetic field is defined as the product of the current generated by the rotating particle times the area enclosed by the rotation. Show that this is equal to μ_{mag} defined in (4.3.11).

4.9. Magnetic Drifts

(a) Cross multiplying (4.3.6) by $\mathbf{B}/qB^2$, obtain the three general guiding center drifts (force, curvature, and grad-B drifts in Table 4.1). This calculation is not straightforward; see Chen (1984).

(b) Show that an expansion of the left-hand side of (4.3.6) gives rise to (4.3.18).

4.10. Calculating the Gradient Drift   For the geometry in Fig. 4.8b, derive (4.3.19), starting from first principles, with the magnetic field $B_z = B_0 + r_c(\partial B/\partial x)\sin\omega_c t$ and $\mathbf{v}$ as given in (4.1.6) ($\phi_0 = 0$). To do this, first find the time average Lorentz force and then use (4.1.19).

4.11. Waves in Magnetized Plasmas   Sketch the wave dispersion ω versus k for the principal waves in an electron plasma (immobile ions) for high densities $\omega_{pe} > \omega_{ce}$, and compare to Fig. 4.9.

4.12. Whistler Waves   The right-hand polarized wave is known as the whistler wave in the frequency range for which $\omega_{ci} \ll \omega \ll \omega_{ce}$. Using these approximations in the dispersion relation, find the dependence of the phase velocity of the wave on the frequency.

4.13. Microwave Diagnostic   Consider a 3-cm-diameter uniform plasma column. It is desired to measure the plasma density either by measuring the perturbation of the resonant frequency of a 6-cm-diameter TM_{010} mode cavity or by measuring the phase shift of the ordinary wave using a $\lambda = 1.5$-cm interferometer.

(a) What is the approximate unperturbed resonant frequency of the cavity?

(b) Using the perturbation formula, calculate the frequency shift due to the plasma column for $n = 10^{10} cm^{-3}$ and $n = 10^{11} cm^{-3}$.

(c) Sketch the cross section of the electric field magnitude for each case.

(d) Find the phase shift for the 1.5-cm interferometer for each case.

(e) Explain which method you would use to find the density if it was expected to lie in the range of each of the two cases.

CHAPTER 5

DIFFUSION AND TRANSPORT

5.1 BASIC RELATIONS

Diffusion and Mobility

We have already seen in Section 4.2 that adding a friction term to the force equation, in a cold uniform plasma with an applied electric field, gives rise to a conductivity. The friction term, arising from collisions with a background species, also leads to diffusion in a nonuniform warm plasma. To see this we start with the steady-state macroscopic force equation (2.3.15),

$$qn\mathbf{E} - \nabla p - mn\nu_m\mathbf{u} = 0 \qquad (5.1.1)$$

where we assume that the background species is at rest and that the momentum transfer frequency ν_m is a constant, independent of the drift velocity $\mathbf{u}$. Taking an isothermal plasma, such that $\nabla p = kT\nabla n$, and solving (5.1.1) for $\mathbf{u}$, we obtain

$$\mathbf{u} = \frac{q\mathbf{E}}{m\nu_m} - \frac{kT}{m\nu_m}\frac{\nabla n}{n} \qquad (5.1.2)$$

Equation (5.1.2) can be written

$$\Gamma = \pm\mu n\mathbf{E} - D\nabla n \qquad (5.1.3)$$

where $\Gamma = n\mathbf{u}$ is the particle flux, and

$$\mu = \frac{|q|}{m\nu_{\mathrm{m}}} \qquad \text{m}^2\text{/V-s} \tag{5.1.4}$$

and

$$D = \frac{kT}{m\nu_{\mathrm{m}}} \qquad \text{m}^2\text{/s} \tag{5.1.5}$$

are the macroscopic *mobility* and *diffusion* constants. These are calculated separately for each species. In (5.1.3), the positive sign is for q positive and the negative sign is for q negative. Using the definition of the mean speed $\bar{v} = (8kT/\pi m)^{1/2}$ and a mean free path (for hard-sphere scattering) $\lambda = \bar{v}/\nu_{\mathrm{m}}$ we can write D as

$$D = \frac{\pi}{8}\lambda^2\nu_{\mathrm{m}} \tag{5.1.6}$$

Notice that D in (5.1.6) is in the form $(\Delta x)^2/\tau$, where Δx is the step length and τ is the time between steps of a *random walk*. This is the basic structure of a diffusion process.

Free Diffusion

From (5.1.3), in the absence of an electric field, we can directly obtain the diffusion law, relating the flux $\Gamma = n\mathbf{u}$ to the density gradient,

$$\Gamma = -D\nabla n \tag{5.1.7}$$

which is called *Fick's law*. Substituting (5.1.7) in the continuity equation (2.3.7) and choosing the source and sink terms to be zero, we obtain

$$\frac{\partial n}{\partial t} + \nabla \cdot \Gamma = 0$$

With D independent of position, we obtain the diffusion equation for a single species:

$$\frac{\partial n}{\partial t} - D\nabla^2 n = 0 \tag{5.1.8}$$

Finally, we note that the *transport coefficients* μ and D are related by the *Einstein relation*:

$$\mu = \frac{|q|}{kT}D \tag{5.1.9}$$

Ambipolar Diffusion

Returning to the more general relation (5.1.3) we consider this to hold separately for electrons and ions. Furthermore, we make the *congruence assumption* that the flux of electrons and ions out of any region must be equal, $\Gamma_e = \Gamma_i$, such that charge does not build up (see Problem 5.9). This is still true in the presence of ionizing collisions, which create equal numbers of both species. Since the electrons are lighter, and would tend to flow out faster (in an unmagnetized plasma), an electric field must spring up to maintain the local flux balance. That is, a few more electrons than ions initially leave the plasma region to set up a charge imbalance and consequently an electric field. Using (5.1.3) for both species, with $\Gamma_e = \Gamma_i = \Gamma$ and $n_e \approx n_i = n$, we have

$$\mu_i n \mathbf{E} - D_i \nabla n = -\mu_e n \mathbf{E} - D_e \nabla n$$

from which we can solve for $\mathbf{E}$ in terms of ∇n:

$$\mathbf{E} = \frac{D_i - D_e}{\mu_i + \mu_e} \frac{\nabla n}{n} \tag{5.1.10}$$

Substituting this value of $\mathbf{E}$ into the common flux relation we have (in the ion equation)

$$\Gamma = \mu_i \frac{D_i - D_e}{\mu_i + \mu_e} \nabla n - D_i \nabla n$$

$$= -\frac{\mu_i D_e + \mu_e D_i}{\mu_i + \mu_e} \nabla n \tag{5.1.11}$$

which is symmetric in the coefficients and (of course) holds for both ions and electrons. Introducing the ambipolar diffusion coefficient

$$D_a = \frac{\mu_i D_e + \mu_e D_i}{\mu_i + \mu_e} \tag{5.1.12}$$

we see that (5.1.11) again has the form of Fick's law $\Gamma = -D_a \nabla n$. Substituting (5.1.11) in the continuity equation, and assuming that all coefficients are independent of position, we obtain

$$\frac{\partial n}{\partial t} - D_a \nabla^2 n = 0 \tag{5.1.13}$$

the ambipolar diffusion equation.

The ambipolar diffusion coefficient can usually be simplified by noting that $\mu_e \gg \mu_i$ in a weakly ionized discharge. Dropping μ_i in the denominator of (5.1.12) we have

$$D_a \approx D_i + \frac{\mu_i}{\mu_e} D_e$$

and using the Einstein relation, we obtain

$$D_a \approx D_i \left(1 + \frac{T_e}{T_i} \right) \tag{5.1.14}$$

From (5.1.14) we see that the ambipolar diffusion is tied to the slower species, in this case the ions, but that it is increased by a term proportional to the ratio of temperatures. Thus, in the usual case in weakly ionized plasmas, in which $T_e \gg T_i$, the ions and electrons both diffuse at a rate that greatly exceeds the ion-free diffusion rate.

Let us note that in the regime where $\mu_e \gg \mu_i$ and $T_e \gg T_i$, the pressure gradient term in (5.1.3) is small compared to the flux and field terms for ions, such that

$$\Gamma_i = \Gamma \approx \mu_i n \mathbf{E} \tag{5.1.15}$$

On the other hand, for electrons the flux term is small compared to the field and pressure gradient terms, such that

$$\Gamma_e = \Gamma = -\mu_e n \mathbf{E} - D_e \nabla n \approx 0 \tag{5.1.16}$$

Hence ion motion is mobility dominated and electron motion is determined by a Boltzmann equilibrium.

In the above calculations we have considered only unmagnetized plasmas. In a magnetic field the motion of electrons is strongly confined perpendicular to the field, as we have already seen in Chapter 4, which can lead to quite different diffusion rates parallel to and perpendicular to the applied magnetic field. We shall discuss this situation in Section 5.4.

5.2 DIFFUSION SOLUTIONS

One-Dimensional Time-Dependent Solution

Solutions to the diffusion equation (5.1.8) or (5.1.13) are easily obtained for spatial variation in one dimension. Because there are no source terms in (5.1.8), the solution must decay in time. For simplicity, taking a slab geometry of width l, we introduce a separation of variables,

$$n(x, t) = X(x)T(t)$$

which when substituted in (5.1.8) gives

$$X \frac{dT}{dt} = DT \frac{d^2 X}{dx^2} \tag{5.2.1}$$

Dividing by XT, we obtain on the left-hand side a function of time alone, and on the right-hand side a function of space alone. Consequently, both must equal a constant which we call $-1/\tau$. The function of T then is determined from

$$\frac{dT}{dt} = -\frac{T}{\tau} \tag{5.2.2}$$

which integrates to

$$T = T_0 e^{-t/\tau} \tag{5.2.3}$$

Similarly, the spatial part is determined by

$$\frac{d^2X}{dx^2} = -\frac{X}{D\tau} \tag{5.2.4}$$

which has a solution of the form

$$X = A \cos \frac{x}{\Lambda} + B \sin \frac{x}{\Lambda} \tag{5.2.5}$$

where $\Lambda = (D\tau)^{1/2}$ is the diffusion length, and A and B are constants. Taking boundary conditions of $X = 0$ at $x = \pm l/2$, then the lowest-order solution is symmetric ($B = 0$) and

$$\Lambda_0 = (D\tau_0)^{1/2} = \frac{l}{\pi}$$

Solving for $\tau = \tau_0$, we have the decay constant

$$\tau_0 = \left(\frac{l}{\pi}\right)^2 \frac{1}{D} \tag{5.2.6}$$

Combining the solutions for T and X, the complete solution is

$$n = n_0 e^{-t/\tau_0} \cos \frac{\pi x}{l} \tag{5.2.7}$$

with τ_0 given from (5.2.6). This gives the decay of the lowest-order mode. For an arbitrary initial value of the density within $-l/2 < x < l/2$, the initial density can be written as a Fourier series, which, with $n(x) = 0$ at $x = -l/2$ and $l/2$, is

$$n = n_0 \left[\sum_{i=0}^{\infty} A_i \cos \frac{(2i+1)\pi x}{l} + \sum_{i=1}^{\infty} B_i \sin \frac{2i\pi x}{l} \right] \tag{5.2.8}$$

Then assuming that each mode decays at its own characteristic rate, the symmetric ith mode has a product solution:

$$n_i = n_0 A_i e^{-t/\tau_i} \cos \frac{(2i+1)\pi x}{l} \tag{5.2.9}$$

where from the diffusion equation, as above, we find

$$\tau_i = \left[\frac{l}{(2i+1)\pi} \right]^2 \frac{1}{D} \tag{5.2.10}$$

From (5.2.10) we see that the higher modes, $i > 0$, decay more rapidly than the lowest mode, which becomes the dominant decay mode after sufficient time.

One-Dimensional Steady-State Solution

The more interesting diffusion solution for analyzing a steady discharge is one without time dependence. In this case it is necessary to either have flow into the region or a source within the region, to balance the diffusion out of the region. The simplest case is with flux entering on one side and leaving on the other. Again taking a slab geometry, we have

$$-D \frac{d^2 n}{dx^2} = 0 \tag{5.2.11}$$

The solution is just a linear decay across the region of interest:

$$n = Ax + B \tag{5.2.12}$$

If we specify that the flux is $\Gamma = \Gamma_0$ at $x = 0$ and the density $n(l/2) = 0$, then

$$n = \frac{\Gamma_0}{D} \left(\frac{l}{2} - x \right) \tag{5.2.13}$$

and the flux $\Gamma = -D\nabla n$ is independent of x.

A more interesting case, physically, is with ionization in the region of interest. In this case the steady-state diffusion equation must be modified to include the source term,

$$-D\nabla^2 n = \nu_{iz} n \tag{5.2.14}$$

where, as in (2.3.8), we have assumed an ionization proportional to $n = n_e$, with ν_{iz} the ionization frequency as defined in Chapter 3. We see that in slab geometry

(5.2.14) is in the same form as (5.2.4), such that the symmetric solution is

$$n = n_0 \cos \beta x \tag{5.2.15}$$

where

$$\beta = \left(\frac{\nu_{iz}}{D} \right)^{1/2} \tag{5.2.16}$$

The flux is

$$\Gamma = -D \frac{dn}{dz} = D n_0 \beta \sin \beta x \tag{5.2.17}$$

and the diffusion velocity is

$$u = \frac{\Gamma}{n} = D\beta \tan \beta x \tag{5.2.18}$$

For ambipolar diffusion, $D = D_a$ and the electric field, given by (5.1.10), points toward the walls, thus confining the more mobile electrons. With boundary conditions $n(l/2) = n(-l/2) = 0$, (5.2.15) gives the spatial dependence:

$$n = n_0 \cos \frac{\pi x}{l} \tag{5.2.19}$$

with

$$\beta = \left(\frac{\nu_{iz}}{D} \right)^{1/2} = \frac{\pi}{l} \tag{5.2.20}$$

as shown in Fig. 5.1. The reader may well ask how it is possible to have a relation of the type (5.2.20) when ν_{iz} and D are both given functions of the medium. The answer is that they are both temperature dependent, with ν_{iz} an exponentially sensitive function of T_e, as we have seen in Chapter 3. Thus, (5.2.20) is an equation for the electron temperature. We shall make this quite explicit in our discharge models in Chapters 10–14.

The boundary conditions $n(\pm l/2) = 0$ that we have specified for the diffusion solutions are not self-consistent. To see this we note from (5.2.17) that the flux at the wall is $\Gamma(l/2) = D n_0 \pi/l$. However, since we can also write

$$\Gamma\left(\frac{l}{2} \right) = n\left(\frac{l}{2} \right) u\left(\frac{l}{2} \right) \tag{5.2.21}$$

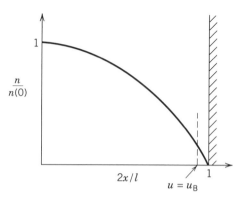

FIGURE 5.1. High-pressure diffusion solution for normalized density $n/n(0)$ versus normalized position $2x/l$.

we see that the finite flux combined with the zero edge density would lead to an infinite macroscopic edge velocity. Clearly, this cannot happen, so that the boundary condition cannot be exact. In fact, in the next chapter, considering the region close to the wall, we find that the velocity at the plasma–sheath edge $x = l'/2$, where $l'/2 = l/2 - s$, with s the sheath thickness, is fixed at a particular value* called the *Bohm velocity* $u_B = (kT_e/M)^{1/2}$. Hence, equating u_B to the diffusion velocity (5.2.18), we obtain

$$u_B = D\beta \tan \frac{\beta l'}{2} \tag{5.2.22}$$

where from (5.2.16), $\beta = (\nu_{iz}/D)^{1/2}$. For a thin sheath, we have $l' \approx l$. Since u_B, ν_{iz}, and D are all functions of the electron temperature alone (with the neutral density specified), (5.2.22) again is an equation for T_e.

The analysis above is easily performed in cylindrical or spherical geometries. We will perform the calculations as they naturally occur in the contexts of the particular discharges under investigation. The cylindrical geometry is typical for the analysis of the positive column of a dc glow discharge, which we analyze in Section 14.2. The spherical geometry is a useful approximation to a small driving electrode in an rf-excited plasma, which we discuss in Section 11.4. Both cylindrical and spherical geometries are useful in analyzing electrostatic probes, which are considered in Section 6.6.

For high pressures with $\lambda_i/l \lesssim T_i/T_e$, ν_m is independent of **u** (see next section), which is required for the diffusion solution (5.2.16) with (5.2.17) to be valid. However, in this high pressure range, for some discharges the electrons are hotter near the plasma edge, so that the assumption that ν_{iz} is proportional to n_e is not valid. We discuss this further in Chapter 10. If, on the other hand, we have $\lambda_i/l \gtrsim T_i/T_e$, then the

---

*The velocity at the plasma-sheath edge is u_B only for $\lambda_i \gg \lambda_{De}$; see (6.5.9) and accompanying discussion.

assumptions of the constant D macroscopic diffusion theory begin to break down, and other approximations must be employed. We discuss some of the most important of these situations in Section 5.3.

5.3 LOW-PRESSURE SOLUTIONS

Variable Mobility Model

Many discharges are run at low pressure where the assumptions used to obtain the solutions in Section 5.2 break down. In particular, at low pressure the effective ion velocity for collision of ions with neutrals is the ion drift velocity $|\mathbf{u}|$ rather than the ion thermal velocity v_{thi}; i.e., for the pressure regime of interest $|\mathbf{u}| \gg v_{\mathrm{thi}}$ over most of the discharge region. In this case the ion neutral collision rate can be written as $v_{\mathrm{m}} \approx |\mathbf{u}_{\mathrm{i}}|/\lambda_{\mathrm{i}}$, where λ_{i} is the ion mean free path. Hence, we can replace the mobility from (5.1.4) by the relation (Smirnov, 1981, Problem 4.5)

$$\mu_{\mathrm{i}} = \frac{2e\lambda_{\mathrm{i}}}{\pi M |\mathbf{u}_{\mathrm{i}}|} \tag{5.3.1}$$

Experimentally, over usual velocity ranges, λ_{i} is found to be reasonably approximated by a constant, and we assume this to be the case for the following analysis (see Fig. 3.15 for some typical data). For the regime of interest here, $\mu_{\mathrm{e}} \gg \mu_{\mathrm{i}}$ and $T_{\mathrm{e}} \gg T_{\mathrm{i}}$, the basic equations can be simplified in a manner similar to that used to to obtain (5.1.15) and (5.1.16). We make the assumption that the ion drift velocity due to the electric field dominates over the velocity due to the pressure gradient, such that

$$\mathbf{u}_{\mathrm{i}} = \mu_{\mathrm{i}} \mathbf{E} \tag{5.3.2}$$

For the electrons we make the opposite assumption, namely that the drift velocity is negligible, to obtain

$$\mathbf{E} = -\frac{kT_{\mathrm{e}}}{e} \frac{\nabla n}{n} \tag{5.3.3}$$

This is equivalent to assuming that the electrons are governed by a Boltzmann distribution, as we have already described in Section 2.4.

With the above assumptions and the steady-state ion continuity equation,

$$\nabla \cdot (n\mathbf{u}_{\mathrm{i}}) = v_{\mathrm{iz}} n \tag{5.3.4}$$

we can derive a differential equation for the density profile. Taking a parallel plane geometry, as in Section 5.2 and solving for u_{i} in terms of $\nabla n/n$ from (5.3.2) and

(5.3.3) we have, for $u_i > 0$,

$$u_i^2 = -\frac{kT_e}{M}\frac{2}{\pi}\frac{\lambda_i}{n}\frac{dn}{dx} \tag{5.3.5}$$

Taking the square root of (5.3.5) and substituting in (5.3.4) we obtain

$$u_B\left(\frac{2\lambda_i}{\pi}\right)^{1/2}\frac{d}{dx}\left(-n\frac{dn}{dx}\right)^{1/2} = \nu_{iz}n \tag{5.3.6}$$

Equation (5.3.6), which is nonlinear, has been solved by Godyak and Maximov (see Godyak, 1986) for the boundary conditions that $u_i = u_B$ at the sheath edge. The solution is

$$\alpha^{2/3}\xi = \frac{1}{2}\ln\left[(1-y^3)^{1/3}+y\right] + \frac{1}{\sqrt{3}}\tan^{-1}\left[\frac{2(y^3-1)^{1/3}-1}{\sqrt{3}}\right] + \frac{\pi}{6\sqrt{3}} \tag{5.3.7}$$

where $\xi = 2x/l$, $y = n/n(0)$, and

$$\alpha = \frac{\nu_{iz}l}{2u_B}\left(\frac{\pi l}{4\lambda_i}\right)^{1/2} \approx 1.25 \tag{5.3.8}$$

Actually, α varies slightly with ν_{iz} as shown in Fig. 5.2a. We see from (5.3.7) that $n/n(0)$ is a function of a single parameter $2\alpha^{2/3}x/l$ where ν_{iz} and therefore α is determined from (5.3.8), which expresses the balance of ionization and loss at the boundary. The result for y is shown in Fig. 5.2b. The density profile is roughly similar to the sinusoidal profile of the simpler case discussed in Section 5.2, but is flatter in the middle and steeper at the edge.

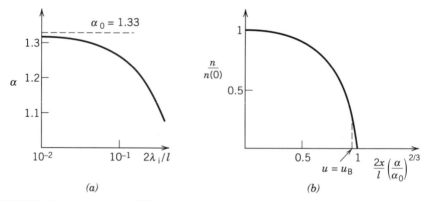

FIGURE 5.2. Low-pressure diffusion solutions for variable mobility model: (a) normalized ionization rate $\alpha = (\nu_{iz}l/2u_B)(\pi l/4\lambda_i)^{1/2}$ versus $2\lambda_i/l$; (b) normalized density $n/n(0)$ versus normalized position $(2x/l)(\alpha/\alpha_0)^{2/3}$ (after Godyak, 1986).

Langmuir Solution

At very low pressures, there is a limiting regime for which the ions are essentially collisionless ($\lambda_i > l$). In this situation we keep the Boltzmann relation for the electrons (5.3.3) but replace the ion drift equation (5.3.2) by an ion velocity governed by energy conservation:

$$\frac{1}{2}Mu_i^2 + e\Phi = 0 \tag{5.3.9}$$

This is equivalent to keeping the inertial and field terms in the force equation (2.3.9). In (5.3.9) we have dropped the negligible ion thermal velocity in the plasma center where the potential Φ is taken to be zero. Note that Φ is negative elsewhere. Solving for $u_i(\Phi)$ from (5.3.9) and eliminating Φ in terms of n using the Boltzmann relation for electrons,

$$n = n_0 e^{\Phi/T_e}$$

(here T_e is in volts), we obtain $u_i(n)$. Substituting this into the continuity equation (5.3.4), we obtain

$$\frac{d}{dx}\left[\left(-\frac{2eT_e}{M}\ln\frac{n}{n_0}\right)^{1/2} n\right] = \nu_{iz}n \tag{5.3.10}$$

Again we are faced with a nonlinear equation which in this case must be solved numerically. This was first done for various geometries in a seminal paper by Tonks and Langmuir (1929), which included matching to the sheath region. They formulated (5.3.10) in terms of the potential, rather than the density, obtaining the integral equation

$$\exp\left(\frac{\Phi(\zeta)}{T_e}\right) = \left(\frac{T_e}{2}\right)^{1/2} \int_0^\zeta \frac{\exp\left(\Phi(\xi')/T_e\right)}{[\Phi(\zeta) - \Phi(\xi')]^{1/2}}\, d\xi' \tag{5.3.11}$$

where $\zeta = x\nu_{iz}/u_B$. This equation has a closed-form solution in terms of Dawson functions, but the solution was originally obtained by Tonks and Langmuir in the form of a power series, and is shown in Fig. 5.3. We note that the variable ζ is a function of the ionization, but ζ/ζ_w is not, where ζ_w is the value of ζ at the plasma edge $x = l'/2$, and so the solution is valid as $\nu_{iz} \rightarrow 0$. The endpoint, where there is a singularity in the derivative of n, occurs at $\zeta_w = 0.572$, $n_s/n_0 = 0.425$, $\Phi/T_e = 0.854$. The solution yields the velocity $u_s \approx 1.3u_B$ at the sheath edge; see also Section 6.2.

The solution in Fig. 5.2b for λ_i/l large does not coincide with the collisionless solution shown in Fig. 5.3. It is possible to construct a heuristic solution that approximates the low-pressure constant λ_i solution for $\lambda_i/l \lesssim 1$, but has a transition to the approximate collisionless solution as $\lambda_i/l \rightarrow \infty$. Godyak (1986) has done this,

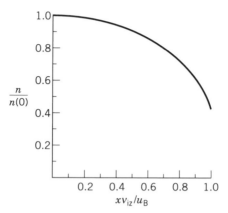

FIGURE 5.3. Free-fall solution: variation of the normalized density $n/n(0)$ versus normalized position xv_{iz}/u_B.

obtaining an approximate result useful for calculations:

$$v_{iz} \approx \frac{2u_B}{l}\left(3 + \frac{l}{2\lambda_i}\right)^{-1/2} \tag{5.3.12}$$

and

$$h_l = \frac{n(l/2)}{n(0)} \approx 0.86\left(3 + \frac{l}{2\lambda_i}\right)^{-1/2} \tag{5.3.13}$$

A similar result for diffusion in a infinitely long cylinder of radius R was obtained (see Godyak, 1986):

$$v_{iz} \approx 2.2\frac{u_B}{R}\left(4 + \frac{R}{\lambda_i}\right)^{-1/2} \tag{5.3.14}$$

and

$$h_R = \frac{n(R)}{n(0)} \approx 0.8\left(4 + \frac{R}{\lambda_i}\right)^{-1/2} \tag{5.3.15}$$

5.4   DIFFUSION ACROSS A MAGNETIC FIELD

We consider diffusion in the presence of magnetic fields, electric fields, and gradients. Generally the species for which the magnetic field is important, in weakly ionized plasmas, is the electrons that have small gyration orbits. To focus our attention we consider a long cylinder, with the magnetic field $\mathbf{B} = \hat{z}B_0$ taken along the cylinder.

The density gradient points radially inward, and the ambipolar electric field, to contain the weakly magnetized ions, also points inward. When an electron gyrating around a line of force suffers a collision, it changes its direction, which would tend to move its center of gyration, on the average, by a gyration radius r_{ce}. This process is random, and therefore diffusive, with r_{ce} replacing λ_e as the diffusion mean free path when $r_{ce} \ll \lambda_e$.

To derive the perpendicular diffusion coefficient, we write the perpendicular component of the fluid equation for either species from (2.3.15):

$$0 = qn(\mathbf{E} + \mathbf{u}_\perp \times \mathbf{B}_0) - kT\nabla n - mn\nu_m\mathbf{u}_\perp$$

where we have again assumed an isothermal plasma and taken ν_m sufficiently large that the inertial (time-derivative) term is negligible. It is convenient to express the vector equation in terms of the rectangular components (taken to be x and y):

$$mn\nu_m u_x = qnE_x - kT\frac{\partial n}{\partial x} + qnu_y B_0 \tag{5.4.1a}$$

and

$$mn\nu_m u_y = qnE_y - kT\frac{\partial n}{\partial y} - qnu_x B_0 \tag{5.4.1b}$$

Using the definitions of μ and D from (5.1.4) and (5.1.5), (5.4.1) can be rewritten:

$$u_x = \pm\mu E_x - \frac{D}{n}\frac{\partial n}{\partial x} + \frac{\omega_c}{\nu_m}u_y \tag{5.4.2a}$$

and

$$u_y = \pm\mu E_y - \frac{D}{n}\frac{\partial n}{\partial y} - \frac{\omega_c}{\nu_m}u_x \tag{5.4.2b}$$

where we have also used the definition of the gyration frequency $\omega_c = qB_0/m$. Equations (5.4.2) may be solved simultaneously for u_x and u_y to obtain

$$\left[1 + (\omega_c\tau_m)^2\right] u_x = \pm\mu E_x - \frac{D}{n}\frac{\partial n}{\partial x} + (\omega_c\tau_m)^2\frac{E_y}{B_0} - (\omega_c\tau_m)^2\frac{kT}{qB_0}\frac{1}{n}\frac{\partial n}{\partial y} \tag{5.4.3a}$$

$$\left[1 + (\omega_c\tau_m)^2\right] u_y = \pm\mu E_y - \frac{D}{n}\frac{\partial n}{\partial y} + (\omega_c\tau_m)^2\frac{E_x}{B_0} + (\omega_c\tau_m)^2\frac{kT}{qB_0}\frac{1}{n}\frac{\partial n}{\partial x}, \tag{5.4.3b}$$

where we have defined $\tau_m \equiv 1/\nu_m$. Dividing by $1 + (\omega_c\tau_m)^2$, we define perpendicular mobility and diffusion coefficients,

$$\mu_\perp = \frac{\mu}{1 + (\omega_c \tau_m)^2} \tag{5.4.4}$$

$$D_\perp = \frac{D}{1 + (\omega_c \tau_m)^2} \tag{5.4.5}$$

and combining (5.4.3a) and (5.4.3b) in vector form, we find

$$\mathbf{u}_\perp = \pm\mu_\perp \mathbf{E} - D_\perp \frac{\nabla n}{n} + \frac{\mathbf{u}_E + \mathbf{u}_D}{1 + (\omega_c \tau_m)^{-2}} \tag{5.4.6}$$

Here $\mathbf{u}_E$ and $\mathbf{u}_D$ are the $E \times B$ *drift* and the *diamagnetic drift* velocities which are perpendicular to the field and the gradients:

$$\mathbf{u}_E = \frac{\mathbf{E} \times \mathbf{B}_0}{B_0^2} \tag{5.4.7}$$

$$\mathbf{u}_D = -\frac{kT}{qB_0^2} \frac{\nabla n \times \mathbf{B}_0}{n} \tag{5.4.8}$$

The drifts perpendicular to the field and gradients are slowed by the collisions, while the mobility and diffusion fluxes parallel to the gradients and perpendicular to the field exist only in the presence of collisions, and are slowed by the presence of the magnetic field. For some plasma discharges the drifts can be important, because they can lead to instabilities with a resulting anomalous transport and they can also lead to large current flows.

The factor $\omega_c \tau_m$ is an important quantity in magnetic confinement, with $\omega_c \tau_m \gg 1$ indicating strong retardation of diffusion. In this limit, dropping the 1, we have

$$D_\perp = \frac{kT}{m\nu_m} \frac{1}{(\omega_c \tau_m)^2} = \frac{kT\nu_m}{m\omega_c^2} \tag{5.4.9}$$

Comparing (5.4.9) with the diffusion coefficient without a magnetic field (or $D = D_\parallel$ parallel to B_0), from (5.1.5) we see that the position of the collision frequency is reversed, with $D_\perp \propto \nu_m$ while $D_\parallel \propto \nu_m^{-1}$. Since $\nu_m \propto m^{-1/2}$ at fixed energy and cross section, we also find $D_\perp \propto m^{1/2}$ and $D_\parallel \propto m^{-1/2}$. This is easily understood in that the lighter electrons move faster without a magnetic field, but are strongly inhibited across the field. We can also understand these relations in terms of random walk distances. As in Section 5.1 we use $\bar{v}^2 = 8kT/\pi m$, and with the mean gyroradius $\bar{r}_c = \bar{v}/\omega_c$ substituted into (5.4.9), we have

$$D_\perp = \frac{\pi}{8} \bar{r}_c^2 \nu_m \tag{5.4.10}$$

Comparing (5.4.10) with (5.1.6), we see that the mean gyration radius has taken the place of the mean free path as the characteristic random walk step.

Ambipolar Diffusion

If plasma can be lost only across the magnetic field, then equating the electron and ion fluxes, as in Section 5.1, leads to a cross-field ambipolar diffusion coefficient as in (5.1.12), except that the quantities refer to the perpendicular mobility and diffusion

$$D_{\perp a} = \frac{\mu_{\perp i} D_{\perp e} + \mu_{\perp e} D_{\perp i}}{\mu_{\perp i} + \mu_{\perp e}}. \tag{5.4.11}$$

If the magnetic field is sufficiently strong that $\mu_{\perp i} \gg \mu_{\perp e}$, reversing the inequality used in Section 5.1, then the simpler form, analogous to (5.1.14), is

$$D_{\perp a} = D_{\perp e}\left(1 + \frac{T_i}{T_e}\right) \tag{5.4.12}$$

where $D_{\perp e}$ is given by (5.4.9). Again the slower diffusion controls the behavior, but in the usual weakly ionized plasma with $T_i \ll T_e$, the ambipolar and electron diffusion coefficients perpendicular to B_0 are not significantly different.

The assumption that the diffusion takes place only across the magnetic field is almost never satisfied. Even for finite length systems in which l (along B_0) $\gg$ d (across B_0), the more rapid diffusion along B_0 is usually important. We therefore consider the regime in which $l \sim d$, as shown in Fig. 5.4. For simplicity, rectangular coordinates are used and the y direction is taken to be uniform and of infinite extent. Since the walls are conducting, it is clear that the fluxes across and along B_0 are coupled, and ambipolarity requires only that the total electron and ion fluxes integrated over the wall surfaces to be equal.

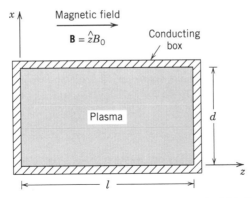

FIGURE 5.4. A plasma-filled conducting box in a dc magnetic field, illustrating the calculation of ambipolar diffusion in a magnetized plasma.

The diffusion is obtained from the continuity equations for electrons and ions:

$$\frac{\partial n}{\partial t} = D_e \frac{\partial^2 n}{\partial z^2} + \mu_e \frac{\partial}{\partial z}(nE_z) + D_{\perp e}\frac{\partial^2 n}{\partial x^2} + \mu_{\perp e}\frac{\partial}{\partial x}(nE_x) \tag{5.4.13}$$

$$\frac{\partial n}{\partial t} = D_i \frac{\partial^2 n}{\partial z^2} - \mu_i \frac{\partial}{\partial z}(nE_z) + D_{\perp i}\frac{\partial^2 n}{\partial x^2} - \mu_{\perp i}\frac{\partial}{\partial x}(nE_x) \tag{5.4.14}$$

Exact two-dimensional solutions to these two coupled nonlinear diffusion equations have not been obtained. Letting $V_{s\perp}$ and $V_{s\parallel}$ be the potential drops across the perpendicular and parallel sheaths, then because the plasma is surrounded by a conducting wall, the potential in the center can be estimated as

$$\Phi \sim V_{s\parallel} + \frac{1}{2}E_z l \sim V_{s\perp} + \frac{1}{2}E_x d$$

Two limiting cases can be considered depending on the size of E_x. For $E_x d \lesssim T_i$, the perpendicular mobility terms in (5.4.13) and (5.4.14) are small compared to the perpendicular diffusion terms. Dropping the mobility terms, as done by Simon (1959), multiplying (5.4.13) by μ_i and (5.4.14) by μ_e and adding the two equations, we obtain

$$\frac{\partial n}{\partial t} = \frac{\mu_i D_e + \mu_e D_i}{\mu_i + \mu_e}\frac{\partial^2 n}{\partial z^2} + \frac{\mu_i D_{\perp e} + \mu_e D_{\perp i}}{\mu_i + \mu_e}\frac{\partial^2 n}{\partial x^2} \tag{5.4.15}$$

Thus, the ambipolar diffusion coefficients are

$$D_{\parallel a} = \frac{\mu_i D_e + \mu_e D_i}{\mu_i + \mu_e} \tag{5.4.16}$$

parallel to the field, and

$$D_{\perp a} = \frac{\mu_i D_{\perp e} + \mu_e D_{\perp i}}{\mu_i + \mu_e} \tag{5.4.17}$$

perpendicular to the field. We see that the parallel diffusion is the same as the case without an applied magnetic field. However, (5.4.17) and (5.4.11) are not the same. Since $\mu_e \gg \mu_i$ and normally $D_{\perp i} \gtrsim D_{\perp e}$, (5.4.17) simplifies to

$$D_{\perp a} \approx D_{\perp i} \tag{5.4.18}$$

With this approximation the diffusion equation (5.4.15) becomes

$$\frac{\partial n}{\partial t} = D_a \frac{\partial^2 n}{\partial z^2} + D_{\perp i}\frac{\partial^2 n}{\partial x^2} \tag{5.4.19}$$

such that the perpendicular loss of ions is by free (not ambipolar) diffusion alone. Physically this corresponds to a situation in which the electrons, flowing along field lines, almost completely remove the negative charge that produces E_x. Since electrons preferentially flow out along the field and ions flow out perpendicular to the field, $\Gamma_i \neq \Gamma_e$ and currents must flow in the wall.

If electron flow along field lines is impeded by inertial or collisional effects or if the axial sheath voltage $V_{s\parallel}$ varies with x, then there can be a substantial ion acceleration potential $E_x d \gtrsim T_i$. In this case the perpendicular ion diffusion term in (5.4.14) is smaller than the mobility term and the preceding derivation of $D_{\perp a}$ is invalid. There is experimental evidence (see Lieberman and Gottscho, 1994, Section VIII.D.2) and also computer simulations (Porteous et al., 1994) that indicate the existence of these radial potentials in magnetized processing discharges such as ECR's (see Section 13.1). Measurements and simulations both show that ions are lost radially from the bulk plasma with a characteristic loss velocity of order the Bohm velocity $u_B = (eT_e/M)^{1/2}$. However, radial expansion of field lines might affect the results. If an electric field exists across field lines with magnitude $E_x \sim T_e/d$, then we can estimate $\Gamma_{\perp i} \sim \mu_{\perp i} n T_e/d$. Then defining $D_{\perp a}$ through $\Gamma_{\perp i} \equiv -D_{\perp a} dn/dx \sim D_{\perp a} n/d$, we obtain

$$D_{\perp a} \sim \mu_{\perp i} T_e \sim D_{\perp i} \frac{T_e}{T_i}$$

in place of (5.4.18). For $d \sim l$, this can lead to substantial perpendicular ion losses in magnetized discharges, as observed in ECR measurements and simulations.

It is well known that plasmas not in thermal equilibrium are subject to insta-bilities. This is a major subject of fully ionized, near collisionless plasmas, and is treated in detail in most texts on plasma physics (see, for example, Chen, 1984). Magnetic field confinement is one source of such disequilibrium that leads to various instabilities which tend to destroy the confinement. Large-amplitude disturbances can lead to turbulent diffusion, which has the upper limit of the *Bohm diffusion* coefficient,

$$D_B = \frac{1}{16} \frac{T_e}{B} \qquad (5.4.20)$$

The scaling with B makes Bohm diffusion increasingly important as a source of cross-field diffusion at high magnetic fields, since from (5.4.10), we see that classical cross-field diffusion scales as $D_\perp \propto 1/B^2$. Bohm diffusion tends to be less important at high collisionality (low temperature and high pressure) both due to the comparative scaling of D_B to $D_\perp$ and also due to the fact that high collisionality tends to inhibit some of the instabilities. We have not considered nonclassical diffusion in this text. The reader wishing to explore the subject further can turn to Chen or other texts on high-temperature plasmas.

5.5   MAGNETIC MULTIPOLE CONFINEMENT

In magnetic multipole confinement, a set of alternating rows of north and south pole permanent magnets is placed around the surface of a discharge chamber. A typical configuration, with the rows arranged around the circumference of a cylindrical chamber, is shown in Fig. 5.5. In some cases, one or both cylindrical endwalls are also covered with rows of magnets. Commonly, each row is composed of a set of many permanent magnets (diameter $\sim$ length $\sim$ 1 inch, $B_0 \sim$ 1 kG). The alternating rows of magnets generate a *line cusp* magnetic configuration in which the magnetic field strength B is a maximum near the magnets and decays with distance into the chamber, as shown in Fig. 5.5. Hence most of the plasma volume can be virtually magnetic field free, while a strong field can exist near the discharge chamber wall, inhibiting plasma loss and leading to an increase in plasma density and uniformity.

Magnetic Fields

The structure of the magnetic field can be understood by unwrapping the circumference to obtain the alternating periodic arrangement of magnet rows in rectangular geometry shown in Fig. 5.6. Assuming that each row of magnets has a width $\Delta \ll d$, the separation of the rows, then B_y at $y = 0$ can be approximated as

$$B_y(x, 0) = B_0 \Delta \sum_{i=-\infty}^{\infty} (-1)^i \delta\left(x - id - \frac{d}{2}\right) \qquad (5.5.1)$$

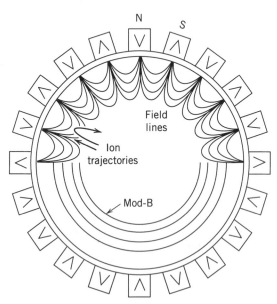

FIGURE 5.5. Magnetic multipole confinement in cylindrical geometry, illustrating the magnetic field lines and the $|\mathbf{B}|$ surfaces near the circumferential walls.

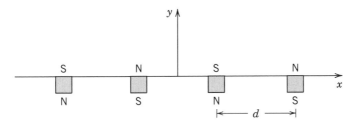

FIGURE 5.6.   Schematic for determining multipole fields in rectangular geometry.

where δ is the Dirac delta function. Introducing the Fourier transform,

$$B_y(x, 0) = \sum_{m=1}^{\infty} A_m \sin \frac{m\pi}{d} x \qquad (5.5.2)$$

and equating (5.5.1) and (5.5.2), then if we multiply by $\sin(\pi x/d)$ and integrate from 0 to d, we obtain the fundamental ($m = 1$) Fourier mode amplitude A_1, such that

$$B_{y1}(x, 0) = \frac{2B_0 \Delta}{d} \sin \frac{\pi x}{d} \qquad (5.5.3)$$

Because $\nabla \cdot \mathbf{B} = 0$ and $\nabla \times \mathbf{B} = 0$ for $y > 0$, B_{y1} satisfies Laplace's equation:

$$\frac{\partial^2 B_{y1}}{\partial x^2} + \frac{\partial^2 B_{y1}}{\partial y^2} = 0 \qquad (5.5.4)$$

The solution to (5.5.4) with boundary conditions that $B_{y1}(x, 0)$ is given by (5.5.3) and that $B_{y1}(x, y \rightarrow \infty)$ is not infinite is

$$B_{y1}(x, y) = \frac{2B_0 \Delta}{d} \sin \frac{\pi x}{d} e^{-\pi y/d} \qquad (5.5.5)$$

From the z component of $\nabla \times \mathbf{B} = 0$, we have

$$\frac{\partial B_{x1}}{\partial y} = \frac{\partial B_{y1}}{\partial x} \qquad (5.5.6)$$

Using (5.5.5) in (5.5.6) and integrating with respect to y, we obtain

$$B_{x1}(x, y) = -\frac{2B_0 \Delta}{d} \cos \frac{\pi x}{d} e^{-\pi y/d} \qquad (5.5.7)$$

The field amplitude is $B_1 = (B_{x1}^2 + B_{y1}^2)^{1/2}$. Using (5.5.5) and (5.5.7), we obtain

$$B_1(x, y) = \frac{2B_0\Delta}{d} e^{-\pi y/d} \tag{5.5.8}$$

showing an exponential decay that is independent of x into the discharge column with decay length d/π. The smooth B_1 surfaces, as well as the alternating B_{y1} and B_{x1} components can be clearly seen in Fig. 5.5. The higher-order Fourier modes with nonzero coefficients ($m = 3, 5, \ldots$) have even shorter decay lengths ($d/3\pi, d/5\pi, \ldots$), and their effect is negligible a short distance from the chamber wall. Thus, we expect this picture to hold at distances significantly greater than d/π within the plasma chamber. Midway between the magnets (at $x = 0, \pm d, \ldots$), the magnetic field is zero at $y = 0$ and rises to a maximum value

$$B_{max} = \frac{\pi^2}{8} \frac{\Delta^2}{d^2} B_0$$

at $y \approx 0.28\, d$, after which it decays exponentially with y. The diffusion across this region is important in determining the confinement properties of the multipoles.

Plasma Confinement

Experimentally (Leung et al., 1975, 1976), multipole fields have been found to have three important effects on low-pressure plasma confinement:

1. Hot electrons, having energies $\gtrsim$ dc sheath potential, can be efficiently confined, provided there is end confinement either with magnetic mirrors, multipoles, or negative electrostatic potentials. These electrons, if created and trapped at low pressures (large mean free path compared to the discharge size) can be the main ionization source for a discharge.
2. Significant (but not large) improvements can be obtained in the confinement of the bulk (low-temperature) plasma in a discharge.
3. Significant improvements in radial plasma uniformity can be obtained.

The effects can, at least partly, be understood in terms of magnetic mirroring in the cusps as governed by (4.3.15). The energetic electrons that are not lost by moving parallel to field lines are mirrored as they move into the higher field near the cusp. Their velocity vectors with respect to the magnetic field at the wall are randomized within the central plasma chamber, where (4.3.15) does not hold. The number of reflections from the cusp then depends on the size of the "loss cone" angle in velocity space compared to the possible solid angle of 4π within which the velocity vector can be found. At lower velocities (or higher pressures), the scattering can take place collisionally on the outward flight, greatly increasing the loss rate. Ambipolar fields also play a part, but in a complicated manner. The improvement in plasma uniformity follows because the diffusion is inhibited in the region of strong magnetic field, as described in Section 5.4. Thus, most of the density gradient occurs at the plasma edge, where the diffusion coefficient is small, leading to a relatively uniform central region.

As an example (Leung et al., 1975), a low-pressure dc argon discharge was created in a 30-cm-diameter, 33-cm-long chamber by primary energetic electrons emitted from a hot filament placed inside the chamber and biased at -60 V. With multipoles and at $p = 0.8$ mTorr, the energetic electrons were confined for up to 70 bounces within the chamber, and the plasma density was increased by approximately a factor of 100. Of this increase, roughly a factor of 30 was measured to be due to the increased confinement of the energetic electrons, and an additional factor of three increase was due to the improvement in confinement for the bulk plasma. However, in most processing discharges the ionization is not produced by a class of very energetic electrons, and the second and third effects listed above are most significant.

A useful concept to discuss confinement is the *effective leak width w* of a line cusp. If there are $\mathcal{N}$ cusps of width w, then the effective circumferential loss width is $\mathcal{N}w$ and the fraction f_{loss} of diffusing electron–ion pairs that will be lost to the wall is

$$f_{\mathrm{loss}} = \frac{\mathcal{N}w}{2\pi R}, \qquad \mathcal{N}w < 2\pi R \qquad (5.5.9)$$

The boundary condition at the wall ($y \approx 0$) for the ambipolar diffusion of plasma within the field-free discharge volume is then

$$\Gamma_{\mathrm{wall}} = f_{\mathrm{loss}} n_s u_B \qquad (5.5.10)$$

We return to the example in Section 5.2 of steady-state diffusion in a plasma slab of length l with an ionization source proportional to the density. The density profile is given by (5.2.15). Equating $\Gamma(l/2)$ in (5.2.17) to Γ_{wall} in (5.5.10), we obtain, for a thin sheath,

$$\frac{f_{\mathrm{loss}} u_B}{D_a \beta} = \tan \frac{\beta l}{2} \qquad (5.5.11)$$

This transcendental equation for β must in general be solved numerically. However, if f_{loss} is not too small, such that the left-hand side of (5.5.11) still remains much greater than unity, then we can approximate $\beta \approx \pi/l$ on the left-hand side to obtain

$$\tan \frac{\beta l}{2} = \frac{f_{\mathrm{loss}} u_B l}{\pi D_a} \qquad (5.5.12)$$

This is the usual regime for most processing discharges. Taking the ratio of $n_s \equiv n(l/2)$ to $n_0 \equiv n(0)$, and using (5.2.15) to substitute for $\tan(\beta l/2)$ in terms of n_s, we find

$$\frac{n_s}{n_0} = \left[1 + \left(\frac{f_{loss} u_B l}{\pi D_a} \right)^2 \right]^{-1/2} \tag{5.5.13}$$

We see that the uniformity of the plasma improves as f_{loss} is reduced below unity by the presence of the multipoles. Since uniformity is often a critical issue plasma processing, multipole confinement may offer a means to control this parameter. A measured density profile with and without multipole confinement is shown in Fig. 12.10, where rf inductive discharges are discussed. As will be shown in Chapter 10 (see (10.2.2) and accompanying discussion), for a fixed absorbed power the plasma density is inversely proportional to the loss area. Hence we would expect $n_0 \propto f_{loss}^{-1}$, when $f_{loss} u_B l / \pi D_a \gg 1$.

Leak Width w

The size of the leak width w is not fully understood. At very low pressures, theoretical calculations, confirmed by measurements (see Hershkowitz et al., 1975), indicate that

$$w \approx 4(\bar{r}_{ce} \bar{r}_{ci})^{1/2} \tag{5.5.14}$$

where $\bar{r}_{ce}$ and $\bar{r}_{ci}$ are the mean electron and ion gyroradii at the location where the magnetic field lines enter the wall. However, the leak width is observed to increase with pressure and is much larger than indicated by (5.5.14) at typical process pressures ($\gtrsim 1$ mTorr). The mechanism for this increase in w is that ions and electrons collisionally diffuse across magnetic field lines, and diffuse or flow along the field lines to the wall. An estimate of the leak width for intermediate pressures is (Matthieussent and Pelletier, 1992)

$$w \approx \frac{2}{\pi} (\bar{r}_{ce} \bar{r}_{ci})^{1/2} \frac{d}{(\lambda_{me} \lambda_{mi})^{1/2}} \tag{5.5.15}$$

where λ_{me} and λ_{mi} are the electron and ion mean free paths. By comparing (5.5.14) and (5.5.15), a heuristic formula valid for low and intermediate pressures can be constructed. The general scalings have been observed experimentally. At some pressure where $w \approx 2\pi R / \mathcal{N}$, f_{loss} given by (5.5.9) rises to unity and the multipoles have little effect on the bulk plasma confinement. Other mechanisms, such as *Bohm diffusion* across magnetic fields due to fluctuating electric fields in the plasma, can also be present and are known to be important for particle losses, e.g., from weakly collisional cusp magnetic fields.

PROBLEMS

5.1. Ambipolar Diffusion Coefficient   Making the assumptions of electric field driven flux for ions and Boltzmann equilibrium for electrons, as in (5.1.15) and

(5.1.16), solve to obtain the ambipolar diffusion coefficient D_a, and compare with (5.1.14).

5.2. Diffusion for Uniform Ionization Rate   A high-pressure, steady-state argon plasma discharge confined between two parallel plates located at $x = \pm l/2$ is created in argon gas at density n_g by uniformly illuminating the region within the plates with ultraviolet radiation. The radiation creates a uniform number G_0 of electron–ion pairs per unit volume per unit time $(m^{-3}\text{-}s^{-1})$ everywhere within the plates. Electrons and ions are lost to the walls by ambipolar diffusion. Assuming that the electron and ion temperatures are uniform and constant with time, with $T_e \gg T_i$, and choosing boundary conditions such that $n(x) \approx 0$ at the walls, find the plasma density $n(x)$ and the peak density n_0 within the plates.

5.3. Ambipolar Diffusion in Parallel Plate Geometry   An rf discharge is ignited between two parallel electrodes located at $x = \pm l/2$. The steady-state diffusion equation

$$\frac{d^2 n}{dx^2} + \beta^2 n = 0$$

with the boundary condition that $n(\pm l/2) = 0$, has the solution $n(x) = n_0 \cos(\pi x/l)$, where $\beta^2 = \nu_{iz}/D_a = (\pi/l)^2$, ν_{iz} is the electron-neutral ionization rate, and $D_a \approx \mu_i T_e$ is the ambipolar diffusion coefficient (T_e is in volts).

(a) Find the steady-state (dc) particle flux $\Gamma(x)$, ambipolar electric field $E(x)$, potential $\Phi(x)$, and total charge density $\rho(x)$. Sketch Γ, E, and Φ for $|x| \leq l/2$.

(b) Plot $\rho(x)/e$ and $n(x)$ on the same graph for $|x| \leq l/2$. Are the ambipolar solutions valid for $\rho(x)/e > n(x)$? Explain your answer.

(c) Find the steady-state diffusion velocity $u(x) = \Gamma(x)/n(x)$. Equating $u(x)$ to the Bohm velocity u_B, find the sheath thickness s (the thickness of the region near the electrode where $u(x) > u_B$). You may assume that $s \ll l/2$.

(d) In view of (b), obtain a condition for your result in (c) to be valid.

5.4. Density at a Sheath Edge   For a constant ambipolar diffusion coefficient D_a and for a diffusion velocity equal to the Bohm velocity u_B at the sheath edge $x = l'/2$, show that the ratio of the density n_{sl} at the sheath edge to the density n_0 in the center of a plane parallel discharge of length l' is

$$\frac{n_{sl}}{n_0} = \left[1 + \left(\frac{1}{\beta} \frac{u_B}{D_a} \right)^2 \right]^{-1/2}$$

where β is given by (5.2.22).

5.5. Diffusion in a Magnetic Field   A plasma is generated in a cylindrical tube of radius R and length l in argon $(M_{Ar}/M_H = 40)$ at $p = 3$ mTorr with a strong magnetic field $B_0 = 1$ kG along the axis of the tube.

(a) Assuming that the ambipolar diffusion coefficient along B_0 has been measured to be $D_a = C/p(\text{Torr})$, with $C = 10^4$ cm^2-Torr/s, and that the ambipolar ion drift velocity corresponds to an energy $\mathcal{E}_\| = 10$ V, calculate the mean free path of argon ions along B_0.

(b) Considering that the transverse ion velocity corresponds to a temperature $T_\perp = 1$ V, calculate the ion gyration radius and determine if the radial diffusion will be significant for $L = 30$ cm and $R = 10$ cm.

5.6. Diffusion in Cylindrical Coordinates

(a) Obtain the steady-state diffusion equation, which is Bessel's equation, for the ambipolar diffusion radially across a long cylinder (the plasma is uniform axially and azimuthally).

(b) The diffusion solution is the lowest-order Bessel function $J_0(\rho)$, where ρ is the argument found from part (a). Using the approximation that $n(r) = 0$ at $r = R$, the cylinder radius, what is the relationship between D_a and ν_{iz}?

(c) What is the relationship between the solution in part (b) and the physical situation in Problem 5.5, in which the diffusion is mostly axial?

5.7. Random Walk Diffusion

In a multiple-mirror device, which has been proposed for confining fusion plasmas, ions are injected into the central magnetic mirror and diffuse through a series of mirrors to the device ends. In the steady state a flux Γ_0 flows out through each half of the machine. The density is a maximum in the center of the machine and falls linearly to n_{min} at each end. The axial diffusion mechanism is that an ion travels an axial distance $l_z = \lambda_i/R_m$, where λ_i is the ion mean free path and $R_m = B_{max}/B_{min}$ is the "mirror ratio." The ion remains trapped in the mirror for a time τ_i before again escaping axially in either direction. Assume $l_z \gg l$, the length between mirrors, and that $\tau_i \gg \bar{v}_i/l_z$, the flight time between mirrors, and that the total device length $2L \gg l_z$.

(a) Derive an approximate one-dimensional diffusion equation for the ion transport in terms of the above parameters (electron effects are neglected), and find the axial diffusion coefficient D_z.

(b) The density falls to n_{min} at $z = \pm L$. Solve the diffusion equation for the central density n_0 as a function of Γ_0, n_{min}, D_z, and L.

5.8. Diffusion in a Magnetized Plasma

Solve (5.4.19) in the steady state with a source term $\nu_{iz}n$ and boundary conditions that $n = 0$ at the rectangular walls $x = \pm d/2$ and $z = \pm l/2$. Find ν_{iz} as a function of D_a, $D_{\perp i}$, d, and l.

5.9. The Congruence Assumption

The congruence assumption $\Gamma_e = \Gamma_i$ is used to derive the ambipolar relation (5.1.11).

(a) Show from particle conservation that $\nabla \cdot \Gamma_e = \nabla \cdot \Gamma_i$.

(b) Show that

$$\nabla \times \Gamma_e = -\mu_e \nabla \Phi \times \nabla n$$
$$\nabla \times \Gamma_i = \mu_i \nabla \Phi \times \nabla n$$

Hence for $\nabla \Phi \times \nabla n = 0$, we find $\nabla \times \Gamma_e = \nabla \times \Gamma_i = 0$.

(c) For $\nabla\Phi \times \nabla n = 0$, from parts (a) and (b), show that $\Gamma_e = \Gamma_i + $ const. The boundary conditions generally set the condition that the constant is zero, and hence $\Gamma_e = \Gamma_i$.

(d) Show that if $n(\mathbf{r}) = n[\Phi(\mathbf{r})]$, e.g., if n is given by the Boltzmann relation (2.4.16), with T_e a constant, then $\nabla\Phi \times \nabla n = 0$.

CHAPTER 6

DC SHEATHS

6.1 BASIC CONCEPTS AND EQUATIONS

At the edge of a bounded plasma a potential exists to contain the more mobile charged species. This allows the flow of positive and negative carriers to the wall to be balanced. In the usual situation of an electropositive plasma, consisting of equal numbers of positive ions and electrons, the electrons are far more mobile than the ions. The plasma will therefore charge positively with respect to a grounded wall. The nonneutral potential region between the plasma and the wall is called a *sheath*.

In a weakly ionized plasma the energy to sustain the plasma is generally heating of the electrons by the source, while the ions are at near equilibrium with the background gas. The electron temperature is then typically of few volts, while the ions are cold. In this situation we may think of monoenergetic ions being accelerated through the sheath potential, while the electron density decreases according to a Boltzmann factor, as described in Section 2.4. The electron density would then decay on the order of a Debye length λ_{De}, to shield the electrons from the wall. However, we cannot linearize the Poisson equation, as we did in deriving λ_{De} in Section 2.4, if we wish to obtain the exact flux balance. Furthermore, we will show that a transition layer or *presheath* must exist between the neutral plasma and the nonneutral sheath in order to maintain the continuity of ion flux, giving rise to an ion velocity at the plasma-sheath edge known as the *Bohm velocity* u_B. The need for this presheath will arise naturally in our derivation in Section 6.2.

If a potential is placed between bounding electrodes, then, while the overall flux balance is maintained, each electrode may separately draw current. The most straight-forward analysis is of a boundary with a large negative potential with respect to the plasma. The simplest example is a uniform ion charge density, or *matrix sheath.* This occurs in the cathode sheath of a dc discharge, for example, considered in Section 14.3. A matrix sheath is also created transiently with a pulsed negative electrode voltage in which the electrons are expelled from a plasma region, leaving a uniform ion density behind. This occurs naturally in *plasma immersion ion implantation,* discussed in Chapter 16. We consider the matrix sheath in Section 6.3.

For a high-voltage sheath, the current to the electrode is almost all ion current. Provided the ion motion in the sheath is collisionless, then the steady self-consistent ion density is not uniform, but rather is described by the *Child–Langmuir law* of *space-charge-limited current* in a planar diode. We also discuss this situation in Section 6.3.

The idealized conditions described in Sections 6.2 and 6.3 are not always met. The temperature of the ions cannot always be ignored with respect to the electron temperature. This situation arises, for example, in highly ionized plasmas. In this case more complicated kinetic treatments are required. In a similar vein, the electron distribution may not be Maxwellian. This may arise due to particular heating or loss mechanisms, which occur, for example, in low-pressure capacitive rf plasmas, discussed in Chapter 11. In this situation the decrease in electron density in the sheath is not given by a Boltzmann factor but must be obtained kinetically. If the neutral gas is electronegative, such that electron attachment is significant, then the negative charges divide between electrons and negative ions. If the fraction of negative ions present becomes large, the mobility of the negative charges can be greatly reduced, changing the conditions at the sheath edge. We consider these various topics, which, in fact, have some unity of analysis, in Section 6.4. Electronegative plasmas are of considerable importance in processing applications, and their analysis is described in Chapter 10.

Other situations that differ from the basic theory arise due to collisional effects in the sheath region. In this case the ion flow is impeded by collisional processes with neutrals, and the transport is mobility rather than inertia limited, similar to that already described in Chapter 5. We discuss two simple limiting collisional cases in Section 6.5. A full treatment, including both inertial and collisional effects, is very complicated, requiring numerical solution of the kinetic equations.

This chapter deals with sheaths that are constant in time. Two other interesting cases are sheaths formed in oscillating rf potentials and sheaths formed transiently by pulsed potentials. In both situations approximate solutions can be obtained if there is a separation of time scales such that electrons respond rapidly to the time variation while ions respond slowly. This separation is characterized by the inequalities

$$f_{pe} \gg \frac{1}{\tau} \gg f_{pi} \qquad (6.1.1)$$

where τ is the time scale of field variation ($\tau = 2\pi/\omega$ for an oscillatory variation) and f_{pe} and f_{pi} are the electron and ion plasma frequencies, respectively. An oscillatory potential applied to an electrode is characteristic of a capacitively excited rf discharge, and we consider this sheath in Chapter 11. The pulsed potential sheath is analyzed in Chapter 16.

The Collisionless Sheath

We use the assumptions (1) Maxwellian electrons at temperature T_e, (2) cold ions ($T_i = 0$), and (3) $n_e(0) = n_i(0)$ at the plasma–sheath interface (interface between essentially neutral and nonneutral regions) at $x = 0$. As shown in Fig. 6.1, we define the zero of the potential Φ at $x = 0$ and take the ions to have a velocity u_s there. Ion energy conservation (no collisions) then gives

$$\frac{1}{2}Mu^2(x) = \frac{1}{2}Mu_s^2 - e\Phi(x) \qquad (6.1.2)$$

The continuity of ion flux (no ionization in the sheath) is

$$n_i(x)u(x) = n_{is}u_s \qquad (6.1.3)$$

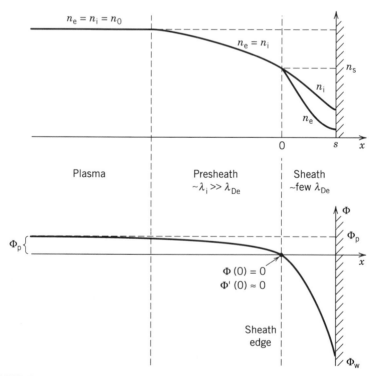

FIGURE 6.1. Qualitative behavior of sheath and presheath in contact with a wall.

where n_{is} is the ion density at the sheath edge. Solving for u from (6.1.2) and substituting in (6.1.3) we have

$$n_i = n_{is} \left(1 - \frac{2e\Phi}{Mu_s^2} \right)^{-1/2} \tag{6.1.4}$$

The electron density is given by the Boltzmann relation

$$n_e(x) = n_{es} e^{\Phi(x)/T_e} \tag{6.1.5}$$

Setting $n_{es} = n_{is} \equiv n_s$ at the sheath edge and substituting n_i and n_e into Poisson's equation

$$\frac{d^2\Phi}{dx^2} = \frac{e}{\epsilon_0}(n_e - n_i)$$

we obtain

$$\frac{d^2\Phi}{dx^2} = \frac{en_s}{\epsilon_0} \left[\exp\frac{\Phi}{T_e} - \left(1 - \frac{\Phi}{\mathcal{E}_s} \right)^{-1/2} \right] \tag{6.1.6}$$

where $e\mathcal{E}_s = \frac{1}{2}Mu_s^2$ is the initial ion energy. Equation (6.1.6) is the basic nonlinear equation governing the sheath potential and ion and electron densities. However, as we shall see in the next section, it has stable solutions only for sufficiently large u_s, created in an essentially neutral *presheath* region.

6.2   THE BOHM SHEATH CRITERION

A first integral of (6.1.6) can be obtained by multiplying (6.1.6) by $d\Phi/dx$ and integrating over x:

$$\int_0^\Phi \frac{d\Phi}{dx}\frac{d}{dx}\left(\frac{d\Phi}{dx}\right) dx = \frac{en_s}{\epsilon_0}\int_0^\Phi \frac{d\Phi}{dx}\left[\exp\frac{\Phi}{T_e} - \left(1 - \frac{\Phi}{\mathcal{E}_s} \right)^{-1/2} \right] dx \tag{6.2.1}$$

Canceling the dx's and integrating with respect to Φ, we obtain

$$\frac{1}{2}\left(\frac{d\Phi}{dx}\right)^2 = \frac{en_s}{\epsilon_0}\left[T_e \exp\frac{\Phi}{T_e} - T_e + 2\mathcal{E}_s\left(1 - \frac{\Phi}{\mathcal{E}_s} \right)^{1/2} - 2\mathcal{E}_s \right] \tag{6.2.2}$$

where we have set $\Phi = 0$ and $d\Phi/dx = 0$ at $x = 0$ corresponding to a field free plasma. Equation (6.2.2) can be integrated numerically to obtain $\Phi(x)$. However, it is apparent that the RHS of (6.2.2) should be positive for a solution to exist. Physically this means that the electron density must always be less than the ion density in the sheath region. Since we expect this to be a problem only for small Φ, we expand the RHS of (6.2.2) to second order in a Taylor series to obtain the inequality

$$\frac{1}{2}\frac{\Phi^2}{T_e} - \frac{1}{4}\frac{\Phi^2}{\mathcal{E}_s} \geq 0 \tag{6.2.3}$$

We see that (6.2.3) is satisfied for $\mathcal{E}_s \geq T_e/2$ or, substituting for $\mathcal{E}_s$,

$$u_s \geq u_B = \left(\frac{eT_e}{M}\right)^{1/2} \tag{6.2.4}$$

This result is known as the *Bohm sheath criterion*. To give the ions this directed velocity u_s, there must be a finite electric field in the plasma over some region, typically much wider than the sheath, called the *presheath* (see Fig. 6.1). Hence the presheath region is not strictly field free, although E is very small there. Since the field at the edge between the sheath and the presheath is not precisely defined, only approximate solutions are obtained by matching sheath to presheath solutions. Nevertheless, if we can make (6.2.4) sharper, by using the equality on the right, this relation is sufficient to obtain quantitative solutions for the plasma equilibrium. The procedure for doing this is to examine the solution in the presheath region of the plasma in the quasineutral approximation $n_i = n_e$, to see how the presheath solution joins with that of the sheath region. We sketch the calculation below.

Presheath Requirements

Setting

$$n_i = n_e \tag{6.2.5}$$

within the presheath and taking the derivative of the logarithm of (6.2.5) we have

$$\frac{1}{n_i}\frac{dn_i}{dx} = \frac{1}{n_e}\frac{dn_e}{dx} \tag{6.2.6}$$

Substituting on the left for the ion current, through the relation $n_i = J_i/eu_i$, this becomes

$$\frac{1}{J_i}\frac{dJ_i}{dx} - \frac{1}{u_i}\frac{du_i}{dx} = \frac{1}{n_e}\frac{dn_e}{dx} \tag{6.2.7}$$

Assuming the Boltzmann form of n_e and rearranging, (6.2.7) becomes

$$\frac{1}{u_i}\frac{du_i}{dx} + \frac{1}{T_e}\frac{d\Phi}{dx} = \frac{1}{J_i}\frac{dJ_i}{dx} \tag{6.2.8}$$

In the presheath we expect the flow to be subsonic, with $u_i < u_B$, and applying this condition to (6.2.8) we have

$$\frac{1}{u_B}\frac{du_i}{dx} + \frac{1}{T_e}\frac{d\Phi}{dx} < \frac{1}{J_i}\frac{dJ_i}{dx} \tag{6.2.9}$$

This is satisfied for either

$$\frac{1}{u_B}\frac{du_i}{dx} + \frac{1}{T_e}\frac{d\Phi}{dx} < 0, \qquad \frac{1}{J_i}\frac{dJ_i}{dx} = 0 \tag{6.2.10}$$

or

$$\frac{1}{u_B}\frac{du_i}{dx} + \frac{1}{T_e}\frac{d\Phi}{dx} > 0, \qquad \frac{1}{J_i}\frac{dJ_i}{dx} > \frac{1}{u_B}\frac{du_i}{dx} + \frac{1}{T_e}\frac{d\Phi}{dx} \tag{6.2.11}$$

Since the ion energy conservation (6.1.2) would make

$$\frac{1}{u_B}\frac{du_i}{dx} + \frac{1}{T_e}\frac{d\Phi}{dx} > 0$$

relations (6.2.10) imply ion friction in the presheath, whereas taking the equality on the right implies current conservation. Relations (6.2.11) imply ionization or geometric contraction. At the sheath–presheath interface there is a transition from subsonic to supersonic ion flow, where the condition of charge neutrality must break down. Putting in specific values of momentum mean free path, ionization, or geometric contraction, the presheath equations can be solved analytically. This has been done, for example, for (a) a geometric presheath with current contraction onto a spherical probe, (b) a plane parallel collisional presheath, and (c) an ionizing presheath with the ionization proportional to n_e. These solutions are plotted in Fig. 6.2. They show quite different behavior in the plasma region: The geometric presheath (a) relaxes to the undisturbed (field free) plasma, the collisional presheath (b) tends to a logarithmic potential shape (indicating that the ion transport requires a residual plasma field), and the ionizing presheath (c) ends with zero field at a finite point representing the midplane of a symmetric plasma. For (b) or (c) the presheath width is of order the mean free path for ion-neutral collisions or for electron-neutral ionization, respectively. Despite the differences, all solutions run quite similarly into the singularity $u_i = u_B$ at the sheath edge. The growing field inhomogeneity approaching this singularity indicates the formation of space charge and the breakdown of the quasineutral approximation. Matching the ion velocity across the sheath–presheath interface then gives us the equality $u_i = u_B$ for the sheath region. Of course, the true behavior is quite complicated at this interface, thus needing a more sophisticated treatment. For more details, including a kinetic treatment, the reader is referred to a review paper by Reimann (1991).

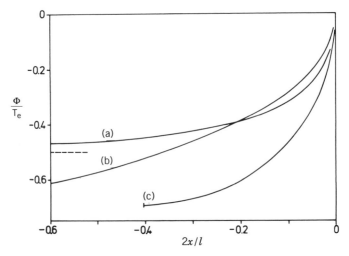

FIGURE 6.2. Φ/T_e versus position within the presheath, showing (a) the geometric presheath, (b) a planar collisional presheath, and (c) a planar ionization presheath. The sheath-presheath edge is at the right (after Riemann, 1991).

The potential drop across the presheath, which accelerates the ions to the Bohm velocity, is given by

$$\frac{1}{2}Mu_B^2 = e\Phi_p$$

where Φ_p is the plasma potential with respect to the potential at the sheath–presheath edge. Substituting for the Bohm velocity from (6.2.4), we find

$$\Phi_p = \frac{T_e}{2} \qquad (6.2.12)$$

This is shown as the dashed line in Fig. 6.1. The ratio of the density at the sheath edge to that in the plasma is then found from the Boltzmann relation

$$n_s = n_b e^{-\Phi_p/T_e} \approx 0.61 n_b \qquad (6.2.13)$$

where n_b is the density where the presheath and bulk plasma join.

Sheath Potential at a Floating Wall

It is quite straightforward to determine the potential drop within the sheath between a plasma and a floating wall. We equate the ion flux (assumed constant through the sheath),

$$\Gamma_i = n_s u_B \qquad (6.2.14)$$

to the electron flux at the wall,

$$\Gamma_e = \frac{1}{4} n_s \bar{v}_e e^{\Phi_w/T_e} \tag{6.2.15}$$

where $\bar{v}_e = (8eT_e/\pi m)^{1/2}$ is the mean electron speed and Φ_w is the potential of the wall with respect to the sheath–presheath edge. We have, after substituting for the Bohm velocity from (6.2.4),

$$n_s \left(\frac{eT_e}{M}\right)^{1/2} = \frac{1}{4} n_s \left(\frac{8eT_e}{\pi m}\right)^{1/2} e^{\Phi_w/T_e} \tag{6.2.16}$$

Solving for Φ_w, we obtain

$$\Phi_w = -T_e \ln \left(\frac{M}{2\pi m}\right)^{1/2} \tag{6.2.17}$$

The wall potential Φ_w is negative and is related linearly to T_e with a factor proportional to the logarithm of the square root of the mass ratio. For hydrogen, for example, $\ln (M/2\pi m)^{1/2} \approx 2.8$, while for argon ($M = 40$ amu) the factor is 4.7. Thus argon ions with initial energy $\mathcal{E}_s = T_e/2$ at the sheath–presheath edge that fall through a collisionless dc sheath to a floating wall would bombard the wall with an energy of $\mathcal{E}_i \approx 5.2T_e$. Of course, electrodes that have potentials on them, either dc or rf, can be bombarded with much higher energy, but these electrodes must draw a substantial net current, as we will show in Section 6.3.

The sheath width s is found by integrating (6.2.2) to obtain $\Phi(x)$ and setting $\Phi(s) = \Phi_w$, with Φ_w given by (6.2.17). The integral must be done numerically. Typical sheath widths are a few electron Debye lengths λ_{De}.

Particle-in-cell simulations can illustrate some of the phenomena we have described, as well as introduce some new features. Figure 6.3 shows a simulation of sheath formation during the decay of a warm, initially uniform density electron–proton plasma between short-circuited parallel plates (no source). The initial plasma parameters are $T_e = T_i = 1$ V and $n_0 = 10^8$ cm^{-3}, with $p = 50$ mTorr, $l = 1$ cm, and an ion-neutral momentum transfer cross section $\sigma_{mi} = 5 \times 10^{-15}$ cm^2. For these parameters, $\lambda_{De} \approx 0.074$ cm, $f_{pe}^{-1} \approx 1.11 \times 10^{-8}$ s, $D_a \approx 1.5 \times 10^5$ cm^2/s, and the fundamental diffusion mode timescale is $\tau_0 \approx 0.68 \times 10^{-6}$ s. The density, field, and potential profiles are shown in (a), (b), and (c) at $t = 5 \times 10^{-8}$ s, after the sheaths have partially formed, but before the decay of the higher-order ($i > 1$) diffusion modes. Hence the ion density in (a) is relatively uniform in the bulk plasma rather than the cosine variation given in (5.2.7), and the steady-state sheaths have not fully formed due to ion transit timescale effects. However, we clearly see the sheath formation. The midpotential variation with time is shown on a short timescale in (d), illustrating its formation with $\Phi_{max} \sim T_e$ as the sheaths form on the very fast electron timescale f_{pe}^{-1}, along with accompanying electron plasma oscillations, as noted previously for Fig. 2.2.

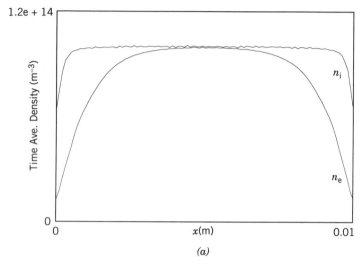

FIGURE 6.3. Particle-in-cell simulation showing sheath formation from warm, initially uniform electron–proton plasma between short-circuited parallel plates: (a) density profiles at time $t = 4 \times 10^{-8}$ s.

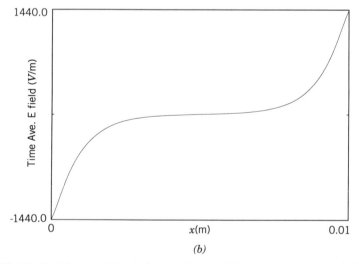

FIGURE 6.3. Particle-in-cell simulation showing sheath formation from warm, initially uniform electron–proton plasma between short-circuited parallel plates: (b) electric field profile.

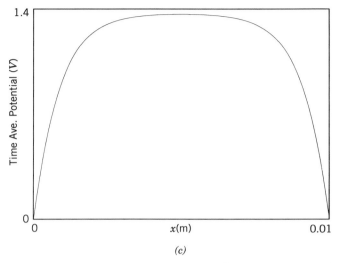

(c)

FIGURE 6.3. Particle-in-cell simulation showing sheath formation from warm, initially uniform electron–proton plasma between short-circuited parallel plates: (c) potential profile.

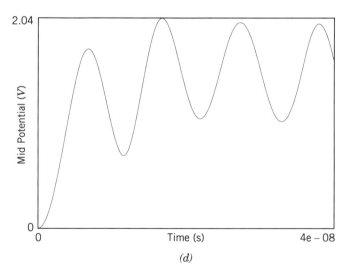

(d)

FIGURE 6.3. Particle-in-cell simulation showing sheath formation from warm, intially uniform electron–proton plasma between short-circuited parallel plates: (d) midpotential versus time.

6.3 THE HIGH-VOLTAGE SHEATH

Matrix Sheath

Sheath voltages are often driven to be very large compared to T_e. The potential Φ in these sheaths is highly negative with respect to the plasma–sheath edge; hence $n_e \sim n_s e^{\Phi/T_e} \to 0$ and only ions are present in the sheath. The simplest high-voltage sheath, with a uniform ion density, is known as a *matrix sheath*. Letting $n_i = n_s = $ const within the sheath of thickness s and choosing $x = 0$ at the plasma–sheath edge, then from (2.2.3),

$$\frac{dE}{dx} = \frac{en_s}{\epsilon_0} \tag{6.3.1}$$

which yields a linear variation of E with x:

$$E = \frac{en_s}{\epsilon_0} x \tag{6.3.2}$$

Integrating $d\Phi/dx = -E$, we obtain a parabolic profile

$$\Phi = -\frac{en_s}{\epsilon_0} \frac{x^2}{2} \tag{6.3.3}$$

Setting $\Phi = -V_0$ at $x = s$, we obtain the matrix sheath thickness

$$s = \left(\frac{2\epsilon_0 V_0}{en_s} \right)^{1/2} \tag{6.3.4}$$

In terms of the electron Debye length $\lambda_{De} = (\epsilon_0 T_e/en_s)^{1/2}$ at the sheath edge, we see that

$$s = \lambda_{De} \left(\frac{2V_0}{T_e} \right)^{1/2} \tag{6.3.5}$$

Hence the sheath thickness can be tens of Debye lengths.

Child Law Sheath

In the steady state, the matrix sheath is not self-consistent since it does not account for the decrease in ion density as the ions accelerate across the sheath. In the limit that the initial ion energy $\mathcal{E}_s$ is small compared to the potential, the ion energy and flux conservation equations (6.1.2) and (6.1.3) reduce to

$$\frac{1}{2} M u^2(x) = -e\Phi(x) \tag{6.3.6}$$

$$en(x)u(x) = J_0 \tag{6.3.7}$$

where J_0 is the constant ion current. Solving for $n(x)$, we obtain

$$n(x) = \frac{J_0}{e}\left(-\frac{2e\Phi}{M}\right)^{-1/2} \tag{6.3.8}$$

Using this in Poisson's equation, we have

$$\frac{d^2\Phi}{dx^2} = -\frac{J_0}{\epsilon_0}\left(-\frac{2e\Phi}{M}\right)^{-1/2} \tag{6.3.9}$$

Multiplying (6.3.9) by $d\Phi/dx$ and integrating from 0 to x, we have

$$\frac{1}{2}\left(\frac{d\Phi}{dx}\right)^2 = 2\frac{J_0}{\epsilon_0}\left(\frac{2e}{M}\right)^{-1/2}(-\Phi)^{1/2} \tag{6.3.10}$$

where we have chosen $d\Phi/dx = -E = 0$ at $\Phi = 0$ ($x = 0$). Taking the (negative) square root (since $d\Phi/dx$ is negative) and integrating again, we obtain

$$-\Phi^{3/4} = \frac{3}{2}\left(\frac{J_0}{\epsilon_0}\right)^{1/2}\left(\frac{2e}{M}\right)^{-1/4}x \tag{6.3.11}$$

Letting $\Phi = -V_0$ at $x = s$ and solving for J_0, we obtain

$$J_0 = \frac{4}{9}\epsilon_0\left(\frac{2e}{M}\right)^{1/2}\frac{V_0^{3/2}}{s^2} \tag{6.3.12}$$

Equation (6.3.12) is the well-known Child law of space-charge-limited current in a plane diode. With fixed spacing s it gives the current between two electrodes as a function of the potential difference between them, and has been traditionally used for electron diodes. However, with J_0 given explicitly as

$$J_0 = en_s u_B \tag{6.3.13}$$

in (6.3.12), we have a relation between the sheath potential, the sheath thickness, and the plasma parameters, which can be used to determine the sheath thickness s. Substituting (6.3.13) in (6.3.12) and introducing the electron Debye length at the sheath edge, we obtain

$$s = \frac{\sqrt{2}}{3}\lambda_{De}\left(\frac{2V_0}{T_e}\right)^{3/4} \tag{6.3.14}$$

Comparing this to the matrix sheath width, we see that the Child law sheath is larger by a factor of order $(V_0/T_e)^{1/4}$. Hence the Child law sheath can be of order of 100 Debye lengths (~ 1 cm) in a typical processing discharge. Since there are no electrons within the sheath to excite the gas, the sheath region appears dark when observed visually.

Inserting (6.3.12) into (6.3.11) yields the potential within the sheath as a function of position

$$\Phi = -V_0 \left(\frac{x}{s}\right)^{4/3} \tag{6.3.15}$$

The electric field $E = d\Phi/dx$ is

$$E = \frac{4}{3}\frac{V_0}{s}\left(\frac{x}{s}\right)^{1/3} \tag{6.3.16}$$

and the ion density $n = (\epsilon_0/e)\, dE/dx$ is

$$n = \frac{4}{9}\frac{\epsilon_0}{e}\frac{V_0}{s^2}\left(\frac{x}{s}\right)^{-2/3} \tag{6.3.17}$$

We see that n is singular as $x \to 0$, a consequence of the simplifying assumption in (6.3.6) that the initial ion energy $\mathcal{E}_s = 0$. The analysis can be carried through for a finite $e\mathcal{E}_s = \frac{1}{2}Mu_B^2$, using (6.1.2), resolving the singularity and yielding $n \to n_s$ as $x \to 0$ (Problem 6.1).

The Child law solution is valid if the sheath potentials are large compared to the electron temperature. It is therefore not appropriate for use where the sheath potential is the potential between a plasma and a floating electrode. However, with some modification, we shall see in Chapter 12 that it is useful in determining the sheath width of an rf-driven discharge. Because the ion motion was assumed collisionless, it is also not appropriate for higher pressure discharges. We shall treat collisional formulations of the sheath region in Section 6.5.

6.4 GENERALIZED CRITERIA FOR SHEATH FORMATION

Using a kinetic treatment without ion collisions, the Bohm criterion for a stable sheath can be generalized to arbitrary ion and electron distributions. First formulated by Boyd and Thompson (1959), a more rigorous and complete treatment in the limit $\lambda_{De} \to 0$ can be found in Riemann (1991). The result is

$$\frac{eT_e}{M}\int_0^\infty \frac{1}{v^2}f(v)\,dv \le T_e\frac{d(n_e + n_-)}{d\Phi}\bigg|_{\Phi=0} \tag{6.4.1}$$

where $f(v)$ is the one-dimensional speed distribution of the positive ions, $n_e + n_-$ is the sum of the densities of the negatively charged species, and Φ is the potential, with $\Phi = 0$ at the sheath–presheath edge. For our previous case of cold ions and

Maxwellian electrons, (6.4.1) becomes

$$\frac{eT_e}{M} \int_0^\infty \frac{1}{v^2} \delta(v - u_s)\, dv \le T_e \frac{d}{d\Phi}\left(e^{\Phi/T_e}\right)\Big|_{\Phi=0} \tag{6.4.2}$$

where $\delta(v - u_s)$ is the Dirac δ function. Evaluating the integral on the left and taking the derivative on the right, we have

$$\frac{eT_e}{M} \frac{1}{u_s^2} \le 1$$

or

$$u_s \ge \left(\frac{eT_e}{M}\right)^{1/2} = u_B$$

which is the Bohm criterion from (6.2.4).

The more general form can be calculated for finite temperature ion distributions, but can lead to mathematical difficulties at low energies due to the average over $1/v^2$. Non-Maxwellian electron distributions, such as power-law distributions that can arise from stochastic rf heating (Chapter 11), can also lead to mathematical singularities. In physical devices, however, collisional processes at low energies generally allow nonsingular solutions to exist.

Electronegative Gases

A physical situation in which (6.4.1) is particularly useful is for electronegative gases in which electron attachment allows a significant number of negative ions to be present. This situation was treated by Boyd and Thompson (1959) and we follow their approach here. The Poisson equation for the potential is

$$\nabla^2\Phi = -\frac{e}{\epsilon_0}(n_+ - n_e - n_-) \tag{6.4.3}$$

where n_+, n_e, and n_- are the positive ion, electron, and negative ion densities, respectively. At the sheath edge we use quasineutrality, $n_{s+} = n_{se} + n_{s-}$, and define the ratio of negative ions to electrons as $\alpha_s \equiv n_{s-}/n_{se}$. Then the quasineutral condition becomes

$$n_{s+} = (1 + \alpha_s)\, n_{se} \tag{6.4.4}$$

If we further consider that the electron and negative ion distributions are Maxwellian, with a temperature ratio $T_e/T_i \equiv \gamma$, then for cold positive ions we can directly repeat the calculation in Section 6.2 to obtain a new Bohm criterion (Problem 6.3). Here we use the more general expression (6.4.1). The Boltzmann relation for electrons and negative ions gives

$$n_e + n_- = n_{se}e^{\Phi/T_e} + \alpha_s n_{se}e^{\gamma\Phi/T_e}$$

which combined with (6.4.4) gives

$$n_e + n_- = \frac{n_{s+}}{1 + \alpha_s}\left(e^{\Phi/T_e} + \alpha_s e^{\gamma\Phi/T_e}\right) \tag{6.4.5}$$

Taking a derivative of (6.4.5) with respect to Φ, and evaluating at $\Phi = 0$, on the RHS of (6.4.1), that equation becomes

$$\frac{eT_e}{M}\int_0^\infty \frac{1}{v^2}f(v)\,dv \leq n_{s+}\left(\frac{1 + \alpha_s\gamma}{1 + \alpha_s}\right) \tag{6.4.6}$$

For cold ions, evaluating the integral as in (6.4.2) and taking the reciprocal we have

$$u_s \geq \left[\frac{eT_e(1 + \alpha_s)}{M(1 + \alpha_s\gamma)}\right]^{1/2} \tag{6.4.7}$$

which is the generalization of the Bohm criterion (6.2.4) for an electronegative plasma. It is immediately apparent that, if γ is large and α_s not too small, the negative ions strongly reduce the velocity required at the sheath edge. However, in this situation the positive ion temperature cannot be ignored and the left-hand side of (6.4.6) must be integrated over the ion distribution.

This is not the end of the story, because the potentials in the bulk plasma and presheath regions will repel the colder negative ions, thus reducing α_s at the sheath edge as compared to $\alpha_b \equiv n_{b-}/n_{be}$ where the presheath and bulk plasma join, thus increasing the importance of the electrons in the sheath region. If Φ_p is the potential at this position with respect to the sheath–presheath edge, then using the Boltzmann relation for both electrons and negative ions, $n_{se} = n_{be}\exp(-\Phi_p/T_e)$, $n_{s-} = n_{b-}\exp(-\gamma\Phi_p/T_e)$, we combine these expressions with the definition of α to obtain

$$\alpha_s = \alpha_b \exp\left[\frac{\Phi_p(1 - \gamma)}{T_e}\right] \tag{6.4.8}$$

We have previously found that for $\alpha_s = 0$ (electrons only) that $\Phi_p/T_e = 1/2$. Using the same argument of conservation of ion energy we obtain (Problem 6.4)

$$\frac{\Phi_p}{T_e} = \frac{1 + \alpha_s}{2(1 + \gamma\alpha_s)} \tag{6.4.9}$$

Substituting (6.4.9) in (6.4.8) we can solve explicitly for α_b:

$$\alpha_b = \alpha_s \exp\left[\frac{(1 + \alpha_s)(\gamma - 1)}{2(1 + \gamma\alpha_s)}\right] \qquad (6.4.10)$$

Considering α_b as the known quantity for an electronegative gas, then α_s must be determined numerically from (6.4.10), and Φ_p from (6.4.9). This was done by Boyd and Thompson with the result shown in Fig. 6.4. The ratio Φ_p/T_e is seen to be

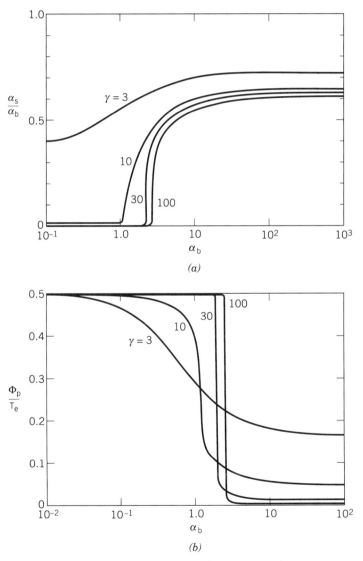

FIGURE 6.4. Negative ion sheath solutions; (a) α_s/α_b and (b) Φ_p/T_e versus α_b, with γ as a parameter (after Boyd and Thompson, 1959).

very nearly $1/2$ for electronegative discharges if $\alpha_b < 2$ and $\gamma > 30$, which hold in weakly electronegative gases under typical discharge operating conditions. As will be seen in Chapter 10, α_b is, in turn, determined from a diffusion solution within the bulk plasma, in terms of $\alpha_0 = n_{0-}/n_{0e}$, the value at the center of the plasma.

6.5   COLLISIONAL SHEATHS

If the mean free path for ion momentum transfer $\lambda_i < s$, the sheath width, then the assumption (6.3.6) of energy conservation, used to derive the Child law, fails. This modifies both the dynamics in the high potential sheath region and the ion velocity at the sheath edge. Consider first the high-voltage sheath region. If the ionization within the sheath is negligible, current continuity still holds, which is expressed as

$$n_i u_i = n_s u_s \tag{6.5.1}$$

where n_s and u_s are the values at the sheath edge. Considering the collisional case, we take

$$u_i = \mu_i E \approx \frac{2e\lambda_i}{\pi M |\mathbf{u}_i|} E \tag{6.5.2}$$

where μ_i is the mobility as defined in (5.3.1). Generally both μ_i and λ_i are functions of the velocity. However, as we have discussed in Section 5.3, depending on the gas pressure and ion velocity, one or the other of these quantities may be relatively independent of velocity. For argon, for example, λ_i is relatively independent of velocity at intermediate pressures and with sheath voltages commonly used for plasma processing. With the assumption of constant λ_i, solving for $u_i > 0$ from (6.5.2) and substituting the result in (6.5.1) we have

$$n_i = \frac{n_s u_s}{(2e\lambda_i E/\pi M)^{1/2}} \tag{6.5.3}$$

Substituting this in Gauss' law (2.2.3), we have

$$\frac{dE}{dx} = \frac{en_s u_s}{\epsilon_0 (2e\lambda_i E/\pi M)^{1/2}} \tag{6.5.4}$$

Separating variables, we can integrate and solve for E to obtain

$$E = \left[\frac{3en_s u_s}{2\epsilon_0 (2e\lambda_i/\pi M)^{1/2}} \right]^{2/3} x^{2/3} \tag{6.5.5}$$

where we have set $E(0) \approx 0$ at the sheath edge. A second integration gives the potential

$$\Phi = -\frac{3}{5} \left(\frac{3}{2\epsilon_0}\right)^{2/3} \frac{(en_s u_s)^{2/3}}{(2e\lambda_i/\pi M)^{1/3}} x^{5/3} \qquad (6.5.6)$$

where we have set $\Phi(0) = 0$. Noting that $en_s u_s = J_0$, the constant current, we can take the 3/2 power of (6.5.6), rearrange, and taking $\Phi = -V_0$ at the electrode position $x = s$, we obtain

$$J_0 = \left(\frac{2}{3}\right)\left(\frac{5}{3}\right)^{3/2} \epsilon_0 \left(\frac{2e\lambda_i}{\pi M}\right)^{1/2} \frac{V_0^{3/2}}{s^{5/2}} \qquad (6.5.7)$$

Equation (6.5.7) gives a collisional form of the Child law for the regime in which λ_i is independent of ion velocity. We note that the current scales the same with voltage, but differently with sheath spacing, than for the collisionless case. For a fixed J_0 and V_0, the sheath width scales as $s \propto \lambda_i^{1/5}$ and therefore weakly decreases as the gas pressure is increased.

Alternatively to our relation (6.5.7), we could equally well have chosen the higher pressure regime to make the calculation, taking ν_{mi} and hence μ_i independent of velocity. In this case, a similar integration procedure leads to the result (Problem 6.5)

$$J_0 = \frac{9}{8} \epsilon_0 \mu_i \frac{V_0^2}{s^3} \qquad (6.5.8)$$

We note here new scalings of J_0 with both V_0 and s.

For the highly collisional case, the velocity at the sheath edge u_s is no longer equal to u_B calculated in Section 6.2. A numerical solution has been obtained (Godyak and Sternberg, 1990a) for u_s/u_B as a function of λ_{De}/λ_i where λ_{De} is the electron Debye length and λ_i is the ion mean free path. A reasonable analytic approximation to this numerical result is

$$\frac{u_s}{u_B} \approx \frac{1}{(1 + \pi\lambda_{De}/2\lambda_i)^{1/2}} \qquad (6.5.9)$$

such that $u_s/u_B \approx (2\lambda_i/\pi\lambda_{De})^{1/2}$ for $\lambda_i \ll \lambda_{De}$. More detailed use of these various relations will be given in Chapter 10, where we calculate discharge equilibria.

6.6 ELECTROSTATIC PROBE DIAGNOSTICS

A metal probe, inserted in a discharge and biased positively or negatively to draw electron or ion current, is one of the earliest and still one of the most useful tools for diagnosing a plasma. These probes, introduced by Langmuir and analyzed in

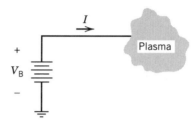

FIGURE 6.5.   Definition of voltage and current for a Langmuir probe.

considerable detail by Mott–Smith and Langmuir (1926) are usually called *Langmuir probes*. As with any other electrode, the probe is surrounded by a sheath, such that its analysis naturally fits into the present chapter. However, unlike large electrode surfaces that are used to control a plasma, probes are usually quite small and under suitable conditions, produce only minor local perturbations of the plasma.

The voltage and current of a probe defined in Fig. 6.5 lead to a typical probe voltage–current characteristic as shown in Fig. 6.6. At the probe voltage $V_B = \Phi_p$, the probe is at the same potential as the plasma and draws mainly current from the more mobile electrons, which is designated as positive current flowing from the probe into the plasma. For increasing V_B above this value, the current tends to saturate at the electron saturation current, but, depending on the probe geometry, can increase due to increasing effective collection area. For $V_B < \Phi_p$, electrons are repelled according to the Boltzmann relation, until at Φ_f the probe is sufficiently negative with respect to the plasma that the electron and ion currents are equal such that $J = 0$. Φ_f is known as the *floating potential*, because it is the potential at which an insulated probe, which cannot draw current, will float. For $V_B < \Phi_f$, the current is increasingly ion current (negative into the plasma), tending to an ion saturation current that may also vary with voltage due to a change of the effective collection area. The magnitude of the

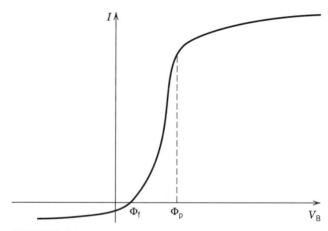

FIGURE 6.6.   Typical I–V_B characteristic for a Langmuir probe.

ion saturation current is, of course, much smaller than the electron saturation current due to the much greater ion mass.

The basic theory for a plane collecting area, based on the sheath calculations of the previous sections, is quite simple. However, to minimally disturb the plasma and also for ease of construction, Langmuir probes are often thin wires with the wire radius $a < \lambda_{De}$. The trajectories of charged particles in the sheath then become important in determining the collected current, and the analysis becomes quite complicated. As the voltage is raised, either to large positive or large negative values with respect to the plasma, the sheath thickness s increases according to Child's law, and consequently the effective collecting area also increases. If $T_i \sim T_e$, then additional complications arise to make calculations very involved. There are also difficulties if the momentum transfer mean free paths λ_i, $\lambda_e \lesssim s$, which can occur in high-pressure discharges. A review of the analysis, including many of these complications, is given by Chen (1965). The extension to $T_i \sim T_e$, which is not usually of great interest in processing discharges, is given in a report by Laframboise (1966).

The story does not end here. More complicated probe configurations, such as double probes and emissive probes, have proven quite useful in various situations. These are also reviewed in Chen, and we consider them below. In an rf-driven plasma an additional complication arises in that the potential of the plasma oscillates with respect to ground. Since we generally wish to use probes in a quasielectrostatic manner, the probe is usually made to oscillate with the plasma to eliminate the effect of the oscillating potential. Detailed consideration of oscillating plasma potentials and methods of using probes in their presence is found in reviews by Godyak (1990a) and Hershkowitz (1989). We summarize some of these considerations below.

Probe theory has generally been developed for plasmas in which the electron distribution is approximated by a Maxwellian. Various deviations from Maxwellian electrons exist in discharges. As described in Chapter 11, sheath heating in a capacitive discharge can result in a high-energy tail to the electron distribution, leaving the bulk electrons considerably colder than they would be in an equilibrium discharge with a Maxwellian distribution. These "two-temperature" distributions modify the results of Langmuir probes. Godyak et al. (1993) have critically examined this phenomenon and argue that the use of standard electron and ion saturation current techniques for analyzing probe data can lead to considerable error in the resulting plasma parameters. They present an alternative technique in which the electron energy distribution function (EEDF) is measured and used directly in calculating the plasma density. We review their arguments and technique below.

Plasma densities obtained from the ion saturation current to probes have been compared with other measurement techniques such as microwaves (see Section 4.6). Generally, the comparisons have indicated that probe-predicted densities, using ion saturation current, are somewhat high when compared under conditions for which the microwave predictions are expected to be highly accurate. This result would generally agree with the arguments presented by Godyak and associates. However, in many situations, the densities obtained by probe and microwave techniques are quite close (e.g., see Fig. 4.13). The accuracy of using the ion saturation current to measure the plasma density depends on the closeness of the electron distribution to an assumed Maxwellian at the probe sheath edge, and therefore to the type of plasma being diagnosed.

Finally, we shall briefly discuss practical probes and circuits for their use. Details of probes and probe circuitry are usually to be found in original articles, references to which can be found in the review articles cited here. Basic information on probes and circuits, beyond that given here, can also be found in the review articles by Chen (1965), Hershkowitz (1989), Godyak (1990a), and Godyak et al. (1992).

Planar Probe With Collisionless Sheath

Consider a flat plate probe with the (two-sided) physical probe area $A \gg s^2$, where s is the sheath thickness, such that the collecting area A is essentially independent of s. As we saw in Section 6.3, if a large voltage is applied to the probe, then $s \gg \lambda_{\mathrm{De}}$, and we find that A is quite large to satisfy the above condition. For this reason we expect that biasing the probe strongly positive to collect only electron current would strongly perturb the plasma. Consider therefore that the probe is biased sufficiently negatively to collect only ion current. From (6.3.13) the current "collected" (see direction in Fig. 6.5) by the probe is

$$I = -I_{\mathrm{i}} = -en_s u_{\mathrm{B}} A \tag{6.6.1}$$

where, as in (6.2.4) with $T_i \ll T_e$, the Bohm velocity u_{B} is given by

$$u_{\mathrm{B}} = \left(\frac{eT_e}{M} \right)^{1/2} \tag{6.6.2}$$

If we know T_e, then the density at the sheath edge n_s is determined from the measurement of I_i. As in (6.2.13), the plasma density in the probe neighborhood is then obtained as

$$n_0 \approx \frac{n_s}{0.61} \tag{6.6.3}$$

Since the electron temperature in most discharges is clamped in the range of 2–5 V by particle balance (see Section 10.1), a reasonable estimate of density can be obtained without knowing T_e. However, by varying the probe voltage, it is also straightforward to measure T_e. Considering that the probe potential is retarding with respect to the plasma potential, then, using Boltzmann's relation as in Section 6.2, the electron component of the probe current is

$$I + I_i = I_e = \frac{1}{4} en_s \bar{v}_e A \exp \left(\frac{V_{\mathrm{B}} - \Phi_{\mathrm{p}}}{T_e} \right) \tag{6.6.4}$$

where $\bar{v}_e = (8eT_e/\pi m)^{1/2}$, and $V_{\mathrm{B}} - \Phi_{\mathrm{p}} < 0$ is the potential between the probe and the plasma. There is an exponential increase in I_e with increasing V_{B} in this range.

Defining an electron saturation current

$$I_{esat} = \frac{1}{4} e n_s \bar{v}_e A \tag{6.6.5}$$

and taking the logarithm of (6.6.4), we have

$$\ln \left(\frac{I_e}{I_{esat}} \right) = \frac{V_B - \Phi_p}{T_e} \tag{6.6.6}$$

From (6.6.6) we see that the inverse slope of the logarithmic electron probe current with respect to V_B (in volts) gives T_e directly in volts.

The above simple interpretation is limited by the dynamic range over which (6.6.4) holds. For I_e too small, adding the measured I_i to I can introduce errors in the determination of I_e. For V_B too large, the Boltzmann exponential no longer is accurate, as electron saturation is approached. The nominal useful range of voltages over which the slope can be measured is then

$$\frac{|\Delta V_B|}{T_e} \approx \ln \left(\frac{\bar{v}_e}{4u_B} \right) = \ln \left(\frac{M}{2\pi m} \right)^{1/2} \tag{6.6.7}$$

which is approximately 4.7 for argon. This range is sufficient, provided there are no geometric complications.

The floating potential Φ_f and the plasma potential Φ_p are often of interest in discharge operation. The floating potential is the potential at which the probe draws equal electron and ion currents. If the plasma is mainly surrounded by a grounded conducting surface, then we would expect the floating potential to lie near this ground, as shown in Fig. 6.6. This follows because the ground is usually not, itself, drawing significant net current, and thus at $V_B = \Phi_f$ the probe behaves as part of the ground. The plasma (space) potential, given by (6.6.7) with $\Phi_p - \Phi_f = \Delta V_B$, can be approximately determined from the knee (point of maximum first derivative) of the electron saturation portion of the I–V_B characteristic of Fig. 6.6. For planar probes the knee is easily recognizable, but the current drawn may be too large, either modifying the plasma or destroying the probe. For cylindrical probes, considered below, the measurement is usually possible, but its accuracy is reduced due to the variation of current with voltage in the electron saturation region.

Non-Maxwellian Electrons

A low-pressure discharge often has an electron energy distribution that departs significantly from a Maxwellian. For example, in Fig. 11.10a (page 359), the electron distribution of a low-pressure rf discharge is given, which can be approximated by a two-temperature Maxwellian. At higher pressures, for which a two-temperature distribution is not evident, high accelerating fields may also result in a non-Maxwellian distribution. For an arbitrary distribution function, the electron current to a planar probe in the retarding potential region $\Phi_p - V_B > 0$ can be written as

$$I_e = eA \int_{-\infty}^{\infty} dv_x \int_{-\infty}^{\infty} dv_y \int_{v_{min}}^{\infty} dv_z \, v_z f_e(\mathbf{v}) \tag{6.6.8}$$

where

$$v_{min} = \left[\frac{2e(\Phi_p - V_B)}{m}\right]^{1/2} \tag{6.6.9}$$

is the minimum velocity along z for an electron at the plasma–sheath edge to reach the probe. For an isotropic distribution we can introduce spherical polar coordinates in velocity to obtain

$$I_e = eA \int_{v_{min}}^{\infty} dv \int_{0}^{\theta_{min}} d\theta \int_{0}^{2\pi} d\phi \, v \cos\theta \, v^2 \sin\theta f_e(v) \tag{6.6.10}$$

where A is the physical collecting area of the probe and where

$$\theta_{min} = \cos^{-1} \frac{v_{min}}{v} \tag{6.6.11}$$

The ϕ and θ integrations are easily done, yielding

$$I_e = \pi eA \int_{v_{min}}^{\infty} dv \, v^3 \left(1 - \frac{v_{min}^2}{v^2}\right) f_e(v) \tag{6.6.12}$$

A transformation of (6.6.12) allows f_e to be obtained directly in terms of the second derivative of I_e with respect to $V = \Phi_p - V_B$. Introducing the change of variable $\mathcal{E} = \frac{1}{2}mv^2/e$, then (6.6.12) becomes

$$I_e = \frac{2\pi e^3}{m^2} A \int_{V}^{\infty} d\mathcal{E} \, \mathcal{E} \left\{ \left(1 - \frac{V}{\mathcal{E}}\right) f_e[v(\mathcal{E})] \right\} \tag{6.6.13}$$

where $v(\mathcal{E}) = (2e\mathcal{E}/m)^{1/2}$. Differentiating I_e we obtain[*]

$$\frac{dI_e}{dV} = -\frac{2\pi e^3}{m^2} A \int_{V}^{\infty} d\mathcal{E} f_e[v(\mathcal{E})] \tag{}$$

---

[*] Note that if $G = \int_{x_1}^{x_2} g(x_1, x) \, dx$ then $\partial G/\partial x_1 = \int_{x_1}^{x_2} (\partial g/\partial x_1) \, dx - g(x_1, x_1)$.

and a second differentiation yields

$$\frac{d^2 I_e}{dV^2} = \frac{2\pi e^3}{m^2} A f_e[v(V)] \tag{6.6.14}$$

It is usual to introduce the electron *energy distribution function* (EEDF) $g_e(\mathcal{E})$ by

$$g_e(\mathcal{E}) \, d\mathcal{E} = 4\pi v^2 f_e(v) \, dv \tag{6.6.15}$$

Using the relation between $\mathcal{E}$ and v, we find

$$g_e(\mathcal{E}) = 2\pi \left(\frac{2e}{m}\right)^{3/2} \mathcal{E}^{1/2} f_e[v(\mathcal{E})] \tag{6.6.16}$$

Using this to eliminate f_e from (6.6.14), we obtain

$$g_e(V) = \frac{2m}{e^2 A} \left(\frac{2eV}{m}\right)^{1/2} \frac{d^2 I_e}{dV^2} \tag{6.6.17}$$

which gives $g_e(V)$ directly in terms of the measured value of $d^2 I_e/dV^2$. The electron energy probability function (EEPF) $g_p(\mathcal{E}) = \mathcal{E}^{-1/2} g_e(\mathcal{E})$ is sometimes introduced instead. For a Maxwellian distribution,

$$g_p(\mathcal{E}) = \frac{2}{\sqrt{\pi}} n_e T_e^{-3/2} e^{-\mathcal{E}/T_e} \tag{6.6.18}$$

such that $\ln g_p$ is linear with $\mathcal{E}$. The electron density n_e and the average energy $\langle \mathcal{E}_e \rangle$ can then be determined as

$$n_e = \int_0^\infty g_e(\mathcal{E}) \, d\mathcal{E} \tag{6.6.19}$$

and

$$\langle \mathcal{E} \rangle = \frac{1}{n_e} \int_0^\infty \mathcal{E} g_e(\mathcal{E}) \, d\mathcal{E} \tag{6.6.20}$$

The effective temperature is defined as $T_{eff} = \frac{2}{3}\langle \mathcal{E} \rangle$. The maximum in the first derivative dI_e/dV_B of the electron current is also a good indicator for the location of the plasma potential Φ_p. The use of (6.6.17), along with (6.6.19) and (6.6.20) to determine n_e and T_{eff} from the probe characteristic has a number of virtues. First, (6.6.19) can be shown to be valid for any isotropic electron velocity distribution. Second, (6.6.17) is valid for any convex probe geometry, planar, cylindrical or spherical (Kagan and Perel, 1964); e.g., $A = 2\pi a d$ for a cylindrical probe of radius a and length d. Third, non-Maxwellian distributions can be measured. Fourth, the result (6.6.17) does not depend on the ratio of probe dimension to Debye length or the ratio T_i/T_e (Godyak, 1990a).

Cylindrical Probe With a Collisionless Sheath

As we have seen in Section 6.3, the sheath thickness s can be quite significant, $s \gg \lambda_{De}$ such that one cannot routinely satisfy $A \gg s^2$. This recognition led to analysis of cylindrical and spherical probes (Mott–Smith and Langmuir, 1926). Because the cylindrical probe, consisting of a simple wire, is much more convenient and consequently almost exclusively used, we concentrate our attention on that geometry. The initial analysis and most subsequent improvements in analysis have concentrated on the pressure range for which the sheath is collisionless, $\lambda_i \gg s$, and we consider that pressure range here.

We consider first the case of a thin wire probe for which $s \gg a$, the probe radius, but take the probe tip length d (collecting part of the wire) to be sufficiently long, $s \ll d$, that an infinite cylinder approximation applies. In the saturation condition, where only a single species is collected, if all the electrons or ions entering the sheath were collected, then the collisionless Child law would predict that $I \propto s \propto |\Phi_p - V_B|^{3/4}$. However, the collisionless trajectories preclude this happening, giving a weaker scaling which we now determine. The geometry is shown in Fig. 6.7. A

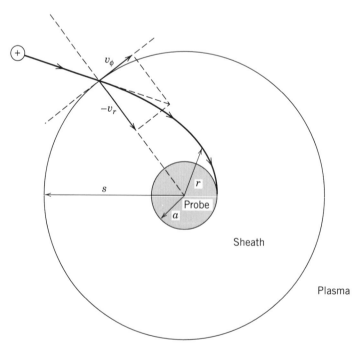

FIGURE 6.7. Ion orbital motion within the sheath of a cylindrical Langmuir probe.

given incoming particle in the attractive central force of the probe has initial velocity components $-v_r$ and v_ϕ in the radial and azimuthal directions at the edge of the sheath $r = s$. At the probe radius $r = a$, the corresponding components are $-v_r'$ and v_ϕ'. For a collisionless sheath we require conservation of energy,

$$\frac{1}{2}m(v_r^2 + v_\phi^2) + e|\Phi_p - V_B| = \frac{1}{2}m(v_r'^2 + v_\phi'^2) \tag{6.6.21}$$

and conservation of angular momentum,

$$sv_\phi = av_\phi' \tag{6.6.22}$$

where m is the mass of the attracted species, either electrons or ions. Solving, we obtain

$$v_\phi' = \frac{s}{a}v_\phi \tag{6.6.23}$$

$$v_r'^2 = v_r^2 + v_\phi^2 + \frac{2e|\Phi_p - V_B|}{m} - \frac{s^2}{a^2}v_\phi^2 \tag{6.6.24}$$

For an ion to reach the probe, $v_r < 0$ and $v_r'^2 > 0$. Setting $v_r'^2 = 0$ in (6.6.24), we obtain

$$v_{\phi 0} = \left(\frac{v_r^2 + 2e|\Phi_p - V_B|/m}{s^2/a^2 - 1} \right)^{1/2} \tag{6.6.25}$$

such that particles only reach the probe if $|v_\phi| \leq v_{\phi 0}$.

The saturation current collected by the probe is found by integrating the radial flux $-n_s v_r$ over the distribution function at the plasma sheath edge, for those particles that reach the probe:

$$I = -2\pi sdn_s e \int_{-\infty}^{0} v_r \, dv_r \int_{-v_{\phi 0}}^{v_{\phi 0}} dv_\phi f(v_r, v_\phi) \tag{6.6.26}$$

where f is the normalized distribution function of electrons or ions. Making the rather strong assumption that the distribution is an isotropic Maxwellian, averaged over the third velocity coordinate, we have

$$f = \frac{m}{2\pi e T_s} \exp\left[-\frac{m(v_r^2 + v_\phi^2)}{2eT_s} \right] \tag{6.6.27}$$

where T_s is the temperature of the collected species at the sheath edge. The integrations can be performed in terms of error functions, but the results, which can be found in the literature quoted above, are not particularly illuminating. However, for large probe voltages we can simplify the evaluation of (6.6.26) by assuming that

$$\frac{a}{s} \ll 1 \tag{6.6.28a}$$

$$v_r^2 \ll \frac{e|\Phi_p - V_B|}{m} \tag{6.6.28b}$$

and

$$v_{\phi 0}^2 \ll \frac{eT_s}{m} \tag{6.6.28c}$$

Then using (6.6.28a) and (6.6.28b) to evaluate (6.6.25), we obtain

$$v_{\phi 0} = \frac{a}{s} \left(\frac{2e|\Phi_p - V_B|}{m} \right)^{1/2}$$

We note that since $s \propto |\Phi_p - V_B|^{3/4}$ for Child's law, (6.6.28c) is well satisfied at high voltages. Using $v_{\phi 0}$ in (6.6.26), with (6.6.27) and the condition (6.6.28c), we integrate to find that

$$I = 2en_s ad \left(\frac{2e|\Phi_p - V_B|}{m} \right)^{1/2} \tag{6.6.29}$$

where I represents either electron or ion saturation current. We see that I is independent of T_s in this limit. Hence a plot of I^2 versus $-V_B$ should be linear, with n_s^2 determined by the slope of this line, independent of T_e and T_i. Expression (6.6.29) is widely used to determine n_s in low-pressure discharges. However, the orbital ion motion is sensitive to ion collisions in the sheath, and orbital motion is destroyed at quite low pressures. In addition, the result (6.6.29) is sensitive to the isotropy of the distribution function at the sheath edge. From Fig. 6.7 it is apparent that significant radial anisotropy will enhance the fraction of particles that are collected. For electrons we might reasonably expect to find an isotropic distribution at the sheath edge, even if it is not Maxwellian. We have seen in Section 6.2 that ions, on the other hand, gain an energy $T_e/2$ in a presheath, which may lead to significant anisotropy. Although we have assumed a collisionless sheath, the presheath is not necessarily so, and presheath collisions will tend to isotropize the distribution of ion velocities. For an alternative distribution at the sheath edge of monoenergetic ions on a cylindrical (isotropic) shell in velocity space, Hershkowitz finds that the coefficient 2, in (6.6.29) is replaced by $\pi/\sqrt{2}$, which is quite similar. A more extreme assumption of anisotropy of f_i, which might be approached at very low pressures, is that the radial ion velocity component is given by the Bohm velocity $u_B = (eT_e/M)^{1/2}$, while the azimuthal component remains Maxwellian at temperature T_i,

$$f_i = \delta(v_r + u_B) \left(\frac{M}{2\pi eT_i} \right)^{1/2} \exp \left(-\frac{Mv_\phi^2}{2eT_i} \right)$$

Using this in (6.6.26) along with the conditions (6.6.28), we integrate to obtain

$$I_i = 2e \left(\frac{2\pi T_e}{T_i} \right)^{1/2} n_s ad \left[\frac{2e(\Phi_p - V_B)}{M} \right]^{1/2} \tag{6.6.30}$$

Comparing (6.6.30) to (6.6.29), we see that n_s is smaller by a factor of $(T_i/2\pi T_e)^{1/2}$ for the same current. We do not expect to find such extreme overestimations of density from the measured orbital ion saturation current, but the sensitivity to the ion velocity distribution suggests that (6.6.29) provides only a semiquantitative estimate of the ion density. Similarly, in low-pressure discharges, the ion drift velocity tends to exceed the ion thermal velocity (see Section 5.3), leading to further modifications in the collected ion current. For a capacitive rf discharge at a pressure $p = 30$ mTorr argon, Godyak et al. (1993) found, by using the measured $I_e(V)$ in the expression (6.6.17) for the energy distribution, a two-temperature distribution, as in Fig. 11.10a (page 359), with $T_{ec} = 0.50$ V and $T_{eh} \approx 3.4$ V. Using the energy distribution (6.6.17) in (6.6.19) and (6.6.20), they found $n_e \approx 4.4 \times 10^9$ cm^{-3} and $T_{eff} \approx 0.67$ V. From the standard Langmuir procedure (6.6.6) applied to the electron current I_e collected by the cylindrical probe, and using the measured electron saturation current (6.6.5) at the plasma potential to find the density, they found $T_{ec} \approx 0.73$ V, $T_{eh} \approx 4.2$ V, and $n_e \approx 3.3 \times 10^9$ cm^{-3}, close to the values determined from the measured energy distribution, as expected for an isotropic distribution. The density determined from the orbital ion current I_i using (6.6.29) was $n_e \approx 1.1 \times 10^{10}$ cm^{-3}, a factor of 2.5 larger than found from the measurement of the electron distribution, as might be expected if the ion distribution had significant anisotropy at the sheath edge.

It should be pointed out, however, that the more accurate calculational procedure, using g_e determined from (6.6.17), is considerably more difficult, experimentally. In particular, taking derivatives of measured quantities results in the introduction of system noise, much of which is intrinsic to the plasma. While averaging procedures can be employed to increase the signal-to-noise ratio, it is all to easy to substitute experimental uncertainty for the uncertainties of the ion orbital theory. Measuring electron saturation current, which does not suffer particularly from the above uncertainties, may be excluded by consideration of the power limits to the probe, unless the measurement system is pulsed, which introduces additional complexities. The experimenter must navigate carefully among these alternatives.

Double Probes and Emissive Probes

Other probe configurations have also been used to measure plasma parameters, with various claims as to accuracy, convenience, etc. Two of the most frequently used alternatives are double probes and emissive probes. Double probes are generally used if there is no well-defined ground electrode in the plasma. A schematic of a double probe is shown in Fig. 6.8a, with a typical probe characteristic in Fig. 6.8b. Since the two probes draw no net current they will both be negative with respect to the plasma. Current flows between the probes if the differential potential $V \neq 0$. As V becomes large, the more negative probe (in this case probe 2) essentially draws ion saturation current, which is just balanced by the net electron current to probe 1. The probe

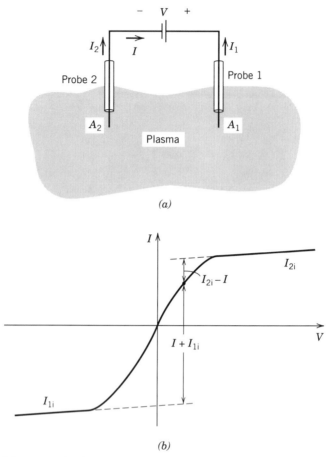

FIGURE 6.8. Schematic of double probe measurement: (a) definition of voltage and currents; (b) typical current–voltage characteristic (Chen, 1965).

system has the advantage that the net current never exceeds the ion saturation current, minimizing the disturbance to the discharge, but has a consequent disadvantage that only the high-energy tail of the electron distribution is collected by either probe. The distribution of these electrons may not be representative of the distribution of bulk electrons in the discharge.

Defining the ion and electron currents to probes 1 and 2 as I_{1i}, I_{1e}, I_{2i}, I_{2e}, then the condition that the system float is

$$I_{1i} + I_{2i} - I_{1e} - I_{2e} = 0 \qquad (6.6.31a)$$

The loop current is

$$I_{2i} - I_{2e} - (I_{1i} - I_{1e}) = 2I \qquad (6.6.31b)$$

Combining (6.6.31*a*) with (6.6.31*b*) we obtain

$$I = I_{1e} - I_{1i} = I_{2i} - I_{2e} \tag{6.6.31c}$$

For the electron current we have

$$I_{1e} = A_1 J_{esat} e^{V_1/T_e}, \quad I_{2e} = A_2 J_{esat} e^{V_2/T_e} \tag{6.6.32}$$

where J_{esat} is the electron random current density and V_1 and V_2 are the probe potentials with respect to the plasma potential. Using $V = V_1 - V_2$ and substituting (6.6.32) into (6.6.31*c*) we obtain

$$\frac{I + I_{1i}}{I_{2i} - I} = \frac{A_1}{A_2} e^{V/T_e} \tag{6.6.33}$$

which generally plots as shown in Fig. 6.8b. For $A_1 = A_2$, then $I_{1i} = I_{2i} \equiv I_i$, such that (6.6.33) simplifies to

$$I = I_i \tanh \left(\frac{V}{T_e} \right) \tag{6.6.34}$$

It is straightforward to fit (6.6.34) to the experimental curve, obtaining both T_e and I_i (and thus n). A simpler procedure can be used to determine T_e. Again taking $A_1 = A_2$ the slope of the I–V plot at the origin ($V = 0$) can be calculated to be

$$\left. \frac{dI}{dV} \right|_{V=0} = \frac{I_i}{T_e} \tag{6.6.35}$$

The details are left to a problem. Note that for cylindrical probes the I_i in either (6.6.34) or (6.6.35) is that obtained by extrapolation, as shown by the dashed lines in Fig. 6.8b.

A hot wire electron-emitting (emissive) probe can be used for a simple measurement of the plasma space potential. Since it works with electron emission, it has the disadvantage of requiring a separate filament circuit carrying high currents, but because it is hot it is less subject to contamination, which can be a serious problem with other probe measurements. The basic idea is very simple. Since the temperature T_w of the electrons emitted from the hot probe wire is related to the wire temperature, we have $T_w \ll T_e$. This results in a sharp change in probe current as the probe potential passes through the plasma potential. This is easily seen from the equations for the electron current. The plasma electron current is approximately (Hershkowitz, 1989)

$$
\begin{aligned}
I_{pe} &= I_{p0} e^{-(\Phi_p - V_B)/T_e} & V_B < \Phi_p \\
&= I_{p0} \left[1 + \frac{(V_B - \Phi_p)^{1/2}}{T_e} \right] & V_B > \Phi_p
\end{aligned}
\tag{6.6.36}
$$

and the emission current is approximately

$$I_{we} = I_{w0}e^{-(V_B-\Phi_p)/T_w}g_w(V_B - \Phi_p) \quad V_B > \Phi_p$$
$$= I_{w0} \quad V_B < \Phi_p$$

(6.6.37)

It can be shown that $g_w \approx [1 + (V_B - \Phi_p)/T_w]^{1/2}$, but this result is not necessary for the argument. Neglecting the small ion current, the total probe current is given by

$$I = I_{pe} - I_{we}$$

(6.6.38)

Without detailed calculation, if we choose $I_{w0} \approx I_{p0}$, then for the case considered with $T_w \ll T_e$, there is a sharp change in I due to the exponential term in I_{we} at $(V_B - \Phi_p) \approx T_w$, which thus gives the plasma potential. The result for a typical case with $T_e = 3$ V and $T_w = 0.3$ V is shown in Fig. 6.9. We note that measuring $V_B - \Phi_p$ also gives an estimate of T_e as obtained from (6.2.17) with $\Phi_w = V_B - \Phi_p$.

Probes in Time-Varying Fields

A widely used discharge for plasma processing is one driven by an rf excited electrode which is capacitively coupled to the plasma. We discuss this discharge in Chapter 11. Because the plasma must remain positive with respect to the rf voltage on the electrode, in order to keep the electron flux equal to the ion flux, the space potential of the plasma also oscillates. In this situation the current drawn to a biased probe,

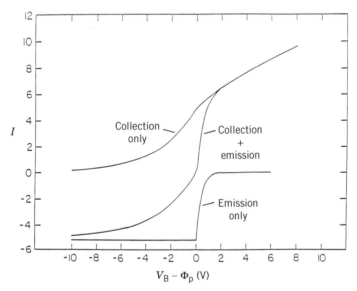

FIGURE 6.9. Typical collecting and emitting current voltage characteristics for an emissive wire probe in a plasma; the electron and wire temperatures are $T_e = 3$ V and $T_w = 0.3$ V (Hershkowitz, 1989).

through a low impedance, is quite different than described in the preceding sections. Although it is possible to interpret the current measurements (see Hershkowitz, 1989), it is also possible to modify the probe circuits so that the I–V characteristic can be interpreted in the normal way. The technique is to place an inductor L in series with the probe such that the probe reactance to ground $\omega L \gg 1/\omega C_s$, the reactance between the probe and the plasma, where ω is the radian rf driving frequency. This may be somewhat hard to achieve if ω is not too high, but can reasonably be obtained at $\omega/2\pi = 13.6$ MHz, a commonly used frequency. If a single frequency is used to drive the discharge (such as 13.6 MHz), a variation on the simple inductance is to include a capacitance C in parallel, such that the parallel LC circuit is in resonance at the desired frequency. This technique, which can also be employed with double probes, is described in the last subsection.

Effect of Collisions and DC Magnetic Fields

Collisions can significantly affect probe diagnostics when the mean free path λ_i becomes of the order of the sheath width. For planar probes with $\lambda_i \ll s$ we can directly use the collisional sheath theory in Section 6.4, just as we used collisionless sheath theory to describe collisionless planar probes. However, in the transition region, even the planar theory is complicated and difficult to use. For other geometries the analysis becomes still more complicated and difficult to interpret. A good account of collisional effects can be found in Chen (1965). A fairly complete theory has been developed for large spherical probes by Su and Lam (1963).

One reason for studying collisional effects is that they also bear on the use of probes in the presence of an applied dc magnetic field. As we have seen in Section 5.4, the electron diffusion across a magnetic field is severely inhibited. For each species (without considering ambipolar effects) the diffusion across the field is related to the diffusion along the field by

$$D_\perp = \frac{D_\parallel}{1 + \omega_c^2 \tau_c^2}$$

where $\omega_c = eB/m$ is the gyration frequency, and τ_c is the mean collision time. For electrons in a gas with $p = 10$ mTorr and $B = 100$ G, we find $\omega_c \tau_c \approx 10^2$. For ions, since ω_c is decreased by m/M and τ_c increased by $(M/m)^{1/2}$, $\omega_c \tau_c \propto (m/M)^{1/2}$ and therefore the ion diffusion is not severely limited. The result is that the probe, drawing electron current, behaves similarly to a plane probe without B but with an effective probe area equal to the probe cross section along the field lines. The ion orbital collection regime (6.6.29) may be used as previously, if the ions have gyroradii large compared to the sheath width. The above simple interpretation of a probe in a B field is limited by a phenomenon called shadowing. Because the probe collects electrons from a thin layer of plasma corresponding to the probe cross section, it acts similarly to a plane probe, as discussed in the first subsection. We indicated there that a large probe can deplete the nearby plasma, thus modifying the plasma it is supposed to measure. This probe shadowing can occur even for small-diameter probes with a magnetic field present. However, the depleted region can be refilled by diffusion across the magnetic field from the neighboring plasma. As one might expect, the calculations can become quite complicated, and the reader is again referred to the review by Chen (1965) for a summary and further references.

As mentioned above, shadowing can also occur for flowing plasma or when electron beams are present. If the plasma is flowing with a velocity of order of the Bohm velocity, then the ion collection can be distorted such that operation in the ion orbital motion region is modified. This is a common situation in low-pressure discharges in which the ion drift velocity typically exceeds the ion thermal velocity. Similarly, if the electrons are streaming through ions with beam velocities comparable to the electron thermal velocity, the electron collection will be distorted.

Probe Construction and Circuits

A basic cylindrical Langmuir probe consists of a thin wire surrounded by a thin insulator that, for dc discharges, may itself be encased in a thin grounded shield. The probe tip usually extends many wire diameters from the insulator. A typical probe, shown in Fig. 6.10, has a tungsten wire probe tip 6.3 mm in length and 38 μm in radius, with a quartz or ceramic capillary sleeve preventing electrical contact between the probe and any conductive material on the probe holder. The insulating

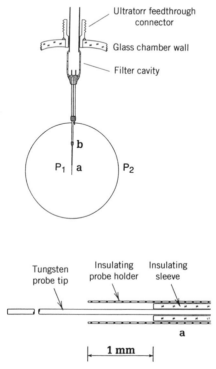

FIGURE 6.10. Construction of a cylindrical probe for rf discharge measurements (Godyak et al., 1992).

holder surrounding the capillary sleeve should have a radius smaller than an electron mean free path to prevent perturbation of the plasma by the probe. To construct other geometries a small plate (plane probe geometry) or sphere (spherical probe geometry) may be attached to the probe tip. Complications include vacuum sealing the probe, allowing the probe tip to be replaceable (tip burnout can be a serious problem), and allowing the probe body to slide through a vacuum seal in order to scan the plasma. Details of various probe designs can be found in the literature; a typical design is shown in Fig. 6.10.

For other types of probes obvious constructional changes are made. The simplest emissive probe construction uses a high-resistivity refractory wire loop tip with the two sides of the loop returning with low resistivity insulated wire through the probe body, where they can be connected to a power source for heating. The heating current is switched off during the measurement. For dense plasmas a single probe can be made emissive by heating from electron current alone, but such probes are more subject to burnout. The simple Langmuir probe may also incorporate some means of heating to drive off impurities which can severely affect current measurements. Double probes are also often constructed with the two probe tips emanating from a single probe body. In this case the wires must be sufficiently far apart that the sheaths surrounding the wires do not interact. For expected plasma parameters, estimates of the sheath widths should be made before designing the probe separation. For measurement of rf plasmas, the inductance required to allow the probe to follow the oscillating plasma space potential may be incorporated into the probe body to minimize stray capacitance. The probe labeled P1 in Fig. 6.10 can be used in this way. In this design, a large circular wire loop P2 is used to establish a ground reference for P1. Note that the probe does not have a grounded shield, which, if present, would greatly increase the stray capacitance of the probe tip to ground.

A simple probe biasing circuit is shown in Fig. 6.11, with the probe voltage V_B given by

$$V_B = V_0 - R_L I \qquad (6.6.39)$$

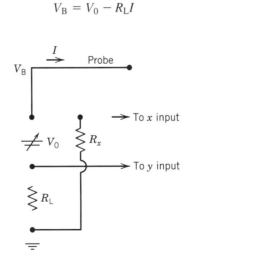

FIGURE 6.11.   Simple Langmuir probe biasing circuit.

where $R_X \gg R_L$ and the current through R_X is neglected. The current I is measured directly from the voltage across R_L, and V_B is measured either directly, as shown, or by measuring V_0 and subtracting $R_L I$. Clearly $R_L \ll \partial I / \partial V_B$ for the measurement technique to work; that is, V_B must be able to be varied by varying V_0. The points labeled y input, measuring I, and x input, measuring V_B, may be the vertical and horizontal inputs on an oscilloscope, x, y recorder, or simply voltmeters. The circuit is usually a little more complicated, since V_0 is not only variable, but must be able to change signs. The voltage can also be swept at a slow rate. For a floating potential measurement $V_0 = 0$, and it then also improves accuracy to make R_L large. Amplifiers may also be used to adjust impedance levels in practical circuits.

For measurement in an rf excited plasma, the probe circuit is more complicated and must explicitly take into account the various capacitances of the probe and probe body. The probe circuit elements, the additional series inductive "choke" element L, and a large bypass capacitor C_{bypass} are shown in Fig. 6.12. Here C_s is the effective capacitance of the probe sheath. The amplitude $\tilde{\Phi}_p - \tilde{V}_{rf}$ of the rf voltage across the probe sheath must satisfy $(\tilde{\Phi}_p - \tilde{V}_{rf})/T_e \ll 1$. In fact, fractional measurement errors appear to be $\lesssim 0.2$ if $(\tilde{\Phi}_p - \tilde{V}_{rf})/T_e \lesssim 1$. Using the voltage divider formula with the impedances $Z_s = (j\omega C_s)^{-1}$ and $Z_L = j\omega L$,

$$\tilde{\Phi}_p - \tilde{V}_{rf} = \tilde{\Phi}_p \frac{Z_s}{Z_L + Z_s}$$

we obtain the criterion

$$\frac{\tilde{\Phi}_p - \tilde{V}_{rf}}{T_e} = \frac{Z_s}{Z_L + Z_s} \frac{\tilde{\Phi}_p}{T_e} \lesssim 1 \qquad (6.6.40)$$

A particular measurement of a probe with a 5-mH inductive choke, at 13.56 MHz gave $|Z_L| = 450$ kΩ and $|Z_s| = 12$ kΩ ($C_s \approx 1$ pF), limiting $\tilde{\Phi}_p/T_e$ to less than $|Z_L/Z_s| \approx 37$. We shall see in Chapter 11 that this may limit the use of a simple blocking inductance in practical discharges. This has led, as mentioned earlier, to the use of resonant blocking circuits at the first (ω) and second (2ω) harmonics of the driving frequency.

The above discussion, and that of the previous subsections, does not include all of the complications that can be encountered in probe diagnostics. The experimenter

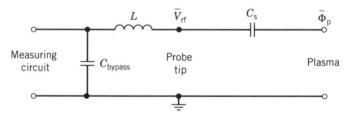

FIGURE 6.12. Probe circuit elements and blocking inductor used to measure the current–voltage characteristics in an rf discharge.

wishing to use probes as a diagnostic tool can proceed from the information given here, but may also wish to look further into the reviews referenced in this section, and also into the original literature referenced in those reviews.

PROBLEMS

6.1. Finite Density for Collisionless Child Law The Child law density (6.3.17) is singular at the sheath edge $x = 0$, while the potential (6.3.15) is not. Assuming that (6.3.15) still holds and that all ions enter the sheath with the Bohm velocity u_B, find a nonsingular expression for $n(x)$ as a function of J_0, u_B, $\Phi(x)$, and other constants. Plot n/n_s versus x/s for $V_0/T_e = 100$. Plot n/n_s given from (6.3.17) on the same graph to compare with your result.

6.2. Ion Transit Time for Child Law Sheath Show that the time for an ion with zero initial energy to transit a collisionless Child law sheath having voltage V_0 and thickness s is $t = 3s/v_0$, where $v_0 = (2eV_0/M)^{1/2}$.

6.3. Bohm Criterion for an Electronegative Plasma Derive the Bohm criterion (6.4.7) for an electronegative plasma with cold positive ions along with electrons and negative ions in Boltzmann equilibrium at temperatures T_e and T_i, respectively, by repeating the calculation leading to (6.2.4) with three species rather than two.

6.4. Potential Across an Electronegative Presheath Show that the potential Φ_p across the presheath in an electronegative plasma is given by (6.4.9).

6.5. Collisional Sheath Law For a high-pressure, high-voltage, collisional sheath, the ion drift velocity can be written as $v_i = \mu_i E$, where $\mu_i = e/M\nu_{mi}$ is the constant ion mobility, with ν_{mi} a constant ion-neutral momentum transfer frequency.

(a) Using particle conservation and Poisson's equation, derive the high-pressure, collisional Child law for ions (6.5.8).

(b) For an argon discharge with $\lambda_i = (330p)^{-1}$ cm, with the pressure p in Torr and $p = 10$ Torr, calculate the sheath thickness s for $n_s = 10^9$ cm^{-3} at the sheath edge, $T_e = 3$ V, $T_i = 0.026$ V, and $V_0 = 300$ V across the sheath. Note for these parameters that $\lambda_i < \lambda_{De}$, such that $u_s < u_B$ from (6.5.9). Compare this s to that obtained for the same discharge parameters from the collisionless Child law.

6.6. Langmuir Probe Calculation A probe whose collecting surface is a square tantalum foil 2×2 mm is found to give a saturation ion current of 100 μA in an argon plasma (atomic mass $= 40$). If $T_e = 2$ V, what is the approximate plasma density? (Assume that the probe can be considered as a plane collector with both sides collecting.) If a bias voltage of -20 V is applied between the probe and ground, calculate the sheath thickness, using the collisionless Child law, to determine if the plane collector assumption is justified.

6.7. Langmuir Probe Theory

    (a) Referring to Fig. 6.7, starting from (6.6.21) and (6.6.22), and using (6.6.26) and (6.6.27), fill in the steps to obtain (6.6.29).

    (b) Starting from (6.6.31) and using (6.6.32), derive (6.6.33) and (6.6.34).

    (c) Verify (6.6.35).

6.8. Emissive Probes   The relation between the floating potential and the probe potential for an emissive probe is found accurately by equating the emission current I_{we} to the plasma electron current I_{pe} to the probe. Taking $g_w = \left[1 + (\Phi_f - \Phi_p)/T_w\right]^{1/2}$ and assuming that $T_e \gg (\Phi_f - \Phi_p)$ in the emission current $I_{we} = I_{w0} \exp\left[-(\Phi_f - \Phi_p)/T_w\right] g_w(\Phi_f - \Phi_p)$, show that

$$\frac{\Phi_f - \Phi_p}{T_w} - \frac{1}{2}\ln\left(1 + \frac{\Phi_f - \Phi_p}{T_w}\right) = \ln\left(\frac{I_{w0}}{I_{pe}}\right)$$

Plot $(\Phi_f - \Phi_p)/T_w$ versus $\ln(I_{w0}/I_{p0})$ for $1 < I_{w0}/I_{p0} < 10$.

CHAPTER 7

CHEMICAL REACTIONS
AND EQUILIBRIUM

7.1 INTRODUCTION

Gas- and surface-phase chemical reactions play a critical role in plasma-assisted materials processing. To see why, consider the typical reactor, shown in Fig. 7.1, that is used to etch an SiO_2 film. A CF_4/O_2 gas mixture is fed into the reactor and rf or microwave energy is applied to form a plasma. Electron impact ionization and dissociation create ions such as CF_3^+, CF_2^+, O_2^+, O^-, F^-, and free radicals such as CF_3, CF_2, O, and F. Chemical reactions in the gas phase and on the SiO_2 surface create additional molecules such as CO, CO_2, SiF_2, and SiF_4. The etch rates, anisotropies, and selectivities depend on the concentrations and energy distributions of all these species.

The concentrations are determined by general chemical reactions such as

$$e + AB \rightarrow AB^+ + 2e \qquad \text{(electron–ion pair production)}$$
$$e + AB \rightarrow e + A + B \qquad \text{(radical production)}$$
$$e + AB \rightarrow A^- + B \qquad \text{(negative ion production)}$$
$$A + B \rightarrow C + D \qquad \text{(gas-phase chemical reactions)}$$
$$\Gamma_i = -D_a \nabla n_i \qquad \text{(ion transport to surfaces)}$$
$$\Gamma_A = -D_A \nabla n_A \qquad \text{(radical transport to surfaces)}$$
$$A(g) + B(s) \rightarrow C(g) \qquad \text{(surface-phase reactions)}$$

The net energy absorbed by these and other reactions must be supplied by the discharge power source. For example, electron energy is lost due to ionization,

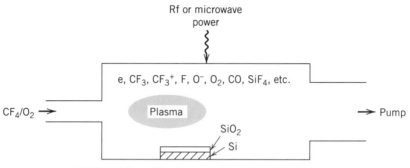

FIGURE 7.1. Typical materials processing reactor.

excitation, elastic scattering, and dissociation (the second reaction listed above). Hence, the discharge model must account for these energy losses. Further, the rates of these reactions depend critically on the energy distributions or temperatures of the reactants. Although thermodynamics determines the energy of reaction and can constrain the extent of reaction, most reactions occurring in typical reactors are far from thermodynamic equilibrium. Then collisions between pairs of species determine the reaction kinetics, including the reaction rates and the steady-state distribution of reactor species.

Another aspect seen in Fig. 7.1 is the dual importance of homogeneous reactions in the gas-phase and heterogeneous reactions of gas-phase species with surfaces (the last reaction listed above). Hence, one must describe not only the properties of a given species, but also possible changes in the phase of that species, e.g., from solid to gas, as well as changes in composition due to chemical reactions.

This and the following two chapters deal with the fundamentals of chemical dynamics. In this chapter we describe the energetics of gas-phase and surface chemical reactions and chemical equilibrium. In Chapter 8, building on the study of atomic collisions in Chapter 3, we describe the fundamentals of molecular collisions, including such processes as dissociation, attachment, and recombination, and introduce appropriate rate constants. In Chapter 9, we introduce the principles of gas-phase and surface chemical kinetics, using the rate constants obtained in the previous chapter. We also describe the principles of surface interactions, including physical and chemical surface processes, the transport of species to surfaces, and surface reactions.

7.2 ENERGY AND ENTHALPY

The state of a system of $\mathcal{M}$ chemical species is uniquely determined by the temperature T, the total volume $\mathcal{V}$, and the number N_j of moles of each species (1 mole $=$ 6.022×10^{23} molecules). This is illustrated for $\mathcal{M} = 3$ in Fig. 7.2, for two states labeled 1 and 2, where the five axes shown in the figure are considered to be mutually perpendicular. State variables, such as the internal energy U, pressure p, entropy S, enthalpy H, and Gibbs free energy G, are then uniquely determined. For example, $U = \frac{3}{2}NRT$, and the equation of state determines $p = NRT/\mathcal{V}$ for a perfect gas. Often

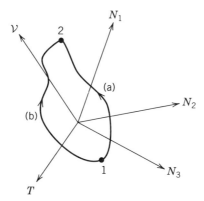

FIGURE 7.2.   State space for a chemically reactive system.

the equations for U, p, S, etc., can be inverted. Hence other combinations of $\mathcal{M} + 2$ variables, such as (U, p, N_j), uniquely specify the state and thus determine T, $\mathcal{V}$, S, etc.

If a chemical system can exchange heat and work, but not matter with its surroundings, and undergo changes in chemical composition, then the first law of thermodynamics states that the increase dU in internal energy is equal to the sum of the heat flow $đQ$ into the system and the work done $đW$ on the system

$$dU = đQ + đW \tag{7.2.1}$$

If neither heat nor work are exchanged with the surroundings then U does not change. Equation (7.2.1) is the law of conservation of energy. Physically, U accounts for the random translational, vibrational, and rotational kinetic energy of the molecules in the system, the potential energies stored in the molecular chemical bonds, and the interaction energies between molecules.

The notation $đ$ is used for the differentials of heat and work because $đQ$ and $đW$ are not, in general, exact differentials. Consider a process leading to a change from an initial state 1 to a final state 2 along two different paths (a) and (b), as shown in Fig. 7.2. For exact differentials, such as dU, the total change is independent of the path:

$$\Delta U_a = \int_{\text{path a}} dU = \Delta U_b = \int_{\text{path b}} dU = U_2 - U_1$$

The differentials of all state variables are exact; e.g., $\Delta p = p_2 - p_1$, $\Delta \mathcal{V} = \mathcal{V}_2 - \mathcal{V}_1$, etc. However, heat and work are not state variables. Hence,

$$\Delta Q_a = \int_{\text{path a}} đQ \neq \Delta Q_b = \int_{\text{path b}} đQ$$

and, similarly, $\Delta W_a \neq \Delta W_b$.

Different kinds of work (mechanical, electrical, etc.) can be done on a system. We are considering here only $p\,dV$ work due to a change of volume V. The work done on a system by its surroundings is found, from Newton's laws, to be

$$dW = -p_{ext}\,dV \tag{7.2.2}$$

where p_{ext} is the pressure of the surroundings. In general, p_{ext} is not equal to the system pressure p. However, if the system is in near equilibrium with its surroundings, then $p_{ext} \approx p$ and $T_{ext} \approx T$. In this case,

$$dW = -p\,dV \tag{7.2.3}$$

If, during a process of change from state 1 to state 2, the system remains in near equilibrium with its surroundings, then the process is called *reversible*. Examples of reversible processes are the slow heating of a gas in a closed container ($V =$ const) or in an open container capped by a piston exerting a constant pressure on the gas. Examples of irreversible processes are the bursting of a balloon ($p \neq p_{ext}$) and the flow of heat in a metal rod heated at one end and cooled at the other ($T_{ext} \neq T$). The *reversible work* done on the system is found by integrating (7.2.3). Substituting (7.2.3) into (7.2.1), we see that

$$dU = dQ - p\,dV \tag{7.2.4}$$

at every point along the path of a reversible process.

The work done is zero for a constant-volume reversible process. Integrating (7.2.4) shows that the increase in internal energy is equal to the total heat flow into the system:

$$U_2 - U_1 = \Delta Q \tag{7.2.5}$$

However, in plasma reactors, most processes occur at constant pressures, not constant volumes. It is useful to introduce a new state variable, the enthalpy

$$H = U + pV \tag{7.2.6}$$

for constant-pressure processes. For example, for a perfect gas, $U = \frac{3}{2}NRT$ and $pV = NRT$, so $H = \frac{5}{2}NRT$. Differentiating H and using (7.2.4), we obtain

$$dH = dQ + V\,dp \tag{7.2.7}$$

Hence, the increase in enthalpy is equal to the total heat flow for constant-pressure processes:

$$H_2 - H_1 = \Delta Q \tag{7.2.8}$$

In general, there is a change of volume for a constant-pressure process. Integrating (7.2.3) yields the total work done on the system:

$$\Delta W = -p(\mathcal{V}_2 - \mathcal{V}_1) \tag{7.2.9}$$

Differentiating (7.2.6) at constant pressure, we obtain

$$\Delta H = \Delta U + p\Delta \mathcal{V} \tag{7.2.10}$$

Hence, the enthalpy change is equal to the sum of the internal energy change and the $p\,d\mathcal{V}$ work done *by* the system on its surroundings. Generally, $|\Delta W| \ll |\Delta U|$ for chemical reactions at the low pressures characteristic of plasma processing discharges; hence $\Delta H \approx \Delta U$.

If a system containing N_j moles of each species undergoes a chemical reaction at constant temperature and pressure, then the N_j's change, and the total enthalpy $H_2(T, p, N_j')$ after the reaction is not the same as the enthalpy $H_1(T, p, N_j)$ before the reaction. By (7.2.8), the excess enthalpy appears as heat. For $\Delta H > 0$, the reaction is called *en*dothermic and heat *en*ters the system. For $\Delta H < 0$, the reaction is called *exo*thermic and heat *ex*its. Although the enthalpy H_f for formation of a particular product species is a function of T, p, and the N_j's, a *standard molar formation enthalpy* $H_f^{\circ}(T_0)$ is tabulated in the thermodynamic literature for a standard temperature and pressure (STP) and for one mole (1 mol) of the product created by the reaction of the most stable natural forms of the elements. The standard pressure, denoted with a superscript $\circ$, is usually taken to be either 1 bar $= 10^5$ Pa in the newer tables or 1 atm $= 760$ Torr $= 1.013$ bar in the older tables; the difference is not significant for our purposes. The standard temperature, denoted T_0, is taken to be 298.15 K $= 25\,^{\circ}$C. An example is the reaction for formation of SiO_2: $Si(s) + O_2 \rightarrow SiO_2(s, \alpha)$; $H_f^{\circ}(T_0) = -910.9$ kJ/mol, where s, l, and g denote solid, liquid, and gas, and α denotes the most stable (α) phase of SiO_2. In older tables, enthalpies are often specified in kcal/mol, where 1 kcal $= 4.184$ kJ. We also note that an energy equivalent voltage of 1 V/molecule corresponds to 96.49 kJ/mol. When considering chemical reactions, only changes in enthalpies are significant. Hence the standard enthalpies of formation of the elements in their most stable state are taken to be zero at all temperatures. Some standard enthalpies of formation are given in Tables 7.1 and 7.2.

The standard enthalpy $H_r^{\circ}(T_0)$ for any chemical reaction can be calculated by subtracting the enthalpies of formation of the reactants from those of the products. For example, consider the etching of one mole of $SiO_2(s)$ by fluorine gas:

$$SiO_2(s) + 2F_2(g) \rightarrow SiF_4(g) + O_2(g) \tag{7.2.11}$$

From Table 7.1, $H_f^{\circ}(T_0) = -910.9$ kJ/mol for one mole of $SiO_2(s)$ and $H_f^{\circ}(T_0) = -1614.9$ kJ/mol for one mole of SiF_4. Hence

$$H_r^{\circ}(T_0) = (1)(-1614.9) - (1)(-910.9) = -704.0 \text{ kJ/mol}$$

and the reaction is exothermic.

TABLE 7.1. Thermodynamic Properties. *Note.* **Substances are in gas phase unless otherwise specified.**

Substance	$H_f^\circ(T_0)$ (kJ/mol)	$G_f^\circ(T_0)$ (kJ/mol)
O	249.2	231.7
O_3	142.7	163.2
H	218.0	203.2
OH	39.0	34.2
H_2O (l)	−285.8	−237.1
H_2O	−241.8	−228.6
F	78.99	61.91
HF	−271.1	−273.2
Cl	121.7	105.7
HCl	−92.3	−95.3
Br	111.9	82.4
Br_2	30.9	3.11
S	278.8	238.3
SF_4	−774.9	−731.6
SF_6	−1209.	−1105.
N	472.7	455.6
C (graphite cr)	0	0
C (diamond cr)	1.90	2.90
CO	−110.5	−137.2
CO_2	−393.5	−394.4
CH_2	390.4	372.9
CH_3	145.7	147.9
CH_4	−74.8	−50.7
CF_3	−477.	−464.
CF_4	−925.	−879.
COF_2	−634.7	−619.2
CH_2F_2	−446.9	−419.2

For $SiO_2(s)$ etching by chlorine gas,

$$SiO_2(s) + 2Cl_2(g) \rightarrow SiCl_4(g) + O_2(g) \qquad (7.2.12)$$

we obtain

$$H_r^\circ(T_0) = (1)(-657) - (1)(-910.9) = 253.9 \text{ kJ/mol}$$

and the reaction is endothermic.

The reactions in plasma processing do not necessarily take place at the standard temperature. To determine the temperature dependence of the enthalpy, we note that at constant pressure and composition, a small heat flow $đQ$ into the system produces a proportionate temperature rise,

$$đQ = C_p \, dT \qquad (7.2.13)$$

TABLE 7.1. (*continued*) **Thermodynamic Properties**

Substance	$H_f^\circ(T_0)$ (kJ/mol)	$G_f^\circ(T_0)$ (kJ/mol)
CHF_3	688.3	−653.9
CCl_4	−102.9	−60.59
$COCl_2$	−218.8	−204.6
CH_3Cl	−80.8	−57.4
CH_2Cl_2	−92.5	−65.9
$CHCl_3$	−103.1	−70.3
C_2H_2	226.7	209.2
C_2H_4	52.3	68.2
C_2H_6	−84.7	−32.8
C_2F_4	−650.6	−615.9
C_2F_6	−1297.	−1213.
Si (cr)	0	0
Si	455.6	411.3
SiO	−99.6	−126.4
SiO_2 (α quartz, cr)	−910.9	−856.6
SiO_2 (amorphous)	−903.5	−850.7
SiH_4	34.3	56.9
SiF	7.1	−24.3
SiF_2	−619.	−628.
SiF_4	−1614.9	−1572.7
$SiCl_2$	−165.6	−177.2
$SiCl_4$ (l)	−687.0	−619.8
$SiCl_4$	−657.	−617.
Si_3N_4 (α, cr)	−743.5	−642.6
SiC (β, cubic)	−65.3	−62.8
Al_2O_3 (α)	−1675.7	−1582.3
AlF_3 (cr)	−1510.4	−1431.1
AlF_3	−1204.6	−1188.2
$AlCl_3$ (cr)	−704.2	−628.8
WF_6	−1721.7	−1632.1

where the constant of proportionality C_p is called the *specific heat at constant pressure*. Since $dH = đQ$ under these conditions, we find that

$$C_p = \left(\frac{\partial H}{\partial T}\right)_{p,N_j} \tag{7.2.14}$$

For a perfect gas, $H = \frac{5}{2}NRT$ and $C_p = \frac{5}{2}RN$. The specific heat for one mole of perfect gas is $C_{pm} \approx 20.8$ J/K-mol. Most substances, including real gases, have $C_{pm} \sim 30 - 100$ J/K-mol. The enthalpy at temperature T can be written as

$$H(T) = H(T_0) + \int_{T_0}^{T} C_p(T')\,dT' \tag{7.2.15}$$

TABLE 7.2. Enthalpies of Formation. *Note.*
All substances are in gas phase.

Substance	$H_f^\circ(T_0)$ (kJ/mol)
CH	595.8
CCl_3	59
CF_2	-194.1
CF_3	-467.4
SiH	377
SiH_2	269.0
SiH_3	194.1
SiF	-19.3
SiF_2	-587.9
SiF_3	$-1025.$
SiCl	195.8
$SiCl_2$	-163.6
$SiCl_3$	-318
$AlCl_3$	-583.2

Since reaction enthalpies are typically hundreds of kilojoules per mole, the integral in (7.2.15) is not too important for temperatures within a few hundred degrees of T_0, as is common in processing discharges.

Similarly, the enthalpy depends only weakly on the pressure. In fact, for a perfect gas, $H = \frac{5}{2}NRT$ and therefore is independent of p. At the low pressures of processing discharges, the pressure dependence is negligible.

The enthalpies associated with breaking chemical bonds to form neutral products are also of interest. The dissociation reaction for the molecule AB,

$$AB(g) \rightarrow A(g) + B(g)$$

where both A and B may be groups of atoms, has a *dissociation enthalpy* $H_{diss}^\circ(T_0)$ for breaking the AB bond. Some bond dissociation enthalpies are given in Table 7.3. A *mean bond dissociation enthalpy*, which is an average of $H_{diss}^\circ(T_0)$ over many different types of molecules containing the bond, can also be defined. For example, $H_{diss}^\circ(T_0) = 492$ kJ/mol for the HO–H bond and 428 kJ/mol for the O–H radical bond; the mean enthalpy of O–H bonds in many different molecules is 463 kJ/mol. The *enthalpy of phase transition* is also of interest, including *sublimation* s → g, *vaporization* l → g, and *melting* s → l; e.g., $H_2O(l) \rightarrow H_2O(g)$ has $H_{vap}^\circ(100\,°C) = 40.66$ kJ/mol. Some enthalpies of formation of gaseous atoms are given in Table 7.4. The data in Tables 7.3 and 7.4 can be used to estimate the enthalpy of formation of various substances (see Problem 7.2). Other enthalpies include *ionization*,

$$A(g) \rightarrow A^+(g) + e$$

TABLE 7.3. Bond Dissociation Enthalpies

Bond	$H^\circ_{\text{diss}}(T_0)$ (kJ/mol)
F–F	158.75
F–Ni	435
F–O	222
F–S	342.7
F–Si	552.7
F–W	548
F–Zn	368
F–C	552
C–C	607
C–H	338.3
C–O	1076.5
C–Si	451.5
Si–O	799.6
Si–Si	326.8
Al–Al	186.2
Al–Cu	216.7
Al–F	663.6
Al–Cl	511.3
Al–O	512.1
O–H	427.5
Si–H	299.2
F–SF$_5$	381.2
F–SF$_4$	222.2
F–SF$_3$	351.9
F–SF$_2$	264.0
F–SF	383.7
S–F	342.7
CF$_2$ = CF$_2$	319.2
CF$_3$ − CF$_3$	413.0

and *electron affinity*,

$$A(g) + e \rightarrow A^-(g)$$

For example, the enthalpy for ionization of Cl is 1251 kJ/mol, corresponding to 12.96 V/atom. The electron affinity enthalpy for Cl$^-$ is -348.6 kJ/mol, corresponding to -3.61 V/atom. The affinity reaction is exothermic for Cl$^-$ production.

7.3 ENTROPY AND GIBBS FREE ENERGY

The second law of thermodynamics asserts that there is a state variable S, the entropy, defined by

$$dS = đQ/T \tag{7.3.1}$$

TABLE 7.4. Enthalpies of Formation of Gaseous Atoms

Element	$H_f^\circ(T_0)$ (kJ/mol)
Si	455.6
C	716.7
Br	111.9
Cl	121.7
F	79.4
H	218.0
Al	329.7
Mo	658.1
O	249.2
S	278.8
W	849.8
Zn	130.42
N	472.7
Cu	341
Ge	328
Ni	425

where $đQ$ is the heat injected into a system by a reversible process. The second law also asserts that, for an irreversible process,

$$dS > đQ/T \tag{7.3.2}$$

The entropy is a measure of the disorder in the system.

Consider a thermally isolated system of chemical species that irreversibly (spontaneously) undergoes a chemical reaction, leading to a change in temperature, pressure, and species concentrations. Since $đQ = 0$ for a thermally isolated system, (7.3.2) shows that the system entropy must increase; i.e., the direction of spontaneous change in a thermally isolated system is to increase the system disorder.

Inserting (7.3.1) into the internal energy change (7.2.4), we find

$$dU = T dS - p dV \tag{7.3.3}$$

Although (7.3.3) was derived for a reversible process, it applies for any process, reversible or irreversible, because dU is an exact differential. Although $đQ = T dS$ and $đW = -p dV$ for a reversible process, and $đQ < T dS$ and $đW > -p dV$ for an irreversible process, the sum $đQ + đW$ is always equal to $T dS - p dV$. Similarly, inserting (7.3.1) into the enthalpy change (7.2.7), we find

$$dH = T dS + V dp \tag{7.3.4}$$

for any process, reversible or irreversible.

Equation (7.3.1) can be used to determine the variation of S with temperature and pressure. Consider a constant pressure process for which the surroundings are heated slowly from T_0 to T_1. Then from the definition of specific heat (7.2.13), a reversible heat

$$đQ = dH(p, T) = C_p(p, T) dT \tag{7.3.5}$$

flows into the system. Inserting (7.3.5) into (7.3.1) and integrating, we obtain

$$S(p_0, T) - S(p_0, T_0) = \int_{T_0}^{T} \frac{C_p(p_0, T')}{T'} dT' \tag{7.3.6}$$

For a perfect gas, $C_p = \frac{5}{2}RN$ and

$$S(p_0, T) - S(p_0, T_0) = \frac{5}{2}RN \ln\left(\frac{T}{T_0}\right) \tag{7.3.7}$$

which gives the temperature variation of the entropy.

Similarly, the change in internal energy for a constant-volume reversible process is, from the internal energy change (7.2.4),

$$đQ = dU(V, T) = C_V(V, T) dT \tag{7.3.8}$$

where

$$C_V = \left(\frac{\partial U}{\partial T}\right)_V \tag{7.3.9}$$

is the *specific heat at constant volume*. Inserting (7.3.8) into (7.3.1) and integrating, we find

$$S(V, T) - S(V, T_0) = \int_{T_0}^{T} \frac{C_V(V, T')}{T'} dT' \tag{7.3.10}$$

For a perfect gas, $C_V = \frac{3}{2}RN$ and

$$S(V, T) - S(V, T_0) = \frac{3}{2}RN \ln\left(\frac{T}{T_0}\right) \tag{7.3.11}$$

Equations (7.3.6) and (7.3.10) can be used to determine the variation of entropy with pressure by considering the two-step reversible process

$$(p_0, T_0) \xrightarrow{p \text{ const}} (p_0, T) \xrightarrow{V \text{ const}} (p, T_0)$$

For a perfect gas, using (7.3.7) and (7.3.11), and noting that $V = NRT/p_0 = NRT_0/p$, we obtain

$$S(p, T_0) - S(p_0, T_0) = -RN \ln \left(\frac{p}{p_0} \right) \tag{7.3.12}$$

which gives the pressure variation of the entropy.

In general, the specific heats are continuous functions of temperature except at isolated values of T where the system undergoes a change of phase (first-order phase transition). At these temperatures the specific heats are singular. An example is shown in Fig. 7.3 for a change of phase of a pure substance from a solid to a liquid to a gas. The third law of thermodynamics states that the entropy of all perfect crystalline compounds may be taken to be zero at $T = 0$. Hence, integrating C_p from 0 to T, including the appropriate δ functions at T_{melt} and T_{vap}, yields the entropy. The *standard molar entropies* $S_m^\circ(T_0)$ J/K-mol of various pure substances and compounds are tabulated in the thermodynamic literature. The *standard reaction entropies* $S_r^\circ(T_0)$ for any reaction are found by subtracting the standard entropies of the reactants from those of the products.

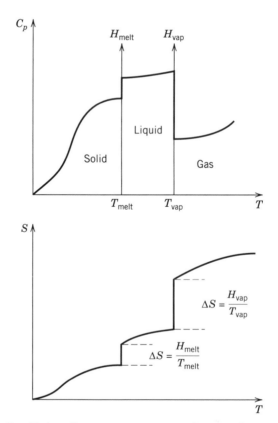

FIGURE 7.3.   Specific heat C_p at constant pressure and entropy S versus temperature T.

Gibbs Free Energy

For a constant-pressure process đQ = dH and the second law, (7.3.1) and (7.3.2) can be written

$$dH - T\,dS \leq 0 \tag{7.3.13}$$

where the equality applies for a reversible process. Introducing a new state variable, the Gibbs free energy

$$G = H - TS \tag{7.3.14}$$

such that

$$dG = dH - T\,dS - S\,dT \tag{7.3.15}$$

and comparing (7.3.13) and (7.3.15) at constant temperature, we see that

$$dG = dH - T\,dS \leq 0 \tag{7.3.16}$$

Hence, for a chemical reaction to proceed spontaneously at constant temperature and pressure, the Gibbs free energy must decrease. Inserting (7.3.4) into (7.3.15), we obtain

$$dG = \mathcal{V}\,dp - S\,dT \tag{7.3.17}$$

If we let $G = G(p, T, N_j)$, where (p, T, N_j), $j = 1, \ldots, \mathcal{M}$, specifies the state of the system, then the differential of G is

$$dG = \left(\frac{\partial G}{\partial p}\right)_{T,\{N_i\}} dp + \left(\frac{\partial G}{\partial T}\right)_{p,\{N_i\}} dT + \sum_{j=1}^{\mathcal{M}} \left(\frac{\partial G}{\partial N_j}\right)_{p,T,\{N_i \neq N_j\}} dN_j \tag{7.3.18}$$

Comparing (7.3.18) with (7.3.17), we see that

$$\mathcal{V} = \left(\frac{\partial G}{\partial p}\right)_{T,\{N_i\}} \tag{7.3.19}$$

$$S = -\left(\frac{\partial G}{\partial T}\right)_{p,\{N_i\}} \tag{7.3.20}$$

and, introducing the *chemical potential*

$$\mu_j = \left(\frac{\partial G}{\partial N_j}\right)_{p,T,\{N_i \neq N_j\}} \tag{7.3.21}$$

we see that

$$\sum_{j=1}^{M} \mu_j \, dN_j = 0 \tag{7.3.22}$$

The chemical potential specifies how G changes as various substances j are added to the system. For a *closed system*, for which heat and work, but not matter, can be exchanged with the surroundings, (7.3.22) must hold; i.e., dG is independent of changes in composition. However, for an *open system*, for which matter can be exchanged with the surroundings, we must write

$$dG = V \, dp - S \, dT + \sum_{j=1}^{M} \mu_j \, dN_j \tag{7.3.23}$$

in place of (7.3.17). We note that V, S and μ_j in (7.3.23) are all functions of the state (p, T, N_j), $j = 1, \ldots, M$. However, for a single substance

$$\mu = \left(\frac{\partial G}{\partial N} \right)_{p,T} \tag{7.3.24}$$

is independent of N. Hence μ is equal to the *molar Gibbs free energy* $G_m(p, T)$ for that substance.

From (7.3.14), the *standard molar Gibbs free energy of formation* of any substance from the elements in their most stable natural states is

$$\mu^\circ(T_0) \equiv G_f^\circ(T_0) = H_f^\circ(T_0) - T_0 S_m^\circ(T_0) \tag{7.3.25}$$

These data are tabulated in the thermodynamic literature, and some selected values are given in Table 7.1. The standard Gibbs free energy $G_r^\circ(T_0)$ for any chemical reaction is found by subtracting the standard Gibbs free energies for formation of the reactants from those of the products. Again, $G_f^\circ(T_0)$ for the elements in their most stable natural state is taken to be zero.

As an example, consider reaction (7.2.12) for etching one mole of $SiO_2(s)$ by chlorine gas. From Table 7.1, we find

$$G_r^\circ(T_0) = (1)(-617.0) - (1)(-856.6) = 239.6 \text{ kJ/mol} \tag{7.3.26}$$

The pressure and temperature variation of μ are found by integrating (7.3.19) and (7.3.20) for one mole of substance from STP at (p°, T_0) to (p, T). First integrating (7.3.20) from (p°, T_0) to (p°, T) and assuming a perfect gas, such that $S(p^\circ, T)$ is found from (7.3.7), we obtain

$$\mu^\circ(T) = \mu^\circ(T_0) + (T - T_0) \left[\frac{5}{2} R - S_m^\circ(T_0) \right] - \frac{5}{2} RT \ln \left(\frac{T}{T_0} \right) \tag{7.3.27}$$

To obtain the pressure variation we integrate (7.3.19) from (p°, T) to (p, T), using $V = RT/p$ for one mole of a perfect gas, to obtain

$$\mu(p, T) = \mu^\circ(T) + RT \ \ln \left(\frac{p}{p^\circ} \right) \tag{7.3.28}$$

For a mixture of perfect gases, p is replaced by the partial pressure p_j in (7.3.28):

$$\mu_j(g) = \mu_j^\circ(T) + RT \ \ln \left(\frac{p_j}{p^\circ} \right) \tag{7.3.29}$$

Introducing the *mole fractions* $x_j = p_j/p = N_j / \sum_{i=1}^{\mathcal{M}_g} N_i$ for the $\mathcal{M}_g$ gas-phase species, we have

$$\mu_j(g) = \mu_j^\circ(T) + RT \ \ln \left(\frac{x_j p}{p^\circ} \right) \tag{7.3.30}$$

The x_j's give the composition dependence. For typical processing discharges, most gases can be considered ideal. For solids or liquids, (7.3.28) is replaced by

$$\mu_j = \mu_j^\circ(T) + RT \ \ln a_j \tag{7.3.31}$$

where $a_j = \gamma_j x_j$, a_j is the *activity*, γ_j is the *activity coefficient*, and x_j is the mole fraction in the solid or liquid phase. For a pure solid or liquid, $x_j = 1$ and γ_j is chosen to be unity at standard pressure p°. Hence $a_j = 1$ and $\mu_j^\circ = \mu_j$ for the pure substance at p°. Integrating (7.3.19) and (7.3.20) for one mole of solid or liquid substance shows that μ_j depends only weakly on p and T for typical values of the molar volume V_m and molar entropy S_m. Assuming that the mutual solubilities of the constituents in the solid or liquid phases are small, then these phases are pure, and the a_j's can be taken to be unity for the solid or liquid reactants and products at the pressures and temperatures in typical processing discharges:

$$\mu_j(s, l) = \mu_j^\circ(T) \tag{7.3.32}$$

7.4 CHEMICAL EQUILIBRIUM

Consider a chemical reaction in a closed system, such as

$$3A + B = 2C + 4D \tag{7.4.1}$$

Letting $\mathcal{J}_1 = A$, $\mathcal{J}_2 = B$, $\mathcal{J}_3 = C$, $\mathcal{J}_4 = D$, etc. denote the species and introducing the *stoichiometric coefficients* $\alpha_1 = -3$, $\alpha_2 = -1$, $\alpha_3 = 2$, $\alpha_4 = 4$, etc., (the α's are negative for reactants and positive for products), the reaction can be written as

$$\sum_j \alpha_j \mathcal{J}_j = 0 \qquad (7.4.2)$$

Let the number of moles of $\mathcal{J}_j$ change by $dN_j = \alpha_j dN$, where dN is the *extent* of the reaction. For reaction at constant pressure and temperature, (7.3.18) shows that

$$dG = \sum_j \alpha_j \mu_j \, dN \qquad (7.4.3)$$

If the reaction (7.4.1) proceeds spontaneously, either to the right (C and D are formed) or to the left (A and B are formed), then the second law (7.3.16) shows that $dG < 0$ and hence G must decrease. Eventually, the system attains a state of equilibrium in which the concentrations of the various species no longer change spontaneously; at this equilibrium state $dG = 0$. Hence, as shown in Fig. 7.4, the equilibrium state is a minimum of G with respect to composition changes. Using (7.4.3), we see that

$$\sum_{j=1}^{\mathcal{M}} \alpha_j \mu_j = 0 \qquad (7.4.4)$$

at equilibrium. Inserting the chemical potentials (7.3.29) for the gas-phase constituents and (7.3.32) for the liquid- and solid-phase constituents into (7.4.4), we obtain

$$-RT \sum_{j=1}^{\mathcal{M}_g} \alpha_j \ln \left(\frac{\bar{p}_j}{p^\circ} \right) = \sum_{j=1}^{\mathcal{M}} \alpha_j \mu_j^\circ (T) \qquad (7.4.5)$$

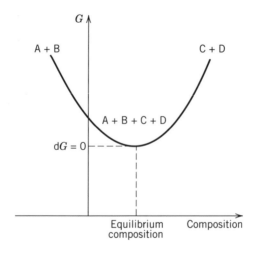

FIGURE 7.4.   Gibbs free energy G versus composition.

where $\bar{p}_j$ is the equilibrium partial pressure of the jth species, and the sum on the left is over the $\mathcal{M}_g$ gas-phase constituents only. The term on the right-hand side of (7.4.5) is the Gibbs free energy $G_r^\circ(T)$ of the reaction. Using this and introducing the *equilibrium constant*

$$\mathcal{K} = \prod_{j=1}^{\mathcal{M}_g} \left(\frac{\bar{p}_j}{p^\circ} \right)^{\alpha_j} \tag{7.4.6}$$

into (7.4.5), we obtain

$$\mathcal{K}(T) = \exp \left[-\frac{G_r^\circ(T)}{RT} \right] \tag{7.4.7}$$

Equations (7.4.6) and (7.4.7) are the fundamental equations of chemical equilibrium. $\mathcal{K}$ can be written in terms of the equilibrium mole fractions $\bar{x}_j = \bar{p}_j/p$ as

$$\mathcal{K} = \mathcal{K}_x \left(\frac{p}{p^\circ} \right)^{\alpha_g} \tag{7.4.8}$$

where

$$\mathcal{K}_x = \prod_{j=1}^{\mathcal{M}_g} \bar{x}_j^{\alpha_j} \tag{7.4.9}$$

and

$$\alpha_g = \sum_{j=1}^{\mathcal{M}_g} \alpha_j \tag{7.4.10}$$

is the sum of the gas-phase stoichiometric coefficients.

As an example, consider the reaction (7.2.12) for the etching of one mole of SiO_2 by Cl_2 gas at STP. The reaction Gibbs free energy is, from (7.3.26), $G_r^\circ(T_0) = 239.6\ kJ/mol$. Using (7.4.7) with $RT = 2.479\ kJ/mol$, we find $\mathcal{K}(T_0) = 1.02 \times 10^{-42}$. Let x_{Cl_2}, x_{O_2}, and x_{SiCl_4} be the gas-phase mole fractions and N_0 be the total number of gas-phase moles in the initial state. Let N be the extent of the reaction to attain the equilibrium state. Then we obtain the following table based on conservation of Cl_2 and O_2 for the reaction (7.2.12):

Species	Initial Moles	Equilibrium Moles
Cl_2	$x_{Cl_2}N_0$	$x_{Cl_2}N_0 - 2N$
O_2	$x_{O_2}N_0$	$x_{O_2}N_0 + N$
$SiCl_4$	$x_{SiCl_4}N_0$	$x_{SiCl_4}N_0 + N$

Using $\alpha_{Cl_2} = -2$, $\alpha_{O_2} = 1$, $\alpha_{SiCl_4} = 1$, we obtain $\alpha_g = 0$ from (7.4.9) and hence $K = K_x$ from (7.4.10). Dividing each element in the third column of the table by the initial number of gas-phase moles, we obtain from (7.4.9) that

$$K = \frac{(x_{O_2} + N/N_0)(x_{SiCl_4} + N/N_0)}{(x_{Cl_2} - 2N/N_0)^2} = 1.02 \times 10^{-42} \tag{7.4.11}$$

If the initial state contains only SiO_2 and Cl_2, then $x_{O_2} = x_{SiCl_4} = 0$ and $x_{Cl_2} = 1$. Then (7.4.11) becomes

$$\left(\frac{N/N_0}{1 - 2N/N_0}\right)^2 = 1.02 \times 10^{-42}$$

from which we obtain $N/N_0 = 1.01 \times 10^{-21} \ll 1$. Hence, only a negligible etching of SiO_2 occurs before equilibrium is obtained.

In contrast, consider reaction (7.2.11) for SiO_2 etching by fluorine gas, for which, using the data from Table 7.1, $G_r^\circ(T_0) = -716.1$ kJ/mol. Using (7.4.7), we obtain $K = 3.2 \times 10^{125} \gg 1$. Hence, almost the entire F_2 gas charge reacts to attain the equilibrium state.

It is necessary to emphasize at this point that thermodynamics has nothing to say about the rate of the reaction to attain the equilibrium state. The reaction timescale might be microseconds or centuries. Rates are typically fast for gas- or liquid-phase reactions due to the high mobilities of the reactants and products, but they can be very slow if one of the reactants or products is a solid. Catalysts can be used to increase the reaction rates without altering the thermodynamic equilibrium. Reaction rates are the provenance of chemical kinetics, which we consider in Chapter 9.

Pressure and Temperature Variations

Changing the reaction pressure and temperature can have a strong effect on the equilibrium. First considering pressure variations, we note from (7.4.7) that K is independent of pressure. However, the mole fractions $\bar{x}$ will generally change as p changes. Inserting (7.4.9) into (7.4.8), we obtain

$$\prod_{j=1}^{\mathcal{M}_g} \bar{x}_j^{\alpha_j} = \left(\frac{p}{p^\circ}\right)^{-\alpha_g} K \tag{7.4.12}$$

Recall that α_g, given by (7.4.10), is the difference between the number of gas-phase product and reactant molecules for the stoichiometric reaction. For $\alpha_g > 0$, there are more gas-phase product molecules than product molecules. If the pressure is decreased, then the right-hand side of (7.4.12) is increased, driving the reaction to the right; i.e., the $\bar{x}_j$'s for the products increase and the $\bar{x}_j$'s for the reactants decrease. Hence, at low pressures, it is desirable to seek reactions having $\alpha_g > 0$. For $\alpha_g < 0$, a decrease in pressure drives the reaction to the left (fewer products, more reactions).

For $\alpha_g = 0$, the $\bar{x}_j$'s are independent of pressure. These variations are summarized in the following table:

p change	$\alpha_g > 0$	$\alpha_g = 0$	$\alpha_g < 0$
$p \downarrow$	products ↑	no change	products ↓
$p \uparrow$	products ↓	no change	products ↑

Equation (7.4.7) shows that the temperature variation of K is specified by the variation of $G_r^\circ(T)/T$, which we can derive as follows: Inserting the entropy (7.3.20) into the definition of G in (7.3.14) we obtain, at constant pressure,

$$G = H + T\frac{\partial G}{\partial T} \tag{7.4.13}$$

Dividing (7.4.13) by T^2 and rearranging, we obtain the *Gibbs–Helmholtz equation*

$$-\frac{H}{T^2} = -\frac{G}{T^2} + \frac{1}{T}\frac{\partial G}{\partial T} = \frac{\partial}{\partial T}\left(\frac{G}{T}\right) \tag{7.4.14}$$

Using (7.4.14) for each reaction species, we find

$$\left(\frac{\partial}{\partial T}\frac{G_r^\circ(T)}{T}\right) = -\frac{H_r^\circ(T)}{T^2} \tag{7.4.15}$$

Substituting (7.4.15) into the derivative of (7.4.7) and then dividing by (7.4.7), we obtain

$$\frac{d}{dT}\ln K(T) = \frac{H_r^\circ(T)}{RT^2} \tag{7.4.16}$$

We see from (7.4.16) that increasing the temperature for an exothermic reaction ($H_r < 0$) drives the reaction toward the left (fewer products, more reactants). Increasing the temperature drives an endothermic reaction toward the right (more products, fewer reactants). Integrating (7.4.16) over a temperature change from T_0 to T_1 and assuming that $H_r^\circ \approx$ const, independent of temperature, we obtain

$$K(T_1) = K(T_0)\exp\left[\frac{H_r^\circ}{R}\left(\frac{1}{T_0} - \frac{1}{T_1}\right)\right] \tag{7.4.17}$$

The following table summarizes the temperature variation:

T change	$H_r < 0$	$H_r > 0$
$T \uparrow$	products ↓	products ↑
$T \downarrow$	products ↑	products ↓

7.5 HETEROGENEOUS EQUILIBRIUM

Equilibrium Between Phases

We consider equilibrium between gas and liquid phases of a pure substance, e.g., H_2O, at constant temperature and pressure. Suppose that N_g moles of gas are in equilibrium with N_l moles of liquid. Let μ_g and μ_l be the chemical potentials of the gas and liquid. If dN moles are transferred from the gas to the liquid, then the Gibbs free energy changes by

$$dG = -\mu_g\, dN + \mu_l\, dN$$

If $\mu_g \neq \mu_l$, then dN can be chosen to make $dG < 0$; hence the system is not in equilibrium. Therefore, in equilibrium,

$$\mu_g = \mu_l = \mu_s \equiv \mu \tag{7.5.1}$$

independent of phase.

Now suppose that T and p are changed slightly so as to remain in equilibrium with N_g and N_l constant. Using the Gibbs free energy change (7.3.17), we obtain

$$d\mu_g = -S_{\text{gm}}\, dT + V_{\text{gm}}\, dp \tag{7.5.2}$$

$$d\mu_l = -S_{\text{lm}}\, dT + V_{\text{lm}}\, dp \tag{7.5.3}$$

where S_{gm} and S_{lm} are the entropy per mole and V_{gm} and V_{lm} are the volume per mole of the gas and liquid phases. Using (7.5.1), we can equate the RHS's of (7.5.2) and (7.5.3) to obtain

$$\frac{dp}{dT} = \frac{\Delta S_m}{\Delta V_m} \tag{7.5.4}$$

where

$$\Delta S_m = S_{\text{gm}} - S_{\text{lm}} \tag{7.5.5}$$

$$\Delta V_m = V_{\text{gm}} - V_{\text{lm}} \tag{7.5.6}$$

From the entropy change (7.3.1) with $dQ = dH$ (see also Fig. 7.3), the change in the molar entropy is

$$\Delta S_m = \frac{H_{\text{vap}}}{T} \tag{7.5.7}$$

Assuming that $V_{\text{lm}} \ll V_{\text{gm}}$ and using the perfect gas law to determine V_{gm}, we have

$$\Delta V_m \approx V_{\text{gm}} = \frac{RT}{p} \tag{7.5.8}$$

Inserting (7.5.7) and (7.5.8) into (7.5.4) yields

$$\frac{dp}{dT} = \frac{H_{vap}}{RT^2} p \qquad (7.5.9)$$

which is known as the *Clausius–Clapeyron equation*. Assuming that H_{vap} varies only weakly with T, we can integrate this to find

$$p_j = p_{0j} \exp\left(-\frac{H_{vap}}{RT}\right) \qquad (7.5.10)$$

where the subscript j denotes a pure substance. Equation (7.5.10) specifies the *vapor pressure* p_j of the gas in equilibrium with the liquid at temperature T. For the two phases to coexist, p_j and T cannot be independently chosen. Conversely, if p_j and T do not satisfy (7.5.10), then one of the phases does not exist.

The preceding analysis can be applied similarly to equilibrium between the gas and solid phases, yielding

$$p_j = p'_{0j} \exp\left(-\frac{H_{subl}}{RT}\right) \qquad (7.5.11)$$

where H_{subl} is the sublimation enthalpy per mole. For most substances $H \gg RT$, and thus p is a strong function of T. Plotting $\ln p_j$ versus $1/RT$ yields a straight line with slope $-H$. In the usual case, the curves (7.5.10) and (7.5.11) intersect at the *triple point* (p_3, T_3), leading to the phase diagram shown in Fig. 7.5. All three phases can coexist only at the triple point.

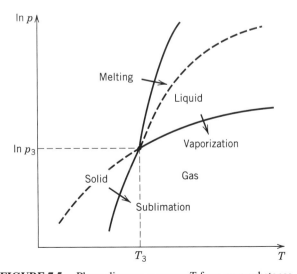

FIGURE 7.5.   Phase diagram p versus T for a pure substance.

TABLE 7.5. Vapor Pressures. *Note.* s, solid phase.

	Temperature (°C)		
Substance	1 Torr	10 Torr	100 Torr
$AlBr_3$	81.3 (s)	118.0	176.1
$AlCl_3$	100.0 (s)	123.8 (s)	152.0 (s)
AlF_3	1238	1324	1422
NH_3	−109.1 (s)	−91.9 (s)	−68.4
Br_2	−48.7 (s)	−25.0 (s)	9.3
Cl_2	−118.0 (s)	−101.6 (s)	−71.7
Cu_2Cl_2	546	702	960
$NiCl_2$	671 (s)	759 (s)	866 (s)
$SiCl_4$	−63.4	−34.4	5.4
SiF_4	−144.0 (s)	−130.4 (s)	−113.3 (s)
H_2O	−17.3 (s)	11.3	51.6
WF_6	−71.4 (s)	−49.2 (s)	−20.3 (s)

As an example, for H_2O, $H_{vap} \approx 40.66$ kJ/mol and $p_j = 1$ atm at 100 °C. This determines p_{0j} in (7.5.10). Table 7.5 gives some vapor pressure data for various substances.

For a mixture of substances, (7.5.10) and (7.5.11) hold for the partial pressures p_j, where the total pressure is the sum of the partial pressures:

$$p = \sum_j p_j \qquad (7.5.12)$$

Referring to Fig. 7.5, we see that if $T > T_3$ and $p > p_j(T)$ for vaporization, then the liquid and gas phases of substance j can coexist; if $T < T_3$ and $p > p_j(T)$ for sublimation, then the solid and gas phases can coexist.

As an application of these ideas, consider an etching process in which the etch product forms on the substrate in liquid form and in equilibrium with the gas phase. Then the product gas equilibrium density is $\bar{n} = p/kT$, where p, the vapor pressure, is given by (7.5.10). Now the flux of product molecules to and from the surface must balance in equilibrium. Using (2.4.10), the flux to the surface is $\Gamma_{in} = \frac{1}{4}\bar{n}\bar{v}$, where $\bar{v} = (8kT/\pi M)^{1/2}$ is the mean speed of the product molecules. Hence the flux from the surface is

$$\Gamma_{out} = \frac{1}{4}\bar{n}\bar{v} \qquad (7.5.13)$$

Now consider the *nonequilibrium* situation in which the product gas is efficiently pumped away, such that the gas density $n \ll \bar{n}$. In this case, $\Gamma_{in} \ll \Gamma_{out}$. However, if the surface remains completely covered with the liquid etch product, then Γ_{out} is still given by (7.5.13). Hence (7.5.13) determines a maximum etch product removal rate due to vapor pressure limitations. The removal rate can be less if the surface coverage is less than 100%, but it can never exceed this rate.

In this example, equilibrium thermodynamics (the vapor pressure p versus T) has been applied to determine an unknown kinetic rate (Γ_{out}) in terms of another known rate (Γ_{in}) for a system *that is not in equilibrium*. This important application of thermodynamics will be elaborated in Chapter 9.

Equilibrium at a Surface

We now consider thermal equilibrium for desorption and adsorption of gas molecules at a surface:

$$A : S = A(g) + S \qquad (7.5.14)$$

where the notation A: S denotes an adsorbed molecule A on the surface S. In almost all cases, adsorption (the reverse reaction) proceeds only if it is exothermic, $H_{ads} < 0$, because the entropy change S_{ads} is almost always negative, due to the binding of the gas molecule to the surface. Consequently, $G_{ads} = H_{ads} - TS_{ads} < 0$ only if $H_{ads} < 0$. Adsorption must be balanced by desorption (the forward reaction, with $G_{desor} = -G_{ads}$) in thermal equilibrium. Let $\bar{n}_A$ (m^{-3}) be the equilibrium gas-phase volume density, n_0' (m^{-2}) be the area density of surface sites, and $\bar{\theta}_A$ be the equilibrium fraction of sites on which molecules have adsorbed, such that the area densities covered and not covered with A molecules are $\bar{n}_{S:A}' = n_0' \bar{\theta}_A$ and $\bar{n}_S' = n_0'(1 - \bar{\theta}_A)$, respectively. Then as was done for pure gas-phase reactions, leading to an equilibrium constant $\mathcal{K}$ given by (7.4.6) and (7.4.7), we can write for reaction (7.5.14),

$$\frac{n_0' \bar{\theta}_A}{\bar{n}_A n_0'(1 - \bar{\theta}_A)} = \frac{\bar{\theta}_A}{\bar{n}_A(1 - \bar{\theta}_A)} = \mathcal{K}_{desor}(T) \qquad (7.5.15)$$

where

$$\mathcal{K}_{desor}(T) = \frac{1}{n^\circ} \exp\left(-\frac{G_{desor}^\circ}{RT}\right) \qquad (7.5.16)$$

and $n^\circ \approx 2.69 \times 10^{19}$cm^{-3} is the gas-phase density at standard temperature and pressure (Loschmidt's number).

Solving (7.5.15) for $\bar{\theta}_A$, we obtain

$$\bar{\theta}_A = \frac{\mathcal{K}\bar{n}_A}{1 + \mathcal{K}\bar{n}_A} \qquad (7.5.17)$$

which is known as the *Langmuir isotherm* because it specifies the equilibrium surface coverage as a function of pressure at fixed temperature. Plotting $\bar{\theta}_A$ versus $\bar{n}_A$ in Fig. 7.6, we see that $\bar{\theta}_A \propto \mathcal{K}\bar{n}_A$ for $\mathcal{K}\bar{n}_A \ll 1$, $\bar{\theta}_A \to 1$ for $\mathcal{K}\bar{n}_A \gg 1$, and $\bar{\theta}_A = 1/2$ at $\mathcal{K}\bar{n}_A = 1$. At fixed gas density, increasing T increases $\mathcal{K}$ and hence reduces $\bar{\theta}_A$. This behavior can be important in determining processing rates due to chemical reactions at surfaces. Although the rate of reaction for an adsorbed molecule A:S generally increases with temperature, the surface coverage decreases. Hence the overall reaction

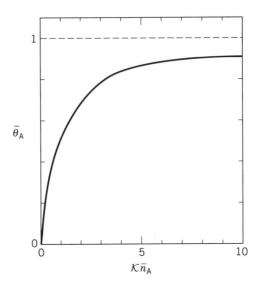

FIGURE 7.6. The Langmuir isotherm.

rate can first increase with T up to some maximum value and then decrease as T is further increased. Such behavior has been observed, for example, for silicon etching using XeF_2 gas.

Now let us consider the desorption and adsorption of two kinds of gas molecules on a surface:

$$A: S = A(g) + S$$

$$B: S = B(g) + S$$

Let $\bar{\theta}_A$ and $\bar{\theta}_B$ be the surface fractions covered with A and B molecules in thermal equilibrium; hence $1 - \bar{\theta}_A - \bar{\theta}_B$ is the surface fraction not covered. In thermal equilibrium, we must have

$$\frac{\bar{\theta}_A}{\bar{n}_A(1 - \bar{\theta}_A - \bar{\theta}_B)} = \mathcal{K}_A \qquad (7.5.18a)$$

$$\frac{\bar{\theta}_B}{\bar{n}_B(1 - \bar{\theta}_A - \bar{\theta}_B)} = \mathcal{K}_B \qquad (7.5.18b)$$

Solving for $\bar{\theta}_A$ and $\bar{\theta}_B$, we obtain

$$\bar{\theta}_A = \frac{\mathcal{K}_A \bar{n}_A}{1 + \mathcal{K}_A \bar{n}_A + \mathcal{K}_B \bar{n}_B} \qquad (7.5.19a)$$

$$\bar{\theta}_B = \frac{\mathcal{K}_B \bar{n}_B}{1 + \mathcal{K}_A \bar{n}_A + \mathcal{K}_B \bar{n}_B} \qquad (7.5.19b)$$

Comparing (7.5.19a) with (7.5.17), we see that the adsorption of B reduces the surface coverage of A. If A reacts at the surface and B does not, then B is an *inhibitor* for the reaction. Sidewalls in etching of silicon trenches are often protected by the use of inhibitors, which are cleared by ion bombardment at the bottom of the trench, thus yielding a low horizontal etch rate at the sidewall and a high vertical etch rate at the bottom.

PROBLEMS

7.1. High Temperature Equilibrium A professor has suggested that hydrogen gas at a high temperature $T = 1100\,°C$ and pressure p can be used to convert a thin layer of a SiO_2 (quartz) wafer to silicon. The reaction is

$$2H_2(g) + SiO_2(s) \rightarrow Si(s) + 2H_2O(g)$$

At STP, $G_f = -228.6\ kJ/mol$ for $H_2O(g)$ and $-856.6\ kJ/mol$ for $SiO_2(s)$. Also, $H_f = -241.8\ kJ/mol$ for $H_2O(g)$ and $-910.9\ kJ/mol$ for $SiO_2(s)$. You may assume that H_f is independent of temperature.

(a) Show that the equilibrium constant for the reaction at $1100\,°C$ is approximately 4.3×10^{-12}.

(b) Find the pressure p of H_2O gas necessary to convert a 1-nm-thick layer of SiO_2 to silicon. The SiO_2 wafer has an exposed area of $78.5\ cm^2$ and is placed in a reaction vessel having a volume of 10L. Note that the density of SiO_2 is $2.65 g/cm^3$.

7.2. Estimating Enthalpies of Formation The enthalpy of formation of $H_f^°(AB)$ of the substance AB can be written in terms of the bond dissociation enthalpy $H_{diss}^°(AB)$ and the enthalpies of formation $H_f^°(A)$ and $H_f^°(B)$ of the gaseous atoms A and B as

$$H_f^°(AB) = H_f^°(A) + H_f^°(B) - H_{diss}^°(AB)$$

This relation can be generalized to substances containing more than one bond.

(a) Using the data in Tables 7.3 and 7.4, estimate $H_f^°(T_0)$ for CF_4, CF_3, CF_2, and CF. Compare your estimates with data given in Tables 7.1 and 7.2.

(b) Using the data in Tables 7.3 and 7.4, estimate $H_f^°(T_0)$ for SiH_4, SiH_3, SiH_2, and SiH. Compare your estimates with data given in Tables 7.1 and 7.2.

(c) Using the data in Tables 7.3 and 7.4, estimate $H_f^°(T_0)$ for TEOS [$Si(OC_2H_5)_4$] and compare your estimate to the measured value of $-1397\ kJ/mol$.

7.3. The Triple Point Find p_3 and T_3 for H_2O by using the partial pressures for vaporization and sublimation (7.5.10) and (7.5.11), and compare to tabulated experimental data. Note that at standard pressure, the enthalpies of melting (at 273 K) and vaporization (at 373 K) are 6.01 and 40.66 kJ/mol, respectively. Assume that the heat capacity of liquid water is $1\ cal/K\text{-}cm^3$ and that the heat capacity of water vapor is given by the ideal gas formula.

7.4. Phase Equilibrium for a Mixture of Pure Substances   A mixture of Cl_2 and $SiCl_4$ is in equilibrium at room temperature $T = 25\,°C$ and $p = 760$ Torr. Find all the phases that exist, and find the vapor pressures of the gas phases of the two substances.

7.5. Thermodynamics and Vapor Pressures   Aluminum at $T = 298$ K (standard temperature) is etched reasonably fast in Cl_2 gas but not in F_2 gas, because the vapor pressure of AlF_3 is very low while that of $AlCl_3$ is reasonably high. The reactions are

$$Al(s) + \frac{3}{2}Cl_2(g) \rightarrow AlCl_3(s)$$

$$Al(s) + \frac{3}{2}F_2(g) \rightarrow AlF_3(s)$$

(a) Show that both reactions are thermodynamically strongly downhill (proceed far to the right) by finding the fraction x_{Cl_2} or x_{F_2} of unreacted Cl_2 or F_2 in equilibrium, given that the initial gas pressure of Cl_2 or F_2 is p_0. (Assume that there is a very large initial supply of aluminum to be etched.) Note that $G_f = -628.8$ kJ/mol for $AlCl_3(s)$ and -1431 kJ/mol for $AlF_3(s)$ at STP.

(b) Estimate the maximum etch rate (Å/min) at 298 K that can be achieved for Cl_2 and F_2 etching of aluminum due to vapor pressure limitations. Note that $H_{vap} = 116$ kJ/mol for $AlCl_3$ and 531 kJ/mol for AlF_3 at STP; the vapor pressure is 760 Torr at $T = 453.2$ K for $AlCl_3$ and 1810 K for AlF_3. The density of solid aluminum is $2.70\,g/cm^3$. (Industrial processes generally require etch rates exceeding 2000 Å/min.)

7.6. Vapor Pressure Data   The vapor pressure data for $NiCl_2(s) \rightarrow NiCl_2(g)$ is given below:

p (Torr)	1	10	100	760
T (°C)	671	759	866	987

Plot $\log p$ versus $1000/T$ (T in kelvins, not degrees Centigrade!) and use this plot to show that the sublimation enthalpy per mole at STP is ≈ 210 kJ/mol.

7.7. Equilibrium for Dissociation on a Surface   For dissociative adsorption in thermal equilibrium with associative desorption,

$$A{:}S + A{:}S = A_2(g) + 2S$$

show that the equilibrium surface coverage is

$$\theta_A = \frac{(\mathcal{K}\bar{n}_{A_2})^{1/2}}{1 + (\mathcal{K}\bar{n}_{A_2})^{1/2}}$$

where $\bar{n}_{A_2}$ is the equilibrium gas-phase density and $\mathcal{K}$ is the equilibrium constant for the reaction.

CHAPTER 8

MOLECULAR COLLISIONS

8.1 INTRODUCTION

Basic concepts of gas-phase collisions were introduced in Chapter 3, where we described only those processes needed to model the simplest noble gas discharges: electron-atom ionization, excitation, and elastic scattering; and ion-atom elastic scattering and resonant charge transfer. In this chapter we introduce other collisional processes that are central to the description of chemically reactive discharges. These include the dissociation of molecules, the generation and destruction of negative ions, and gas-phase chemical reactions.

Whereas the cross sections have been measured reasonably well for the noble gases, with measurements in reasonable agreement with theory, this is not the case for collisions in molecular gases. Hundreds of potentially significant collisional reactions must be examined in simple diatomic gas discharges such as oxygen. For feedstocks such as CF_4/O_2, SiH_4/O_2, etc., the complexity can be overwhelming. Furthermore, even when the significant processes have been identified, most of the cross sections have been neither measured nor calculated. Hence, one must often rely on estimates based on semiempirical or semiclassical methods, or on measurements made on molecules analogous to those of interest. As might be expected, data are most readily available for simple diatomic (H_2, N_2, O_2, CO, NO) and polyatomic (CO_2, NH_3, CH_4) gases.

8.2 MOLECULAR STRUCTURE

The energy levels for the electronic states of a single atom were described in Chapter 3. The energy levels of molecules are more complicated for two reasons. First, molecules have additional vibrational and rotational degrees of freedom due to the motions of their nuclei, with corresponding quantized energies $\mathcal{E}_v$ and $\mathcal{E}_J$. Second, the energy $\mathcal{E}_e$ of each electronic state depends on the instantaneous configuration of the nuclei. For a diatomic molecule, $\mathcal{E}_e$ depends on a single coordinate R, the spacing between the two nuclei. Since the nuclear motions are slow compared to the electronic motions, the electronic state can be determined for any fixed spacing. We can therefore represent each quantized electronic level for a frozen set of nuclear positions as a graph of $\mathcal{E}_e$ versus R, as shown in Fig. 8.1. For a molecule to be stable, the ground (minimum energy) electronic state must have a minimum at some value $\bar{R}_1$ corresponding to the mean intermolecular separation (curve 1). In this case, energy must be supplied in order to separate the atoms ($R \rightarrow \infty$). An excited electronic state can either have a minimum ($\bar{R}_2$ for curve 2) or not (curve 3). Note that $\bar{R}_2$ and $\bar{R}_1$ do not generally coincide. As for atoms, excited states may be short-lived (unstable to electric dipole radiation) or may be metastable. Various electronic levels may tend to the same energy in the unbound ($R \rightarrow \infty$) limit.

For diatomic molecules, the electronic states are specified first by the component (in units of $\hbar$) Λ of the total orbital angular momentum along the internuclear axis, with the symbols Σ, Π, Δ, and Φ corresponding to $\Lambda = 0$, ± 1, ± 2, and ± 3, in

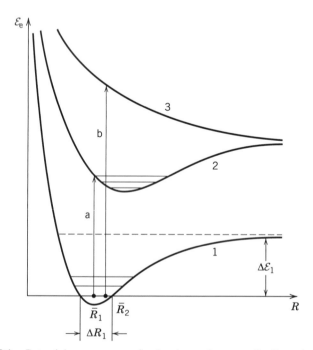

FIGURE 8.1. Potential energy curves for the electronic states of a diatomic molecule.

analogy with atomic nomenclature. All but the Σ states are doubly degenerate. For Σ states, $+$ and $-$ superscripts are often used to denote whether the wave function is symmetric or antisymmetric with respect to reflection at any plane through the internuclear axis. The total electron spin angular momentum S (in units of $\hbar$) is also specified, with the multiplicity $2S + 1$ written as a prefixed superscript, as for atomic states. Finally, for homonuclear molecules (H_2, N_2, O_2, etc.) the subscripts g or u are written to denote whether the wave function is symmetric or antisymmetric with respect to interchange of the nuclei. In this notation, the ground states of H_2 and N_2 are both singlets, $^1\Sigma_g^+$, and that of O_2 is a triplet, $^3\Sigma_g^-$. For polyatomic molecules, the electronic energy levels depend on more than one nuclear coordinate, so Fig. 8.1 must be generalized. Furthermore, since there is generally no axis of symmetry, the states cannot be characterized by the quantum number Λ, and other naming conventions are used. Such states are often specified empirically through characterization of measured optical emission spectra. Typical spacings of low-lying electronic energy levels range from a few to tens of volts, as for atoms.

Vibrational and Rotational Motion

Unfreezing the nuclear vibrational and rotational motions leads to additional quantized structure on smaller energy scales, as illustrated in Fig. 8.2. The simplest (harmonic oscillator) model for the vibration of diatomic molecules leads to equally spaced quantized, nondegenerate energy levels

$$e\mathcal{E}_v = \hbar\omega_{\text{vib}}\left(v + \frac{1}{2}\right) \tag{8.2.1}$$

where $v = 0, 1, 2, \ldots$ is the vibrational quantum number and ω_{vib} is the linearized vibration frequency. Fitting a quadratic function

$$e\mathcal{E}_v = \frac{1}{2}k_{\text{vib}}(R - \bar{R})^2 \tag{8.2.2}$$

near the minimum of a stable energy level curve such as those shown in Fig. 8.1, we can estimate

$$\omega_{\text{vib}} \approx \left(\frac{k_{\text{vib}}}{m_{\text{Rmol}}}\right)^{1/2} \tag{8.2.3}$$

where k_{vib} is the "spring constant" and m_{Rmol} is the reduced mass of the AB molecule. The spacing $\hbar\omega_{\text{vib}}$ between vibrational energy levels for a low-lying stable electronic state is typically a few tenths of a volt. Hence for molecules in equilibrium at room temperature (0.026 V), only the $v = 0$ level is significantly populated. However, collisional processes can excite strongly nonequilibrium vibrational energy levels.

We indicate by the short horizontal line segments in Fig. 8.1 a few of the vibrational energy levels for the stable electronic states. The length of each segment gives the

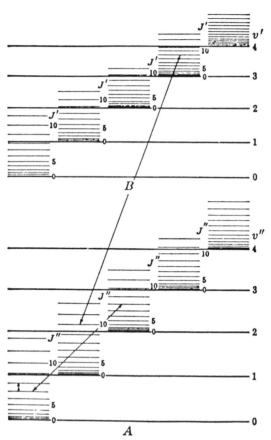

FIGURE 8.2. Vibrational and rotational levels of two electronic states A and B of a molecule; the three double arrows indicate examples of transitions in the pure rotation spectrum, the rotation-vibration spectrum, and the electronic spectrum (after Herzberg, 1971).

range of classically allowed vibrational motions. Note that even the ground state $(v = 0)$ has a finite width ΔR_1 as shown, because from (8.2.1), the $v = 0$ state has a nonzero vibrational energy $\frac{1}{2}\hbar\omega_{\text{vib}}$. The actual separation ΔR about $\bar{R}$ for the ground state has a Gaussian distribution, and tends toward a distribution peaked at the classical turning points for the vibrational motion as $v \to \infty$. The vibrational motion becomes anharmonic and the level spacings tend to zero as the unbound vibrational energy is approached ($\mathcal{E}_v \to \Delta\mathcal{E}_1$). For $\mathcal{E}_v > \Delta\mathcal{E}_1$, the vibrational states form a continuum, corresponding to unbound classical motion of the nuclei (breakup of the molecule). For a polyatomic molecule there are many degrees of freedom for vibrational motion, leading to a very complicated structure for the vibrational levels.

The simplest (dumbbell) model for the rotation of diatomic molecules leads to the nonuniform quantized energy levels

$$e\mathcal{E}_J = \frac{\hbar^2}{2I_{\text{mol}}}J(J + 1) \tag{8.2.4}$$

where $I_{mol} = m_{Rmol}\bar{R}^2$ is the moment of inertia and $J = 0, 1, 2, \ldots$ is the rotational quantum number. The levels are degenerate, with $2J + 1$ states for the Jth level. The spacing between rotational levels increases with J (see Fig. 8.2). The spacing between the lowest ($J = 0$ to $J = 1$) levels typically corresponds to an energy of 0.001–0.01 V; hence, many low-lying levels are populated in thermal equilibrium at room temperature.

Optical Emission

An excited molecular state can decay to a lower energy state by emission of a photon or by breakup of the molecule. As shown in Fig. 8.2, the radiation can be emitted by a transition between electronic levels, between vibrational levels of the same electronic state, or between rotational levels of the same electronic and vibrational state; the radiation typically lies within the optical, infrared, or microwave frequency range, respectively. Electric dipole radiation is the strongest mechanism for photon emission, having typical transition times of $t_{rad} \sim 10^{-9}$ s, as obtained in (3.4.13). The selection rules for electric dipole radiation are

$$\Delta\Lambda = 0, \pm 1 \tag{8.2.5a}$$

$$\Delta S = 0 \tag{8.2.5b}$$

In addition, for transitions between Σ states the only allowed transitions are

$$\Sigma^+ \rightarrow \Sigma^+ \quad \text{and} \quad \Sigma^- \rightarrow \Sigma^- \tag{8.2.6}$$

and for homonuclear molecules, the only allowed transitions are

$$g \rightarrow u \quad \text{and} \quad u \rightarrow g \tag{8.2.7}$$

Hence homonuclear diatomic molecules do not have a pure vibrational or rotational spectrum. Radiative transitions between electronic levels having many different vibrational and rotational initial and final states give rise to a structure of emission and absorption bands within which a set of closely spaced frequencies appear. These give rise to characteristic molecular emission and absorption bands when observed using low-resolution optical spectrometers. As for atoms, metastable molecular states having no electric dipole transitions to lower levels also exist. These have lifetimes much exceeding 10^{-6} s; they can give rise to weak optical band structures due to magnetic dipole or electric quadrupole radiation.

Electric dipole radiation between vibrational levels of the same electronic state is permitted for molecules having permanent dipole moments. In the harmonic oscillator approximation, the selection rule is $\Delta v = \pm 1$; weaker transitions $\Delta v = \pm 2, \pm 3, \ldots$ are permitted for anharmonic vibrational motion.

The preceding description of molecular structure applies to molecules having arbitrary electronic charge. This includes neutral molecules AB, positive molecular ions AB^+, AB^{2+}, etc., and negative molecular ions AB^-. The potential energy curves for the various electronic states, regardless of molecular charge, are commonly plotted on the same diagram. Figures 8.3 and 8.4 give these for some important electronic states of H_2^-, H_2, and H_2^+, and of O_2^-, O_2, and O_2^+. Examples of both attractive (having a potential energy minimum) and repulsive (having no minimum) states can be seen. The vibrational levels are labeled with the quantum number v for the attractive levels. The ground states of both H_2^+ and O_2^+ are attractive; hence these molecular ions are stable against *autodissociation* ($AB^+ \rightarrow A + B^+$ or $A^+ + B$). Similarly, the ground states of H_2 and O_2 are attractive and lie below those of H_2^+ and O_2^+; hence they are stable against autodissociation and *autoionization* ($AB \rightarrow AB^+ + e$). For some molecules, e.g., diatomic argon, the AB^+ ion is stable but the AB neutral is not stable. For all molecules, the AB ground state lies below the AB^+ ground state and is stable

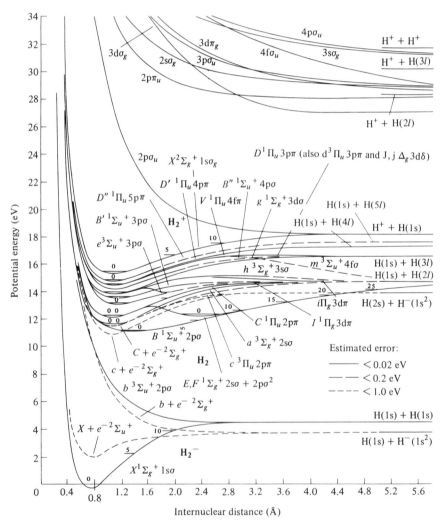

FIGURE 8.3. Potential energy curves for H_2^-, H_2, and H_2^+. (From Jeffery I. Steinfeld, *Molecules and Radiation: An Introduction to Modern Molecular Spectroscopy*, 2d ed. © MIT Press, 1985.)

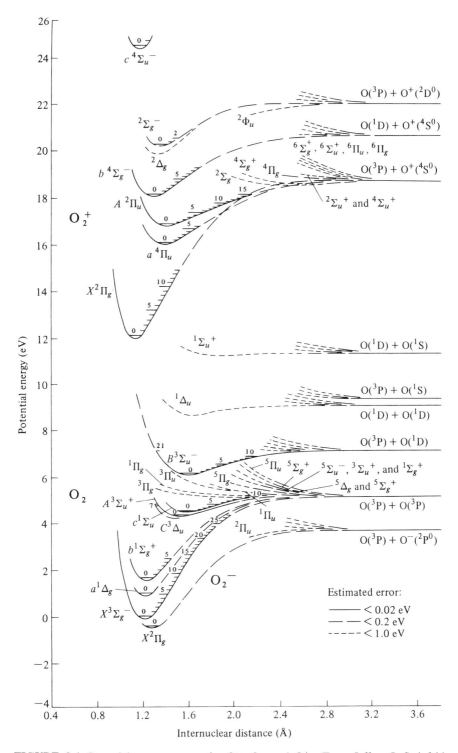

FIGURE 8.4. Potential energy curves for O_2^-, O_2, and O_2^+. (From Jeffery I. Steinfeld, *Molecules and Radiation: An Introduction to Modern Molecular Spectroscopy*, 2d ed. © MIT Press, 1985.)

against autoionization. Excited states can be attractive or repulsive. A few of the attractive states may be metastable; some examples are the $^3\Pi_u$ state of H_2 and the $^1\Delta_g$, $^1\Sigma_g^+$ and $^3\Delta_u$ states of O_2.

Negative Ions

Recall from Section 7.2 that many neutral atoms have a positive electron affinity $\mathcal{E}_{aff}$; i.e., the reaction

$$A + e \rightarrow A^-$$

is exothermic with energy $\mathcal{E}_{aff}$ (in volts). If $\mathcal{E}_{aff}$ is negative, then A^- is unstable to *autodetachment*, $A^- \rightarrow A + e$. A similar phenomenon is found for negative molecular ions. A stable AB^- ion exists if its ground (lowest energy) state has a potential minimum that lies below the ground state of AB. This is generally true only for strongly electronegative gases having large electron affinities, such as O_2 ($\mathcal{E}_{aff} \approx$ 1.463 V for O atoms) and the halogens ($\mathcal{E}_{aff} > 3$ V for the atoms). For example, Fig. 8.4 shows that the $^2\Pi_g$ ground state of O_2^- is stable, with $\mathcal{E}_{aff} \approx 0.43$ V for O_2. For weakly electronegative or for electropositive gases, the minimum of the ground state of AB^- generally lies above the ground state of AB, and AB^- is unstable to autodetachment. An example is hydrogen, which is weakly electronegative ($\mathcal{E}_{aff} \approx 0.754$ V for H atoms). Figure 8.3 shows that the $^2\Sigma_u^+$ ground state of H_2^- is unstable, although the H^- ion itself is stable. In an electropositive gas such as N_2 ($\mathcal{E}_{aff} \lesssim 0$), both N_2^- and N^- are unstable.

8.3 ELECTRON COLLISIONS WITH MOLECULES

The interaction time for the collision of a typical (1–10 V) electron with a molecule is short, $t_c \sim 2a_0/v_e \sim 10^{-16}$–$10^{-15}$ s, compared to the typical time for a molecule to vibrate, $\tau_{vib} \sim 10^{-14}$–10^{-13} s. Hence for electron collisional excitation of a molecule to an excited electronic state, the new vibrational (and rotational) state can be determined by freezing the nuclear motions during the collision. This is known as the *Franck-Condon principle* and is illustrated in Fig. 8.1 by the vertical line a, showing the collisional excitation at fixed R to a high quantum number bound vibrational state and by the vertical line b, showing excitation at fixed R to a vibrationally unbound state, in which breakup of the molecule is energetically permitted. Since the typical transition time for electric dipole radiation ($\tau_{rad} \sim 10^{-9}$–10^{-8} s) is long compared to the dissociation ($\sim$ vibrational) time τ_{diss}, excitation to an excited state will generally lead to dissociation when it is energetically permitted. Finally, we note that the time between collisions $\tau_c \gg \tau_{rad}$ in typical low-pressure processing discharges. Summarizing the ordering of timescales for electron–molecule collisions, we have

$$t_{at} \sim t_c \ll t_{vib} \sim t_{diss} \ll \tau_{rad} \ll \tau_c$$

Dissociation

Electron impact dissociation,

$$e + AB \rightarrow A + B + e$$

of feedstock gases plays a central role in the chemistry of low-pressure reactive discharges. The variety of possible dissociation processes is illustrated in Fig. 8.5. In collisions a or a′, the $v = 0$ ground state of AB is excited to a repulsive state of AB. The required threshold energy $\mathcal{E}_{thr}$ is $\mathcal{E}_a$ for collision a and $\mathcal{E}_{a'}$ for collision a′, and it leads to an energy after dissociation lying between $\mathcal{E}_a - \mathcal{E}_{diss}$ and $\mathcal{E}_{a'} - \mathcal{E}_{diss}$ that is shared among the dissociation products (here, A and B). Typically, $\mathcal{E}_a - \mathcal{E}_{diss} \sim$ few volts; consequently, hot neutral fragments are typically generated by dissociation processes. If these hot fragments hit the substrate surface, they can profoundly affect the process chemistry. In collision b, the ground state AB is excited to an attractive state of AB at an energy $\mathcal{E}_b$ that exceeds the binding energy $\mathcal{E}_{diss}$ of the AB molecule, resulting in dissociation of AB with fragment energy $\mathcal{E}_b - \mathcal{E}_{diss}$. In collision b′, the excitation energy $\mathcal{E}_{b'} = \mathcal{E}_{diss}$, and the fragments have low energies; hence this process creates fragments having energies ranging from essentially thermal energies up to $\mathcal{E}_b - \mathcal{E}_{diss} \sim$ few volts. In collision c, the AB atom is excited to the bound excited state AB* (labeled 5), which subsequently radiates to the unbound AB state (labeled 3), which then dissociates. The threshold energy required is large, and the fragments are hot. Collision c can also lead to dissociation of an excited state by a radiationless transfer from state 5 to state 4 near the point where the two states cross:

$$AB^*(bound) \rightarrow AB^*(unbound) \rightarrow A + B^*$$

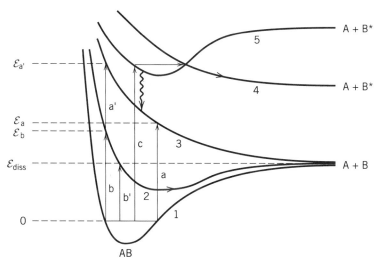

FIGURE 8.5. Illustrating the variety of dissociation processes for electron collisions with molecules.

The fragments can be both hot and in excited states. We discuss such radiationless electronic transitions in the next section. This phenomenon is known as *predissociation*. Finally, a collision (not labeled in the figure) to state 4 can lead to dissociation of AB^*, again resulting in hot excited fragments.

The process of electron impact excitation of a molecule is similar to that of an atom, and, consequently, the cross sections have a similar form. A simple classical estimate of the dissociation cross section for a level having excitation energy U_1 can be found by requiring that an incident electron having energy W transfer an energy W_L lying between U_1 and U_2 to a valence electron. Here, U_2 is the energy of the next higher level. Then integrating the differential cross section dI (given in (3.4.20) and repeated here),

$$dI = \pi \left(\frac{e^2}{4\pi\epsilon_0} \right)^2 \frac{1}{W} \frac{dW_L}{W_L^2} \tag{3.4.20}$$

over W_L, we obtain

$$
\begin{aligned}
\sigma_{\text{diss}} &= 0 & W &< U_1 \\
&= \pi \left(\frac{e^2}{4\pi\epsilon_0} \right)^2 \frac{1}{W} \left(\frac{1}{U_1} - \frac{1}{W} \right) & U_1 &< W < U_2 \\
&= \pi \left(\frac{e^2}{4\pi\epsilon_0} \right)^2 \frac{1}{W} \left(\frac{1}{U_1} - \frac{1}{U_2} \right) & W &> U_2
\end{aligned}
\tag{8.3.1}
$$

Letting $U_2 - U_1 \ll U_1$ and introducing voltage units $W = e\mathcal{E}$, $U_1 = e\mathcal{E}_1$ and $U_2 = e\mathcal{E}_2$, we have

$$
\begin{aligned}
\sigma_{\text{diss}} &= 0 & \mathcal{E} &< \mathcal{E}_1 \\
&= \sigma_0 \frac{\mathcal{E} - \mathcal{E}_1}{\mathcal{E}_1} & \mathcal{E}_1 &< \mathcal{E} < \mathcal{E}_2 \\
&= \sigma_0 \frac{\mathcal{E}_2 - \mathcal{E}_1}{\mathcal{E}} & \mathcal{E} &> \mathcal{E}_2
\end{aligned}
\tag{8.3.2}
$$

where

$$\sigma_0 = \pi \left(\frac{e}{4\pi\epsilon_0 \mathcal{E}_1} \right)^2 \tag{8.3.3}$$

We see that the dissociation cross section rises linearly from the threshold energy $\mathcal{E}_{\text{thr}} \approx \mathcal{E}_1$ to a maximum value $\sigma_0(\mathcal{E}_2 - \mathcal{E}_1)/\mathcal{E}_{\text{thr}}$ at $\mathcal{E}_2$ and then falls off as $1/\mathcal{E}$. Actually, $\mathcal{E}_1$ and $\mathcal{E}_2$ can depend on the nuclear separation R. In this case, (8.3.2) should be averaged over the range of R's corresponding to the ground state vibrational energy, leading to a broadened dependence of the average cross section on energy $\mathcal{E}$. The maximum cross section is typically of order 10^{-15} cm^2. Typical rate constants for a single dissociation process with $\mathcal{E}_{thr} \gtrsim T_e$ have an Arrhenius form

$$K_{diss} \propto K_{diss0} \exp\left(-\frac{\mathcal{E}_{thr}}{T_e}\right) \tag{8.3.4}$$

where $K_{diss0} \sim 10^{-7}$ cm^3/s. However, in some cases $\mathcal{E}_{thr} \lesssim T_e$. For excitation to an attractive state, an appropriate average over the fraction of the ground-state vibration that leads to dissociation must be taken.

Dissociative Ionization

In addition to normal ionization,

$$e + AB \rightarrow AB^+ + 2e$$

electron–ion collisions can lead to dissociative ionization

$$e + AB \rightarrow A + B^+ + 2e$$

These processes, common for polyatomic molecules, are illustrated in Fig. 8.6. In collision a having threshold energy $\mathcal{E}_{iz}$, the molecular ion AB$^+$ is formed. Collisions b and c occur at higher threshold energies $\mathcal{E}_{diz}$ and result in dissociative ionization, leading to the formation of fast, positively charged ions and neutrals. These cross sections have a similar form to the Thompson ionization cross section for atoms.

Dissociative Recombination

The electron collision,

$$e + AB^+ \rightarrow A + B^*$$

illustrated as d and d$'$ in Fig. 8.6, destroys an electron–ion pair and leads to the production of fast excited neutral fragments. Since the electron is captured, it is not available to carry away a part of the reaction energy. Consequently, the collision cross section has a resonant character, falling to very low values for $\mathcal{E} < \mathcal{E}_d$ and $\mathcal{E} > \mathcal{E}_{d'}$. However, a large number of excited states A* and B* having increasing principal quantum numbers n and energies can be among the reaction products. Consequently, the rate constants can be large, of order 10^{-7}–10^{-6} cm^3/s. Dissociative recombination to the ground states of A and B cannot occur because the potential energy curve for AB$^+$ is always greater than the potential energy curve for the repulsive state of AB. Two-body recombination for atomic ions or for molecular ions that do not subsequently dissociate can only occur with emission of a photon:

$$e + A^+ \rightarrow A + h\nu.$$

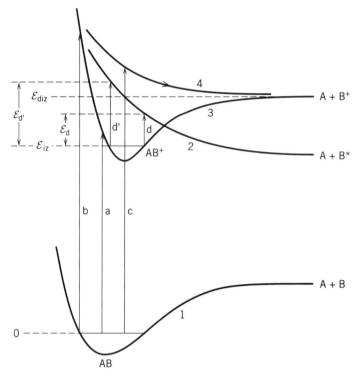

FIGURE 8.6. Illustrating dissociative ionization and dissociative recombination for electron collisions with molecules.

As shown in Section 9.2, the rate constants are typically three to five orders of magnitude lower than for dissociative recombination.

Example of Hydrogen

The example of H_2 illustrates some of the inelastic electron collision phenomena we have discussed. In order of increasing electron impact energy, at a threshold energy of ~ 8.8 V, there is excitation to the repulsive $^3\Sigma_u^+$ state followed by dissociation into two fast H fragments carrying ~ 2.2 V/atom. At 11.5 V, the $^1\Sigma_u^+$ bound state is excited, with subsequent electric dipole radiation in the UV region to the $^1\Sigma_g^+$ ground state. At 11.8 V, there is excitation to the $^3\Sigma_g^+$ bound state, followed by electric dipole radiation to the $^3\Sigma_u^+$ repulsive state, followed by dissociation with ~ 2.2 V/atom. At 12.6 V, the $^1\Pi_u$ bound state is excited, with UV emission to the ground state. At 15.4 V, the $^2\Sigma_g^+$ ground state of H_2^+ is excited, leading to the production of H_2^+ ions. At 28 V, excitation of the repulsive $^2\Sigma_u^+$ state of H_2^+ leads to the dissociative ionization of H_2, with ~ 5 V each for the H and H^+ fragments.

Dissociative Electron Attachment

The processes,

$$e + AB \rightarrow A + B^-$$

produce negative ion fragments as well as neutrals. They are important in discharges containing atoms having positive electron affinities, not only because of the production of negative ions, but because the threshold energy for production of negative ion fragments is usually lower than for pure dissociation processes. A variety of processes are possible, as shown in Fig. 8.7. Since the impacting electron is captured and is not available to carry excess collision energy away, dissociative attachment is a resonant process that is important only within a narrow energy range. The maximum cross sections are generally much smaller than the hard-sphere cross section of the molecule. Attachment generally proceeds by collisional excitation from the ground AB state to a repulsive AB^- state, which subsequently either autodetaches or dissociates. The attachment cross section is determined by the balance between these processes. For most molecules, the dissociation energy $\mathcal{E}_{diss}$ of AB is greater than the electron affinity $\mathcal{E}_{affB}$ of B, leading to the potential energy curves shown in Fig. 8.7a. In this case, the cross section is large only for impact energies lying between a minimum value $\mathcal{E}_{thr}$, for collision a, and a maximum value $\mathcal{E}'_{thr}$ for collision a'. The fragments are hot, having energies lying between minimum and maximum values $\mathcal{E}_{min} = \mathcal{E}_{thr} + \mathcal{E}_{affB} - \mathcal{E}_{diss}$ and $\mathcal{E}_{max} = \mathcal{E}'_{thr} + \mathcal{E}_{affB} - \mathcal{E}_{diss}$. Since the AB^- state lies above the AB state for $R < R_x$, autodetachment can occur as the molecules begin to separate: $AB^- \rightarrow AB + e$. Hence the cross section for production of negative ions can be much smaller than that for excitation of the AB^- repulsive state. As a crude estimate, for the same energy, the autodetachment rate is $\sqrt{M_R/m} \sim 100$ times the dissociation rate of the repulsive AB^- molecule, where M_R is the reduced mass. Hence only one out of 100 excitations lead to dissociative attachment.

Excitation to the AB^- bound state can also lead to dissociative attachment, as shown in Fig. 8.7b. Here the cross section is significant only for $\mathcal{E}_{thr} < \mathcal{E} < \mathcal{E}'_{thr}$, but the fragments can have low energies, with a minimum energy of zero and a maximum energy of $\mathcal{E}'_{thr} + \mathcal{E}_{affB} - \mathcal{E}_{diss}$. Collision b,

$$e + AB \rightarrow AB^{-*}$$

does not lead to production of AB^- ions because energy and momentum are not generally conserved when two bodies collide elastically to form one body (see Problem 3.12). Hence the excited AB^{-*} ion separates,

$$AB^{-*} \rightarrow e + AB$$

unless vibrational radiation or collision with a third body carries off the excess energy. These processes are both slow in low pressure discharges (see Section 9.2).

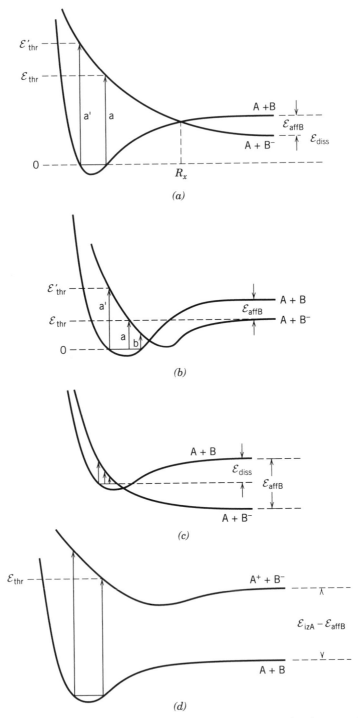

FIGURE 8.7. Illustrating a variety of electron attachment processes for electron collisions with molecules; (a) capture into a repulsive state; (b) capture into an attractive state; (c) capture of slow electrons into a repulsive state; (d) polar dissociation.

At high pressures (say, atmospheric), three-body attachment to form AB^- can be very important.

For a few molecules, such as some halogens, the electron affinity of the atom exceeds the dissociation energy of the neutral molecule, leading to the potential energy curves shown in Fig. 8.7c. In this case the range of electron impact energies $\mathcal{E}$ for excitation of the AB^- repulsive state includes $\mathcal{E} = 0$. Consequently, there is no threshold energy, and very slow electrons can produce dissociative attachment, resulting in hot neutral and negative ion fragments. The range of R's over which autodetachment can occur is small; hence the maximum cross sections for dissociative attachment can be as high as 10^{-16} cm^2.

A simple classical estimate of electron capture can be made using the differential scattering cross section for energy loss (3.4.20), in a manner similar to that done for dissociation. For electron capture to an energy level $\mathcal{E}_1$ that is unstable to autodetachment, and with the additional constraint for capture that the incident electron energy lie within $\mathcal{E}_1$ and $\mathcal{E}_2 = \mathcal{E}_1 + \Delta\mathcal{E}$, where $\Delta\mathcal{E}$ is a small energy difference characteristic of the dissociative attachment timescale, we obtain, in place of (8.3.2),

$$
\begin{aligned}
\sigma_{\text{att}} &= 0 & \mathcal{E} &< \mathcal{E}_1 \\
&= \sigma_0 \frac{\mathcal{E} - \mathcal{E}_1}{\mathcal{E}_1} & \mathcal{E}_1 &< \mathcal{E} < \mathcal{E}_2 \\
&= 0 & \mathcal{E} &> \mathcal{E}_2
\end{aligned}
\tag{8.3.5}
$$

where

$$
\sigma_0 \approx \pi \left(\frac{m}{M_R}\right)^{1/2} \left(\frac{e}{4\pi\epsilon_0\mathcal{E}_1}\right)^2
\tag{8.3.6}
$$

The factor of $(m/M_R)^{1/2}$ roughly gives the fraction of excited states that do not autodetach. We see that the dissociative attachment cross section rises linearly at $\mathcal{E}_1$ to a maximum value $\sigma_0 \Delta\mathcal{E}/\mathcal{E}_1$ and then falls abruptly to zero.

As for dissociation, $\mathcal{E}_1$ can depend strongly on the nuclear separation R, and (8.3.5) must be averaged over the range of $\mathcal{E}_1$'s corresponding to the ground state vibrational motion; e.g., from $\sim \mathcal{E}_{\text{thr}}$ to $\sim \mathcal{E}'_{\text{thr}}$ in Fig. 8.7a. Because generally $\Delta\mathcal{E} \ll \mathcal{E}'_{\text{thr}} - \mathcal{E}_{\text{thr}}$, we can write (8.3.5) in the form

$$
\sigma_{\text{att}} \approx \pi \left(\frac{m}{M_R}\right)^{1/2} \left(\frac{e}{4\pi\epsilon_0}\right)^2 \frac{(\Delta\mathcal{E})^2}{2\mathcal{E}_1^3} \delta(\mathcal{E} - \mathcal{E}_1)
\tag{8.3.7}
$$

where δ is the Dirac delta function. Using (8.3.7), the average over the vibrational motion can be performed, leading to a cross section that is strongly peaked lying between $\mathcal{E}_{\text{thr}}$ and $\mathcal{E}'_{\text{thr}}$. We leave the details of the calculation to a problem.

Polar Dissociation

The process,

$$e + AB \rightarrow A^+ + B^- + e$$

produces negative ions without electron capture. As shown in Fig. 8.7d, the process proceeds by excitation of a polar state A^+B^- of AB^* that has a separated atom limit of A^+ and B^-. Hence at large R, this state lies above the $A + B$ ground state by the difference between the ionization potential of A and the electron affinity of B. The polar state is weakly bound at large R by the Coulomb attraction force, but is repulsive at small R. The maximum cross section and the dependence of the cross section on electron impact energy are similar to that of pure dissociation. The threshold energy $\mathcal{E}_{thr}$ for polar dissociation is generally large.

The measured cross section for negative ion production by electron impact in O_2 is shown in Fig. 8.8. The sharp peak at 6.5 V is due to dissociative attachment. The variation of the cross section with energy is typical of a resonant capture process. The maximum cross section of $\sim 10^{-18}$ cm^2 is quite low because autodetachment from the repulsive O_2^- state is strong, inhibiting dissociative attachment. The second gradual maximum near 35 V is due to polar dissociation; the variation of the cross section with energy is typical of a nonresonant process.

Metastable Negative Ions

In some complex molecules a negative ion state lies at an energy very close to but just above the ground state. In this case, pure attachment of electrons having nearly

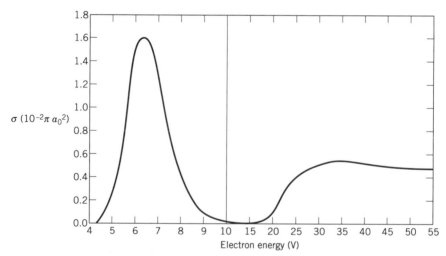

FIGURE 8.8. Cross section for production of negative ions by electron impact in O_2 (Rapp and Briglia, 1965).

zero energy can occur at low pressures. A good example is SF_6 where the SF_6^- state lies about 0.1 V above the SF_6 state, leading to the process

$$e + SF_6 \rightarrow SF_6^-$$

The negative ion is unstable to autodetachment and may also be unstable to autodissociation, but in some complex molecules, such as SF_6, these processes are weak, leading to lifetimes for the SF_6^- metastable ion in excess of 10^{-6} s. The cross section is sharply resonant with a maximum value of order 10^{-15} cm^2. For very low electron energies, this process might be important in low-pressure SF_6 discharges.

Electron Impact Detachment

The processes

$$e + A^- \rightarrow A + 2e$$
$$e + AB^- \rightarrow AB + 2e$$

can be important in destroying atomic or molecular negative ions. The process is similar to electron neutral ionization, with the electron affinity $\mathcal{E}_{aff}$ of A or AB playing the role of the ionization potential. However, the peak in the cross section is shifted to energies of order 10–$20\,\mathcal{E}_{aff}$ due to the repulsive Coulomb force between the incident electron and the negative ion. The maximum cross section per valence electron is smaller than the Thomson result (3.4.21), with $\mathcal{E}_{aff}$ replacing $\mathcal{E}_{iz}$, due to the same effect.

Vibrational and Rotational Excitations

Vibrational and rotational energy levels are separated by energies of order $\mathcal{E}_v \sim 0.2$ V and $\mathcal{E}_J \sim 0.01$ V, respectively. Classically, slow electrons are unlikely to excite ground state molecules to higher vibrational or rotational levels because an electron having energy $\mathcal{E}$ transfers an energy $\sim (2m/M)\mathcal{E} \ll \mathcal{E}_v, \mathcal{E}_J$ in an elastic collision with a heavy particle. However, it is found experimentally that there can be significant excitations when $\mathcal{E} \sim \mathcal{E}_v$ or $\mathcal{E}_J$, respectively. For vibrational excitations, the cross sections are generally sharply peaked, indicating that a resonant (electron capture) process is involved. A common mechanism is a two step process in which the electron is first captured by the $v = 0$ AB ground state to form an unstable negative molecular ion:

$$e + AB(v = 0) \rightarrow AB^-$$

The AB^- ion is unstable, but its lifetime for decay (typically 10^{-15}–10^{-10} s) can be comparable to or larger than its vibrational (or autodissociation) timescale (10^{-14} s). Eventually, the unstable negative ion undergoes autodetachment to an excited vibrational state of AB:

$$AB^- \rightarrow AB(v > 0) + e$$

For N_2, the N_2^- ground state is attractive (has a potential energy minimum), lies about 2.3 V above the ground state, and has a lifetime of about 10^{-14} s. Hence the cross section for vibrational excitation of N_2 is strongly peaked about 2.3 V. The maximum cross section is large, about 5×10^{-16} cm^2. For O_2, the $v' = 0$ to $v' = 3$ states of the $^2\Pi_g$ ground state of O_2^- lie below the $v = 0$ ground state of O_2 and do not autodetach. The set of O_2^- states with $v' > 3$ lie above the O_2 ground state and can autodetach. The lifetimes of these states are long; e.g., 10^{-10} s for $v' = 4$. Excitation of these states by electron impact leads to a series of 8–10 peaks for the total vibrational cross section lying between 0.3 and 2.5 V, with the energy-integrated cross section associated with each peak in the range 10^{-19}–10^{-18} cm^2.

Pure rotational excitation by electron impact can be a resonant process as for vibrational excitation, or can be a nonresonant process in which the electron interacts with the permanent dipole moment of the molecule (or with the quadrupole moment for a homonuclear diatomic molecule). Angular momentum is transferred to the molecule; hence the angular momentum of the electron must change. The cross sections for $J \rightarrow J'$ are of order 10^{-18}–10^{-16} cm^2 at energies a few times the rotational energy level difference.

Neither vibrational nor rotational cross sections have been especially well measured or calculated for most molecules. This is unfortunate because electron impact excitations to higher vibrational (and, to a lesser extent, rotational) levels can be an important source of electron energy loss in low-pressure discharges, particularly for the lower range of electron temperatures ($\lesssim 2$ V) in these discharges. We consider these energy losses further in Section 8.5.

Elastic Scattering

Elastic scattering of electrons by atoms was described in Section 3.3. For slow electrons, polarization scattering dominates, and the cross sections typically vary as $1/v$, with v the incident electron velocity, as described by the Langevin cross section (3.3.13). In some cases, however, a relatively constant cross section is found at low energies (see Fig. 3.9). For molecules having a permanent dipole moment, scattering by the resulting $1/r^3$ potential can also be significant, and the Langevin cross section is increased (Su and Bowers, 1973). We consider this process in Section 8.4.

8.4 HEAVY-PARTICLE COLLISIONS

Heavy particle energies in a discharge range from room temperature (~ 0.026 V) for most ions and neutrals in the bulk plasma, to a few volts for ion and neutral fragments newly created by dissociation processes, to hundreds of volts for ions in rf discharge sheaths. In all cases, however, the heavy particle velocities are much smaller than the characteristic velocities of orbital electron motion in an atom or molecule. The time $t_c \sim 2a_0/v_i$ for a collision between two slowly moving heavy particles is $\sim 10^{-13}$ s for room-temperature energies and is 10^{-15}–10^{-14} s for fast moving particles. These times are comparable to the molecular vibration timescale and are much longer than the timescale $t_{at} \sim 10^{-16}$–10^{-15} s for electron motion in the molecule. Hence we have the ordering for heavy-particle collisions,

$$t_{at} \ll t_c \sim \tau_{vib} \ll \tau_{rad} \ll \tau_c$$

where, as previously, τ_{rad} is the timescale for electric dipole radiation and τ_c is the mean free time between collisions. Because $t_{at} \ll t_c$, we expect that as two heavy particles approach each other, the electronic states and their corresponding energy levels will adiabatically vary, in a manner described by the variation of the potential energy with nuclear separation R shown in Fig. 8.1 and in succeeding figures. During a collision, two heavy particles move toward smaller separations along the potential energy curve, reflect at some minimum radius R_{min} corresponding to their center of mass energy, and retrace the incoming trajectory along the same curve to larger separations. This corresponds to an elastic scattering between heavy particles without a change of electronic state.

If two potential energy curves cross or nearly touch at some separation R_x, then a change of electronic state can occur with a very small energy transfer as the collision passes through R_x. A small energy transfer is required classically because the energy transferred by a heavy particle of energy $\mathcal{E}$ to an orbital electron is $\sim (2m/M)\mathcal{E}$, which is much less than the typical energy (1–10 V) required for electronic excitations of the molecule. The condition for a change of state between two electronic energy levels separated by an energy $\Delta\mathcal{E}$ during a heavy-particle collision can be estimated by requiring that the collision time $\sim R_x/v_i$ be shorter than the characteristic time $\sim \hbar/e\Delta\mathcal{E}$ for the orbital electron to change its state:

$$\frac{R_x}{v_i} \lesssim \frac{\hbar}{e\Delta\mathcal{E}} \tag{8.4.1}$$

which yields

$$\Delta\mathcal{E} \lesssim \frac{\hbar v_i}{eR_x} \tag{8.4.2}$$

This is known as the *adiabatic Massey criterion*. In practical units, we find

$$\Delta\mathcal{E} \lesssim \frac{1}{6R'_x} \left(\frac{\mathcal{E}}{A_R}\right)^{1/2} \tag{8.4.3}$$

where $\Delta\mathcal{E}$ is in volts, $\mathcal{E}$ is the center of mass energy in volts, R'_x is the nuclear separation in units of the Bohr radius, and A_R is the reduced mass in atomic mass units (amu).

For example, letting $\mathcal{E} = 1$ V, $A_R = 8$, and $R'_x = 2$, we find that $\Delta\mathcal{E} \lesssim 0.03$ V for a nonadiabatic transition to occur. Hence the states must cross or nearly touch.

Resonant and Nonresonant Charge Transfer

For some processes, such as resonant charge transfer,

$$A^+ + A \rightarrow A + A^+$$

which was described in Section 3.4, the two states have exactly the same energy, such that $\Delta\mathcal{E} \equiv 0$ for all separations. From the present point of view, the transition is very likely even at large separations, leading to a large cross section of the high- or low-energy form (3.4.33) or (3.4.37).

Nonresonant charge transfer between atoms,

$$A^+ + B \rightarrow A + B^+$$

is illustrated in Fig. 8.9 for the reactions between N^+ and O and between O^+ and N. Since the ionization potentials of N and O are 14.53 and 13.61 V, respectively, the separated $N^+ + O$ level is 0.92 V higher than the $N + O^+$ level. At the crossing separation R_x between the attractive $N^+ + O$ and the repulsive $O^+ + N$ level, a change of state corresponding to a transfer of charge can occur. Collision a-x-b in Fig. 8.9 for the exothermic reaction

$$N^+ + O \rightarrow N + O^+$$

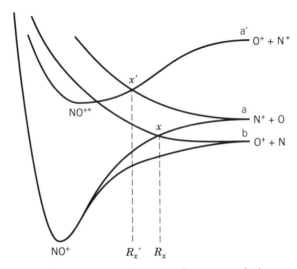

FIGURE 8.9. Illustrating nonresonant charge transfer processes for heavy particle collisions.

does not have a threshold energy, and the N and O^+ products share an increase in kinetic energies of 0.92 V; hence for slow (thermal) collisions of N^+ and O, the charge transfer products are fast. The cross section is of order the resonant cross section (3.4.33) or (3.4.37). The inverse reaction,

$$O^+ + N \rightarrow O + N^+$$

is endothermic with a threshold energy of 0.92 V; hence the rate constant for charge transfer collisions of O^+ and N at thermal energies is very small. However, if either the O^+ ion or the N atom is in an excited atomic state, then the reaction a'-x'-a, e.g.,

$$O^+ + N^* \rightarrow O + N^+$$

has no threshold, and the cross section can be large at thermal energies. Consequently, excited atoms and molecules (particularly metastables) can be important in charge transfer processes.

Similar collisions can occur between atoms and molecules. The ionization potential of O_2 is 12.2 V, so the cross section for the reaction

$$O^+ + O_2 \rightarrow O + O_2^+$$

does not have a threshold and can be expected to be large, while the cross section for the endothermic reverse reaction,

$$O_2^+ + O \rightarrow O_2 + O^+$$

has a threshold energy of 1.4 V; hence it is very unlikely for collisions between thermal particles. As for collisions between atoms, excited O_2^+ and/or O atom charge transfer collisions can have no threshold. In fact, a proper combination of excited electronic and vibrational states can have $\Delta \mathcal{E} \approx 0$, leading to a large (resonant) cross section.

The charge transfer cross section between O_2 molecules,

$$O_2^+ + O_2 \rightarrow O_2 + O_2^+$$

is resonant if the molecules have the same vibrational and rotational states after the collision, but this is not very likely. However, we may expect any energy change due to the change in vibrational and rotational quantum numbers to be small, leading to a near resonant cross section.

Charge transfer processes between negative ions and neutrals can be important in electronegative discharges. For example, in oxygen discharges, we have

$$O_2^- + O \rightarrow O_2 + O^-$$
$$O^- + O_2 \rightarrow O + O_2^-$$
$$O^- + O \rightarrow O + O^-$$
$$O_2^- + O_2 \rightarrow O_2 + O_2^-$$

Since the electron affinities of O_2 and O are 0.43 and 1.463 V, respectively, the first reaction has no threshold energy, while the second reaction has a threshold energy of 1.03 V. Hence we expect a large cross section for the first reaction, but the second reaction is very unlikely for thermal particles. The last two processes are resonant or near resonant and have large cross sections.

Positive-Negative Ion Recombination

This process,

$$A^- + B^+ \rightarrow A + B^*$$

is a type of charge transfer and can be the dominant mechanism for the loss of negative ions in a low-pressure discharge. The potential energy diagram is shown in Fig. 8.10. The separated $A^- + B^+$ state lies below the separated $A + B^+$ state by the electron affinity $\mathcal{E}_{affA}$ of A and lies above the separated $A + B^*$ state. The $A^- + B^+$ potential energy falls as the nuclear separation decreases because of the attractive Coulomb force between the A^- and the B^+ ions. The energy level difference between the separated $A^- + B^+$ and $A + B^*$ states is of order

$$\Delta \mathcal{E} \sim \frac{\mathcal{E}_{izB}}{n^2} - \mathcal{E}_{affA} \tag{8.4.4}$$

where $\mathcal{E}_{izB}$ is the ionization potential of B and n is the principal quantum number of the excited state B^*. For $\mathcal{E}_{affA} \approx 1$ V and $\mathcal{E}_{izB} \approx 14$ V, we find that $\Delta \mathcal{E}$ is small for $n \approx 3$–4. Since $\Delta \mathcal{E}$ can be quite small, the separation R_x at the crossing can be large, and positive–negative ion recombination can have a large near-resonant cross section. A crude classical estimate of σ_{rec} can be found by putting $n \sim 3$–4 in (3.4.28), to obtain

$$\sigma_{rec} \sim 3000\text{--}10000 \; \pi a_0^2 \tag{8.4.5}$$

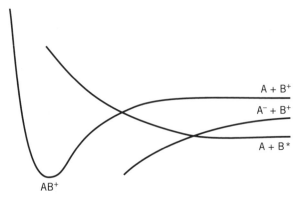

FIGURE 8.10. Illustrating positive-negative ion recombination for heavy particle collisions.

However, this does not expose the energy dependence. To estimate this for attractive Coulomb collisions with $\mathcal{E} \ll \mathcal{E}_{\text{affA}}$, we write conservation of angular momentum and energy during a collision as

$$v_i b = v_{\max} b_0 \tag{8.4.6}$$

$$\frac{1}{2} m_R v_{\max}^2 \approx \frac{e^2}{4\pi\epsilon_0 b_0} \tag{8.4.7}$$

where v_i and b are the initial velocity and impact parameter in the center of mass system, $v_{\max}$ is the velocity at the distance of closest approach b_0, and m_R is the reduced mass. Solving (8.4.6) and (8.4.7) for b, we obtain an estimate for the cross section

$$\sigma_{\text{rec}} \approx \pi b^2 = \pi \frac{e}{4\pi\epsilon_0 \mathcal{E}} b_0 \tag{8.4.8}$$

where $e\mathcal{E} = \frac{1}{2} m_R v_i^2$. We can crudely estimate the value of b_0 for a significant probability of transition to be $b_0 \approx R_x$, where for $\Delta\mathcal{E} \approx \mathcal{E}_{\text{affA}}$,

$$b_0 \approx R_x \approx \frac{e}{4\pi\epsilon_0 \mathcal{E}_{\text{affA}}} \tag{8.4.9}$$

Substituting (8.4.9) in (8.4.8), we obtain

$$\sigma_{\text{rec}} \approx \pi \left(\frac{e}{4\pi\epsilon_0} \right)^2 \frac{1}{\mathcal{E}\mathcal{E}_{\text{affA}}} \tag{8.4.10}$$

We see that $\sigma_{\text{rec}} \propto 1/\mathcal{E}$, where $\mathcal{E}$ is the collision energy in the center of mass system. Hence for collisions between heavy particles at thermal energies, the cross sections are very large. If we put $\mathcal{E} \approx 0.026$ V and $\mathcal{E}_{\text{affA}} \approx 1$ V, then (8.4.10) yields a value of σ_{rec} in the range given by (8.4.5).

Actually, b_0 is more properly determined from a consideration of quantum mechanical electron tunneling. This was done in Section 3.4, to obtain the result (3.4.36) for b_0, which scales as $b_0 \propto \mathcal{E}_{\text{affA}}^{-1/2}$; hence $\sigma_{\text{rec}} \propto \mathcal{E}_{\text{affA}}^{-1/2}$, not $\propto \mathcal{E}_{\text{affA}}^{-1}$, as in (8.4.10). The reader should consult Smirnov (1982) for further details.

Associative Detachment

This process,

$$A^- + B \rightarrow AB + e$$

proceeds by formation of an unstable AB^- state that autodetaches. Figure 8.11a gives a potential energy diagram illustrating this process. At low energies, the collision partners move along path a-b-c of the attractive AB^- state 2, which autodetaches at

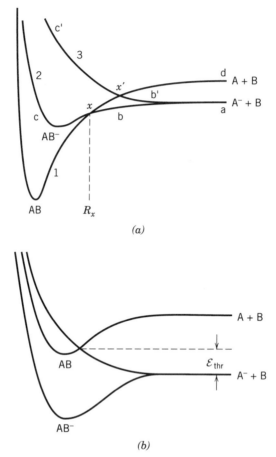

FIGURE 8.11. Illustrating associative detachment processes for heavy particle collisions; (a) the AB^- ground state lies above the AB ground state; (b) the AB^- ground state lies below the AB ground state.

c to the AB ground state 1, often falling into a highly excited vibrational state. If the collision partners follow path a-b'-a along the repulsive AB^- state 2, then there is mainly elastic scattering with little detachment. If the two AB^- states have equal statistical weight, then roughly half the collisions will lead to associative detachment. At higher energies, the path a-b'-c' can result in autodetachment from the repulsive state 3, instead of elastic scattering.

At thermal energies, the interaction between the negative ion and neutral is dominated by the polarization force, and the cross section for associative detachment will tend toward half the Langevin value (for a statistical weight of $\frac{1}{2}$):

$$\sigma_{adet} \approx \frac{1}{2}\sigma_L \tag{8.4.11}$$

where σ_L is given in (3.3.13). At higher energies, where the trajectories are practically straight lines, the cross section will be of order $\frac{1}{2}\pi R_x^2$. Finally, at energies higher than the electron affinity $\mathcal{E}_{affA}$ of A, the process

$$A^- + B \rightarrow A + B + e$$

can occur, as shown by the path a-b'-x'-d, leading to detachment of the electron from A^- by collision with B.

If the AB^- attractive ground state lies below the AB ground state, as shown in Fig. 8.11b, then associative detachment from the ground state of AB^- cannot occur. However, at high energies, $\mathcal{E} > \mathcal{E}_{thr}$, associative detachment from the repulsive AB^- state is possible, and at still higher energies, detachment from A^- due to collision with B can occur.

Associative detachment reactions in an oxygen discharge include

$$O^- + O \rightarrow O_2 + e$$
$$O^- + O_2 \rightarrow O_3 + e$$
$$O_2^- + O \rightarrow O_3 + e$$
$$O_2^- + O_2 \rightarrow O_4 + e \rightarrow 2O_2 + e$$

For oxygen, the O_2^- ground state lies below the O_2 ground state and is stable against autodetachment. However, there are a large number of shallow attractive O_2^- electronic states that lie above the O_2 ground state, and hence are subject to autodetachment. Consequently, there is a large rate constant for associative detachment of O^- on O (the first reaction listed above); at thermal energies, $K_{adet} \sim 5 \times 10^{-10}$ cm^3/s. The importance of the second and third reactions listed above can be understood by noting that ozone (O_3) has a dissociation energy of only 1.04 V. Because the electron affinity of O is 1.463 V, the potential energy diagram for the second reaction is similar to that shown in Fig. 8.11b, and the reaction has a very small rate constant at thermal energies, of order 5×10^{-15} cm^3/s. Since $\mathcal{E}_{aff} \approx 0.43$ V for O_2, the third reaction has a potential energy diagram similar to that shown in Fig. 8.11a, and the rate constant is large at thermal energies, of order 1.5×10^{-10} cm^3/s. The fourth reaction requires a threshold energy equal to the electron affinity of O_2, $\mathcal{E}_{aff} \approx 0.43$ V, and is not very likely at thermal energies.

Transfer of Excitation

Ionization or excitation by impact of ground state atoms or molecules,

$$A + B \rightarrow A^+ + B + e$$
$$A + B \rightarrow A^* + B$$

is improbable because, as we have already seen, the potential energy curve for the A+B state is widely separated from the potential energy curves of the A^++B and

A*+B states. Classically, as noted earlier, only a very small fraction, $\sim 2m/M$, of the initial kinetic energy can be transferred to an orbital electron. However, transfer of energy from an excited electronic state to another excited (or ionized) state can be accomplished if the potential energy curves cross or nearly touch at some nuclear separation R_x. Examples of processes of this type include

$$A + B^* \rightarrow A^+ + B + e$$
$$A + B^* \rightarrow AB^+ + e$$
$$A + B^* \rightarrow A^* + B$$

Some examples of potential energy curves for these processes are given in Fig. 8.12.

The first process is illustrated in Fig. 8.12a. For the reaction to proceed at thermal energies, the excitation energy of B should equal or exceed the ionization potential of A, as shown for the path a-x-d in the figure. When the excited atom is metastable, then this process is known as *Penning ionization*. This is the most important case because the metastable atom density can be significant in many discharges. The most effective

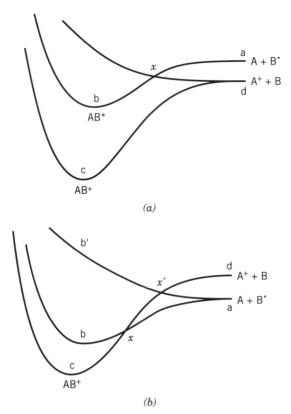

FIGURE 8.12. Illustrating transfer of excitation for heavy particle collisions; (a) Penning ionization; (b) associative ionization.

metastable atom is helium with 19.82 V for the 2^3S state and 20.6 V for the 2^1S state. Because the valence electrons in excited ($n > 1$) states have large radii, $a \sim a_0 n^2$ from (3.4.3), the maximum cross sections can be very large: $\sigma_{max} \sim 10^{-15}$ cm^2 for He(2^3S) ionization of Ar, and $\sim 1.4 \times 10^{-14}$ cm^2 for Hg.

For the second process, called *associative ionization*, to proceed at thermal energies, the sum of the excitation energy of B and the dissociation energy of the AB$^+$ ion should exceed the ionization potential of A, as illustrated in Fig. 8.12b. If the bound AB$^+$ ground state is lower than the AB* bound state, then the path a-x-b has no threshold energy and leads to formation of an unstable AB* molecule that can decay to the ground state at c by electron emission. This leads to associative ionization, which can have a large maximum cross section $\sim 10^{-15}$ cm^2. This path is also possible for the potential energy diagram of Fig. 8.12a. The path a-x'-b' along the repulsive AB* curve, with subsequent decay to the AB$^+$ ground state, can also lead to associative ionization above a fairly high threshold energy. If the bound molecular complex AB* has a very short lifetime, then the electron can be emitted near the points x and x' along the a-x-c and a-x'-c paths, again resulting in associative ionization. Finally, the path a-x'-d leads to production of A$^+$ at a threshold energy equal to the difference between the ionization potential of A and the excitation energy of B.

The third process listed above, *transfer of excitation*, proceeds along the path a-x-d shown in Fig. 8.12a, with replacement of the AB$^+$ ground state by an AB* excited state and replacement of the separated A$^+$+B state by the separated state A*+B. Because there is no emitted electron to carry away the excess energy, the process is highly resonant. The energy uncertainty of the A+B* and A*+B levels is of order $\hbar / \tau_{rad}$, where τ_{rad} is the lifetime of the excited states, and the excitation energies of the A* and B* states must coincide to within this uncertainty. An important example of transfer occurs in the He–Ne gas laser, where the transfers

$$\text{He}(^1S) + \text{Ne} \rightarrow \text{He} + \text{Ne}(5s)$$

$$\text{He}(^3S) + \text{Ne} \rightarrow \text{He} + \text{Ne}(4s)$$

are near resonant, resulting in a population inversion for the 4s and 5s levels of neon and subsequent laser action.

Rearrangement of Chemical Bonds

Exothermic chemical reactions between ions and neutrals of the form

$$\text{AB}^+ + \text{CD} \rightarrow \text{AC}^+ + \text{BD}$$

$$\rightarrow \text{ABC}^+ + \text{D}$$

$$\rightarrow \text{etc.}$$

result in rearrangements of chemical bonds. For thermal collisions, the collision is dominated by the polarization force, and the maximum rate constant for reactions of this type might be expected to be the Langevin value (3.3.17). However, the thermal rate constants are often considerably smaller than this, indicating that the collision complex does not live long enough to allow for efficient bond rearrangement. An exception occurs for exothermic *proton abstraction* processes,

$$AH^+ + B \rightarrow BH^+ + A$$

which have rate constants close to the Langevin value for thermal collisions. An example of bond rearrangement in oxygen discharges is the exothermic reaction

$$O^+ + O_3 \rightarrow O_2^+ + O_2$$

which has a rate constant of $\sim 10^{-10}$ cm^3/s.

Exothermic neutral–neutral bond rearrangements,

$$AB + CD \rightarrow AC + BD$$
$$\rightarrow ABC + D$$
$$\rightarrow \text{etc.}$$

generally have rate constants $\sim 10^{-11}$ cm^3/s, one or two orders of magnitude smaller than the Langevin value. The maximum cross sections are of order the gas kinetic value $\pi(a_1 + a_2)^2$, where a_1 and a_2 are the mean radii of the reactants. Generally, even exothermic reactions are impeded by energy barriers, such that many such reactions have an Arrhenius form

$$K(T) = K_0 \exp\left(-\frac{\mathcal{E}_a}{T}\right) \tag{8.4.12}$$

with the preexponential factor K_0 and the activation energy $\mathcal{E}_a$ roughly independent of temperature T. An example in oxygen discharges is

$$O + O_3 \rightarrow 2O_2$$

with $K \approx 2 \times 10^{-11} \exp(-0.2/T)$ cm^3/s.

Ion-Neutral Elastic Scattering

If the molecular ion has a permanent dipole moment p_d, then the polarization scattering and the Langevin capture cross section are increased due to the additional interaction potential $U \propto p_d/r^2$. The increase in the Langevin rate constant has been calculated by Su and Bowers (1973) for thermal collisions of ions and neutrals, with the result

$$K = \left(\frac{\pi q^2}{\epsilon_0 m_R}\right)^{1/2} \left[\alpha_p^{1/2} + C p_d \left(\frac{2}{\pi k T}\right)^{1/2}\right] \tag{8.4.13}$$

where the first term in square brackets gives the Langevin rate constant (3.3.15) and the second term gives the increase due to the permanent dipole moment. The quantity C is a parameter between 0 and 1 that describes the effectiveness of the charge "locking" in the dipole, and is a function of T and $p_d/\alpha_p^{1/2}$ alone. At $T = 300$ K, C is plotted against $p_d/\alpha_d^{1/2}$ in Fig. 8.13.

Three-Body Processes

We have said little in this and the previous section about three-body reactions such as electron–ion recombination

$$e + A^+ (+e) \rightarrow A (+e)$$

attachment

$$e + A (+M) \rightarrow A^- (+M)$$

association

$$A^+ + B (+M) \rightarrow AB^+ (+M)$$

and positive-negative ion recombination

$$A^- + B^+ (+M) \rightarrow AB (+M)$$

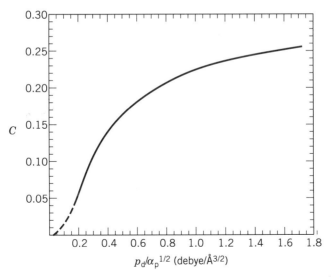

FIGURE 8.13. A plot of the dipole locking constant C; 1 debye $\approx 3.34 \times 10^{-30}$ C-m (Su and Bowers, 1973).

Here, A or B can be any atom or molecule, and M can be any atom or molecule including A or B. In most cases, for the densities of interest in low pressure materials processing discharges, these processes proceed by a series of two body reactions in which the third body (shown in parentheses for the reactions listed above) absorbs the excess reaction energy. For low densities of the third body, the equivalent two-body rate constants (cm^3/s) for three-body processes are proportional to the density of the third body, and are generally smaller than the rate constants for two-body processes. We will show this and consider other aspects of three-body processes in Chapter 9.

8.5   REACTION RATES AND DETAILED BALANCING

As described in Section 3.5, the cross sections must be averaged over the energy distributions of the colliding particles in order to determine the reaction rates. For a general reaction of A and B particles,

$$A + B \rightarrow \text{products}$$

the number of A and B particles reacting per unit volume per unit time is

$$\frac{dn_A}{dt} = \frac{dn_B}{dt} = K_{AB} n_A n_B \tag{8.5.1}$$

where the two-body rate constant K_{AB} is a function of the particle energy distributions but is independent of their densities.

We described the averaging required for electron collisions with heavy particles in Section 3.5. Here we consider the case of heavy-particle collisions. If A and B are unlike particles (of different species) that each have a Maxwellian distribution with a common temperature T, then the averaging yields

$$K_{AB}(T) = \langle \sigma_{AB} v_R \rangle = \int_0^\infty f_m v_R \sigma_{AB}(v_R) 4\pi v_R^2 \, dv_R \tag{8.5.2}$$

where

$$f_m = \left(\frac{m_R}{2\pi kT} \right)^{3/2} \exp\left(-\frac{m_R v_R^2}{2kT} \right) \tag{8.5.3}$$

and m_R is the reduced mass. If A and B are like particles, e.g., for the collision of two ground-state oxygen atoms, then

$$K_{AA}(T) = \frac{1}{2} \langle \sigma_{AA} v_R \rangle = \frac{1}{2} \int_0^\infty f_m v_R \sigma_{AA}(v_R) 4\pi v_R^2 \, dv_R \tag{8.5.4}$$

The reason for the factor of $\frac{1}{2}$ in (8.5.4) can be understood by numbering the A and B particles within a unit volume. For unlike particles, the collisions of A1 with B2 and A2 with B1 are different collisions, whereas for like particles, the collisions of A1 with A2 and A2 with A1 are the same collision and must not be counted twice.

Temperature Dependence

For thermal collisions ($T \sim 300$ K) with a constant cross section σ_0 near zero velocity, as for hard-sphere collisions, the averages in (8.5.2) and (8.5.4) are easily done, yielding

$$K_{AB} = \sigma_0 \bar{v}_R \tag{8.5.5}$$

$$K_{AA} = \frac{1}{2}\sigma_0 \bar{v}_R \tag{8.5.6}$$

where $v_R = (8kT/\pi m_R)^{1/2}$. Hence K_{AB} and K_{AA} vary weakly as $\sqrt{T}$. For the polarization interaction, with $\sigma \propto 1/v_R$, we have already seen for the Langevin rate constant (3.3.15) that K is independent of T.

Consider now a process that has a threshold energy $\mathcal{E}_{thr}$. The variation of the cross section with energy near the threshold can be estimated from conservation of angular momentum and energy,

$$v_R b = v_\theta b_0 \tag{8.5.7}$$

$$e\mathcal{E} = \frac{1}{2}m_R v_R^2 \approx \frac{1}{2}m_R v_\theta^2 + e\mathcal{E}_{thr} \tag{8.5.8}$$

where b_0 is the effective radius for the reaction and v_θ is the angular component of the velocity. The influence of the interaction potential has been neglected in (8.5.8). Solving (8.5.7) for v_θ, substituting this into (8.5.8), and solving for $\sigma \approx \pi b^2$, we obtain

$$\begin{aligned} \sigma &= 0 & \mathcal{E} < \mathcal{E}_{thr} \\ &= \sigma_0\left(1 - \frac{\mathcal{E}_{thr}}{\mathcal{E}}\right) & \mathcal{E} > \mathcal{E}_{thr} \end{aligned} \tag{8.5.9}$$

where $\sigma_0 = \pi b_0^2$. We see that the cross section rises linearly just above the threshold energy and tends to a maximum value σ_0 for large $\mathcal{E}$. The rise is linear rather than abrupt because the centrifugal energy $\frac{1}{2}m_R v_\theta^2$ is not available to excite the reaction. Many cross sections display this linear rise.

Inserting (8.5.9) into either (8.5.2) or (8.5.4) and integrating, we obtain

$$K_{AB} = \sigma_0 \bar{v}_R\, e^{-\mathcal{E}_{thr}/T} \tag{8.5.10}$$

$$K_{AA} = \frac{1}{2}\sigma_0 \bar{v}_R\, e^{-\mathcal{E}_{thr}/T} \tag{8.5.11}$$

respectively, which have an Arrhenius form, with the preexponential factor varying weakly as $\sqrt{T}$.

The Principle of Detailed Balancing

The cross sections and rate constants for forward and reverse reactions are related by the principle of detailed balancing, which expresses the time reversibility of the equations of motion for a collision. Hence, knowledge of the cross section for a two-body reaction allows one to determine the properties of the reverse reaction. The cross section $\sigma(v_R)$ for the inelastic reaction (endothermic with threshold energy $\mathcal{E}_a$),

$$A + B \rightarrow C + D$$

is related to the cross section $\sigma'(v_R')$ for the reverse reaction,

$$C + D \rightarrow A + B$$

by (Smirnov, 1981, Appendix A2)

$$m_R^2 g_A g_B v_R^2 \sigma(v_R) = m_R'^2 g_C g_D v_R'^2 \sigma'(v_R') \tag{8.5.12}$$

where

$$\frac{1}{2} m_R v_R^2 = \frac{1}{2} m_R' v_R'^2 + e\mathcal{E}_a \tag{8.5.13}$$

m_R and m_R' are the reduced masses for particles A and B, and C and D, respectively, and the g's are the degeneracies of the energy levels of the particles; for example, $g_e = 2$ for a free electron (the two spin states have the same energy), and $g_O = 5$ for the O($^3P_2$) ground state (the five m_J values $2, 1, 0, -1, -2$, have the same energy). We can integrate (8.5.12) over a Maxwellian distribution of v_R to obtain (Problem 8.9)

$$\frac{K(T)}{K'(T)} = \left(\frac{m_R'}{m_R} \right)^{3/2} \frac{\bar{g}_C \bar{g}_D}{\bar{g}_A \bar{g}_B} e^{-\mathcal{E}_a/T} \tag{8.5.14}$$

which expresses the ratio of the rate constants for the forward and reverse reactions in terms of a ratio of reduced masses and energy level degeneracies times a Boltzmann factor. We have written $\bar{g}$ rather than g in (8.5.14) because we are generally more interested in the rate constants for a group of closely spaced energy levels for each particle, rather than for a single level. For example, we specify the ground state of an oxygen atom as O(^3P), which comprises three closely spaced levels: the $^3P_1$ and $^3P_0$ levels lie 0.020 and 0.028 V above the $^3P_2$ level, respectively. We can apply (8.5.14) to this case if we interpret the g's as $\bar{g}$'s, the *statistical weights*, or mean number of occupied states, for the group of levels.

The ratio of statistical weights can be evaluated by assuming that the A, B, C, and D particles are all in thermal equilibrium at temperature T. Generally, for an atom or atomic ion somewhat above room temperature, the electronic states within the fine structure of a group of energy levels are all occupied; consequently, $\bar{g}_{at}$ is equal to the total degeneracy g_{at} of the group of levels. For example, the O(^{3}P) ground-state triplet has five states for $^3P_2$, three states for $^3P_1$, and one state for $^3P_0$, for a total degeneracy $\bar{g}_O = 9$. At room temperature and below, $\bar{g}_O < g_O$ because the upper levels do not have a high probability of being occupied (Problem 8.11). Typically, $\bar{g}_{at} \sim 1\text{--}10$ for ground-state atoms or atomic ions.

For molecules at thermal energies (0.026 V) and above, in addition to the electronic degeneracy $\bar{g}_{at}$ of the molecular level, many rotational states and some vibrational states can be occupied. The energy of a molecule in a vibrational–rotational state (v, J) above the $(0, 0)$ ground state is, summing the vibrational and rotational energies in (8.2.1) and (8.2.4),

$$e\mathcal{E} = \hbar\omega_{vib}\left(v + \frac{1}{2}\right) + eB_{rot}J(J + 1)$$

where ω_{vib} is the vibrational frequency and $B_{rot} = \hbar^2/2eI_{mol}$ is the rotational energy constant of the molecule. In thermal equilibrium, the mean number of levels occupied for a heteronuclear diatomic molecule can be shown to be (Problem 8.12)

$$\bar{g}_{rot}\bar{g}_{vib} = \frac{T}{B_{rot}} \frac{1}{1 - e^{-\hbar\omega_{vib}/eT}} \qquad (8.5.15)$$

For a homonuclear diatomic molecule, $\bar{g}_{rot}$ must be divided by two because the two states with the molecule rotated by 180° are identical. For polyatomic molecules, $\bar{g}_{vib}$ consists of a product of factors, one for each vibrational degree of freedom. The statistical weight of the molecule is then $\bar{g}_{mol} = \bar{g}_{at}\bar{g}_{vib}\bar{g}_{rot}$. For typical diatomic molecules at room temperature, $\bar{g}_{mol} \sim 10^2\text{--}10^3$.

Although the statistical weights in (8.5.14) are determined for thermal equilibrium, the ratio of statistical weights is the same for a system that is not in thermal equilibrium. The only assumption required is that the distribution of v_R (and, consequently, v_R') be Maxwellian. This is because each rate constant in (8.5.14) depends only on the collision dynamics (the cross section) and the assumed velocity distribution (a Maxwellian). Consequently, the ratio of rate constants must be the same whether or not the particles are in thermal equilibrium.

As will be shown in Section 9.1, (see (9.1.13)), the right-hand side of (8.5.14) is the equilibrium constant $\mathcal{K}(T)$, as given in (7.4.7), for the reaction of A+B to form C+D. Writing the Gibbs free energy of reaction, G_r°, in terms of the enthalpy and entropy of reaction using the definition of G (7.3.14) and substituting this into the expression for $\mathcal{K}$ (7.4.7), we obtain

$$\frac{K(T)}{K'(T)} = \mathcal{K}(T) = e^{-G_r^\circ/RT} = e^{S_r^\circ/R}e^{-H_r^\circ/RT} \qquad (8.5.16)$$

The terms exponential in S_r° and H_r° on the right-hand side of (8.5.16) are equal, in (8.5.14), to the product of mass and statistical weight factors, and to the exponential energy factor, respectively. If G_r° is known, then K' can be determined if K is known, and vice versa. This relationship will be elaborated in Chapter 9. Let us note some examples where (8.5.14) can be applied. The rate constant for deexcitation of an excited state

$$A + B^* \rightarrow A + B$$

can be determined from the rate constant for collisional excitation of that state:

$$A + B \rightarrow A + B^*$$

Here A can be an electron, atom, or molecule, and B can be an atom or molecule. The rate constant for associative ionization

$$A + B^* \rightarrow AB^* + e$$

can be determined from the rate constant for dissociative recombination

$$e + AB^* \rightarrow A + B^*$$

Relations similar to (8.5.14) can be found for reactions that change the number of particles, such as

$$e + A \rightarrow e + e + A^+$$
$$AB + M \rightarrow A + B + M$$

These relations connect the two-body rate constants to the three-body rate constants for the reverse reactions.

Finally, let us note that detailed balancing is not as useful to determine rate constants as might first be imagined, because the "forward" and "reverse" reactions of interest may not actually be inverses. For example, electron excitation to B^*

$$e + B \rightarrow B^* + e$$

often proceeds by a compound process of excitation to a higher level or set of levels, followed by radiative decay:

$$e + B \rightarrow B^{*2} + e$$
$$B^{*2} \rightarrow B^* + \hbar\omega$$

The reverse reaction of interest might be direct deexcitation of B^* to the ground state:

$$e + B^* \rightarrow B + e.$$

These two processes are not inverses, and are not connected by detailed balancing. Similarly, excitation of a molecule

$$e + AB \rightarrow AB^* + e$$

may be to a high vibrational state $v' \gg 0$, while deexcitation

$$e + AB^* \rightarrow AB + e$$

is from the ground vibrational state $v' = 0$. The reader should consult other sources (e.g., Smirnov, 1981) for further discussion of these methods for the determination of rate constants.

A Data Set for Oxygen

To illustrate the complexity of molecular processes, we give some data for oxygen, which is a simple diatomic gas that has been particularly well studied. This data set will be used throughout this book to illustrate various features of chemically reactive discharges. In an oxygen discharge, there can be significant ground-state concentrations of O, O_2, O_3, O^+, O_2^+, O_4^+, O_3^-, O_2^-, O^-, and electrons, as well as metastable states such as the ^1D and ^1S states of O and the $^1\Delta_g$ and $^1\Sigma_g^+$ states of O_2. Some basic constants for some of these species are given in Table 8.1. The cross sections for binary processes among these species have mostly not been carefully measured or calculated. To give an example of some of the best data, some cross sections for electron impact excitation of O_2, useful for determining the energy losses, are given in Fig. 8.14. These include momentum transfer, rotational and vibrational excitation, two- and three-body attachment, $^1\Delta_g$ and $^1\Sigma_g^+$ metastable excitation, excitations to states involving energy losses of approximately 4.5, 6.0, 8.4, 10.0, and 14.7 V, and ionization with an energy loss of 12.06 V. The momentum transfer cross section is also given. The identification of the energy losses with specific processes such as dissociation, attachment, etc. is uncertain. Using these data, the energy loss $\mathcal{E}_c$ per e–O_2^+ pair created in oxygen has been determined and plotted in Fig. 3.17.

TABLE 8.1. Basic Constants for Oxygen Discharges

State	$\mathcal{E}_{diss}$ (V)	$\mathcal{E}_{iz}$ (V)	Lifetime (s)	α_p (a_0^3)
$O(^3P)$	—	13.61	—	5.4
$O^-(^2P)$	—	1.463	—	—
$O^*(^1D)$	—	11.64	147.1	—
$O_2(^3\Sigma_g^-)$	5.12	12.14	—	10.6
$O_2^+(^2\Pi_g)$	6.59	—	—	—
$O_2^-(^2\Pi_g)$	4.06	0.44	—	—
$O_2^*(^1\Delta_g)$	4.14	11.16	2700	—
O_3	1.05	12.67	—	—
O_3^-	1.69	2.10	—	—

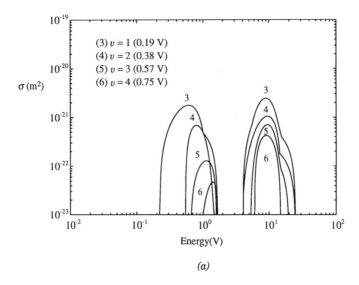

(a)

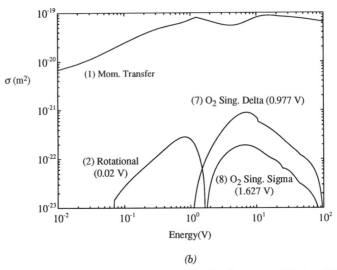

(b)

FIGURE 8.14. Cross sections for electron excitation of O_2 (Lawton and Phelps, 1978; Phelps, 1985; compiled by Vahedi, 1994).

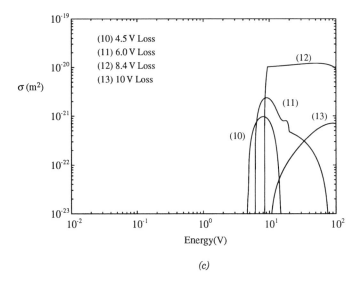

(c)

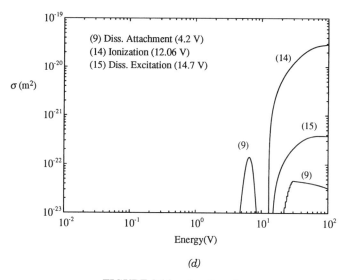

(d)

FIGURE 8.14.   (*continued*)

Similar cross-section sets have been compiled for electron collisions in many reactive gases of interest for materials processing by Hayashi (1987).

Table 8.2 gives some rate constants for a restricted set of two body reactions of interest in modeling low-pressure oxygen discharges. These include reactions among ground states O, O_2, O_3, O^+, O_2^+, O^-, O_2^-, O_3^-, and electrons, and metastable states $O^*(^1D)$ and $O_2^*(^1\Delta_g)$. Electrons are assumed to have a Maxwellian distribution in the range $1 < T_e < 7$ V, and the heavy particles are assumed to be Maxwellian at a common temperature T near room temperature 0.026 V. A first set of reactions is given involving just the species O, O_2, O_2^+, O^-, and electrons, because these often suffice for the simplest discharge models. Additional sets of reactions give added complexity as additional species are added to the model. A key task of the modeler is to choose the set of reactions appropriate to the parameter range of interest.

Finally, Table 8.3 gives some rate constants for three-body reactions. These processes are described in Section 9.2.

8.6   OPTICAL EMISSION AND ACTINOMETRY

Optical diagnostics are powerful tools for the noninvasive measurement of the properties of chemically complex discharges. A wide variety of optical diagnostic techniques are currently in use. A relatively simple technique is that in which the wavelength-resolved optical emission is measured. More complex and expensive schemes, such as laser-induced fluorescence (LIF) and optogalvanic techniques, in which laser beam probes are used to excite specific optical transitions whose subsequent emission or other response is measured, have also been widely used. Infrared emission and absorption techniques are also receiving increasing attention. We refer the reader to review articles by Donnelly (1989), Manos and Dylla (1989), and Selwyn (1993), and research articles referenced therein, for a detailed exposition of the subject.

In this section we discuss the simplest technique of optical emission and actinometry (defined below) to illustrate the usefulness of optical diagnostics. Small variations in discharge operation due to contamination, aging, vacuum leaks, etc., can produce large changes in emission. Hence process reproducibility is often monitored, and even actively controlled, by measurement of emission. Detection of the endpoint for a materials process, particularly an etch, is also conveniently accomplished using optical emission. In this case, an emission line associated with an etch product can be monitored; a sharp decrease in the emission intensity versus time generally signals the completion of the etch process. Optical emission can be spatially resolved perpendicular to the line of sight, but generally is a spatial average along the line of sight. As will be shown below, the emission intensity is a convolution of the species density for the optical wavelength being monitored, the electron distribution function, and the cross section for electron impact excitation of the optical level. For example, a qualitative measure of the importance of F atoms can be obtained by monitoring the 7037-Å F-atom emission line as the discharge power and pressure are varied. A quantitative measure of relative F-atom density can be found by using a tracer gas, such as argon, and measuring the intensity of both an F-atom and an Ar-atom

TABLE 8.2. Selected Second-Order Reaction Rate Constants for Oxygen Discharges

Number	Reaction	Rate Constant (cm^3/s)	Source
Reactions among e, O_2, O_2^+, and O^-			
1	$e + O_2$ momentum transfer	$4.7E{-}8T_e^{0.5}$	a
2	$e + O_2 \rightarrow O^- + O$	$8.8E{-}11\exp(-4.4/T_e)$	b
3	$e + O_2 \rightarrow 2O + e$	$4.2E{-}9\exp(-5.6/T_e)$	d
4	$e + O_2 \rightarrow O_2^+ + 2e$	$9.0E{-}10T_e^{0.5}\exp(-12.6/T_e)$	d
5	$e + O^- \rightarrow O + 2e$	$2.0E{-}7\exp(-5.5/T_e)$	b
6	$e + O_2^+ \rightarrow 2O$	$5.2E{-}9/T_e$	k
7	$O^- + O_2^+ \rightarrow O + O_2$	$(0.96, 2)E{-}7(300/T)^{0.5}$	e,k
8	$O^- + O \rightarrow O_2 + e$	$(1.4, 5)E{-}10$	e,k
9	$O^- + O_2^+ \rightarrow 3O$	$1E{-}7$	k
Addition of O^+			
10	$e + O_2 \rightarrow O^- + O^+ + e$	$7.1E{-}11T_e^{0.5}\exp(-17/T_e)$	r
11	$e + O_2 \rightarrow O + O^+ + 2e$	$5.3E{-}10T_e^{0.9}\exp(-20/T_e)$	b
12	$e + O \rightarrow O^+ + 2e$	$9.0E{-}9T_e^{0.7}\exp(-13.6/T_e)$	d
13	$O^- + O^+ \rightarrow 2O$	$(2.7, 2)E{-}7(300/T)^{0.5}$	e,k
14	$O^+ + O_2 \rightarrow O + O_2^+$	$2.0E{-}11(300/T)^{0.5}$	e
Addition of metastable $O_2^*(^1\Delta_g)$; see note f below			
15	$e + O_2 \rightarrow O_2^* + e$	$1.7E{-}9\exp(-3.1/T_e)$	b
16	$e + O_2^* \rightarrow e + O_2$	$5.6E{-}9\exp(-2.2/T_e)$	b
17	$O_2^* + O_2 \rightarrow 2O_2$	$2.2E{-}18(T/300)^{0.8}$	ke
18	$O_2^* + O \rightarrow O_2 + O$	$(1.0, 7)E{-}16$	e,k
Addition of metastable $O(^1D)$			
19	$e + O_2 \rightarrow O + O^* + e$	$5.0E{-}8\exp(-8.4/T_e)$	d
20	$e + O \rightarrow O^* + e$	$4.2E{-}9\exp(-2.25/T_e)$	d
21	$e + O^* \rightarrow e + O$	$8E{-}9$	d
22	$e + O^* \rightarrow O^+ + 2e$	$9.0E{-}9T_e^{0.7}\exp(-11.6/T_e)$	d
23	$O^* + O \rightarrow 2O$	$8.0E{-}12$	e
24	$O^* + O_2 \rightarrow O + O_2$	$(6.4, 7.0)E{-}12\exp(67/T)$	k,e
25	$O^* + O_2 \rightarrow O + O_2^*$	$1.0E{-}12$	e
Addition of selected reactions for O_2^- and O_3			
26	$O^- + O_2 \rightarrow O_3 + e$	$5E{-}15$	k
27	$e + O_3 \rightarrow O_2^- + O$	$1E{-}9$	k
28	$O^- + O_2^* \rightarrow O_3 + e$	$3E{-}10$	k
29	$O^- + O_2^* \rightarrow O_2^- + O$	$1E{-}10$	k
30	$O_2^- + O_2^+ \rightarrow 2O_2$	$2E{-}7(300/T)^{0.5}$	k
31	$O_2^- + O^+ \rightarrow O_2 + O$	$2E{-}7(300/T)^{0.5}$	k
32	$O_3 + O_2 \rightarrow O_2 + O + O_2$	$7.3E{-}10\exp(-11400/T)$	e
33	$O_3 + O \rightarrow 2O_2$	$1.8E{-}11\exp(-2300/T)$	e

Note. T_e in volts and T in kelvins. Two values from different sources are sometimes given in parentheses. The notation E–8 means 10^{-8}.

[a] Based on cross sections of Phelps (1985).

[b] Based on data of Eliasson and Kogelschatz (1986).

[d] Based on data of Lee et al. (1994).

[e] Eliasson and Kogelschatz (1986).

[f] Reactions 1–4, 10–11 for O_2^* have activation energies reduced by ~ 1 V.

[k] Kossyi et al.(1992).

[r] Based on Rapp and Briglia (1965).

TABLE 8.3. Selected Third-Order Reaction Rate Constants for Oxygen Discharges

Number	Reaction	Rate Constant (cm^6/s)	Source
Reactions among e, O_2, O_2^+, *and* O^-			
1	$e + e + O_2^+ \rightarrow e + O_2$	$1E{-}19(0.026/T_e)^{4.5}$	ke
2	$e + O_2^+ + O_2 \rightarrow O_2 + O_2$	$6E{-}27(0.026/T_e)^{1.5}$, $1E{-}26$	k,e
3	$e + O + O_2 \rightarrow O^- + O_2$	$1E{-}31$	ke
4	$O^- + O_2^+ + O_2 \rightarrow O + O_2 + O_2$	$2E{-}25(300/T)^{2.5}$	k
5	$O + O + O_2 \rightarrow O_2 + O_2$	$2.45E{-}31T^{-0.63}$	k
		$1.3E{-}32(300/T)\exp(-170/T)$	e
6	$O + O + O \rightarrow O_2 + O$	$6.2E{-}32\exp(-750/T)$	e
Addition of O^+			
7	$e + e + O^+ \rightarrow e + O$	$1E{-}19(0.026/T_e)^{4.5}$	ke
8	$e + O^+ + O_2 \rightarrow O + O_2$	$6E{-}27(0.026/T_e)^{1.5}$, $1E{-}26$	k,e
9	$O^- + O^+ + O_2 \rightarrow O_2 + O_2$	$2E{-}25(300/T)^{2.5}$, $2E{-}25$	k,e
10	$O^- + O^+ + M \rightarrow O + O + M$	$2E{-}25(300/T)^{2.5}$	k
11	$O^+ + O + O_2 \rightarrow O_2^+ + O_2$	$1E{-}29$	ke
Addition of metastable $O(^1D)$			
12	$O + O^* + O_2 \rightarrow O_2 + O_2$	$9.9E{-}33$	e
Addition of selected reactions for metastable $O_2^*(^1\Delta_g)$, O_2^-, *and* O_3			
13	$e + O_2 + O_2 \rightarrow O_2^- + O_2$	$1.4E{-}29(0.026/T_e)$ $\times \exp(100/T - 0.061/T_e)$	k
14	$e + O_2 + O \rightarrow O_2^- + O$	$1E{-}31$	k
15	$O^- + O_2^+ + O_2 \rightarrow O_3 + O_2$	$2E{-}25(300/T)^{2.5}$	k,e
16	$O + O_2 + O_2 \rightarrow O_3 + O_2$	$6.9E{-}34(300/T)^{1.25}$, $6.4E{-}35\exp(663/T)$	k,e
17	$O + O_2 + O \rightarrow O_3 + O$	$2.15E{-}34\exp(345/T)$	e
18	$e + O_2^* + O_2 \rightarrow O_2^- + O_2$	$1.9E{-}30$	e
19	$e + O_2^* + O \rightarrow O_2^- + O$	$1E{-}31$	e
20	$O_2^- + O^+ + M \rightarrow O_3 + M$	$2E{-}25(300/T)^{2.5}$	e
21	$O_2^- + O_2^+ + O_2 \rightarrow O_2 + O_2 + O_2$	$2E{-}25(300/T)^{2.5}$	e

Note. T_e in volts and T in kelvins; M denotes either O_2 or O. Two values from different sources are sometimes given. The notation E–19 means 10^{-19}.

^{e}Eliasson and Kogelschatz (1986).

^{k}Kossyi et al.(1992).

emission line. This widely used comparison technique is called *optical actinometry*. More sophisticated measurements, at finer wavelength resolution, can be used to determine ion and neutral energies. Time-resolved emission measurements can be used to determine both volume and surface rate constants.

Optical Emission

Figure 8.15 illustrates the electron impact excitation of the ground state of atom A to an excited state A^*, followed by subsequent emission at frequency ω to some lower energy state A_f. The emission wavelength is

$$\lambda = \frac{2\pi c}{\omega} \tag{8.6.1}$$

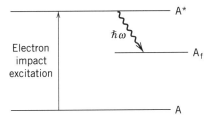

FIGURE 8.15.   Energy level diagram for emission of radiation from an excited state.

where $\hbar\omega = e(\mathcal{E}_{A^*} - \mathcal{E}_f)$. The usual wavelengths are in the optical band, $\lambda \sim 2000 -$ 8000 Å. The emission is sharply peaked about λ, with a small intrinsic linewidth due to the spontaneous emission rate from level A^*, $\Delta\lambda^* \sim 10^{-3}$ Å. The Doppler-broadened linewidth due to a velocity distribution of ions or neutrals is wider,

$$\frac{\Delta\lambda}{\lambda} = \frac{v_{\text{thi}}}{c} \tag{8.6.2}$$

For 1-V argon atoms, $\Delta\lambda \sim 0.025$ Å.

Let n_A be the concentration of the free radical A and let I_λ (in watts) be the optical emission intensity, integrated over the linewidth. The emission due to excitation from the ground state A can be written as

$$I_\lambda = \alpha_{\lambda A} n_A \tag{8.6.3}$$

where

$$\alpha_{\lambda A} = k_D(\lambda) \int_0^\infty 4\pi v^2 dv \, Q_{A^*}(p, \, n_e)\sigma_{\lambda A}(v)vf_e(v) \tag{8.6.4}$$

Here f_e is the electron distribution function, $\sigma_{\lambda A}$ is the cross section for emission of a photon of wavelength λ due to electron impact excitation of A, Q_{A^*} is the quantum yield for photon emission from the excited state ($0 \leq Q_{A^*} \leq 1$), and k_D is the detector response constant. For low-pressure discharges and excited states having short lifetimes, $Q_{A^*} \approx 1$. Q_{A^*} is generally less than unity for metastable states, due to collisional or electric field deexcitation, ionization, or other processes that depopulate the state without emission of a photon. We note that the cross section $\sigma_{\lambda A}$ differs from the cross section σ_{A^*} for excitation of A to level A^*, because spontaneous emission to more than one lower lying level can occur. The two cross sections are related by

$$\sigma_{\lambda A} = b_\lambda \sigma_{A^*} \tag{8.6.5}$$

where b_λ is the *branching ratio* for emission of a photon of wavelength λ from the excited state A^*.

Typically $\sigma_{\lambda A}$ is known but f_e is not; i.e., f_e is not generally a single-temperature Maxwellian. As discharge parameters (pressure, power, driving frequency, length) are varied, f_e changes shape as shown in Chapter 11, Fig. 11.10. In particular, the high-energy tail of the distribution, near the excitation energy E_{A^*}, can vary strongly as discharge parameters are changed. Consequently $\sigma_{\lambda A}$ changes and I_λ given by (8.6.3) is not proportional to n_A. This limits the usefulness of a measurement of I_λ, which provides only qualitative information on the radical density n_A.

Optical Actinometry

An inert tracer gas of known concentration n_T can be added to the feedstock to provide quantitative information on the radical density n_A. We choose an excited state T^* of the tracer T that has nearly the same excitation threshold energy, $\mathcal{E}_{T^*} \approx \mathcal{E}_{A^*} \approx \mathcal{E}_*$. The cross sections $\sigma_{\lambda A}(v)$ and $\sigma_{\lambda' T}(v)$ for photon emission of λ (from A) and λ' (from T) are sketched in Fig. 8.16. A typical form for the multiplicative factor $v^3 f_e(v)$ in the integrand of (8.6.4) is also shown, with the overlap shown as the shaded area. For the tracer gas,

$$I_{\lambda'} = \alpha_{\lambda' T} n_T \tag{8.6.6}$$

with

$$\alpha_{\lambda' T} = k_D(\lambda') \int_0^\infty 4\pi v^2 dv\, Q_{T^*}(p, n_e)\sigma_{\lambda' T}(v)v f_e(v) \tag{8.6.7}$$

Since, from Fig. 8.16 there is only a small range of overlap of f_e with σ, we can replace the cross sections with values near the threshold: $\sigma_{\lambda' T} \approx C_{\lambda' T}(v - v_{thr})$ and $\sigma_{\lambda A} \approx C_{\lambda A}(v - v_{thr})$, where the C's are proportionality constants. We then take the ratio of (8.6.3) and (8.6.6) to obtain

$$n_A = C_{AT}\, n_T \frac{I_\lambda}{I_{\lambda'}} \tag{8.6.8}$$

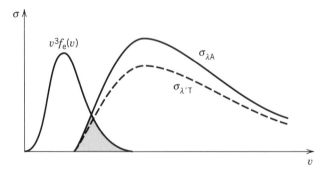

FIGURE 8.16. Overlap of excitation cross sections and electron velocity distribution.

where

$$C_{AT} = \frac{k_D(\lambda)}{k_D(\lambda')} \frac{Q_{A^*}}{Q_{T^*}} \frac{C_{\lambda A}}{C_{\lambda' T}} \qquad (8.6.9)$$

It is often possible to choose $\lambda' \approx \lambda$ such that $k_D(\lambda) \approx k_D(\lambda')$, and also to choose $Q_{A^*} \approx Q_{T^*}$. Hence the constant of proportionality $C_{AT} \approx C_{\lambda A}/C_{\lambda' T}$ is related to the threshold behavior of the two cross sections. If n_T is known and I_λ and $I_{\lambda'}$ are measured, an absolute value of n_A can be determined. Even if C_{AT} is not known, the relative variation of n_A with variation of discharge parameters can be found. For F-atom actinometry, a common choice for the tracer gas is argon with $\lambda' = 7504$ Å; the cross section has a threshold energy of 13.5 V. For F atoms, $\lambda = 7037$ Å is commonly chosen, with a threshold energy of 14.5 V. Typically, n_T is chosen to be 1–5%, of the feedstock gas density.

O Atom Actinometry

To illustrate both the utility and the pitfalls of optical actinometry, we consider O atoms with argon as the tracer gas. Figure 8.17 shows data (Walkup et al., 1986) for n_O for an O_2/CF_4 feedstock mix with 2–3% argon added as a tracer gas. The data were taken in a 13.56-MHz capacitive rf discharge. The oxygen radical density n_O was determined actinometrically using O atom emission at two different wavelengths, $\lambda = 7774$ Å ($3p^5P \rightarrow 3s^5S$ transition) and $\lambda = 8446$ Å ($3p^3P \rightarrow 3s^3S$ transition), each ratioed to the argon emission at wavelength $\lambda' = 7504$ Å. The actinometric measurements were compared with a more accurate (and much more expensive)

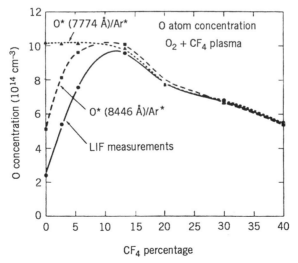

FIGURE 8.17. Comparison of actinometric measurements with a two photon LIF measurement of oxygen atom density in an O_2/CF_4 discharge (Walkup et al., 1986).

determination of n_O using two photon LIF (see Walkup et al., 1986). It can be seen that the 8446/7504-Å actinometric measurement tracks the two photon LIF measurement fairly well as the %CF$_4$ is varied. However, the 7774/7504-Å measurement yields a saturation of n_O rather than a decrease as the CF$_4$ concentration is lowered below 20%, contrary to the LIF measurement.

To understand this behavior, we first note that emission of a photon of wavelength λ can occur due to processes other than excitation from the ground state A. For example, the *dissociative excitation* process

$$e + O_2 \rightarrow O + O^* + e \rightarrow 2O + e + \hbar\omega \tag{8.6.10}$$

can compete with the direct excitation process

$$e + O \rightarrow O^* + e \rightarrow O + e + \hbar\omega \tag{8.6.11}$$

such that the measured emission intensity

$$I_\lambda = \alpha_{\lambda O} n_O + \alpha_{\lambda O_2} n_{O_2}, \tag{8.6.12}$$

has a component proportional to the feedstock density n_{O_2} as well as the radical density n_O. The actinometric measurement of n_O will fail if $\alpha_{\lambda O} n_O \lesssim \alpha_{\lambda O_2} n_{O_2}$, which is the case for the 7774 Å measurement.

Using a high-resolution monochrometer or spectrometer, the radiation due to direct and dissociative excitation can be distinguished. Because dissociative excitation generally results in excited neutral fragments having many volts of energy, the radiation is Doppler broadened according to (8.6.2) and can therefore be distinguished from the much sharper linewidth for radiation produced by direct excitation of a room temperature atom. Subtracting the emission intensity in the broadened tail from the total intensity allows the intensity due to direct excitation alone to be determined. However, other processes can also increase or decrease I_λ. These include radiative transitions from higher-energy excited states to A*, electron impact excitation of metastable states to A*, and collisional and electric field quenching of A*. These can invalidate an actinometric measurement unless the optical transition and discharge operating regime have been selected to minimize their effects.

PROBLEMS

8.1. Vibration and Dissociation of H$_2$

(a) By fitting $\mathcal{E}_e$ for the H$_2(^1\Sigma_g^+)$ ground state in Fig. 8.3 to a parabolic function of $R - \bar{R}$ and using (8.2.2) and (8.2.3), estimate the spring constant k_{vib} and the vibration period $\tau_{vib} = 2\pi/\omega_{vib}$.

(b) From the potential energy curve for the H$_2(^3\Sigma_u^+)$ repulsive state in Fig. 8.3, estimate the timescale τ_{diss} for dissociation of the molecule after electron impact excitation to this excited state.

(c) For excitation of $H_2(^3\Sigma_u^+)$ from the ground vibrational state of $H_2(^1\Sigma_g^+)$, estimate the threshold energy for dissociation and the minimum and maximum energies of the dissociated H atoms.

8.2. Metastable Molecular States In order of increasing energy, the five lowest attractive states of O_2 are $^3\Sigma_g^-$, $^1\Delta_g$, $^1\Sigma_g^+$, $^3\Delta_u$, and $^3\Sigma_u^+$ (see Fig. 8.4). Which of these states are metastable? (give the reasons).

8.3. Dissociation Cross Section for O_2

(a) Using (8.3.2) and the potential energy curves in Fig. 8.4, estimate the cross section $\sigma_{diss}(\mathcal{E})$ for electron impact dissociation of O_2 at the equilibrium nuclear separation $\bar{R}$ to form ground state O atoms. Assume that the dissociation results from direct excitation of the repulsive $^1\Pi$ and $^3\Pi$ energy level curves, and do not average over the vibrational motion. Plot $\sigma_{diss}(\mathcal{E})$ versus $\mathcal{E}$ using linear scales.

(b) Approximating $\sigma_{diss}(\mathcal{E})$ by

$$\sigma_{diss} = 0 \qquad\qquad \mathcal{E} < \mathcal{E}_{thr}$$
$$= \sigma_{max}\mathcal{E}_{thr}/\mathcal{E} \qquad \mathcal{E} > \mathcal{E}_{thr}$$

then integrate $\sigma_{diss}(\mathcal{E})$ over a Maxwellian electron distribution (T_e in the range 2–7 V) to determine the rate constant $K_{diss}(T_e)$. Compare your result to that given in Table 8.2.

8.4. Dissociative Attachment of O_2

(a) For dissociative attachment to a single molecular level having $\mathcal{E}_{att} \sim 4$ V, estimate the rate constant $K_{att}(T_e)$ for T_e's in the range of 2–7 V by integrating (8.3.7) over a Maxwellian electron distribution.

(b) Suppose $\mathcal{E}_{att}$ varies linearly with nuclear separation R over the range of ground-state vibrational motions

$$\mathcal{E}_{att}(R) = \bar{\mathcal{E}}_{att} + \Delta\mathcal{E}_{thr}x$$

where $x = (R - \bar{R})/\Delta R$ has a Gaussian distribution

$$f(x) = \frac{e^{-x^2}}{\sqrt{\pi}}$$

Average (8.3.7) over the vibrational motion and plot your result for $\bar{\sigma}_{att}$ versus $\mathcal{E}$ for $\bar{\mathcal{E}}_{att} = 4$ V, $\Delta\mathcal{E}_{thr} = 1$ V, and $\Delta\mathcal{E}_{att} = 0.2$ V. On the same graph, plot σ_{att} from (8.3.5) with $\mathcal{E}_{att} = 4$ V.

(c) Using detailed balancing (8.5.14), estimate the rate constant for associative detachment

$$O^- + O \to O_2 + e$$

using your result in (a). You will need to use (8.5.15) to estimate the statistical weight of O_2; $\hbar\omega_{vib}/e \approx 0.192$ V and $B_{rot} \approx 1.79 \times 10^{-4}$ V for O_2.

8.5. Polar Ionization of O_2   Interpreting the second (higher energy) peak in Fig. 8.8 as the cross section for polar ionization of O_2, estimate the rate constant for this process for T_e's in the range 2–7 V by fitting the cross section in the energy region above threshold to the form (8.5.9) and then using (8.5.10). Compare your result to that given in Table 8.2.

8.6. Positive Charge Transfer in O_2 Discharges   For thermal ($T \sim$ room temperature) ground state-particles:

(a) Estimate the reaction rate constant for the resonant reaction

$$O^+ + O \rightarrow O + O^+$$

using (3.4.37) and the data in Table 3.2.

(b) Estimate the reaction rate constant for the near-resonant reaction

$$O_2^+ + O_2 \rightarrow O_2 + O_2^+$$

using (3.4.37) and the data in Table 3.2.

(c) Estimate the reaction rate constant for the exothermic (1.4-V) reaction

$$O^+ + O_2 \rightarrow O + O_2^+$$

using (3.4.37) and the data in Table 3.2.

(d) The reaction

$$O_2^+ + O \rightarrow O_2 + O^+$$

has a threshold energy of 1.4 V. Estimate the reaction rate constant using detailed balancing (8.5.14) and your result in (c). To simplify the calculations, you may assume that ω_{vib} and B_{rot} are the same for both molecules and that the fine structure of the atoms is equally occupied. Note that the ground state of O^+ is ^4S.

8.7. Negative Charge Transfer in O_2 Discharges   For thermal ($T \sim$ room temperature) ground-state particles:

(a) Estimate the reaction rate constant for the resonant reaction

$$O^- + O \rightarrow O + O^-$$

using (3.4.37) and the data in Table 3.2.

(b) Estimate the reaction rate constant for the near-resonant reaction

$$O_2^- + O_2 \rightarrow O_2 + O_2^-$$

using (3.4.37) and the data in Table 3.2.

(c) Estimate the reaction rate constant for the exothermic (1.0-V) reaction

$$O_2^- + O \rightarrow O_2 + O^-$$

using (3.4.37) and the data in Table 3.2.

(d) The reaction

$$O^- + O_2 \rightarrow O + O_2^-$$

has a threshold energy of 1.0 V. Estimate the reaction rate constant using detailed balancing (8.5.14) and your result in (c). To simplify the calculations, you may assume that ω_{vib} and B_{rot} are the same for both molecules and that the fine structure of the atoms is equally occupied.

8.8. Positive–Negative Ion Recombination   For thermal particles at temperature T (near room temperature), estimate the rate constant for the reaction

$$O_2^+ + O^- \rightarrow O_2^* + O$$

by integrating the classical cross section (8.4.10) over a Maxwellian distribution of relative velocities. Compare your answer (both magnitude and scaling with T) with that given in Table 8.2.

8.9. Detailed Balancing   For a Maxwellian distribution of relative velocities v_R, integrate the relation (8.5.12) for detailed balancing of the cross sections for forward and reverse reactions using the energy conservation relation (8.5.13), to obtain the relation (8.5.14) for detailed balancing between the rate constants.

8.10. Application of Detailed Balancing

(a) For a Maxwellian electron distribution at temperature T_e, the direct electron collisional excitation of an atom B having statistical weight $\bar{g}_B$ to an excited state having energy $\mathcal{E}$ and statistical weight $\bar{g}_*$ is measured to have an Arrhenius form $K_{ex} = K_0 \exp(-\mathcal{E}_a/T_e)$, where $\mathcal{E}_a \neq \mathcal{E}$ is the activation energy. Using detailed balancing, find the rate constant K_q for quenching (electron collisional deexcitation) of B* to the ground state B.

(b) Apply your formula to determine the rate constant for

$$e + O(^1D) \rightarrow O(^3P) + e$$

using the data in Tables 8.1 and 8.2. Compare your result to that given in Table 8.2.

(c) If $\mathcal{E}_a$ is markedly different from $\mathcal{E}$, then is your result in (a) correct? Explain your answer.

8.11. Statistical Weights

(a) The $^3P_1$ and $^3P_0$ levels of an oxygen atom lie at energies 0.020 and 0.028 V above the $^3P_2$ ground-state level. Assuming that the probability that a level is occupied is given by a Boltzmann factor $e^{-\mathcal{E}/T_e}$, find the statistical weight of $O(^3P)$ at room temperature (0.026 V) and at twice room temperature.

(b) The ground-state of N and O^+ is ^4S. Find the statistical weight if all levels in the fine structure are equally occupied.

(c) The ground-state level of fluorine and chlorine atoms is $^2P_{3/2}$; the $^2P_{1/2}$ levels lie 0.050 and 0.109 V above the ground state, respectively. Find the statistical weights of $F(^2P)$ and $Cl(^2P)$ at room temperature.

(d) The ground-state level of an argon atom is $^1S_0$. Find its statistical weight.

(e) The vibrational and rotational energy constants for $O_2(^3\Sigma_g^-)$, $O_2^+(^2\Pi_g)$, and $O_2^-(^2\Pi_g)$ are $\hbar\omega_{vib}/e = 0.196, 0.236$, and 0.136 V and $B_{rot} = 1.79 \times 10^{-4}$, 2.09×10^{-4}, and 1.45×10^{-4} V, respectively. Find the statistical weights of these molecules at room temperature (0.026 V).

8.12. Statistical Weight for Molecules

(a) Show that

$$\bar{g}_{vib} = \frac{1}{1 - \exp(-\hbar\omega_{vib}/eT)}$$

by summing the probability $\exp(-\hbar\omega_{vib}v/eT)$ over the $v = 0$ to $v = \infty$ vibrational levels.

(b) Show that at temperatures $T \gg B_{rot}$, the mean number of rotational states occupied is $\bar{g}_{rot} = T/B_{rot}$ by summing the probability $\exp[-B_{rot}J(J + 1)/T]$ over the $J = 0$ to $J = \infty$ levels. *Hint*: Convert the sum over J to an integral over dJ, and recall that the degeneracy of level J is $2J + 1$.

8.13. Negative Ions in an O_2 Discharge   Negative ions in a discharge are generally created and lost only through processes in the plasma volume because the plasma potential is positive with respect to all wall surfaces; hence, the negative ions are electrostatically trapped. Use the rate constants given in Table 8.2 to perform the following:

(a) For an oxygen discharge containing room temperature O_2, O_2^+ and O^- and electrons at temperature T_e, obtain the condition on T_e for dissociative attachment to dominate over polar ionization for production of O^- by electron impact on O_2.

(b) Show that O_2^+–O^- recombination dominates over electron detachment for destruction of O^-.

CHAPTER 9

CHEMICAL KINETICS AND SURFACE PROCESSES

9.1 ELEMENTARY REACTIONS

In this chapter we describe aspects of gas-phase and surface chemical kinetics that are important to materials processing. We first introduce the concept of *elementary reactions*, give the definition of the appropriate rate constants, and show their connection to the equilibrium constants for the reactions. Section 9.2 deals with gas-phase kinetics. We introduce first-, second-, and third-order kinetics, and the concept of a rate-limiting step. Although some examples of time-varying kinetics are given, the main applications are to the steady state. Third-order kinetics are described with emphasis on three-body recombination and three-body chemical reactions which, at the low pressures of interest, can often be considered to be a series of two or more one- or two- body reactions. In Sections 9.3 and 9.4 we turn to surface processes and reaction kinetics. The various physical and chemical processes of interest for processing are described in Section 9.3. Section 9.4 deals with heterogeneous reactions on the surface and between the surface and the gas phase. The surface reaction mechanisms for most plasma processes are not well understood or characterized experimentally. Some simple models of surface reactions are introduced, but these, for the most part, should not be regarded as correctly representing the actual plasma induced reactions at substrate surfaces. Rather, they are intended to provide some insight into the more complicated processes that go on in actual surface processing.

Consider stoichiometric reactions such as

$$3A + 2B \rightarrow C + 2D \tag{9.1.1}$$

$$A + B \rightarrow C + D \tag{9.1.2}$$

$$A \rightarrow B + C \tag{9.1.3}$$

etc., where A, B, C, and D are molecules. A reaction is called *elementary* if it proceeds in one step directly as written, i.e., in a simultaneous "collision" of all the reactant molecules for (9.1.1) and (9.1.2), or by a single "decomposition" for (9.1.3). The first reaction is not elementary because it is very unlikely for five particles to simultaneously collide. The second and third reactions might or might not be elementary. If two reactant molecules A and B collide to immediately ($\Delta t \sim t_c$) form two product molecules C and D, then the reaction is elementary. An example from Chapter 8 is

$$O^+ + O_2 \rightarrow O + O_2^+$$

Similarly, if an A molecule suddenly decomposes, then the reaction is elementary. An example is

$$A^* \rightarrow A + \hbar\omega$$

On the other hand, the reaction

$$Cl_2 + H_2 \rightarrow 2HCl$$

having the form (9.1.2), is known not to be elementary. There is no way of knowing from the stoichiometric equations (9.1.2) or (9.1.3) whether a reaction is elementary; additional information is needed. A significant effort in chemical kinetics has been to determine the set of elementary reactions into which a given stoichiometric reaction can be decomposed.

The most important elementary reactions are *unimolecular*

$$A \rightarrow products$$

and *bimolecular*

$$A + B \rightarrow products$$

At high pressures, some *termolecular* gas-phase reactions

$$A + B + C \rightarrow products$$

are elementary; however, in low-pressure discharges, almost all gas-phase termolecular reactions with significant reaction rates are complex.

The reaction rate R for a gas-phase reaction is defined in terms of the stoichiometric coefficients α_j for the reaction introduced in Section 7.4. Recall that these are negative for reactants and positive for products. We define R as

$$R = \frac{1}{\alpha_j}\frac{dn_j}{dt}, \qquad \text{for all } j \tag{9.1.4}$$

where n_j is the volume density (m^{-3}) of molecules of the jth substance. For (9.1.1), for example, this yields

$$R = -\frac{1}{3}\frac{dn_A}{dt} = -\frac{1}{2}\frac{dn_B}{dt} = \frac{dn_C}{dt} = \frac{1}{2}\frac{dn_D}{dt}$$

For surface reactions, n_j is replaced by the area density n_j' (m^{-2}) on the surface. In general, R is a complicated function of the n_j's of the reactants. However, for elementary reactions, R has the following simple forms:

$$A \rightarrow \text{products}$$

$$R = -\frac{dn_A}{dt} = K_1 n_A \tag{9.1.5}$$

$$A + A \rightarrow \text{products}$$

$$R = -\frac{1}{2}\frac{dn_A}{dt} = K_2 n_A^2 \tag{9.1.6}$$

$$A + B \rightarrow \text{products}$$

$$R = -\frac{dn_A}{dt} = -\frac{dn_B}{dt} = K_2 n_A n_B \tag{9.1.7}$$

$$A + A + A \rightarrow \text{products}$$

$$R = -\frac{1}{3}\frac{dn_A}{dt} = K_3 n_A^3 \tag{9.1.8}$$

$$A + A + B \rightarrow \text{products}$$

$$R = -\frac{1}{2}\frac{dn_A}{dt} = -\frac{dn_B}{dt} = K_3 n_A^2 n_B \tag{9.1.9}$$

$$A + B + C \rightarrow \text{products}$$

$$R = -\frac{dn_A}{dt} = -\frac{dn_B}{dt} = -\frac{dn_C}{dt} = K_3 n_A n_B n_C \tag{9.1.10}$$

The quantities K_1 (s^{-1}), K_2 (m^3/s), and K_3 (m^6/s) are the *first-* , *second-* , and *third-order rate constants*. They are functions of temperature but are independent of the densities.

Relation to Equilibrium Constant

Let us consider the two opposing elementary reactions

$$A + B \overset{K_2}{\underset{K_{-2}}{\rightleftharpoons}} C + D$$

The rate at which C is created by the forward reaction is $K_2 n_A n_B$, and the rate at which C is destroyed by the reverse reaction is $K_{-2} n_C n_D$. In thermal equilibrium (reactants and products at temperature T), the rates must balance:

$$K_2 \bar{n}_A \bar{n}_B = K_{-2} \bar{n}_C \bar{n}_D$$

or

$$\frac{K_2(T)}{K_{-2}(T)} = \frac{\bar{n}_C \bar{n}_D}{\bar{n}_A \bar{n}_B} \tag{9.1.11}$$

But from the condition for thermal equilibrium (7.4.6), we find

$$\frac{\bar{n}_C \bar{n}_D}{\bar{n}_A \bar{n}_B} = \mathcal{K}(T) \tag{9.1.12}$$

Substituting this into (9.1.11), we obtain

$$\frac{K_2(T)}{K_{-2}(T)} = \mathcal{K}(T) \tag{9.1.13}$$

Although (9.1.13) was derived for thermal equilibrium between A, B, C, and D, it is also true for a system that is not in thermal equilibrium. The only requirement is that the distribution of relative velocities of the colliding particles be Maxwellian at temperature T. As was noted in Section 8.5, this is because the rate constants K_2 and K_{-2} depend only on the reactant particle collision dynamics and the relative velocity distribution. Therefore, (9.1.13) gives an important relation between the rate constants for the forward and reverse reactions. If the equilibrium constant is known, then K_{-2} can be determined if K_2 is known, and vice versa.

The relation (9.1.13) is just another form of detailed balancing (8.5.14), which was described in Section 8.5 from the point of view of microscopic two-body collision dynamics. However, detailed balancing holds for all opposing pairs of elementary reactions, as is obvious from the derivation presented here. Thus, for the opposing reactions

$$A \overset{K_1}{\underset{K_{-2}}{\rightleftharpoons}} B + C$$

we find

$$\frac{K_1(T)}{K_{-2}(T)} = \frac{\bar{n}_B \bar{n}_C}{\bar{n}_A} = \mathcal{K}(T) \tag{9.1.14}$$

and for

$$A + B \underset{K_{-3}}{\overset{K_2}{\rightleftharpoons}} C + D + E$$

we find

$$\frac{K_2(T)}{K_{-3}(T)} = \frac{\bar{n}_C \bar{n}_D \bar{n}_E}{\bar{n}_A \bar{n}_B} = \mathcal{K}(T) \tag{9.1.15}$$

etc.

9.2 GAS-PHASE KINETICS

Materials processing reactions in the gas phase are almost never elementary, but consist of a complex set of opposing, consecutive, and parallel reactions. For example, for F-atom etching of silicon in a CF_4 discharge, F atoms are created and destroyed by consecutive opposing reactions of the form

$$e + CF_x \rightleftharpoons CF_{x-1} + F + e, \qquad x = 1, 2, 3, 4$$

Most processing is done in steady state; i.e., the processing time is long compared to the reaction or transport times for the gas-phase species of interest. In steady state, there is a constant flow of feedstock gas and a constant discharge power, and the gas-phase species are continuously pumped away or deposited on surfaces. In steady state, all gas-phase densities are constant, independent of time. However, these densities cannot be determined from equilibrium thermodynamics because the system is not in thermal equilibrium. If the reaction rate constants (K's) are known, then the densities can be found by solving the rate equations for particle conservation for each species. Since the reaction set is often very complex, the set of rate equations must generally be solved numerically. However, insight can be developed by considering simplified reaction sets under both time-varying and steady-state conditions, which we do here.

A complete self-consistent discharge model cannot be developed without considering the full set of particle and energy conservation equations. As will be shown in Chapter 10, the full set of equations determines not only the particle densities, but also the electron temperature, and hence the self-consistent rate constants for the discharge equilibrium, which are, in many cases, functions of the electron temperature.

First-Order Consecutive Reactions

Consider the consecutive time-varying first-order reactions

$$A \xrightarrow{K_A} B \xrightarrow{K_B} C \tag{9.2.1}$$

with no sources or sinks. The rate equations are

$$\frac{dn_A}{dt} = -K_A n_A \tag{9.2.2}$$

$$\frac{dn_B}{dt} = K_A n_A - K_B n_B \tag{9.2.3}$$

$$\frac{dn_C}{dt} = K_B n_B \tag{9.2.4}$$

We let $n_A = n_{A0}$ and $n_B = n_C = 0$ at $t = 0$. Then (9.2.2) can be integrated to obtain

$$n_A = n_{A0}\, e^{-K_A t} \tag{9.2.5}$$

Substituting this into (9.2.3) and integrating, we obtain

$$n_B = n_{A0} \frac{K_A}{K_B - K_A} \left(e^{-K_A t} - e^{-K_B t} \right) \tag{9.2.6}$$

This procedure can be repeated to find $n_C(t)$ by substituting (9.2.6) into (9.2.4) and integrating. However, summing (9.2.2)–(9.2.4) yields

$$n_A(t) + n_B(t) + n_C(t) = \text{const} = n_{A0}. \tag{9.2.7}$$

Substituting (9.2.5) and (9.2.6) into (9.2.7), we obtain

$$n_C = n_{A0} \left[1 + \frac{1}{K_A - K_B} \left(K_B e^{-K_A t} - K_A e^{-K_B t} \right) \right] \tag{9.2.8}$$

The time variation of the densities is sketched in Fig. 9.1 for the two cases of (a) $K_A \ll K_B$ and (b) $K_B \ll K_A$. For (a), we obtain the approximate variation

$$n_C = n_{A0} \left(1 - e^{-K_A t} \right) \tag{9.2.9}$$

For case (b), after a short initial transient time $t \sim K_A^{-1}$, we obtain the approximate variation

$$n_C = n_{A0} \left(1 - e^{-K_B t} \right) \tag{9.2.10}$$

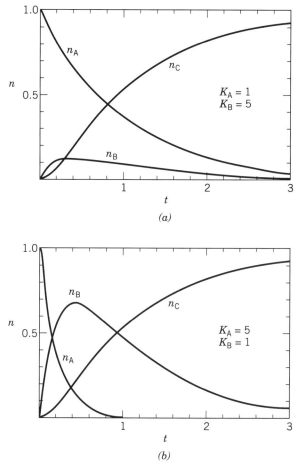

FIGURE 9.1. Transient kinetics for gas phase reaction A → B → C; (a) $K_A = 1$, $K_B = 5$; (b) $K_A = 5$, $K_B = 1$.

In both cases, the rate of formation of the product species C is governed by the *smallest* rate constant. In general, for a series of many consecutive elementary reactions, the reaction with the smallest rate constant limits the overall rate of product formation. The consecutive reaction with the smallest rate constant is called the *rate-limiting step*.

In case (a) ($K_B \gg K_A$), species B is created from A at a slow rate K_A and is immediately converted into C. Hence we should expect that after a short transient time n_B decays with t at a rate K_A, such that $n_B \ll n_A$ and that $dn_B/dt \sim K_A n_B \ll K_A n_A$ at all times. Therefore, dn_B/dt can be set to zero in (9.2.3) to obtain the approximate solution

$$n_A = n_{A0}\, e^{-K_A t}$$

$$n_B \approx \frac{K_A}{K_B} n_{A0}\, e^{-K_A t} \qquad (9.2.11)$$

$$n_C \approx n_{A0}\left(1 - e^{-K_A t}\right)$$

Species B is known as a *reactive intermediate*, and setting $dn_B/dt \approx 0$ is known as the *steady-state approximation for reactive intermediates*.

In case (b) ($K_A \gg K_B$), A creates B before B creates C. Hence there are, approximately, two *uncoupled* first-order reactions having solutions

$$n_A = n_{A0}\, e^{-K_A t}$$

$$n_B \approx n_{A0}\left(1 - e^{-K_A t}\right) \qquad (9.2.12)$$

$$n_C \approx 0$$

for $0 < t < \bar{t}$, and

$$n_A \approx 0$$

$$n_B \approx n_{A0}\, e^{-K_B t} \qquad (9.2.13)$$

$$n_C \approx n_{A0}\left(1 - e^{-K_B t}\right)$$

for $t > \bar{t}$, where $\bar{t} = (K_A K_B)^{-1/2}$ is the characteristic time that divides the fast and slow timescales. The fast reaction in which A is first converted to B is known as a *preequilibrium reaction* for the formation of the product C.

For reaction (9.2.1) in the steady state with a source G ($m^{-3}\text{-}s^{-1}$) for A, and adding a loss term $-K_C n_C$ for C, the rate equations become

$$\frac{dn_A}{dt} = G - K_A n_A = 0$$

$$\frac{dn_B}{dt} = K_A n_A - K_B n_B = 0 \qquad (9.2.14)$$

$$\frac{dn_C}{dt} = K_B n_B - K_C n_C = 0$$

Here, K_C could represent a first order rate constant for loss of C to the surfaces or to the vacuum pump. Solving these equations yields $n_A = G/K_A$, $n_B = G/K_B$, and $n_C = G/K_C$.

Opposing Reactions

Consider the two opposing steady-state reactions

$$A \underset{K_{-A}}{\overset{K_A}{\rightleftharpoons}} B \underset{K_{-B}}{\overset{K_B}{\rightleftharpoons}} C \tag{9.2.15}$$

with $n_A = n_{A0}$ and no sources or sinks. Then

$$\frac{dn_A}{dt} = -K_A n_A + K_{-A} n_B = 0 \tag{9.2.16}$$

$$\frac{dn_B}{dt} = K_A n_A - K_{-A} n_B - K_B n_B + K_{-B} n_C = 0 \tag{9.2.17}$$

$$\frac{dn_C}{dt} = K_B n_B - K_{-B} n_C = 0 \tag{9.2.18}$$

with the solution

$$n_B = \frac{K_A}{K_{-A}} n_{A0} = \bar{n}_B$$

and

$$n_C = \frac{K_B}{K_{-B}} n_B = \bar{n}_C$$

which are the solutions in thermal equilibrium. For opposing elementary reactions with no sources or sinks, the thermal equilibrium solutions must be obtained. However, now consider (9.2.15) with a source G for A and an added first-order loss $-K_C n_C$ for C. Solving (9.2.16)–(9.2.18) under these conditions, we find

$$\frac{n_B}{n_A} = \frac{K_A}{K_{-A} + \dfrac{K_B K_C}{K_{-B} + K_C}}$$

$$\frac{n_C}{n_B} = \frac{K_B}{K_{-B} + K_C} \tag{9.2.19}$$

with $n_C = G/K_C$. We see that n_B/n_A and n_C/n_B are both depressed below their thermal equilibrium values (K_A/K_{-A} and K_B/K_{-B} respectively) by the presence of the source and sink. This situation holds for most low-pressure processing discharges; i.e., the species densities are not in thermal equilibrium.

Bimolecular Association With Photon Emission

Consider the association reaction

$$A + B \longrightarrow AB \qquad (9.2.20)$$

On a molecular level, this reaction cannot occur because energy and momentum cannot be simultaneously conserved in the collision (see Problem 3.12). However, there are many examples known of such stoichiometric reactions; e.g., the associative attachment

$$e + SF_6 \longrightarrow SF_6^-$$

mentioned in Section 8.3.

To understand how a reaction like (9.2.20) can arise, let us note that the molecular reaction

$$A + B \xrightarrow{K_2} AB^* \qquad (9.2.21)$$

can occur, leading to an unstable molecular state. If energy is not taken from AB^*, then it immediately dissociates:

$$AB^* \xrightarrow{K_{-1}} A + B \qquad (9.2.22)$$

One possible mechanism for loss of energy from AB^* is photon emission. This suggests the complex reaction

$$A + B \underset{K_{-1}}{\overset{K_2}{\rightleftharpoons}} AB^* \xrightarrow{K_1} AB + \hbar\omega \qquad (9.2.23)$$

for production of AB. The steady-state rate equations are

$$\frac{dn_A}{dt} = \frac{dn_B}{dt} = -K_2 n_A n_B + K_{-1} n_{AB^*} + G = 0$$

$$\frac{dn_{AB^*}}{dt} = K_2 n_A n_B - K_{-1} n_{AB^*} - K_1 n_{AB^*} = 0 \qquad (9.2.24)$$

$$\frac{dn_{AB}}{dt} = K_1 n_{AB^*} - K_{1w} n_{AB} = 0$$

where, to obtain a steady state, a net input source G and a first-order loss term for AB having rate constant K_{1w} have been added, with $G = K_{1w} n_{AB} = K_1 n_{AB^*}$. The solution of (9.2.24) is

$$n_{AB^*} = \frac{K_2}{K_{-1} + K_1} n_A n_B \qquad (9.2.25)$$

Hence the rate of production of n_{AB} has the form

$$R = K_1 n_{AB^*} = \frac{K_1 K_2}{K_{-1} + K_1} n_A n_B \tag{9.2.26}$$

of a second-order elementary reaction (9.1.7) with rate constant

$$K_2' = \frac{K_1 K_2}{K_{-1} + K_1} \tag{9.2.27}$$

However, this reaction is not elementary.

We can estimate K_2' from the rate constants in (9.2.24). The characteristic time for dissociation of an unstable AB^* molecular state was found in Chapter 8 to be 10^{-13}–10^{-12} s, so that $K_{-1} \approx 10^{12}$–10^{13} s^{-1}. The radiative lifetime for electric dipole radiation was found to be 10^{-9}–10^{-8} s, so that $K_1 \approx 10^{8}$–10^{9} s^{-1}. Using these estimates in (9.2.27), we find $K_2' \approx 10^{-5}$–$10^{-3} K_2$. Therefore, the rate constant for the association reaction (9.2.20) due to photon emission is small; consequently, such reactions are usually not important in low-pressure discharges.

Three-Body Association

A second mechanism for the association reaction (9.2.20) is collision with a third body,

$$A + B + M \longrightarrow AB + M \tag{9.2.28}$$

Here M can be A or B or any other molecule in the system. However, simultaneous collisions of three bodies are very rare at low pressures. This suggests the complex reaction

$$A + B \underset{K_{-1}}{\overset{K_2}{\rightleftharpoons}} AB^* \tag{9.2.29}$$

$$AB^* + M \overset{K_{2M}}{\longrightarrow} AB + M \tag{9.2.30}$$

The rate equations are

$$\frac{dn_A}{dt} = -K_2 n_A n_B + K_{-1} n_{AB^*} + G = 0$$

$$\frac{dn_{AB^*}}{dt} = K_2 n_A n_B - K_{-1} n_{AB^*} - K_{2M} n_{AB^*} n_M = 0 \tag{9.2.31}$$

$$\frac{dn_{AB}}{dt} = K_{2M} n_{AB^*} n_M - K_{1w} n_{AB} = 0$$

with the solution

$$G = K_{1w}n_{AB} = K_{2M}n_{AB^*}n_M$$

$$n_{AB^*} = \frac{K_2}{K_{-1} + K_{2M}n_M}n_An_B \tag{9.2.32}$$

Hence the rate of production of n_{AB} has the form

$$R = K_{2M}n_{AB^*}n_M = \frac{K_2K_{2M}n_M}{K_{-1} + K_{2M}n_M}n_An_B \tag{9.2.33}$$

This rate depends in a complicated way on the third-body density. In the low- and high-pressure limits we find

$$R = \frac{K_2K_{2M}}{K_{-1}}n_An_Bn_M \qquad K_{2M}n_M \ll K_{-1} \tag{9.2.34a}$$

$$= K_2n_An_B \qquad K_{2M}n_M \gg K_{-1} \tag{9.2.34b}$$

Therefore, at low pressure, reaction (9.2.28) looks like an elementary three-body reaction:

$$A + B + M \xrightarrow{K_3'} AB + M \tag{9.2.35}$$

with rate constant

$$K_3' = \frac{K_2K_{2M}}{K_{-1}} \tag{9.2.36}$$

The equivalent second-order rate constant K_2' for the reaction

$$A + B \xrightarrow{K_2'} AB \tag{9.2.37}$$

at low pressures is then

$$K_2' = K_2\frac{K_{2M}n_M}{K_{-1}} \tag{9.2.38}$$

Consider the ratio K_2'/K_2 for neutral particle collisions at thermal energies (300 K). From Section 8.4 we have the estimate $K_{2M} \approx \sigma_{el}\bar{v} \approx 10^{-11}-10^{-10}$ cm^3/s, and $K_{-1} \approx 10^{12}-10^{13}$ s^{-1} for dissociation of the unstable AB* molecule. Hence from (9.2.38), we have

$$K_2' \approx (10^{-24} - 10^{-22})\,n_M\,K_2$$

where n_M is in cm^{-3}. At $p = 1$ Torr, $K_2' \approx (10^{-7} - 10^{-6}) K_2$; consequently, three-body processes involving neutrals are weak in low-pressure discharges. However, let us note that three-body processes can be important for other applications. For example, the three-body reaction

$$O + O_2 + O_2 \longrightarrow O_3 + O_2$$

can be the most significant source of ozone in high-pressure discharges ($p \gtrsim 1$ atm), and is known to be the most important source in the earth's ionosphere. Some three-body rate constants in oxygen discharges are given in Table 8.3.

Three-Body Positive–Negative Ion Recombination

Three-body processes involving charged particles can have rate constants that are much higher than three-body processes involving only neutrals. Consider positive–negative ion recombination at thermal energies as an example:

$$A^+ + B^- + M \longrightarrow AB + M$$

The basic theory of this process was first developed by Thomson (1924), and can be understood as follows. Let the positive and negative ions approach each other to within a critical radius b_0 such that the Coulomb interaction energy is equal to the mean kinetic energy

$$\frac{e^2}{4\pi\epsilon_0 b_0} = \frac{3}{2} kT \tag{9.2.39}$$

If during the time the ions are within the critical radius one of them collides with a neutral molecule M, then with high probability energy is transferred from the ion to the neutral, and the ions become bound to each other. From this description, estimates of the rate constants for the elementary reactions in (9.2.29) and (9.2.30) are

$$K_2 \approx \pi b_0^2 \bar{v}_i \tag{9.2.40}$$

$$K_{-1} \approx \frac{\bar{v}_i}{b_0} \tag{9.2.41}$$

$$K_{2M} \approx (\sigma_{M+} \bar{v}_{M+} + \sigma_{M-} \bar{v}_{M-}) \tag{9.2.42}$$

where $\bar{v}_i$ and $\bar{v} \approx \bar{v}_i$ are the mean speeds of relative motion of the ion pair and the ion–neutral pairs, respectively, and σ_{M+} and σ_{M-} are the cross sections for energy transfer from ions to neutrals. Then from (9.2.38), the equivalent two-body rate constant is

$$K_2' = K_3' n_M \approx \pi b_0^3 K_{2M} n_M \tag{9.2.43}$$

where b_0 is found from (9.2.39):

$$b_0 = \frac{2}{3} \frac{e^2}{4\pi\epsilon_0 kT} \tag{9.2.44}$$

We note from (9.2.42) and (9.2.43) that $K_2' \propto T^{-5/2}$. At room temperature (300 K), we find $b_0 \approx 550$ Å, a very large critical radius. Consider the example of an estimate of K_2' for the reaction

$$O_2^+ + O^- + O_2 \longrightarrow O_3 + O_2$$

In this case, near-resonant charge transfer of O_2^+ on O_2 dominates in (9.2.42), and we estimate from the polarization rate constant (3.3.17) with $\alpha_R \approx 10.6$ and $A_R = 16$ that $K_{2M} \approx 7 \times 10^{-10}$ cm^3/s. Then (9.2.43) yields $K_2' \approx 3.7 \times 10^{-25} n_M$. At 1 Torr, $n_M \approx 3.3 \times 10^{16}$ /cm^3, such that $K_2' \approx 1.2 \times 10^{-8}$ cm^3/s, a very respectable rate constant. Consequently, three-body positive–negative ion recombination can be quite important for processing discharges at pressures $p \gtrsim 1$ Torr. Some rate constants for this process in oxygen discharges are given in Table 8.3.

At very high pressures, the ion–neutral mean free path becomes smaller than the critical radius, leading to multiple ion–neutral collisions within the critical sphere, and the preceding analysis of the mechanism is not correct. This regime is not of interest for low-pressure processing. The reader is referred to Smirnov (1982) for further information.

Three-Body Electron–Ion Recombination

For this process,

$$e + A^+ + e \longrightarrow A + e \tag{9.2.45}$$

with $T_e \gg T$, we have, in place of (9.2.40)–(9.2.42), the rate constants

$$K_2 \approx \pi b_0^2 \bar{v}_e \tag{9.2.46}$$

$$K_{-1} \approx \frac{\bar{v}_e}{b_0} \tag{9.2.47}$$

$$K_{2M} \approx \sigma_1 \bar{v}_e \tag{9.2.48}$$

where

$$b_0 = \frac{2}{3} \frac{e^2}{4\pi\epsilon_0 kT_e} \tag{9.2.49}$$

and

$$\pi b_0^2 \approx \sigma \tag{9.2.50}$$

is the cross section for a single electron–electron Coulomb collision (see Section 3.3) that transfers an energy $\sim \frac{3}{2}T_e$. Substituting (9.2.46)–(9.2.48) into (9.2.38), we obtain

$$K_2' \approx \pi^2 b_0^5 \bar{v}_e n_e \tag{9.2.51}$$

which scales as $K_2' \propto T_e^{-9/2} n_e$. A calculation shows that this process is not important in processing discharges with $T_e \gtrsim 1$ V and $n_e \lesssim 10^{13}$ /cm^3.

9.3 SURFACE PROCESSES

Physical and chemical surface processes are central to plasma processing. For example, in F-atom etching of silicon in a CF_4 discharge, the F atoms created in the gas phase are transported to and successively fluorinate the surface through reactions such as

$$F(g) + Si: F_x(s) \rightleftharpoons SiF_{x+1}(s), \qquad x = 0, 1, 2, 3$$

with production of etch products,

$$Si: F_x(s) \rightleftharpoons SiF_x(g), \qquad x = 2, 4$$

which are transported back into the gas phase. In addition, adsorption–desorption reactions such as

$$F(g) + S \rightleftharpoons F: S$$

play a critical role in determining gas-phase species concentrations. Finally, the discharge equilibrium itself is affected by surface processes such as positive ion neutralization and secondary electron emission at surfaces. For these reasons, the gas-phase and surface reaction sets are coupled, with the coupling being strong at low gas pressures. In this section, we describe some important physical processes, primarily involving positive ions, and some important physical and chemical processes involving neutrals.

Positive Ion Neutralization and Secondary Electron Emission

The strongly exothermic neutralization reaction

$$e + A^+ \longrightarrow A$$

is forbidden in the gas phase because energy and momentum cannot be conserved for the formation of one body from two. However, at the surface, the three-body neutralization reaction

$$e + A^+ + S \longrightarrow A + S$$

is fast. For positive ion energies at the surface in the range 10–1000 V, typical of processing discharges, essentially all positive ions are immediately neutralized at the surface.

To understand the neutralization mechanism, the confinement of electrons in a solid must be briefly described. Figure 9.2a shows the energy versus position near a metal surface. The electrons in the metal fill a set of closely spaced energy levels (conduction band) up to a maximum energy (from the bottom of the conduction band) called the *Fermi energy* $\mathcal{E}_F$. The Fermi energy lies below the $\Phi = 0$ potential energy level for a free electron by an energy equal to the *work function* $\mathcal{E}_\phi$. Hence electrons at $\mathcal{E}_F$ are confined within the solid by a potential barrier of height $\mathcal{E}_\phi$.

A simple classical estimate of $\mathcal{E}_\phi$ for a metal is that it is the work done in moving an electron initially located at a distance $x = a_{\text{eff}}$ from a perfectly conducting surface to $x = \infty$, where a_{eff} is a distance of order an atomic radius. As shown in Fig. 9.3a, the force F_x acting on an electron $-e$ located at x can be found using the method of

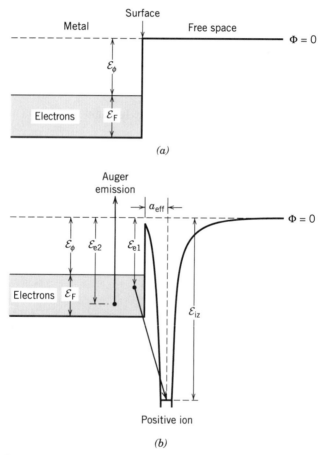

FIGURE 9.2. Illustrating ion neutralization and secondary emission at a metal surface; (a) the work function $\mathcal{E}_\phi$ and the Fermi energy $\mathcal{E}_F$; (b) Auger emission due to electron tunneling.

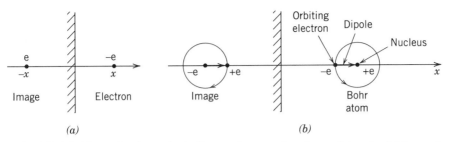

FIGURE 9.3. Illustrating the method of images for a metal surface to determine (a) the work function and (b) the van der Waals force.

images (Ramo et al, 1984), with the image charge $+e$ located at $-x$. From Coulomb's law, we have

$$F_x = -\frac{e^2}{4\pi\epsilon_0(2x)^2} \tag{9.3.1}$$

The work done in moving the electron from a_{eff} to ∞ is then

$$e\mathcal{E}_\phi = -\int_{a_{\mathrm{eff}}}^{\infty} F_x\,dx$$

which yields, upon integration,

$$\mathcal{E}_\phi = \frac{e}{16\pi\epsilon_0 a_{\mathrm{eff}}} \tag{9.3.2}$$

Letting $a_{\mathrm{eff}} = a_0$, the Bohr radius, we obtain $\mathcal{E}_\phi \approx 6.8$ V. Work functions for most materials are in the range 4–6 V, although the alkalis and alkali earths are lower. There is a rough correlation $\mathcal{E}_\phi \propto \mathcal{E}_{iz}^{1/2}$, where $\mathcal{E}_{iz}$ is the ionization potential of the metal atoms. This can be understood from (9.3.2) because the atomic radius $a_{\mathrm{eff}} \propto \mathcal{E}_{iz}^{-1/2}$, as given in (3.4.9).

Now consider the approach of a positive ion to within an atomic radius a_{eff} of the surface. As shown in Fig. 9.2b, this creates a deep potential well very near the surface that is separated from the surface by a narrow potential barrier of width $\sim a_{\mathrm{eff}}$. An electron with energy $\mathcal{E}_e$ from within the conduction band can tunnel through the barrier into the positive ion to neutralize it. There are two possibilities:

1. The electron enters an excited state

$$e + A^+ + S \longrightarrow A^* + S$$

where $\mathcal{E}_* \approx \mathcal{E}_{iz} - \mathcal{E}_e$. If the excited state in not metastable, it radiates a photon in a transition to the ground state or to a metastable state. Hence positive ion neutralization at the surface can create metastables as well as *recombination radiation.*

2. The electron enters the ground state of the atom, and a *second* electron from within the conduction band absorbs the excess energy of neutralization. This mechanism, called *Auger neutralization,* is a nonradiative transition involving two electrons. The electron that enters the ground state of the atom loses an energy $\Delta\mathcal{E} = \mathcal{E}_{iz} - \mathcal{E}_{e1}$, which the second electron gains. If $\Delta\mathcal{E} < \mathcal{E}_{e2}$ for the second electron, then it remains trapped within the solid. However, if $\Delta\mathcal{E} > \mathcal{E}_{e2}$, then the second electron is released from the solid and is free to move away from the surface. This process is called *Auger emission,* or, more commonly, *secondary emission.*

From Fig 9.2b, the condition for release of the second electron is most easily met if both electrons come from the top of the conduction band: $\mathcal{E}_{e1} = \mathcal{E}_{e2} = \mathcal{E}_{\phi}$. The condition for emission is then

$$\mathcal{E}_{iz} \geq 2\mathcal{E}_{\phi} \tag{9.3.3}$$

The released electron has kinetic energy $\mathcal{E}_{max} = \mathcal{E}_{iz} - 2\mathcal{E}_{\phi}$. The minimum kinetic energy is $\mathcal{E}_{min} = \mathcal{E}_{iz} - 2\mathcal{E}_{\phi} - 2\mathcal{E}_F$ (both electrons come from the bottom of the conduction band), or zero if this is negative. Equation (9.3.3) shows that secondary emission is favored for noble gas ions ($\mathcal{E}_{iz}$ is high) and for alkali or alkali earth solids ($\mathcal{E}_{\phi}$ is low). Because the electron tunneling time (see Section 3.4) is short compared to the ion collision time with the surface, the secondary emission process is practically independent of ion kinetic energy, and depends only on the atomic ion species and the near-surface composition of the solid.

Although neutralization and secondary emission have been described for metals, essentially the same processes occur for ions incident on semiconducting and insulating surfaces. As mentioned in Section 3.5, secondary emission is usually characterized by the *secondary emission coefficient* γ_{se}, which is the number of secondary electrons created per incident ion. An empirical expression is (Raizer, 1991)

$$\gamma_{se} \approx 0.016\,(\mathcal{E}_{iz} - 2\mathcal{E}_{\phi}) \tag{9.3.4}$$

provided $\mathcal{E}_{iz} > 2\mathcal{E}_{\phi}$. Metastables produce Auger electron emission very efficiently; the condition for emission is $\mathcal{E}_* > \mathcal{E}_{\phi}$. In addition to Auger emission, secondary electrons can be created by kinetic ejection for ion (or neutral) impact energies $\gtrsim 1$ kV. These heavy particle energies are not common in processing discharges except for ion implantation applications (see Chapter 16).

Although (9.3.4) provides a rough estimate, the actual value of γ_{se} depends sensitively on surface conditions, morphology, impurities, and contamination. Some measured values of $\mathcal{E}_{\phi}$ and γ_{se} for ions incident on atomically clean surfaces are given in Table 9.1. However, surfaces are never atomically clean in processing applications. Secondary emission is an important process in dc discharges, which are described in Chapter 14.

TABLE 9.1. Work Functions and Secondary Emission Coefficients

Solid	Work Function (V)	Ion	Energy (V)	γ_{se}
Si(100)	4.90	He$^+$	100	0.168
		Ar$^+$	10	0.024
			100	0.027
Ni(111)	4.5	He$^+$	100	0.170
		Ar$^+$	10	0.034
			100	0.036
Mo	4.3	He$^+$	100	0.274
		Ar$^+$	100	0.115
		N$_2^+$	100	0.032
		O$_2^+$	100	0.026
W	4.54	He$^+$	100	0.263
		Ar$^+$	10	0.096
			100	0.095
		H$_2^+$	100	0.029
		N$_2^+$	100	0.025
		O$_2^+$	100	0.015

Source. After Konuma (1992).

Apart from neutralization and Auger emission, heavy particles (ions and neutrals) have much the same behavior when they impact surfaces. At low (thermal) energies, physi- and chemisorption and desorption can occur. At higher energies (tens of volts), molecules can fragment into atoms. At still higher energies (hundreds of volts), atoms can be sputtered from the surface, and at still higher energies (thousands of volts), implantation is important.

Adsorption and Desorption

Adsorption and desorption are very important for plasma processing because, in many cases, one or the other of these reactions is the rate limiting step for a surface process. Adsorption,

$$A + S \longrightarrow A:S$$

is the reaction of a molecule with a surface. Desorption is the reverse reaction. Adsorption is due to the attractive force between an incoming molecule and a surface. There are two kinds of adsorption. *Physisorption* is due to the weak attractive van der Waals force between a molecule and a surface. We can understand how this force arises by considering the example of a Bohr atom near a metal surface. As shown in Fig. 9.3b, the Bohr model gives rise to an oscillating dipole moment $p_{dx}(t) \approx a_0 e \cos \omega_{at} t$ normal to the surface as the electron orbits the nucleus. The force F_x acting on the dipole can be found using the method of images. For a dipole $+p_{dx}(t)$ at x, there is an image dipole $+p_{dx}(t)$ at $-x$, and the force is attractive

(Problem 9.5),

$$F_x = -\frac{6\langle p_{dx}^2(t)\rangle}{4\pi\epsilon_0(2x)^4} \approx -\frac{3a_0^2 e^2}{4\pi\epsilon_0(2x)^4}$$

The van der Waals interaction potential is found from $F_x = -e\,dV/dx$ to be

$$V(x) = -\frac{a_0^2 e}{64\pi\epsilon_0 x^3} \tag{9.3.5}$$

When the atom comes to within a distance of order $d \sim 1\text{--}3$ Å from the surface, then the Coulomb clouds of the atom and surface interact, leading to a repulsive force. Hence, a shallow potential well is formed near the surface. Letting $d \sim 1\text{--}3$ Å at equilibrium, the well depth is estimated from (9.3.5) to be $\mathcal{E}_{physi} \sim 0.01\text{--}0.25$ V. Hence physisorption is exothermic with $|\Delta H| \sim 1\text{--}25$ kJ/mol. The vibration frequency ω_{vib} for a molecule trapped in the well can be estimated assuming a harmonic oscillator potential,

$$\frac{1}{2}M\omega_{vib}^2 d^2 \sim e\mathcal{E}_{physi} \tag{9.3.6}$$

which yields $\omega_{vib} \sim 10^{12}\text{--}10^{13}\ \text{s}^{-1}$. Physisorbed molecules are often so weakly bound to the surface that they can diffuse rapidly along the surface.

   Chemisorption is due to the formation of a chemical bond between the atom or molecule and the surface. The reaction is strongly exothermic with $|\Delta H| \sim 40\text{--}400$ kJ/mol, corresponding to a potential well depth $\mathcal{E}_{chemi} \sim 0.4\text{--}4$ V. The minimum of the well is typically located a distance $d \sim 1\text{--}1.5$ Å from the surface. Chemisorption of a molecule having multiple (double, triple, etc.) bonds can occur with the breaking of one bond as the molecule bonds to the surface,

$$A{=}B + S \longrightarrow AB{:}S$$

Molecules that are single bonded are often torn apart as they bond to the surface,

$$AB + S \longrightarrow A{:}S + B{:}S$$

This process is called *dissociative chemisorption* and requires two adsorption sites. Physi- and chemisorption are often found in the same system, with different regimes favored depending on the surface temperature and the form of the potential energy curves. Figure 9.4 gives three examples. In (a), the A+B dissociated chemisorbed state combines with the AB physisorbed state to give a minimum potential energy curve (solid line) that is everywhere negative. AB molecules at low energies incident on the surface can easily pass through the physisorbed region and enter the dissociated chemisorbed state. In (b), there is a potential barrier $\mathcal{E}_{ads}$ to chemisorption, but incident AB molecules can be trapped in the physisorbed state. If the barrier is

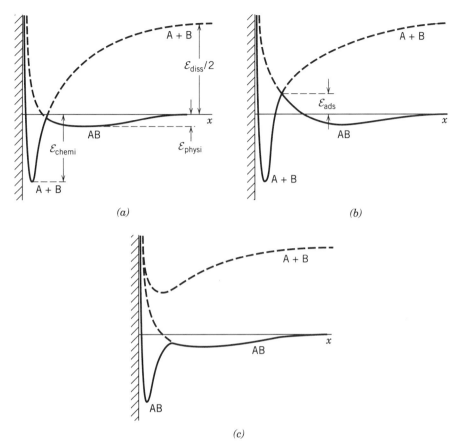

FIGURE 9.4. Schematic diagrams of the potential energy near a surface for adsorption; (a) dissociative chemisorption; (b) physisorption; and (c) molecular chemisorption.

low, then thermal molecules can be first physisorbed and later pass into the lower energy, dissociated chemisorbed state. In (c), there is molecular chemisorption but not dissociative adsorption because the A+B chemisorbed state lies everywhere above the AB state.

Molecules that impinge on a surface cannot be adsorbed unless they lose energy in the collision with the surface. The normal component of the energy loss must be sufficient to trap the molecule in the adsorption well. Let $\Gamma_A = \frac{1}{4}\bar{v}_A n_{AS}$ be the flux of molecules incident on the surface, where $\bar{v}_A$ is mean speed of the molecule and n_{AS} is the gas phase volume density of molecules at the surface. Then the flux of molecules that are chemisorbed can be written as

$$\Gamma_{ads} = s\Gamma_A = \frac{1}{4}s\bar{v}_A n_{AS} \qquad (9.3.7)$$

which defines the *sticking coefficient s*. In general, s is a function of the surface coverage θ (fraction of sites covered with adsorbate) and the gas and surface temperatures. If the gas and surface are in thermal equilibrium at temperature T, then the surface coverage $\bar{\theta}(T)$ is determined and $s = \bar{s}(T)$, the equilibrium *thermal sticking coefficient*. A common assumption for s for nondissociative adsorption for systems not in thermal equilibrium is Langmuir kinetics,

$$s(\theta, T) = s_0(T)(1 - \theta) \tag{9.3.8}$$

where s_0 is the initial or *zero coverage sticking coefficient*, and $1 - \theta$ is the fraction of the surface not covered with adsorbate. Langmuir kinetics is often found to underestimate the sticking coefficient for chemisorption at intermediate values of θ, because molecules that impact sites already filled with adsorbate can be trapped by physisorption and diffuse along the surface to vacant sites, where they chemisorb. Generally, chemisorption ceases after all active sites have been filled; this roughly corresponds to a monolayer of coverage. Continued adsorption is only by the much less tightly bonded physisorption mechanism. Many monolayers can be physisorbed and, in fact, continuous condensation of adsorbate can occur. Usually, however, nonactive surfaces, e.g., reactor walls, come to an equilibrium where physisorption and desorption balance; hence the *net* flux of molecules to these surfaces is zero. The kinetics of physi- and chemisorption are treated in Section 9.4.

The temperature variation of s_0 depends on whether there is an energy barrier to chemisorption (Fig. 9.4b) or not (Fig. 9.4a). If there is no barrier, then s_0 can be near unity at low temperatures and decreases with increasing T because the fraction of incident molecules that lose sufficient energy to trap decreases as T increases. If there is an activation barrier of height $\mathcal{E}_{ads}$, then very little sticking can occur until $T \sim \mathcal{E}_{ads}$. Then s_0 has an Arrhenius form,

$$s_0 = s_{00}(T) e^{-\mathcal{E}_{ads}/T} \tag{9.3.9}$$

where the preexponential factor s_{00} decreases as T increases, as for the case with no barrier. Measured sticking coefficients at $T = 0.026$ V (300 K) vary over a wide range 10^{-6}–1 and strongly depend on crystal orientation and surface roughness, with s_0 increasing as the roughness increases (Morris et al, 1984). For many surfaces, the active sites for sticking are at surface imperfections such as steps, kinks, vacancies, and dislocations. Chemically reactive gases, and especially radicals, usually stick with high probability $s_0 \sim 0.1$–1 on transition metals (Fe, Ni, etc.). Sticking probabilities can be lower for other surfaces. For example, $s_0 \sim 1$ for H on Si, but s_0 is a few percent for H_2 on Si, and $s_0 \sim 10^{-4}$–10^{-3} for O_2 on Si (Joyce and Foxon, 1984).

Desorption,

$$A: S \longrightarrow A + S$$

is the reverse reaction to adsorption. In thermal equilibrium, the two reactions must balance. The (first-order) desorption rate constant can be shown to have an Arrhenius form (Zangwill, 1988)

$$K_{\text{desor}} = K_0\, e^{-\mathcal{E}_{\text{desor}}/T} \tag{9.3.10}$$

where $\mathcal{E}_{\text{desor}}$ is the depth ($\mathcal{E}_{\text{chemi}}$ or $\mathcal{E}_{\text{physi}}$) of the potential well from the zero of potential energy. A crude classical estimate is that K_0 is the number of attempted escapes per second from the adsorption well; hence

$$K_0 \sim \omega_{\text{vib}}/2\pi \tag{9.3.11}$$

where ω_{vib} is the vibration frequency of the adsorbed molecule, as estimated in (9.3.6). However, K_0 cannot be estimated accurately from ω_{vib} in this way. A more precise estimate from transition rate theory (Zangwill, 1988) is that

$$K_0 \approx \bar{s}(T)\frac{eT}{h}\frac{\bar{g}_{\text{esc}}}{\bar{g}_{\text{ads}}}$$

$$\approx 6\times 10^{12}\,\bar{s}(T)\frac{\bar{g}_{\text{esc}}}{\bar{g}_{\text{ads}}}\quad s^{-1} \tag{9.3.12}$$

where $\bar{g}_{\text{esc}}/\bar{g}_{\text{ads}}$ is a ratio of statistical weights for escaping and trapped molecules. For physisorption the ratio of weights is usually large and $K_0 \sim 10^{14}$–10^{16} s^{-1}. For chemisorption the ratio of weights is often smaller and $K_0 \sim 10^{13}$–10^{15} s^{-1}. For activated adsorption, $\bar{s}$ also has an Arrhenius dependence (see (9.3.9)); therefore,

$$K_{\text{desor}} \propto e^{-(\mathcal{E}_{\text{ads}}+\mathcal{E}_{\text{desor}})/T}$$

Associative desorption,

$$2A{:}\,S \longrightarrow A_2 + 2S$$

the reverse of dissociative adsorption, also has a (second order) rate constant on the surface with the Arrhenius form (9.3.10). The classical estimate of the preexponential factor is that it is the number of collisions per second per unit area on the surface between two adsorbed atoms:

$$K_0 \sim \bar{d}\left(\frac{\pi eT}{M_R}\right)^{1/2} \tag{9.3.13}$$

where $\bar{d}$ is the mean diameter for a collision, and $(\pi eT/M_R)^{1/2}$ is the characteristic collision velocity. Typically, $K_0 \sim 10^{-3}$–1 cm^2/s.

Fragmentation

Ionic and neutral molecules that have sufficient impact energy can fragment into atoms that are reflected or adsorbed when they hit a surface. The threshold energy for fragmentation is of order of the energy of the molecular bond. At energies four or five times the threshold energy, over half of the molecules typically fragment. Since molecular bond energies are in the range 1–10 V, and ion-bombarding energies at surfaces are often considerably higher (particularly at capacitively driven electrodes), molecular ions often fragment when they hit surfaces.

Sputtering

At energies above 20–30 V, heavy particles can sputter atoms from a surface. Usually ions are the impacting species. The sputtering yield γ_{sput} (atoms sputtered per incident ion) increases rapidly with energy up to a few hundred volts, where the yield becomes significant for processing applications, with 500–1000 V argon ions the usual projectile for physical sputtering. For these energies, the bombarding ion transfers energy to many target atoms, which in turn collide with other atoms in the solid. Most of the atoms in this collision cascade are trapped in the solid, but one or several can escape from the surface. Over a broad energy range, the yield in this cascade regime can be shown to be independent of the projectile energy and the target atom density, and to scale roughly as (Feldman and Mayer, 1986)

$$\gamma_{sput} \propto \frac{1}{\mathcal{E}_t} \frac{M_i}{M_i + M_t} \qquad (9.3.14)$$

where $\mathcal{E}_t$ is the surface binding energy (roughly, the enthalpy of vaporization) of the target, and M_i and M_t are the incident ion and target atom masses.

In addition to the dependence (9.3.14), measured high fluence sputtering yields have a periodic variation of peaks and valleys versus projectile atomic number, which are not seen in low fluence measurements. These are due to changes in the surface layer due to implantation or deposition of the projectile ion. The peaks are for sputtering by the noble gases and are believed to be due to gas agglomeration and bubble formation in the target material. The valleys are due to buildup of a surface layer which blocks sputtering of the target, e.g., for carbon or calcium projectiles. Some measured sputtering yields for argon ion bombardment at 600 V are given in Table 9.2 (Konuma, 1992). Application of physical sputtering to thin-film deposition is described in Chapter 16.

Above a few hundred volts, there is a significant chance that ions will be implanted in the solid. This process becomes increasingly important above 1 kV. These energies are not common in processing discharges, but can be accessed by applying extremely high dc or pulsed voltages to an electrode immersed in a plasma. This application, plasma-immersion ion implantation, is also described in Chapter 16.

9.4   SURFACE KINETICS

A general reaction set for a surface process is illustrated in Fig. 9.5. This might apply to the etching of a carbon substrate in an oxygen discharge, with O the etchant and

**TABLE 9.2. Measured Sputtering
Yields for Ar$^+$ at 600 V**

Target	γ_{sput}
Al	0.83
Si	0.54
Fe	0.97
Co	0.99
Ni	1.34
Cu	2.00
Ge	0.82
W	0.32
Au	1.18
Al_2O_3	0.18
SiO_2	1.34
GaAs	0.9
SiC	1.8
SnO_2	0.96

Source. After Konuma (1992).

CO the etch product. The etchant atoms diffuse or flow to the surface (rate constant K_a), where they are adsorbed (K_b) and react (K_c) to form the product, which then desorbs (K_d) and diffuses or flows into the gas phase (K_e). In addition, etchants can desorb without reaction (K_f and K_g for normal or associative desorption), and etch products in the gas phase can adsorb back onto the surface (K_h). More complicated reactions can also occur, e.g., to form CO_2 product.

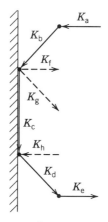

FIGURE 9.5. Illustrating the processes that can occur for reaction of an etchant with a surface.

TABLE 9.3. Gas Kinetic Cross Sections in Units of 10^{-15} cm^2

	He	Ar	H$_2$	N$_2$	O$_2$	CO	CO$_2$
He	1.6	2.9	2.2	3.1	2.9	3.0	3.6
Ar		5.0	3.7	5.4	5.2	5.3	5.7
H$_2$			2.7	3.8	3.7	3.9	4.5
N$_2$				5.2	4.1	5.1	6.8
O$_2$					4.9	4.8	5.9
CO						5.0	6.3
CO$_2$							7.8

Source. Smirnov (1977, Appendix 1).

Diffusion of Neutral Species

Charged particle diffusion was the subject of Chapter 5. Neutral species also diffuse. The diffusion coefficient for A molecules due to collisions with B molecules is

$$D_{AB} = \frac{eT}{M_R \nu_{AB}} \tag{9.4.1}$$

where M_R is the reduced mass and

$$\nu_{AB} = n_B \sigma_{AB} \bar{v}_{AB}$$

is the collision frequency for a constant cross-section (hard-sphere) process, with $\bar{v}_{AB} = (8eT/\pi M_R)^{1/2}$ the mean speed of relative motion. Inserting ν_{AB} into D_{AB} yields[*]

$$D_{AB} = \frac{\pi}{8} \lambda_{AB} \bar{v}_{AB} \tag{9.4.2}$$

where $\lambda_{AB} = 1/n_B \sigma_{AB}$ is the mean free path. The cross section can be estimated from

$$\sigma_{AB} \approx \pi(r_A + r_B)^2 \tag{9.4.3}$$

where r_A and r_B are the mean radii of the molecules. Some gas kinetic cross sections are given in Table 9.3 (Smirnov, 1977). Cross sections are typically in the range 2–6×10^{-15} cm^2. For self-diffusion of A molecules due to collisions with A molecules, $M_R = M_A/2$ in (9.4.2).

Rate Constant for Diffusion

If $\lambda_{AB} \lesssim l_{eff}$, the characteristic size of the plasma, then gas-phase molecules created within the plasma are transported to the surfaces by diffusion. If n_{A0} and n_{AS} are the

---

[*]The exact result from kinetic theory is 3/4 of this value (McDaniel, 1964, p. 50).

densities in the bulk and at the surface, then an estimate of the flux is

$$\Gamma_{AS} \approx D_{AB} \frac{n_{A0} - n_{AS}}{\frac{1}{2} l_{eff}} \tag{9.4.4}$$

Multiplying (9.4.4) by the surface area A, dividing by the volume $\mathcal{V}$, and estimating $l_{eff} \approx \mathcal{V}/A$, we obtain the average volume loss rate for diffusion,

$$\frac{d\bar{n}_A}{dt} \approx -K_{loss}(n_{A0} - n_{AS}) \tag{9.4.5}$$

with

$$K_{loss} \approx \frac{2D_{AB}}{l_{eff}^2} \quad s^{-1} \tag{9.4.6}$$

the rate constant for diffusive loss.

K_{loss} can be estimated more precisely if the discharge geometry is specified. For example, consider the creation, diffusion, and loss of a species A in symmetric one-dimensional slab geometry. We assume that the rate of production G_A within the slab is uniform, that A diffuses due to collisions with B, and that there is a net loss of A to the walls. The diffusion equation,

$$-D_{AB} \frac{d^2 n_A}{dx^2} = G_A \tag{9.4.7}$$

has the solution

$$n_A = n_{A0} - (n_{A0} - n_{AS}) \frac{4x^2}{l^2} \tag{9.4.8}$$

where n_{A0} is the density in the center of the slab, $x = 0$, and n_{AS} is the density at the walls, $x = \pm l/2$. The net flux of A molecules at $x = l/2$ is

$$\Gamma_{AS} = -D_{AB} \left(\frac{dn_A}{dx} \right)_{l/2} = \frac{4D_{AB}}{l} (n_{A0} - n_{AS}) \tag{9.4.9}$$

Integrating (9.4.7) over the slab volume, we note that

$$\Gamma_{AS} = G_A \frac{l}{2} \tag{9.4.10}$$

which determines n_{A0} given n_{AS}. Multiplying (9.4.9) by the surface to volume ratio $2A/l$ and introducing the average density,

$$\bar{n}_A = \frac{2}{3}n_{A0} + \frac{1}{3}n_{AS} \tag{9.4.11}$$

we obtain the average volume loss rate (9.4.5) with

$$K_{loss} = \frac{8D_{AB}}{l^2} = \frac{\pi \lambda_{AB} \bar{v}_{AB}}{l^2} \quad s^{-1} \tag{9.4.12}$$

the first-order rate constant for diffusive loss. To complete the diffusion calculation, as in Chapter 5, we need to determine the relation between n_{A0} and n_{AS} from (9.4.5). This requires a calculation of the fluxes entering and leaving the surface, which we do below.

Let us first estimate K_{loss} for O atoms diffusing through O_2 molecules in a reactor with $l = 10$ cm. Let $\sigma_{AB} \approx 3 \times 10^{-15}$ cm^2 and $p = 10$ mTorr ($n_g \approx 3.3 \times 10^{14}$ cm^{-3}). Then $\lambda_{AB} \approx 1$ cm. At 300 K, $\bar{v}_{AB} \approx 7.7 \times 10^4$ cm/s. Then $K_{loss} \approx 2.4 \times 10^3$ s^{-1}. In many cases this is a sufficiently high rate that the rate-limiting step for loss of etchant is not gas-phase diffusion; for these cases, we find that $n_{AS} \approx n_{A0}$.

At very low pressures, $\lambda_{AB} \gtrsim l_{eff}$, the volume loss to the walls is no longer diffusive; the molecules flow freely to the walls. The characteristic rate of loss is determined by the mean speed of the molecules and the distance they travel:

$$K_{loss} \sim \frac{\bar{v}_A}{2l_{eff}} \tag{9.4.13}$$

Adsorption and Desorption

Consider the opposing reactions for nondissociative adsorption and desorption of A molecules on a surface,

$$A + S \underset{K_d}{\overset{K_a}{\rightleftharpoons}} A:S$$

Let n_0' be the area density of adsorption sites and $n_{A:S}' = n_0'\theta$ be the density of sites covered with adsorbed molecules. Assuming Langmuir kinetics, such that the flux of A adsorbing on the surface is proportional to $1 - \theta$, the fraction of sites not covered with adsorbate, we can write

$$\Gamma_{ads} = K_a n_{AS} n_0'(1 - \theta) \tag{9.4.14}$$

where n_{AS} is the gas phase density at the surface. Equating this to the flux of desorbing molecules

$$\Gamma_{desor} = K_d n_0'\theta \tag{9.4.15}$$

we can solve for θ to obtain

$$\theta = \frac{\mathcal{K}n_{AS}}{1 + \mathcal{K}n_{AS}} \tag{9.4.16}$$

where

$$\mathcal{K} = \frac{K_a}{K_d} \tag{9.4.17}$$

This is the Langmuir isotherm for thermal equilibrium (7.5.17).

Dissociative Adsorption and Associative Desorption

Consider now the opposing reactions

$$A_2(g) + 2S \underset{K_d}{\overset{K_a}{\rightleftharpoons}} 2A:S$$

Because two sites are required for adsorption, the molecular flux adsorbed is

$$\Gamma_{ads} = K_a n_{A_2S} n_0'^2 (1 - \theta)^2 \tag{9.4.18}$$

and the molecular flux desorbed is

$$\Gamma_{desor} = K_d n_0'^2 \theta^2 \tag{9.4.19}$$

Equating fluxes and solving for θ, we obtain the isotherm

$$\theta = \frac{(\mathcal{K}n_{A_2S})^{1/2}}{1 + (\mathcal{K}n_{A_2S})^{1/2}} \tag{9.4.20}$$

where $\mathcal{K} = K_a/K_d$. For θ small, we see that $\theta \propto n_{A_2S}^{1/2}$ for dissociative adsorption, a slower variation than for normal adsorption.

Physical Adsorption

While the density of available sites is usually fixed at some n_0' for chemisorption, many monolayers can be physisorbed. Let n_i' be the area density of sites having a thickness of i physisorbed atoms. Then equating the adsorption to desorption flux for these sites,

$$K_a n_i' n_{AS} = K_d n_{i+1}' \tag{9.4.21}$$

we obtain

$$n_{i+1}' = \beta n_i' \tag{9.4.22}$$

where $\beta = K_a n_{AS}/K_d = \mathcal{K} n_{AS}$. Hence, by induction,

$$n_i' = n_0' \beta^i \tag{9.4.23}$$

The total number of physisorbed molecules per unit area is given by

$$n_T' = \sum_{i=1}^{\infty} i n_i' = n_0' \sum_{i=1}^{\infty} i \beta^i$$

$$= n_0' \frac{\beta}{(1 - \beta)^2} \tag{9.4.24}$$

and the number of sites covered per unit area is

$$n_C' = \sum_{i=1}^{\infty} n_i' = n_0' \frac{1}{1 - \beta} \tag{9.4.25}$$

For $\beta \lesssim 1$, many monolayers can be adsorbed. The condition $\beta = 1$ signals the onset of continuous condensation. The combination of physi- and chemisorption can also be analyzed, leading to the so-called BET isotherm (see Atkins, 1986, p. 779).

Reaction With a Surface

Consider the reaction set where A is adsorbed on the surface S ($\equiv$ B) and reacts directly with the surface to form the gas-phase product AS ($\equiv$ AB):

$$A(g) + S \underset{K_d}{\overset{K_a}{\rightleftharpoons}} A{:}S$$

$$A{:}S \xrightarrow{K_r} AS(g)$$

The surface coverage θ is found from the conservation of adsorbed sites,

$$\frac{dn_{A:S}'}{dt} = K_a n_{AS} n_0'(1 - \theta) - K_d n_0' \theta - K_r n_0' \theta = 0 \tag{9.4.26}$$

Solving for θ, we obtain

$$\theta = \frac{1}{1 + (K_d + K_r)/K_a n_{AS}} \tag{9.4.27}$$

The reaction rate ($\text{cm}^{-2}\text{-s}^{-1}$) for production of AB is then

$$R_{AB} = K_r n_0' \theta \tag{9.4.28}$$

The net flux of A molecules to the surface is found from (9.4.5),

$$\Gamma_A = K_{loss}(n_{A0} - n_{AS}) \tag{9.4.29}$$

where K_{loss} is the rate constant for loss of A from the gas-phase volume. Setting $R_{AB} = \Gamma_A$ leads to a quadratic equation which can be solved to determine n_{AS} in terms of n_{A0} and the rate constants. Finally, n_{A0} itself is set for the given generation rate G_A of A molecules by equating $\Gamma_A A = G_A \mathcal{V}$ and using (9.4.29). Rather than carry out this procedure here, we look at the low- and high-pressure limits.

(a) Low Pressure $K_a n_{AS} \ll K_d + K_r$ Then (9.4.27) reduces to

$$\theta = \frac{K_a n_{AS}}{K_d + K_r} \ll 1 \tag{9.4.30}$$

Equating $R_{AB} = \Gamma_A$,

$$K_{loss}(n_{A0} - n_{AS}) = \frac{K_a K_r n_0'}{K_d + K_r} n_{AS} \tag{9.4.31}$$

we solve for n_{AS} to obtain

$$n_{AS} = \frac{K_{loss}}{K_{loss} + \dfrac{K_a K_r n_0'}{K_d + K_r}} n_{A0} \tag{9.4.32}$$

The reaction rate is found by substituting (9.4.32) into (9.4.30) and (9.4.28) to obtain

$$R_{AB} = K_{AB} n_{A0} \tag{9.4.33}$$

where

$$K_{AB} = \frac{1}{\dfrac{K_d + K_r}{K_a K_r n_0'} + \dfrac{1}{K_{loss}}} \tag{9.4.34}$$

is a composite first-order rate constant. First-order kinetics are typical for surface reactions at low pressures, such as are found in processing discharges.

(b) High Pressure $K_a n_{AS} \gg K_d + K_r$ In this regime $\theta \approx 1$ and $R_{AB} = K_r n_0'$, independent of n_{A0}. This zeroth-order kinetics is not uncommon at high pressures. This regime generally is not of interest for low-pressure processing.

Reactions on a Surface

A common reaction mechanism on the surface, called *Langmuir–Hinshelwood kinetics,* involves the reaction of two adsorbed species:

$$A(g) + S \underset{K_{d1}}{\overset{K_{a1}}{\rightleftharpoons}} A\!:\!S$$

$$B(g) + S \underset{K_{d2}}{\overset{K_{a2}}{\rightleftharpoons}} B\!:\!S$$

$$A\!:\!S + B\!:\!S \overset{K_r}{\longrightarrow} AB(g) + 2S$$

For ease of analysis let the reaction itself be the rate-limiting step. Then the surface concentrations of A and B are the thermal equilibrium values, from (7.5.19),

$$\theta_A = \frac{\mathcal{K}_A n_{AS}}{1 + \mathcal{K}_A n_{AS} + \mathcal{K}_B n_{BS}} \tag{9.4.35a}$$

$$\theta_B = \frac{\mathcal{K}_B n_{BS}}{1 + \mathcal{K}_A n_{AS} + \mathcal{K}_B n_{BS}} \tag{9.4.35b}$$

where $\mathcal{K}_A = K_{a1}/K_{d1}$ and $\mathcal{K}_B = K_{a2}/K_{d2}$. The rate of production of AB(g) is then

$$R_{AB} = K_r n_0'^2 \theta_A \theta_B \tag{9.4.36}$$

At low pressures, $\mathcal{K}_A n_{AS} \ll 1$ and $\mathcal{K}_B n_{BS} \ll 1$, the kinetics is second order,

$$R_{AB} = K_r \frac{K_{a1} K_{a2}}{K_{d1} K_{d2}} n_0'^2 n_{AS} n_{BS} \tag{9.4.37}$$

As previously, n_{AS} and n_{BS} are related to n_{A0} and n_{B0} by using (9.4.5).

A second reaction mechanism, called *Eley–Rideal kinetics,* involves the reaction of adsorbed A directly with an impinging gas-phase molecule B:

$$A(g) + S \underset{K_{d1}}{\overset{K_{a1}}{\rightleftharpoons}} A\!:\!S$$

$$B(g) + S \underset{K_{d2}}{\overset{K_{a2}}{\rightleftharpoons}} B\!:\!S$$

$$A\!:\!S + B(g) \overset{K_r}{\longrightarrow} AB(g) + S$$

Again assuming that the reaction itself is the rate-limiting step, then

$$R_{AB} = K_r n_0' \theta_A n_{BS} \tag{9.4.38}$$

which, at low pressures, reduces again to second-order kinetics,

$$R_{AB} = K_r \frac{K_{a1}}{K_{d1}} n_0' n_{AS} n_{BS} \tag{9.4.39}$$

PROBLEMS

9.1. Complex Reaction for Ozone   Consider the loss of ozone in a dilute, low-pressure O_3/O_2 gas mixture at standard (room) temperature due to the reactions

$$O_3 + O_2 \underset{K_3}{\overset{K_2'}{\rightleftharpoons}} O + O_2 + O_2$$

$$O + O_3 \xrightarrow{K_2} O_2 + O_2$$

(a) Find the reaction rate R (cm^{-3}-s^{-1}) for destruction of ozone based on the above reaction set. Estimate R using the data in Tables 8.2 and 8.3 for $n_{O_2} = 3.3 \times 10^{16}$ cm^{-3} and $n_{O_3} = 3.3 \times 10^{14}$ cm^{-3}.

(b) The reverse reaction,

$$O_2 + O_2 \xrightarrow{K_{-2}} O_3 + O$$

is not listed in Table 8.2. Find the rate constant K_{-2} for this reaction using (9.1.13) and the data in Table 8.2. The standard Gibbs free energies for formation of O and O_3 are 231.75 and 163.16 kJ/mol respectively, and the standard enthalpies of formation of O and O_3 are 249.17 and 142.7 kJ/mol, respectively.

9.2. Reaction Rate Calculations

(a) Consider the kinetics of a stable molecule A that "spontaneously" decomposes into molecules B and C,

$$A \xrightarrow{K_1} B + C$$

Determine the conditions for this to happen and obtain the first-order rate constant K_1 by considering the elementary reactions

$$A + A \underset{K_{-2}}{\overset{K_2}{\rightleftharpoons}} A^* + A$$

$$A^* \xrightarrow{K_{1*}} B + C$$

Assume that the last reaction is rate limiting.

(b) Consider the first-order reaction chain

$$A \xrightarrow{K_{AB}} B \xrightarrow{K_{BC}} C \xrightarrow{K_{CD}} D \xrightarrow{K_{DE}} E$$

Assuming that the concentration $n_A = n_{A0}$ and that all other n's are zero at time $t = 0$, and that $C \longrightarrow D$ is the rate-limiting reaction, then find an approximate expression for $n_E(t)$. Sketch on the same graph the time-varying behavior of n_A, n_B, n_C, n_D, and n_E.

9.3. Stepwise Ionization   Ionization can occur as a two-step process involving excited atoms:

$$e + A \xrightarrow{K_{ex}} e + A^*$$

$$e + A^* \xrightarrow{K_{iz*}} 2e + A^+$$

Competing reactions for loss of A^* are collisional deexcitation

$$e + A^* \xrightarrow{K_{dex}} e + A$$

and first-order losses

$$A^* \xrightarrow{K_1} A$$

where K_1 is the total first-order rate constant for loss of A^* due to radiative emission and to deexcitation at the reactor walls. Let $\mathcal{E}_{iz}$ and $\mathcal{E}_{iz*}$ be the ionization potentials of A and A^*, respectively, and let the statistical weights of A and A^* be the same. Assume that $T_e \ll \mathcal{E}_{iz*}$, $\mathcal{E}_{iz}$ and that $K_{iz*} \ll K_{dex}$.

(a) From detailed balance (8.5.14), show that

$$K_{ex} = K_{dex} \exp\left(-\frac{\mathcal{E}_{iz} - \mathcal{E}_{iz*}}{T_e}\right)$$

(b) Find n_{A^*} as a function of n_e, n_A, and the rate constants.

(c) Using the Thomson ionization rate constants (3.5.4) show that the ratio of two-step to single-step ionization rates is

$$\frac{R_{iz*}}{R_{iz}} = \frac{n_e K_{dex}}{n_e K_{dex} + K_1} \frac{\mathcal{E}_{iz}^2}{\mathcal{E}_{iz*}^2}$$

Hence, two-step ionization is *always* more important than single-step ionization for thermal equilibrium $K_1 \ll n_e K_{dex}$.

(d) Estimate the ratio R_{iz*}/R_{iz} for a typical low-pressure processing discharge. Is two-step ionization important or not?

9.4. Three-Body Recombination

(a) Estimate the rate constant (9.2.38) for K_2' at low pressures for the three-body recombination reaction

$$e + A^+ + M \longrightarrow A + M$$

by modifying the analysis done for positive–negative ion recombination leading to (9.2.40)–(9.2.42). You should obtain the scaling $K_2' = n_M K_3' \propto T_e^{-3/2}$

(b) Compare your result in (a) with the tabulated data in Table 8.3.

9.5. Dipole–Dipole Force   Consider two electric dipoles p_{d1} and p_{d2} oriented along x and separated by a distance r. Each dipole can be regarded as a pair of point charges $+q$ and $-q$ separated by a small distance $d \ll r$: $p_{d1} = p_{d2} = qd$. Using Coulomb's law for the electrostatic force on a point charge due to another point charge, show that the net force on dipole p_{d1} due to dipole p_{d2} is attractive and has a magnitude

$$F_x = \frac{6 p_{d1} p_{d2}}{4\pi\epsilon_0 r^4}$$

9.6. Diffusion Loss in an Asymmetric Discharge   Consider the creation, diffusion and loss of a species A in an *asymmetric* one-dimensional slab geometry, with a uniform rate of production G_A within the slab. Assume that one of the two electrode surfaces is *inactive*, such that the net flux of A to this surface is zero. The other electrode is *active*, such that a net flux Γ_{AS} of A is lost to this surface. Let n_{A0} and n_{AS} be the maximum density and the density at the active surface, respectively.

(a) Find $n_A(x)$ within the slab in terms of G_A and Γ_{AS}. Hence, find n_{A0} and n_{AS}.

(b) Show that the first-order rate constant for diffusive loss is $2D_{AB}/l^2$, where l is the width of the slab.

9.7. Normal and Dissociative Adsorption

(a) Consider the steady-state *chemisorption* and desorption reactions at a reactor wall at room temperature (300 K):

$$A(g) + S \underset{K_d}{\overset{K_a}{\rightleftharpoons}} A : S$$

$$A : S + A : S \xrightarrow{K_{d2}} A_2(g) + 2S$$

Make the following assumptions: $s_0 \approx 1$, the chemisorption well depth for atoms is $\mathcal{E}_{desor} = 3$ V, and the molecular dissociation energy is $\mathcal{E}_{diss} = 5$ V. Note that the activation energies for desorption of atoms and molecules are $\mathcal{E}_{desor}$ and $2\mathcal{E}_{desor} - \mathcal{E}_{diss}$, respectively (see Fig. 9.4). Also use as typical parameters: $\bar{v}_A \approx 8 \times 10^4$ cm/s, $n_0' \approx 10^{15}$ cm^{-2}, $n_{AS} \approx 10^{13}$ cm^{-3}, and use the preexponential factors for normal and associative desorption of 10^{14} s^{-1} and 0.1 cm^2/s, respectively. For these parameters, show that the reactor walls are completely passivated; i.e., $\theta \approx 1$.

(b) Show that for chemisorption with these parameters, virtually all atoms desorb as molecules rather than as atoms.

(c) Find the ratio Γ_{A_2}/Γ_A of the desorbing molecular flux Γ_{A_2} to the flux Γ_A of atoms incident on the surface, and show that this ratio is very small.

(d) Now consider *physisorption* of A along with desorption of A and A_2 on a completely passivated wall (no chemisorption). Make the same assumptions as in (a), except let the physisorption well depth be $\mathcal{E}_{desor} \approx 0.2$ V; hence the activation energy for desorption of atoms is $\mathcal{E}_{desor}$, but the activation energy for desorption of molecules is zero (this reaction is now exothermic). Use a preexponential factor for normal desorption of 10^{15} s^{-1} and an associative desorption rate constant of 0.1 cm^2/s. For these parameters, show that the surface coverage for physisorption is *very small*; $\theta \ll 1$.

(e) Show that for physisorption with these parameters virtually all atoms desorb as atoms rather than as molecules.

(f) Find the ratio Γ_{A_2}/Γ_A of the desorbing molecular flux Γ_{A_2} to the flux Γ_A of atoms incident on the surface, and show that this ratio is very small.

Note that in view of your results in (c) and (f), the surface recycles most reactive atoms back into the discharge as atoms. This is typical for fluorine atoms.

CHAPTER 10

PARTICLE AND ENERGY BALANCE IN DISCHARGES

10.1 INTRODUCTION

For low-pressure discharges the plasma is not in thermal equilibrium and the electrical power is coupled most efficiently to plasma electrons. In the bulk plasma, energy is transferred inefficiently from electrons to ions and neutrals by weak collisional processes; for ions, energy can also be coupled by weak ambipolar electric fields. The fraction of energy transferred by elastic collision of an electron with a heavy ion or neutral is of order $2m/M \sim 10^{-4}$, where m and M are the electron and heavy-particle masses. Hence the electron temperature T_e greatly exceeds the ion and neutral temperatures, T_i and T, respectively, in the bulk; typically $T_e \sim 2$–5 V, whereas T_i and T are a few times room temperature (0.026 V). However, dissociation and excitation processes (see Section 8.3) can create a subgroup of relatively high-energy heavy particles. Also, the ambipolar electric fields accelerate positive ions toward the sheath edge, and typically, the ions in the bulk acquire a directed energy at the sheath edge of order $T_e/2$.

At low pressures the mean free path for ionizing electrons, with energies of 10–15 V, can be comparable to the discharge dimensions. Hence, even if the electrical power is deposited in a small volume within an unmagnetized discharge, the electron–neutral ionization rate ν_{iz} can be relatively uniform, since the ionization occurs on the distance scale of this mean free path. In magnetized plasmas, on the other hand, the ionization may be highly nonuniform as the magnetized electrons have trouble crossing field lines, so ionization along a magnetic flux tube might be uniform but significant radial nonuniformities may persist. In addition, the propagation and absorption of exciting electromagnetic fields can depend on the charge density distribution. In some instances, the density profile can steer power into regions of higher or lower density and make the plasma more or less uniform, respectively (see Chapter 13). At intermediate pressures, the mean free path for ionizing electrons is usually small compared to the discharge dimensions. Hence, even for a uniform T_e, the ionization rate ν_{iz} is nonuniform, with $\nu_{iz} = K_{iz} n_e$, with n_e varying within the plasma according to a diffusion solution.

The electron distribution function f_e need not be Maxwellian. However, insightful estimates of source operation can be obtained by approximating f_e to be Maxwellian, with T_e and the various electron collisional rates assumed to be uniform within the bulk plasma. Electron–neutral collisional processes are important not only for particle production (ionization, dissociation), but also for other collisional energy losses (excitation, elastic scattering). Ion–neutral collisions (charge transfer, elastic scattering) are also important in determining particle production, plasma transport, and ion energy distributions at a substrate surface. The myriad of collisional processes that can occur in molecular feedstock gas mixtures can obscure the fundamental principles of particle and energy balance. Consequently, a noble gas, such as argon, is often used as a reference for describing discharge operation. Although this provides some understanding of plasma properties, it provides little understanding of gas and surface chemistry, which are critical to most processing applications. Furthermore, most process gases are molecular and electronegative (containing some negative ions), leading to significant differences in plasma properties compared to argon. To obtain insight into the more complicated plasma and chemical phenomena that occur in typical materials processing discharges, we examine some properties of electronegative oxygen discharges.

We consider particle and energy balance for two basic types of discharges in this chapter, electropositive and electronegative. In electropositive discharges, there are only two species that are normally considered, electrons and one positive ion species. The rather simple diffusion analysis of Sections 5.2–5.3, or some relatively straightforward modification of it, is usually adequate to treat the particle transport. Similarly, sheath dynamics is treated as in Sections 6.2–6.3. In Section 6.4, we also included a negative ion species, in preparation for a treatment of particle transport in electronegative plasma. In electronegative plasma at least one additional ion species with negative charge is present. Although charge neutrality still holds within the bulk plasma, the low-mobility, low-temperature, negative ions may constitute most of the negative charge, thus profoundly influencing the dynamics. In addition, where for low-pressure electropositive plasmas we usually neglect volume electron–ion recombination, it is generally not possible to neglect recombination of negative and positive ions because this process has a very large rate constant (see Section 8.4). This latter effect makes the diffusion equation fundamentally nonlinear, and therefore much more difficult to solve.

In Section 10.2 we present the simpler treatment of electropositive plasmas, where approximate analytic solutions for the equilibrium can be obtained. For that case the equations for particle balance and the equation for energy balance decouple, the former giving the electron temperature, and the latter giving the electron and ion densities. In Section 10.3 we extend the discussion to electronegative plasmas, which are considerably more complicated. Simplifying assumptions must be made to specify an ambipolar diffusion coefficient for the positive ions, and the resulting equations are fundamentally nonlinear. The particle and energy balance equations are also coupled, further complicating the analysis. As we shall see, rather strong simplifying assumptions are required to obtain analytic solutions.

In any case, these models for the plasma equilibrium are not complete. The voltage across a plasma sheath cannot be specified independently of the heating mechanism and the power absorbed by the plasma. To obtain a complete heating model we must specify the method of sustaining the plasma from an external energy source, and determine how that source transfers energy to the electrons and (indirectly) to the ions. We consider various ways of transferring energy from fields to plasma discharges in Chapters 11–14. The resulting electron heating mechanisms are of the following types:

- Ohmic heating
- Stochastic heating
- Resonant wave–particle interaction heating
- Secondary electron emission heating

Ohmic heating is present in all discharges due to the transfer of energy gained from the acceleration of electrons in electric fields to thermal electron energy through collisional processes. Ohmic heating is particularly important at high pressures at which the collision frequency is high, where it can be the dominant heating mechanism.

Stochastic electron heating has been found to be a powerful mechanism in capacitive rf discharges. Here electrons impinging on the oscillating sheath edge suffer a change of velocity upon reflection back into the bulk plasma. As the sheath moves into the bulk, the reflected electrons gain energy; as the sheath moves away, the electrons lose energy. However, averaging over an oscillation period, there is a net energy gain. Stochastic heating is often the dominant heating mechanism in low pressure capacitive discharges. We shall discuss this heating mechanism in detail in Chapter 11 and consider a similar mechanism for inductive discharges in Chapter 12.

Wave–particle interactions are a fundamental method of transferring energy from fields to electrons and are an important mechanism of electron heating in high-density discharges such as electron cyclotron resonance, helicon, and surface wave sources. The heating can involve both collisional (ohmic) and collisionless energy transfer. We consider these processes in Chapter 13.

Secondary emission heating does not play a central role in most low-pressure discharges. At high pressures, especially in dc and capacitive rf discharges, secondary emission can play a crucial role in plasma production and can also contribute substantially to electron heating. We consider some of these effects in Chapters 11 and 14. The plasma heating mechanism often defines the type of plasma that is generated, as will be seen in the following chapters.

10.2 ELECTROPOSITIVE PLASMA EQUILIBRIUM

Basic Properties

We consider the example of argon discharges. The most important rate constants for electron collisions in argon are K_{iz}, K_{ex}, and K_{el} for electron–neutral ionization, excitation, and momentum transfer. These are given in Fig. 3.16 as a function of T_e. Using these rate constants, the collisional energy $\mathcal{E}_c$ lost per electron–ion pair created was defined in (3.5.8) and is plotted versus T_e in Fig. 3.17. The most important cross sections for ion–neutral collisions in argon are for resonant charge transfer and elastic scattering. As shown in Fig. 3.15, the cross section for resonant charge transfer of Ar^+ on Ar somewhat exceeds that for elastic scattering. The combined ionic momentum transfer cross section σ_i for these two processes is large ($\sigma_i \approx 10^{-14}$ cm^2) and relatively constant for the (thermal) ion energies of interest. The corresponding ion–neutral mean free path is given in (3.5.7): $\lambda_i = 1/n_g\sigma_i$, where n_g is the neutral argon density.

In addition to collisional energy losses, $\mathcal{E}_c$, electrons and ions carry kinetic energy to the walls. For Maxwellian electrons, the mean kinetic energy lost per electron lost is $2T_e$. The mean kinetic energy lost per ion lost is $\mathcal{E}_i$, which is mainly due to acceleration in the dc potential across the sheath. Evaluation of $\mathcal{E}_i$ depends on a calculation of the sheath potential, which can vary greatly depending on the method of plasma production. Summing the three contributions yields the total energy lost per ion lost from the system:

$$\mathcal{E}_T = \mathcal{E}_c + 2T_e + \mathcal{E}_i \tag{10.2.1}$$

At pressures for which the ion loss velocity is the Bohm velocity u_B (see (6.5.9) and accompanying discussion), the overall discharge power balance for a cylindrical plasma having radius R and length l can be written in terms of $\mathcal{E}_T$ as

$$P_{abs} = en_s u_B A \mathcal{E}_T \tag{10.2.2}$$

where P_{abs} is the power absorbed by the plasma, n_s is the ion density at the plasma–sheath edge, and A is the area for particle loss. The Bohm velocity is relatively constant for a given ion mass and for the typical limited range of T_e's of 2–5 V. Hence n_s is controlled by $\mathcal{E}_T$, A, and P_{abs}.

The relation between the density n_s at the sheath edge and the density n_0 at the plasma center is complex because the ambipolar transport of ions and electrons spans the regime $\lambda_i \sim R$, l, depending on the pressure and the values for R and l. As discussed in Chapter 5, there are three regimes.

(a) Low pressure: $\lambda \gtrsim (R, l)$ This *Langmuir* regime was described Section 5.3. The ion transport is collisionless and well described by an ion free-fall profile (Fig. 5.3 in plane parallel geometry) within the bulk plasma. This profile is relatively flat near the plasma center and dips near the sheath edge, with $n_s/n_0 \approx 0.5$ for $R \gg l$ (planar geometry) and $n_s/n_0 \approx 0.4$ for $l \gg R$ (infinite cylinder geometry).

(b) Intermediate pressures: $(R, \ l) \gtrsim \lambda_i \gtrsim (T_i/T_e)(R, \ l)$ In this regime, also described in Section 5.3, the transport is diffusive. However, the ion drift velocity u_i much exceeds the ion thermal velocity within most of the bulk plasma, leading to a nonlinear diffusion equation with the solution (5.3.7) for the density profile in plane parallel geometry shown in Fig. 5.2b. Again the profile is relatively flat in the center and steep near the sheath edge. As discussed in Section 5.3, joining the collisionless (low-pressure) and collisional (intermediate-pressure) results leads to the estimates (5.3.13) and (5.3.15), repeated here:

$$h_l \equiv \frac{n_{sl}}{n_0} \approx 0.86 \left(3 + \frac{l}{2\lambda_i} \right)^{-1/2} \tag{10.2.3}$$

at the axial sheath edge, and

$$h_R \equiv \frac{n_{sR}}{n_0} \approx 0.80 \left(4 + \frac{R}{\lambda_i} \right)^{-1/2} \tag{10.2.4}$$

at the radial sheath edge.

(c) High pressures: $\lambda_i \lesssim (T_i/T_e)(R, \ l)$ In this regime, described in Section 5.2, the transport is diffusive, and the density profile is well described by a J_0 Bessel function variation along r and a cosine variation along z. For this highly collisional case the assumption of a relatively uniform density within the plasma bulk, falling sharply near the sheath edge, is not good. In this regime, for the planar case and assuming that the ionization rate is proportional to n_e, as shown in Problem 5.4,

$$h_l = \frac{n_{sl}}{n_0} = \left[1 + \left(\frac{l}{\pi} \frac{u_B}{D_a} \right)^2 \right]^{-1/2} \tag{10.2.5}$$

where u_B is the Bohm velocity* and D_a is the ambipolar diffusion coefficient. The cylindrical configuration can also be analyzed in a straightforward way, as described below. However, as pointed out in Section 5.2, the ionization may not be proportional to n_e. For example, in the planar capacitive rf discharge described in Chapter 11, stochastic heating produces hotter electrons primarily near the plasma edge, while at higher pressures where ohmic heating dominates, a nonuniform plasma also favors heating near the edge (see Problem 4.5). For edge heating in the regime $\lambda_i/l \lesssim T_i/T_e$, the mean free path of the energetic (ionizing) electrons is generally less than the discharge length. In this situation, the hot electrons perform most of the ionization in the edge region, leading to diffusion solutions with flatter density profiles. For further consideration of this regime in ohmically heated capacitive rf discharges, the reader is referred to Godyak (1986, Sections 5.4 and 5.5).

---

*For $\lambda_i \lesssim \lambda_{De}$, u_B must be replaced by u_s defined in (6.5.9).

Uniform Density Discharge Model

We consider a simple cylindrical discharge model to estimate the plasma parameters and their variation with power, pressure, and source geometry. The electron temperature T_e, the ion-bombarding energy $\mathcal{E}_i$, the plasma density n_0, and the ion current density J_i are the most significant quantities for plasma processing applications. We assume a near uniform (in the bulk) cylindrical plasma with (10.2.3) and (10.2.4) giving the ratios of sheath to bulk density. We assume Maxwellian electrons absorbing an electrical power P_{abs}.

We first determine T_e by equating the total surface particle loss to the total volume ionization,

$$n_0 u_B (2\pi R^2 h_l + 2\pi R l h_R) = K_{iz} n_g n_0 \pi R^2 l \tag{10.2.6}$$

Equation (10.2.6) can be rewritten

$$\frac{K_{iz}(T_e)}{u_B(T_e)} = \frac{1}{n_g d_{eff}} \tag{10.2.7}$$

where

$$d_{eff} = \frac{1}{2} \frac{Rl}{Rh_l + l h_R} \tag{10.2.8}$$

is an effective plasma size, and the T_e dependence of K_{iz} and u_B is explicitly shown. Given n_g and d_{eff}, we can solve (10.2.7) for T_e, obtaining, for argon with K_{iz} in Fig. 3.16, the result for T_e, for typical plasma pressures and sizes, shown in Fig. 10.1. We see that T_e varies over a narrow range between 2 and 5 V. We note that T_e is determined by particle conservation alone, and is independent of the plasma density and therefore the input power.

The ion-bombarding energy $\mathcal{E}_i$ is the sum of the ion energy entering the sheath and the energy gained by the ion as it traverses the sheath. The ion velocity entering the sheath is u_B, corresponding to a directed energy of $T_e/2$. The sheath voltage V_s takes various forms depending on whether significant rf or dc currents are drawn to the surface, as described in Sections 6.2 and 6.3. For an insulating wall, the ion and electron fluxes must balance in the steady state, leading to (6.2.17), which for $V_s \equiv -\Phi_w$ is

$$V_s = \frac{T_e}{2} \ln\left(\frac{M}{2\pi m}\right) \tag{10.2.9}$$

or $V_s \approx 4.7\, T_e$ for argon. Accounting for the initial energy, we obtain $\mathcal{E}_i \approx 5.2\, T_e$. At an undriven conducting wall, the fluxes need not balance, although the integrated fluxes (particle currents) must balance. However, if the fluxes are not too dissimilar, then (10.2.9) remains a good estimate due to the logarithmic dependence of V_s on the ratio of fluxes.

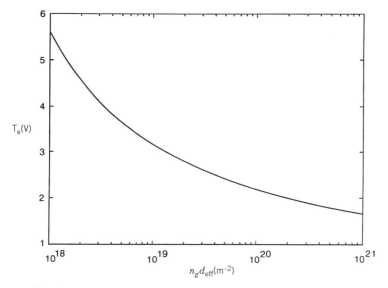

FIGURE 10.1. T_e versus $n_g d_{eff}$ for Maxwellian electrons in argon.

A *high-voltage sheath* exists at the negatively driven electrode (cathode) surface of a dc discharge, as described in Section 6.3, with the sheath voltage

$$V_s \approx V_{dc} \tag{10.2.10}$$

where V_{dc} is the cathode–anode voltage. Similar high-voltage sheaths exist near capacitively driven electrode surfaces. For a symmetrically driven capacitive rf discharge,

$$V_s \approx 0.4\,V_{rf} \tag{10.2.11}$$

where V_{rf} is the driving voltage across the electrodes. For a strongly asymmetrically driven discharge,

$$V_s \approx 0.8\,V_{rf} \tag{10.2.12}$$

at the driven (powered) electrode. More precise calculations of the coefficients in (10.2.10)–(10.2.12) are given in Chapter 14 for dc discharges and in Chapter 11 for capacitive rf discharges. The ion kinetic energy lost at a surface is then

$$\mathcal{E}_i = V_s + \frac{1}{2}T_e \tag{10.2.13}$$

where V_s is given by one of (10.2.9)–(10.2.12). We see from the above discussion that estimating ion energy is not so simple as it depends not only on electron temperature but also on source geometry and the application of bias voltages. Undriven sheath thicknesses s rarely exceed a few Debye lengths λ_{De}; hence such sheaths are less than a millimeter thick in typical discharges. The thickness of a high-voltage sheath follows that of a Child law, with s given by (6.3.14) with $V_0 \sim V_s$. For typical dc or capacitive rf discharges, s is of the order of 0.5 cm.

Finally, we estimate the plasma density n_0. Accounting for possibly different values of n_s at the axial and radial sheath edge [(10.2.3) and (10.2.4)], we introduce an effective area A_{eff} using $\int n_s dA = n_0 A_{eff}$ and solve (10.2.2) to obtain

$$n_0 = \frac{P_{abs}}{eu_B A_{eff} \mathcal{E}_T} \tag{10.2.14}$$

where

$$A_{eff} = 2\pi R(Rh_l + lh_R) \tag{10.2.15}$$

For a specified P_{abs}, and T_e determined from Fig. 10.1, we obtain n_0 from (10.2.14). Note that n_0 is determined by the total power balance in the discharge and is a function of pressure only through the dependence of h_l and h_R on p and through the weaker dependence of T_e on p.

Example 1   Consider a cylindrical discharge having low-voltage sheaths at all surfaces, with V_s given by (10.2.9). Let $R = 0.15$ m, $l = 0.3$ m, $n_g = 3.3 \times 10^{19}$ m^{-3} ($p = 1$ mTorr at 298 K), and $P_{abs} = 800$ W. At 1 mTorr, $\lambda_i \approx 0.03$ m from (3.5.7). Then from (10.2.3) and (10.2.4) $h_l \approx h_R \approx 0.3$, from (10.2.8) $d_{eff} \approx 0.17$ m. From Fig. 10.1, $T_e \approx 4.1$ V, and from Fig. 3.17, $\mathcal{E}_c \approx 42$ V. Using (10.2.1) with $\mathcal{E}_i \approx 5.2\,T_e \approx 21$ V, we find $\mathcal{E}_T \approx 72$ V. The Bohm velocity is $u_B \approx 3.1 \times 10^3$ m/s, and $A_{eff} \approx 0.13$ m^2 from (10.2.15). Substituting these values into the energy balance (10.2.14) yields $n_0 \approx 1.8 \times 10^{17}$ m^{-3}, corresponding to a flux at the axial boundary $\Gamma_{il} = n_0 h_l u_B \approx 1.7 \times 10^{20}$ m^{-2}-s^{-1} or an ion current density of $J_{il} \approx 2.7$ mA/cm^2.

Example 2   If a strong dc magnetic field is applied along the cylinder axis, then particle loss to the circumferential wall is inhibited. For the parameters of Example 1, in the limit of no radial loss, a calculation similar to that in Example 1 yields $n_0 \approx 5.8 \times 10^{17}$ m^{-3}, and $J_{il} \approx 7.8$ mA/cm^2. There is a significant increase in charge density and ion flux due to the magnetic field confinement. The details of the calculation for this example and for Example 3 are left to a problem.

Example 3   Consider the parameters of Example 2 for a symmetrically driven rf discharge with high-voltage sheaths $V_s \approx 500$ V at each of the cylinder endwalls. There is a large increase in $\mathcal{E}_i$ and therefore in $\mathcal{E}_T$, which leads to a significant reduction in n_0 and J_{il}; $n_0 \approx 8.1 \times 10^{16}$ m^{-3} and $J_{il} \approx 1.0$ mA/cm^2.

The last two examples illustrate an important difference between discharges having high-voltage sheaths over a significant fraction of the surface area and discharges having low-voltage sheaths at all surfaces. The densities are significantly lower and the ion-bombarding energies are significantly higher for the same input power and geometry for the high-voltage case than for the low-voltage case. Consequently, low-pressure discharges naturally divide into two types: high- and low-density discharges.

(a) Low density discharges   These discharges have high-voltage sheaths over a significant surface area. We discuss the important cases of capacitive rf discharges in Chapter 11 and dc discharges in Chapter 14.

(b) High density discharges   These discharges have low-voltage sheaths near almost all surfaces. We discuss the cases of rf driven inductive and helical resonator discharges in Chapter 12; and helicon, ECR, and surface wave discharges in Chapter 13. The ion bombarding energy $\mathcal{E}_i$ in high-density discharges is often too low for the materials process of interest. In this case, the substrate surface is often capacitively driven by an additional rf power supply to increase $\mathcal{E}_i$. In this way, the desired ion-bombarding energy at an rf powered substrate holder can be obtained. The additional ion energy flux $en_s u_B \mathcal{E}_i$ striking the wafer holder is supplied by the rf power source driving the holder. The independent control of the ion energy and the ion flux hitting the substrate is a highly desirable feature of high-density (low sheath voltage) discharges.

It should be noted that V_s was arbitrarily chosen to be 500 V in Example 3. In general, as mentioned in Section 10.1, it is not possible to choose the power absorbed P_{abs} and the discharge voltage V_{rf} (or V_{dc}) independently, as was done in Example 3. Therefore, for capacitive rf and dc discharges, the preceding analysis is not complete. We elaborate this in Chapter 11, where we determine the I–V characteristic for capacitive rf discharges and complete the analysis presented here.

Nonuniform Discharge Model

At relatively high pressures, $\lambda_i \lesssim (T_i/T_e)l$, the ambipolar diffusion profile in one dimensional slab geometry was obtained, in Section 5.2, by solving the ion conservation equation,

$$\frac{d\Gamma_i}{dx} = K_{iz} n_g n_e \tag{10.2.16}$$

where $n_e = n_i$ and

$$\Gamma_i = -D_a \frac{dn_i}{dx} \tag{10.2.17}$$

to obtain the density n_i and particle flux Γ_i. These results, repeated here, are

$$n_i(x) = n_0 \cos \beta x \tag{10.2.18}$$

$$\Gamma_i(x) = D_a \beta n_0 \sin \beta x \tag{10.2.19}$$

with $\beta = (\nu_{iz}/D_a)^{1/2}$ $(\nu_{iz} = K_{iz}n_g)$, and D_a is the constant ambipolar diffusion coefficient. The simplest assumption made to obtain a solution is that $n_i \approx 0$ at $x = \pm l/2$, which gives $\beta = \pi/l$. This is reasonable because $\lambda_i \ll l$. Integrating (10.2.16) from $x = 0$ to $x = l/2$ and assuming that K_{iz} is independent of x, we obtain

$$\Gamma_i \left(\frac{l}{2}\right) = K_{iz}n_g \int_0^{l/2} n_i(x)\, dx \tag{10.2.20}$$

Equation (10.2.20) expresses the overall particle conservation in a nonuniform plasma slab, in analogy to (10.2.6), which expresses this same conservation for a uniform cylinder of plasma. Substituting (10.2.18) and (10.2.19) with $x = l/2$ in (10.2.20) and performing the integration, we obtain

$$\frac{\pi}{l}D_a = \frac{l}{\pi}K_{iz}n_g \tag{10.2.21}$$

Since

$$D_a = \frac{eT_e}{Mn_gK_{mi}} \tag{10.2.22}$$

where $K_{mi}(T_i)$ is the ion–neutral momentum transfer rate constant, and substituting $u_B = (eT_e/M)^{1/2}$, (10.2.21) can be rewritten

$$\frac{[K_{mi}K_{iz}(T_e)]^{1/2}}{u_B(T_e)} = \frac{\pi}{n_g l} \tag{10.2.23}$$

Equation (10.2.23) is analogous to (10.2.7) in that it determines T_e for a given n_g and l.

Similarly, the total power absorbed by a unit area of the discharge can be written as

$$S_{abs} = 2\Gamma_i \left(\frac{l}{2}\right) e(\mathcal{E}_e + \mathcal{E}_i) + 2 \int_0^{l/2} e\mathcal{E}_c K_{iz}n_g n_e(x)\, dx \tag{10.2.24}$$

Using (10.2.20) to eliminate the integral in (10.2.24), we find

$$S_{abs} = 2\Gamma_i \left(\frac{l}{2}\right) e\mathcal{E}_T \tag{10.2.25}$$

Substituting (10.2.19) with $\sin(\beta l/2) = 1$ in (10.2.25), and solving for n_0, we obtain

$$n_0 = \frac{S_{abs}l}{2\pi D_a e\mathcal{E}_T} \tag{10.2.26}$$

which is analogous to (10.2.14).

This procedure can be generalized to a three-dimensional nonuniform discharge provided the basic diffusion equation can be solved. For example, for a constant D_a, the diffusion equation

$$-D_a \nabla^2 n_i = \nu_{iz} n_e \tag{10.2.27}$$

with $n_e = n_i$, can be expressed in cylindrical coordinates, with azimuthal symmetry, as

$$\frac{d^2 n_i}{dr^2} + \frac{1}{r}\frac{dn_i}{dr} + \frac{d^2 n_i}{dz^2} + \frac{\nu_{iz}}{D_a} n_i = 0 \tag{10.2.28}$$

With no axial variation ($d^2 n_i / dz^2 = 0$), (10.2.28) is Bessel's equation, with solution

$$n = n_0 J_0(\beta r) \tag{10.2.29}$$

where J_0 is the zero-order Bessel function. Making the same assumptions as in slab geometry that $n_s \approx n(R) \approx 0$ gives

$$\beta = \frac{\chi_{01}}{R} \tag{10.2.30}$$

where $\chi_{01} \approx 2.405$ is the first zero of the Bessel function. If there is variation in z also, then the variables can be separated, in the usual way, by assuming a product solution, to give

$$n_i(r, z) = n_0 J_0\left(\frac{\chi_{01} r}{R}\right) \cos\left(\frac{\pi z}{l}\right) \tag{10.2.31}$$

The particle flux is

$$\Gamma_{iz}(r) = -D_a \frac{\partial n_i}{\partial z} = \frac{\pi D_a}{l} n_0 J_0\left(\frac{\chi_{01} r}{R}\right) \tag{10.2.32}$$

at the endwall, $z = l/2$, and is

$$\Gamma_{ir}(z) = -D_a \frac{\partial n_i}{\partial r} = \frac{\chi_{01} D_a}{R} n_0 J_1(\chi_{01}) \cos\left(\frac{\pi z}{l}\right) \tag{10.2.33}$$

at the radial boundary, $r = R$. Letting $\Gamma_i = -D_a \nabla n_i$ and integrating (10.2.27) over the volume yields

$$\oint_S \Gamma_i \cdot dS = K_{iz} n_g \int n_i 2\pi r \, dr \, dz \tag{10.2.34}$$

The right- and left-hand sides can be evaluated using (10.2.31) and (10.2.32)–(10.2.33), respectively, to obtain the overall particle conservation in a plasma cylinder.

All the plasma parameters can be estimated in a manner completely analogous to the procedure used for a uniform density plasma cylinder. We leave the details to a problem.

10.3   ELECTRONEGATIVE PLASMA EQUILIBRIUM

The addition of a negative ion species greatly complicates the analysis of particle and energy balance in low-pressure discharges, for the following reasons:

1. An additional particle conservation equation is required for the negative ions.
2. The expression for the Bohm velocity, which signals the end of the plasma and the beginning of the sheath, is modified by the presence of negative ions near the plasma–sheath edge. This modification was described in Section 6.4.
3. The diffusive loss is, in general, no longer ambipolar. This means that a set of three nonlinear diffusion equations for positive ions, negative ions, and electrons must be solved simultaneously.
4. Under certain conditions, the diffusive loss is ambipolar. This leads to a single diffusion equation, greatly simplifying the analysis. However, the resulting equation is still nonlinear.

The latter complication is crucial, making analysis difficult and still an active area of research. If the diffusion is not ambipolar, then the analysis is intractable. Of course, the equations can always be solved numerically, but this gives little insight.

Ambipolar Diffusion

As in electropositive plasmas, for each charged species we can write a flux equation

$$
\begin{aligned}
\Gamma_+ &= -D_+ \nabla n_+ + n_+ \mu_+ E \\
\Gamma_- &= -D_- \nabla n_- - n_- \mu_- E \\
\Gamma_e &= -D_e \nabla n_e - n_e \mu_e E
\end{aligned}
\tag{10.3.1}
$$

where all symbols have their usual meaning. In equilibrium the sum of the currents must balance:

$$
\Gamma_+ = \Gamma_- + \Gamma_e
\tag{10.3.2}
$$

We also have

$$
n_+ = n_- + n_e
\tag{10.3.3}
$$

which is the usual quasineutral plasma approximation.

Using these five equations and introducing the ratio of negative ion to electron density, $\alpha = n_-/n_e$, we can eliminate Γ_-, Γ_e, n_+, n_e, and E to obtain an equation in Γ_+, α, and the gradients:

$$\Gamma_+ = -\frac{(\mu_e + \mu_-\alpha)D_+ + \mu_+(1 + \alpha)D_e\left(\dfrac{\nabla n_e}{\nabla n_+}\right) + \mu_+(1 + \alpha)D_-\left(\dfrac{\nabla n_-}{\nabla n_+}\right)}{\mu_e + \mu_-\alpha + \mu_+(1 + \alpha)}\nabla n_+.$$

$$(10.3.4)$$

We have factored out ∇n_+ to put the equation formally in the usual form of an ambipolar diffusion coefficient

$$\Gamma_+ = -D_{a+}\nabla n_+ \qquad (10.3.5)$$

but we note that this is not equivalent to D_a for electropositive plasmas because D_{a+} is a function of n_e, n_- and position, both through α and through the gradients.

The form (10.3.4) was derived by Rogoff (1985) and also, implicitly, by Thompson (1959). However, because (10.3.5) depends explicitly on the other variables it cannot be independently solved for n_+, but is coupled to n_- and n_e. In this form three diffusion equations must be written for the three variables and solved simultaneously. This is clearly a difficult numerical procedure. Thompson (1959) attempted to circumvent this difficulty by assuming that both negative species are in Boltzmann equilibrium, and thus the logarithmic gradients are related by their temperature ratio. Although this assumption is valid for the mobile electrons, it is not necessarily valid for the negative ions. The validity of the approximation will be examined in an example. Introducing the temperature ratio

$$\gamma = \frac{T_e}{T_i} \qquad (10.3.6)$$

where T_i is the common temperature of both ionic species (typically $\gamma \sim 100 \gg 1$), the Boltzmann relation (2.4.15) applied to the negative species yields

$$\frac{\nabla n_-}{n_-} = \gamma\frac{\nabla n_e}{n_e} \qquad (10.3.7)$$

Using (10.3.7) together with

$$\nabla n_+ = \nabla n_- + \nabla n_e$$

we obtain the ratios

$$\frac{\nabla n_e}{\nabla n_+} = \frac{1}{1 + \gamma\alpha}, \qquad \frac{\nabla n_-}{\nabla n_+} = \frac{\gamma\alpha}{1 + \gamma\alpha} \qquad (10.3.8)$$

Substituting (10.3.8), together with the Einstein relations from (5.1.9),

$$\frac{D_-}{D_+} = \frac{\mu_-}{\mu_+}, \qquad \frac{D_e}{D_+} = \gamma\frac{\mu_e}{\mu_+}$$

into (10.3.4), we obtain, after a little algebra,

$$D_{a+} = D_+ \frac{(1 + \gamma + 2\gamma\alpha)\left[1 + \alpha(\mu_-/\mu_e)\right]}{(1 + \gamma\alpha)\left[1 + (\mu_+/\mu_e)(1 + \alpha) + \alpha(\mu_-/\mu_e)\right]} \qquad (10.3.9)$$

which is the form given by Thompson. We note immediately, since μ_-/μ_e, $\mu_+/\mu_e \ll 1$, that for all reasonable cases the second parentheses in both the numerator and denominator are approximately equal to one, yielding

$$D_{a+} \approx D_+ \frac{1 + \gamma + 2\gamma\alpha}{1 + \gamma\alpha} \qquad (10.3.10)$$

Thompson plotted D_{a+} from (10.3.9) with α as a parameter. The structure is easily seen from the simpler form (10.3.10). For $\alpha \gg 1$, γ cancels out such that $D_{a+} \approx 2D_+$. When α decreases below 1, but $\gamma\alpha \gg 1$, $D_{a+} \approx D_+/\alpha$ such that D_{a+} increases with decreasing α. For $\gamma\alpha < 1$, $D_{a+} \approx \gamma D_+$, which is the usual ambipolar diffusion coefficient without negative ions. For plasmas in which $\alpha \gg 1$ in the center of the discharge, the entire transition region takes place over a small range of $1/\gamma < \alpha < 1$ near the discharge surface, such that the simpler value of

$$D_{a+} = 2D_+ \qquad (10.3.11)$$

holds over most of the plasma, except very near the surface where $n_- = 0$.

Oxygen Discharge

To focus attention on the basic principles of electronegative discharge equilibrium, we consider an oxygen plasma in which only three charged species are considered: O_2^+, created by electron impact ionization, O^-, created by dissociative attachment, and electrons. Negative ions are trapped within the discharge by the positive potential of the plasma with respect to all wall surfaces, and are assumed to be lost only by recombination with positive ions in the volume. Positive ions are lost to the walls by diffusion and in the volume by recombination with negative ions. The simplified set of volume reactions is

$$
\begin{array}{ll}
e + O_2 \rightarrow O_2^+ + 2e & \text{(ionization)} \\
e + O_2 \rightarrow O^- + O & \text{(dissociative attachment)} \\
O_2^+ + O^- \rightarrow O_2 + O & \text{(recombination)} \\
O_2^+ + O_2 \rightarrow O_2^+ + O_2 & \text{(elastic scattering)} \\
O_2^+ + O_2 \rightarrow O_2 + O_2^+ & \text{(charge transfer)} \\
O^- + O_2 \rightarrow O^- + O_2 & \text{(elastic scattering)}
\end{array}
$$

The latter three reactions lead to ion–neutral momentum transfers that result in effective diffusion coefficients for positive and negative ion species. The reaction rate constants are approximated to be

$$K_{iz} = 2.13 \times 10^{-14} \exp\left(-\frac{14.5}{T_e}\right) \text{ m}^3/\text{s} \tag{10.3.12a}$$

$$K_{att} = 7.89 \times 10^{-17} \exp\left(-\frac{3.07}{T_e}\right) \text{ m}^3/\text{s} \tag{10.3.12b}$$

$$K_{rec} = 1.4 \times 10^{-13} \text{ m}^3/\text{s} \tag{10.3.12c}$$

$$K_{mi} = 3.95 \times 10^{-16} \text{ m}^3/\text{s} \tag{10.3.12d}$$

Uniform Density Model

Let us first consider a uniform density model to illustrate the complexity of the analysis and the coupling between particle and energy balance equations in these discharges. A simple model of this type is called a zero-dimensional model, because all spatial variations are ignored. It is often used to provide a first estimate of the plasma parameters in complicated discharges. However, as will be shown later, the spatial variation can be quite important. Letting V and S be the volume and surface area of the plasma (sheath thicknesses are assumed to be small) and Γ_{+s} be the average positive ion flux normal to the surface, the equations for conservation of positive ions, negative ions, and energy within the volume are, in the steady state,

$$K_{iz}n_e n_g V - K_{rec}n_+ n_- V - \Gamma_{+s}S = 0 \tag{10.3.13}$$

$$K_{att}n_e n_g V - K_{rec}n_+ n_- V = 0 \tag{10.3.14}$$

$$P_{abs} = e\mathcal{E}_c K_{iz}n_e n_g V + \Gamma_{+s}Se(\mathcal{E}_e + \mathcal{E}_i) \tag{10.3.15}$$

where n_g is the O_2 gas density. For $\alpha_0 \gtrsim 1$, we estimate Γ_{+s} using a simple diffusion model with a diffusion coefficient (10.3.11):

$$\Gamma_{+s} \approx -2D_+ \nabla n_+ \approx \frac{2D_+ n_+}{d_{eff}} \tag{10.3.16}$$

where d_{eff} is the effective diffusion length in the bulk plasma and

$$2D_+ = \frac{2eT_i}{MK_{mi}n_g} \tag{10.3.17}$$

is the diffusion coefficient. For a slab geometry with $\lambda_i \ll l$, we shall see in the following subsection that an approximate parabolic solution gives $d_{eff} = l/4$. We note that $2D_+ \ll D_a$, the usual ambipolar diffusion coefficient in an electropositive plasma. Hence positive ion losses to the walls are much lower in an electronegative plasma than in an electropositive plasma.

Equations (10.3.13)–(10.3.15), along with the quasineutrality condition $n_+ = n_- + n_e$ given in (10.3.3), are four equations that must be simultaneously solved to determine n_+, n_-, n_e, and T_e, for the specified pressure (n_g), absorbed power (P_{abs}) and geometry ($\mathcal{V}$, S, and d_{eff}). We can solve for n_- and n_e in terms of n_+, using (10.3.3) and (10.3.14) to obtain the nonlinear relations

$$n_- = \frac{K_{att} n_g}{K_{att} n_g + K_{rec} n_+} n_+ \tag{10.3.18}$$

$$n_e = \frac{K_{rec} n_+}{K_{att} n_g + K_{rec} n_+} n_+ \tag{10.3.19}$$

which gives the simple form for the ratio

$$\alpha \equiv \frac{n_-}{n_e} = \frac{K_{att} n_g}{K_{rec} n_+} \tag{10.3.20}$$

Also, eliminating $n_+ n_-$ from (10.3.13) and (10.3.14), we find a linear expression for n_e in terms of n_+:

$$n_e n_g = \frac{2 D_+ n_+}{K_{iz} - K_{att}} \frac{S}{\mathcal{V} d_{eff}} \tag{10.3.21}$$

Substituting (10.3.19) into (10.3.21), we obtain

$$\frac{K_{rec} n_+ n_g}{K_{att} n_g + K_{rec} n_+} = \frac{2 D_+}{(K_{iz} - K_{att})} \frac{S}{\mathcal{V} d_{eff}} \tag{10.3.22}$$

Unlike the electropositive plasma in which the density cancels, to give (10.2.7), an equation for T_e alone, (10.3.22) depends on n_+. It is therefore necessary to use the equation for the absorbed power (10.3.15) to eliminate n_+. Substituting (10.3.21) into (10.3.15), we obtain

$$P_{abs} = 2 D_+ n_+ e \left(\mathcal{E}_c \frac{K_{iz}}{K_{iz} - K_{att}} + \mathcal{E}_e + \mathcal{E}_i \right) \frac{S}{d_{eff}} \tag{10.3.23}$$

Solving (10.3.23) for n_+ and substituting this into (10.3.22), we obtain a single implicit equation to determine T_e of the form

$$f\left(T_e, \, n_g, \, \frac{S}{\mathcal{V} d_{eff}}, \, \frac{P_{abs}}{\mathcal{V}}, \, \mathcal{E}_i, \, T_i \right) = 0 \tag{10.3.24}$$

where the T_e dependence arises from the rate constants, $\mathcal{E}_c$, $\mathcal{E}_e$, and, in some cases, $\mathcal{E}_i$. After obtaining T_e, the quantities n_+, n_-, and n_e are found from (10.3.23), (10.3.18), and (10.3.19), respectively.

The general solution is clearly very complicated in that T_e is not a function of $n_g d_{eff}$ alone, as it is for the electropositive case, but now depends separately on n_g and the geometry, as well as the discharge power P_{abs}. Some general features of the equilibrium should be noted. From (10.3.21) we see that $K_{iz} \geq K_{att}$ for an equilibrium to exist; this guarantees enough ionization to supply both volume and surface losses for the positive ions. To get a further feeling for the behavior of the plasma we examine limiting cases, depending on the ratio of volume to surface losses for the positive ions.

(a) Volume Loss Dominates   In this case we can drop $\Gamma_{+s} S$ in (10.3.13) and subtracting (10.3.14) we obtain $K_{iz} = K_{att}$. Using (10.3.12) we find that $T_e = 2.04$ V as a lower limit to the temperature. For this case the power balance (10.3.15), together with (10.3.13), gives

$$P_{abs} \approx \mathcal{V} e \mathcal{E}_c K_{rec} n_+^2 \tag{10.3.25}$$

which determines $n_+ \approx n_-$, and, from (10.3.14)

$$n_e \approx \frac{K_{rec} n_+^2}{K_{att} n_g} \tag{10.3.26}$$

We see from (10.3.26) and (10.3.25) that this solution holds at high pressure (n_g large) and low power (n_+ small) where $\alpha \gg 1$.

(b) Surface Losses Dominate   In this case (10.3.13) gives

$$K_{iz} n_e n_g \mathcal{V} = \Gamma_{+s} S \tag{10.3.27}$$

Inserting (10.3.27) into the power balance (10.3.15) to eliminate K_{iz} and using (10.3.16) for Γ_{+s}, we obtain n_+ as a function of p and P_{abs}. Then (10.3.20) determines α, with $n_e = n_+/(1 + \alpha)$ and $n_- = \alpha n_e$. Solving (10.3.27) then yields T_e. However, the approximation of (10.3.16) is good only for $\alpha \gg 1$, as seen from (10.3.10). As will be seen in the following subsection, this may not be consistent with (10.3.27). Either $\alpha_0 \gtrsim 1$, in which case $D_{a+} \approx 2D_+$ as assumed, or $\alpha_0 \lesssim 1$, in which case the negative ions are localized in the center of the discharge, such that $D_{a+} \approx D_a$ over most of the discharge region. Therefore, we write the flux more generally as

$$\Gamma_{+s} \approx \frac{D_{a+} n_+}{d_{eff}} \tag{10.3.28}$$

Using this in (10.3.27), we obtain

$$K_{iz} n_e n_g \mathcal{V} \approx \frac{2D_{a+}}{d_{eff}} n_+ S \tag{10.3.29}$$

However, if the negative ion profile is highly peaked within a small region in the center of the discharge, then the uniform model clearly cannot be used to determine

n_-, as in (10.3.18). A nonuniform description based on a solution to the appropriate diffusion equation is required.

Nonuniform Electronegative Equilibrium

As shown in Fig 10.2, the structure of an electronegative discharge can be nonuniform, consisting of two plasma regions: a central electronegative region and an edge electropositive region. We consider first the electronegative region alone. For simplicity we restrict our attention to the one-dimensional plane-parallel geometry. We have the positive ion diffusion equation

$$-\frac{d}{dx}\left[D_{a+}(\alpha)\frac{dn_+}{dx}\right] = K_{iz}n_g n_e - K_{rec}n_+ n_- \qquad (10.3.30)$$

Equation (10.3.30) cannot be solved directly, because it is a function of n_- and n_e as well as n_+, all functions of position. Assuming a Boltzmann equilibrium, we can integrate (10.3.7) to obtain

$$\frac{n_e}{n_{e0}} = \left(\frac{n_-}{n_{-0}}\right)^{1/\gamma} \qquad (10.3.31)$$

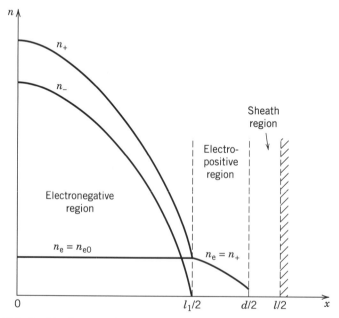

FIGURE 10.2. Positive ion, negative ion, and electron densities versus position for a plane parallel electronegative discharge, showing the electronegative, electropositive, and sheath regions.

from which we can eliminate n_e. We can then use charge neutrality (10.3.3) to eliminate n_-, such that (10.3.30) becomes

$$-\frac{d}{dx}\left[D_{a+}(n_+)\frac{dn_+}{dx}\right] = K_{iz}n_gn_e(n_+) - K_{rec}n_+n_-(n_+) \quad (10.3.32)$$

where $D_{a+}(n_+)$ is a rather complicated function of n_+ and one arbitrary constant $\alpha_0 = n_{-0}/n_{e0}$, the ratio of n_- to n_e at the plasma center. It is now possible to integrate (10.3.32) numerically, given the arbitrary constants. From (10.3.31) we can set $n_e \approx n_{e0}$ in (10.3.32) provided $n_-/n_{-0} \gtrsim 2^{-\gamma}$; i.e., in the region containing at least some negative ions. We use the relation for particle balance of negative ions to complete the set of equations. Assuming that the negative ion flux goes to zero at $x = l_1/2$, we then have the equations: positive ion particle balance,

$$-D_{a+}\frac{dn_+}{dx}\bigg|_{x=l_1/2} = \int_0^{l_1/2} K_{iz}n_gn_{e0}\,dx - \int_0^{l_1/2} K_{rec}n_+n_-(n_+)\,dx \quad (10.3.33)$$

negative ion particle balance,

$$\int_0^{d/2} K_{att}n_gn_e\,dx - \int_0^{l_1/2} K_{rec}n_+n_-(n_+)\,dx = 0 \quad (10.3.34)$$

and energy balance for the electrons in the bulk plasma (neglecting electron kinetic energy lost to the surface)

$$S_e = 2e\mathcal{E}_c \int_0^{d/2} K_{iz}n_gn_e\,dx \quad (10.3.35)$$

where $\mathcal{E}_c(T_e)$, the electron energy lost per electron–positive ion pair created, is a known function of T_e. We note that negative ions created over the whole plasma move into the negative ion region, so the integral on the LHS of (10.3.34) extends over the central electronegative region and the electropositive edge regions. We specify the bulk plasma electron collisional power losses per unit area, S_e, rather than the total power per unit area absorbed, S_{abs}, for simplicity of analysis and because the latter depends on evaluation of the ion energy lost at the surface, $\mathcal{E}_i$, which varies with the type of discharge employed (see Section 10.2). Given the electronegative plasma length l_1, and the power S_e, the three equations can be simultaneously solved for the three unknowns T_e, α_0, and n_{+0}. However, $l_1/2$ is not exactly known, but depends on the Bohm flux condition (6.4.7) at the plasma–sheath edge,

$$-D_{a+}\frac{dn_+}{dx}\bigg|_{x=d/2} = n_+\left(\frac{d}{2}\right)u_B(T_e, T_i, \alpha) \quad (10.3.36)$$

which indicates where the sheath begins. The Bohm velocity in (10.3.36) has a more general form than the usual expression $u_B = (eT_e/M_+)^{1/2}$ since negative ions may be present when (10.3.36) is satisfied. In this case, as described below, there is no electropositive region and $l_1 = d$.

There are actually three different electronegative density regimes depending on neutral pressure and applied power. (1) At low pressure and high power, α_0 is small. The negative ion density becomes quite small, $\alpha < \gamma^{-1}$, well within the plasma volume, such that much of the edge region behaves essentially electropositively, and the diffusion is not well described by $D_{a+} \approx 2D_+$ as given in (10.3.11). (2) In the opposite limit of high pressure and low power, $\alpha_0 \gg 1$ and a significant density of negative ions may exist where (10.3.36) is satisfied, giving no electropositive region, $D_{a+} \approx 2D_+$, and a significantly depressed Bohm velocity. (3) We might expect a large intermediate region to exist where the central α may be quite large but the edge α is essentially zero, giving $D_{a+} \approx 2D_+$ in the electronegative region, but allowing the usual Bohm velocity to be used at the sheath edge. We examine this further when we compare the solutions of (10.3.33)–(10.3.36) to plasma simulations. We find that there is usually a significant edge region in which the plasma is essentially electropositive.

Since for α_0 large, $D_{a+} = 2D_+$ over most of the plasma, we might ask whether a simpler solution with $D_{a+} = 2D_+$ and $n_e = n_{e0}$ might be adequate to describe the bulk plasma. Numerical calculations indicate that this is reasonable. However, there is not much gain over solving the more complete equations, because the nonlinearity in the last term of (10.3.32) does not permit an analytical solution for n_+ to be given explicitly.

Approximate Solutions

Consider the simpler problem in which α is sufficiently large that $D_{a+} \approx 2D_+$, but the effect of recombination can be neglected in determining the spatial distribution. The diffusion equation (10.3.32) then takes the simple form

$$-2D_+ \frac{d^2 n_+}{dx^2} = K_{iz} n_g n_{e0} \tag{10.3.37}$$

where (10.3.31) allows us to set $n_e \approx n_{e0}$. In this approximation $n_+(x)$ has a parabolic solution of the form (see Fig. 10.2)

$$\frac{n_+}{n_{e0}} = \alpha_0 \left(1 - \frac{4x^2}{l_1^2} \right) + 1, \qquad \frac{-l_1}{2} < x < \frac{l_1}{2} \tag{10.3.38}$$

where $l_1/2$ is the nominal position where $\alpha = 0$. We would not normally expect the Bohm flux condition to be met within the validity of this solution, so the $\alpha \gg 1$ solution must be matched to an $\alpha = 0$ electropositive solution which in turn determines the position of the plasma edge satisfying (10.3.36).

We further simplify our analysis by assuming that n_{e0} is known. The absorbed power S_e is then obtained a posteriori from (10.3.35). If S_e is specified rather than n_{e0}, then n_{e0} can be obtained iteratively, as is done for temperature, as described below.

Substituting (10.3.38) in (10.3.33) and (10.3.34) and integrating, we obtain, respectively,

$$K_{iz}n_g\frac{l_1}{2} = K_{rec}n_{e0}\left(\frac{8}{15}\alpha_0^2 + \frac{2}{3}\alpha_0\right)\frac{l_1}{2} + \frac{8D_+\alpha_0}{l_1} \tag{10.3.39}$$

$$K_{att}n_g d = K_{rec}n_{e0}\left(\frac{8}{15}\alpha_0^2 + \frac{2}{3}\alpha_0\right)l_1 \tag{10.3.40}$$

where the integration is over the electronegative region except on the LHS of (10.3.40), and where we have integrated an assumed constant electron density over the whole plasma. For this assumption to be valid, we require that l_1 be a substantial fraction of d. However, even if $l_1 \ll d$, the left hand side of (10.3.40) decreases by only a factor of $2/3$. At $x = l_1/2$, the electronegative solution can be matched to the usual electropositive solution

$$n_+ = n_e = \bar{n}\cos[\beta(x - x_0)]$$

as in (5.2.15). To obtain an algebraic solution we approximate the cosine with a parabola (see Fig. 10.2),

$$n_+ = n_2\left(1 - \frac{4x^2}{l_2^2}\right), \qquad \frac{l_1}{2} < x < \frac{d}{2} \tag{10.3.41}$$

such that the densities and fluxes between the electronegative and electropositive regions can be matched to obtain the simple conditions

$$n_2\left(1 - \frac{l_1^2}{l_2^2}\right) = n_{e0} \tag{10.3.42}$$

and

$$\frac{2D_a n_2 l_1}{l_2^2} = \frac{4D_+\alpha_0 n_{e0}}{l_1} \tag{10.3.43}$$

Requiring that the diffusion flux at $x = d/2$ be equal to the Bohm flux $n_+(d/2)u_B$, we obtain

$$\frac{4n_2 D_a d}{l_2^2} = n_2\left(1 - \frac{d^2}{l_2^2}\right)u_B \tag{10.3.44}$$

We can now eliminate the intermediate variables n_2 and l_2 from (10.3.43) and (10.3.44) and use the Bohm flux condition to solve for l_1:

$$l_1^2 = \frac{2D_+\alpha_0}{D_a + 2D_+\alpha_0}\left(\frac{4D_a d}{u_B} + d^2\right) \tag{10.3.45}$$

In most cases $d^2 \gg 4 D_a d / u_B$ in (10.3.45). Equation (10.3.45) can be used, together with (10.3.39) and (10.3.40) to obtain the three variables of our problem α_0, T_e, and l_1.

The above equations are solved readily by noting that K_{iz} is a strong exponential function of T_e, such that the temperature is essentially clamped by the particle balance of positive ions. We can therefore take the temperature as given from (10.3.39) and solve for α_0 in (10.3.40). With this value of α_0, l_1 is obtained from (10.3.45) and a new temperature obtained from (10.3.39) after substituting the initial values of α_0 and l_1. In this way the complete solution is obtained by iteration. The important scaling is obtained from (10.3.40). Solving for α_0 we obtain

$$\alpha_0 = -\frac{5}{8} + \sqrt{\left(\frac{5}{8}\right)^2 + \frac{15}{8}\frac{K_{att}}{K_{rec}}\frac{n_g}{n_{e0}}\frac{d}{l_1}} \tag{10.3.46}$$

For large α_0 this reduces to

$$\alpha_0 \approx \left(\frac{15}{8}\frac{K_{att}}{K_{rec}}\frac{n_g}{n_{e0}}\frac{d}{l_1}\right)^{1/2} \tag{10.3.47}$$

Substituting $n_{e0} = n_{+0}/\alpha_0$ into (10.3.47) and solving for α_0 yields

$$\alpha_0 = \frac{15}{8}\frac{K_{att}}{K_{rec}}\frac{n_g}{n_{+0}}\frac{d}{l_1} \tag{10.3.48}$$

which, apart from the numerical factor, agrees with the result (10.3.20) for the uniform density model. From (10.3.45) we see that l_1/d varies slowly with parameters. Then from (10.3.47) or from (10.3.48) we again see the essential scaling of α_0 which increases as n_g with increasing pressure, and, from (10.3.25), decreases as $S_e^{1/2}$ with increasing power.

The condition that the negative ions are in Boltzmann equilibrium can be checked by comparing the particle flux for n_-

$$\Gamma_-(x) = \int_0^x K_{att} n_g n_{e0}\, dx - \int_0^x K_{rec} n_+ n_-\, dx \tag{10.3.49}$$

with the terms on the right in the flux equation

$$\Gamma_- = -D_- \nabla n_- - n_- \mu_- E \tag{10.3.50}$$

Since Boltzmann equilibrium is calculated by setting the RHS of (10.3.50) equal to zero, the condition is that

$$|\Gamma_-| \lesssim \left| D_- \frac{dn_-}{dx} \right| \tag{10.3.51}$$

where Γ_- is obtained from (10.3.49). For our model in the negative ion region, n_+ is given by (10.3.38), $n_- = n_+ - n_{e0}$, and $K_{att}n_g$ in (10.3.49) is given by (10.3.40). Making these substitutions and performing the integration in (10.3.49) we obtain

$$\Gamma_- = K_{rec}n_{e0}^2 \left[\left(\frac{7}{15}\alpha_0^2 - \frac{1}{3}\alpha_0 \right) x + \frac{1}{3}\left(2\alpha_0^2 + \alpha_0 \right)\frac{x^3}{l_1^2} - \frac{1}{5}\alpha_0^2\frac{x^5}{l_1^4} \right] \quad (10.3.52)$$

For $\alpha \gtrsim 1$ and $x \ll l_1$, (10.3.52) is approximately

$$\Gamma_- \approx \frac{7}{15}K_{rec}n_{e0}^2\alpha_0^2 x \quad (10.3.53)$$

Substituting (10.3.53) in (10.3.51) and evaluating dn_-/dx we obtain

$$\frac{7}{15}K_{rec}n_{e0}^2\alpha_0^2 x \lesssim 2n_{e0}\alpha_0 D_-\frac{x}{l_1^2} \quad (10.3.54)$$

We see that the factor $n_{e0}\alpha_0 x$ cancels in (10.3.54) such that the Boltzmann condition is

$$n_{-0} \lesssim \frac{30}{7}\frac{D_-}{K_{rec}l_1^2} \quad (10.3.55)$$

independent of x, provided $x \ll l_1$. We check to see if (10.3.55) is satisfied, for a particular example, below.

Simulation Results

Figure 10.3 shows results of simulations of parallel-plate capacitive rf discharge in oxygen. The simulations (Lichtenberg et al., 1994) were done with one-dimensional dynamics using a particle-in-cell code that included the most important collisional processes using Monte Carlo methods. The parameters are $p = 50$ mTorr ($n_g = 1.6 \times 10^{21}$ m^{-3}) at low power ($n_{e0} = 2.4 \times 10^{15}$ m^{-3}) with $l = 4.5$ cm and a 13.56 MHz constant current source. We see the general features of an electronegative core plasma surrounded by an electropositive halo. The parabolic variation of n_- and n_+ and the flat profile for n_e within the electronegative core are clearly seen for this high α_0 case ($\alpha_0 \approx 8$).

Before comparing the analytic model with the simulation, we first check to see that the basic assumption of Boltzmann equilibrium for the negative ions is satisfied. Substituting the numbers we obtain

$$\frac{7K_{rec}n_{e0}\alpha_0}{30D_-/l_1^2} = 0.47$$

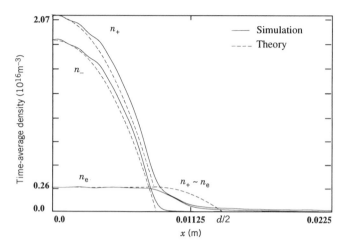

FIGURE 10.3. Simulation results and approximate analytical solutions for a 13.56 MHz plane parallel electronegative discharge in oxygen; $p = 50$ mTorr ($n_g = 1.6 \times 10^{21}$ m^{-3}), $l = 4.5$ cm, at low power ($n_{e0} = 2.4 \times 10^{15}$ m^{-3}).

which reasonably satisfies the inequality. Similarly we can check that the parabolic approximation is justified in the electronegative region by comparing the two terms on the right-hand side of (10.3.30). We find the ionization term to be significantly larger than the recombination term, so justifying the approximation of (10.3.37).

Returning to our comparison, the dashed lines show the results for n_- and n_e found from an analytic model, in which α_0 is matched to the simulation. The α_0 predicted by the model is approximately 45% larger than that found in the simulation. This difference may be due to the different manner in which K_{att} is computed in the model and in the simulation. The model used is somewhat more complicated than that given here because of an additional complication that is important for low-pressure capacitively coupled rf discharges, both electropositive and electronegative, and which may also be present in other types of discharges. In both experiments and simulations the electron distribution is found to be non-Maxwellian with higher temperature tails. Because of the strong exponential dependence of the ionization cross sections, the higher temperature tails may dominate the positive ion particle balance, while only having a modest effect on the negative ion balance. Assuming this situation to hold, the electron distribution can be separated into a two-component Maxwellian, leading to a modification of the analysis presented here. The analytic result gives $T_{eh} = 3.3$ V, with $T_{eh}/T_{ew} = 1.9$ and $n_{eh}/n_{ew} = 1/4$, where the subscripts h and w stand for the hot and warm components of the electron distribution. This gives an average temperature $T_e = 2.05$ V, which is in good agreement with the average temperature found in the simulation of $T_e = 2$ V. For details of the complete model and of the comparison with the simulations, the reader should consult Lichtenberg et al. (1994); the simulations are described in Vahedi (1993).

Although the analytic results do not agree as closely with the simulation if a single temperature Maxwellian is used, the general form of the solution and the scaling with pressure and power are preserved. We leave a calculation, using the simplified equations in the text, for a problem.

PROBLEMS

10.1. Low-Pressure Equilibrium

(a) Using the method outlined in Example 1, calculate d_{eff}, A_{eff}, T_e, $\mathcal{E}_c$, $\mathcal{E}_i$, $\mathcal{E}_T$, and u_B, for Example 2. Confirm that $n_0 \approx 5.8 \times 10^{17}$ m^{-3} and $J_{il} \approx 7.8$ mA/cm^2.

(b) Repeat for Example 3, confirming that $n_0 \approx 7.3 \times 10^{16}$ m^{-3} and $J_{il} \approx 1.0$ mA/cm^2.

10.2. High-Pressure Argon Discharge   Consider a cylindrical argon plasma of radius $R = 5$ cm, length $l = 30$ cm, pressure $p = 20$ mTorr, and absorbed microwave power $P_{abs} = 500$ W. Assume that the ionization rate is $\nu_{iz} n(r, z)$ and that there is diffusive loss to the cylinder side and end walls with a constant axial and radial (ambipolar) diffusion coefficient $D_a \approx (T_e/T_i)^{1/2} \lambda_i u_B$, where λ_i is the ion neutral mean free path for argon ions in argon and u_B is the Bohm velocity. Take $T_i = 0.026$ V.

(a) Assuming that $\lambda_i \ll R, l$ and that the plasma density $n \approx 0$ at the cylinder side and end walls, show that

$$n(r, z) \approx n_0 J_0 \left(\frac{\chi_{01} r}{R} \right) \cos \left(\frac{\pi z}{l} \right)$$

where $\chi_{01} \approx 2.405$ is the first zero of the zero order Bessel function $J_0(\chi)$.

(b) Determine T_e (V) by equating the total (axial + radial) particle loss rate to the total particle creation rate. (Integrate the particle flux $-D_a \nabla n$ over the wall area to obtain the former, and integrate $\nu_{iz} n(r, z)$ over the cylinder volume to obtain the latter.)

(c) At high pressures, the ion-bombarding energy is due to the sheath voltage V_s, given by (10.2.9), that develops at the walls. Assuming that the sheath thickness $s \ll \lambda_i$, equate ion and electron fluxes at the walls to show that $\mathcal{E}_i \approx 5.2 T_e$.

(d) From energy balance, estimate the central density n_0 (cm^{-3}) and the total current I_z (amperes) incident on one end wall.

10.3. Ambipolar Diffusion in Electronegative Plasmas

(a) Starting from Eqs (10.3.1)–(10.3.3), derive (10.3.4).

(b) Assuming the validity of the Boltzmann relation (10.3.7), derive the ambipolar diffusion coefficient (10.3.9).

10.4. Approximate Solution for Electronegative Equilibrium   For the case of $p = 50$ mTorr and $n_{e0} = 2.4 \times 10^{15}$ m^{-3}, and taking $d = 2.6$ cm, use the equations for the approximate parabolic solutions in both the electronegative and electropositive regions to find α_0 and l_1. Assume an initial value of $T_e = 3$ V and iterate your solution once. Compare your results with Fig 10.3 and comment.

10.5. High-Density Oxygen Discharge Model   Consider a high-density oxygen discharge slab model (thickness $l = 10$ cm) at low pressures. Assume that the only volume reactions are 3, 4, 11, and 12 in Table 8.2 for generation of O, O_2^+, and O^+ due to electron impact. Assume further that K_O, $K_{O_2^+}$, and K_{O^+} are the first order rate constants for loss of O, O_2^+ and O^+ to the vacuum pump and/or to the walls. Let $K_O = S_p/V = 30$ s^{-1}, where S_p is the pumping speed and V is the discharge volume. Let $K_{O_2^+} = 2u_{BO_2^+}/l$ and $K_{O^+} = 2u_{BO^+}/l$. Assume that all heavy particles are at 300 K, that there are no other sources for generation or loss of O, O_2^+, and O^+, and that O^- generation is negligible.

(a) Estimate the first-order rate constant for loss of O atoms to the walls due to recombination for a recombination probability of 10^{-4}, and compare this with the value of K_O given above due to the vacuum pump.

(b) Write the steady-state rate equations for n_O, $n_{O_2^+}$, and n_{O^+}.

(c) Find an expression for n_O/n_{O_2} as a function of n_e and the rate constants K_3, K_4, K_{11}, K_{12}, and K_O. For $K_{12}n_e \gg K_O$, show that

$$\frac{n_O}{n_{O_2}} = \frac{2K_3 + K_{11}}{K_{12}}$$

independent of n_e. Evaluate n_O/n_{O_2} and the condition on n_e to achieve this high-density limit for an electron temperature $T_e = 3$ V. Show that $n_O \gg n_{O_2}$.

(d) Find an expression for $n_{O^+}/n_{O_2^+}$ in terms of n_e and the rate constants. In the high-density limit, show that

$$\frac{n_{O^+}}{n_{O_2^+}} = \frac{2K_{O_2^+}}{K_{O^+}} \frac{K_3 + K_{11}}{K_4}$$

Evaluate this in the high-density limit for $T_e = 3$ V, and show that $n_{O^+} \gg n_{O_2^+}$.

(e) Consider now the volume reactions 2 and 13 in Table 8.2 for O^- generation and loss. Show that in the high-density limit, the negative ion density is given by

$$n_{O^-} = \frac{K_2 K_{O^+}}{2K_{13}(K_3 + K_{11})}$$

Evaluate this and show that $n_{O^-} \ll n_e$.

CHAPTER 11

CAPACITIVE DISCHARGES

As discussed in the previous chapter, a complete description of a plasma discharge requires a choice of heating mechanisms and a geometric configuration that sustains those mechanisms. One of the most widely used types of low-pressure discharges is sustained by radio-frequency (rf) currents and voltages, introduced through a capacitive sheath. The currents in the main body of the plasma lead to ohmic heating, while the voltage across the sheath leads to stochastic sheath heating. The complete self-consistent model of such a discharge is quite complicated, even in the simplest plane-parallel geometry. This leads to various simplifying assumptions in order to obtain analytic solutions in which the various scalings of plasma parameters with control parameters are explicit.

In Section 11.1 we describe a uniform, symmetric model for a capacitive parallel plate rf discharge in the regime normally used for reactive ion etching. Because simplifying assumptions are made, the model cannot be used to predict the quantitative behavior of "real" discharges. The model is qualitatively correct and is introduced to describe the underlying physics and the procedures for the analysis of real discharges. The model was developed by V. A. Godyak and his collaborators in the 1970s and is summarized in a monograph by Godyak (1986), where references to his work can be found. In Section 11.2 we consider sheath and plasma nonuniformities in symmetric discharges and develop formulae from which more realistic calculations can be made. In Section 11.3 we give comparisons to symmetric experiments and computer simulations and describe some model limitations. Most discharges are asymmetric because more electrode surfaces are naturally grounded than driven. This leads to a dc bias voltage on the driven electrode with respect to ground. We describe asymmetric models in Section 11.4. One approach to improve the performance of capacitive discharges involves application of a dc magnetic field lying in the plane of the driven electrode. These discharges are known as *magnetically enhanced reactive ion etchers* (MERIEs) or *rf magnetrons*. In Section 11.5, we return to the homogeneous model of Section 11.1 to describe some aspects of these more complicated discharges. Capacitive discharges are commonly driven by 50-Ω rf power sources, usually at 13.56 MHz, although lower (and sometimes higher) frequencies are also used. For efficient power transfer, the power source must drive the discharge through a matching network. We describe matching network operation and rf power measurement techniques in Section 11.6. This chapter is concerned primarily with low-pressure plasmas in which secondary electrons play only a minor role. However, at the end of Section 11.3 we briefly discuss the higher pressure regime, in a subsection on the *role of secondaries*. Throughout the chapter we consider only electropositive plasmas.

11.1   HOMOGENEOUS MODEL

Figure 11.1a shows the basic model. A sinusoidal current $I_{rf}(t)$, having complex representation $I_{rf} = \mathrm{Re}\, \tilde{I}_{rf} e^{j\omega t}$, flows across discharge plates a and b. Here we take $\tilde{I}_{rf} = I_1$, a real number. The plates are separated by a distance l and each has a cross sectional area A. A gas having neutral density n_g is present between the plates. In response to the current flow, a discharge plasma forms between the plates, accompanied by a voltage $V(t)$ across the plates and a power flow $P(t)$ into the plasma. The plasma has an ion density $n_i(\mathbf{r}, t)$ and an electron temperature $T_e(\mathbf{r}, t)$. Because of quasineutrality, $n_e \approx n_i$ almost everywhere except within the oscillating sheaths

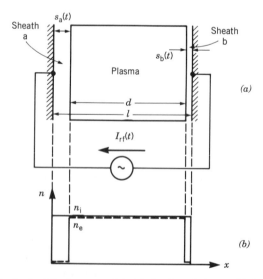

FIGURE 11.1. The basic rf discharge model: (a) sheath and plasma thicknesses; (b) electron and ion densities.

near the plates, where $n_e < n_i$. The instantaneous sheath thickness is $s(t)$ and its time-averaged value is $\bar{s}$. Typically, $\bar{s} \ll l$.

The state of the discharge is specified once a complete set of control parameters is given. The remaining plasma and circuit parameters are then specified as functions of the control parameters. A convenient choice for the control parameters is I_{rf}, ω, n_g, and l. Given these, we develop the basic model to determine n_e, T_e, s, $\bar{s}$, V, and P. The choice of control parameters is not unique. We choose I rather than V or P, in this section, for ease of analysis.

In general, the discharge parameters n_e, n_i, and T_e are complicated functions of position and time. We assume the following to simplify the analysis:

(a) The ions respond only to the time-averaged potentials. This is a good approximation provided

$$\omega_{pi}^2 \ll \omega^2$$

where ω_{pi} is the ion plasma frequency.

(b) The electrons respond to the instantaneous potentials and carry the rf discharge current. This is a good approximation provided

$$\omega_{pe}^2 \gg \omega^2 \left(1 + \frac{\nu_m^2}{\omega^2} \right)^{1/2}$$

where ω_{pe} is the electron plasma frequency and ν_m is the electron–neutral collision frequency for momentum transfer.

(c) The electron density is zero within the sheath regions. This is a good approximation provided $\lambda_{De} \ll \bar{s}$, where λ_{De} is the electron Debye length. This holds if $T_e \ll \bar{V}$, where $\bar{V}$ is the dc voltage across the sheath.

(d) There is no transverse variation (along the plates). This is a good approximation provided $l \ll \sqrt{A}$. Since the divergence of Maxwell's equation $\nabla \times \mathbf{H} = \mathbf{J} + \epsilon_0 \partial \mathbf{E}/\partial t$ is zero, we see that, at any instant of time, the sum of the conduction current $\mathbf{J}$ and the displacement current $\epsilon_0 \partial \mathbf{E}/\partial t$ within the discharge is then independent of x.

These assumptions hold both for the uniform model of this section, and for the inhomogeneous model of Section 11.2. For the simplified model in this section we also assume the following:

(e) The ion density is uniform and constant in time everywhere in the plasma and sheath regions: $n_i(\mathbf{r}, t) = n = \text{const}$. The electron and ion density profiles for the simplified model are shown in Fig. 11.1b, corresponding to the position of the plasma as shown in Fig. 11.1a.

As we shall see in Section 11.2, the variation of the ion density in the sheath, which we obtain from a Child law calculation as in Section 6.3, considerably modifies the results obtained here using the approximation (e).

Plasma Admittance

The admittance of a bulk plasma slab of thickness d and cross-sectional area A is $Y_p = j\omega\epsilon_p A/d$, where

$$\epsilon_p = \epsilon_0 \left[1 - \frac{\omega_{pe}^2}{\omega(\omega - j\nu_m)} \right] \tag{11.1.1}$$

is the plasma dielectric constant given by (4.2.8). We show below that, within the uniform ion density approximation

$$d = l - 2\bar{s} = \text{const} \tag{11.1.2}$$

independent of time. We then find that (see Problem 11.1)

$$Y_p = j\omega C_0 + \frac{1}{j\omega L_p + R_p} \tag{11.1.3}$$

where $C_0 = \epsilon_0 A/d$ is the vacuum capacitance, $L_p = \omega_{pe}^{-2} C_0^{-1}$ is the plasma inductance, and $R_p = \nu_m L_p$ is the plasma resistance. This form for Y_p represents the series combination of L_p and R_p in parallel with C_0. By assumption (b), the displacement current that flows through C_0 is much smaller than the conduction current that flows through L_p and R_p. The sinusoidal current

$$I_{rf}(t) = \text{Re}\,\tilde{I}_{rf} e^{j\omega t} \tag{11.1.4}$$

that flows through the plasma bulk produces a voltage across the plasma

$$V_p(t) = \text{Re}\,\tilde{V}_p e^{j\omega t} \tag{11.1.5}$$

where $\tilde{V}_p = \tilde{I}_{rf}/Y_p$ is the complex voltage amplitude. We see that the plasma voltage is linear in the applied current and that there is no harmonic generation (multiples of ω) or dc component of V_p.

Sheath Admittance

In contrast to the plasma, the current that flows through the two sheaths is almost entirely displacement current; i.e., it is due to a time-varying electric field. This is true because the conduction current in a discharge is carried mainly by electrons, and the electron density is approximately zero within the time-varying sheath. We will see that the conduction current carried by the steady flow of ions across the sheath to the plates is much smaller than the displacement current.

(a) Displacement Current   The electric field $\mathbf{E} = \hat{x}E$ within sheath *a* (see Fig. 11.1) is given by Poisson's equation

$$\frac{dE}{dx} = \frac{en}{\epsilon_0}, \qquad x \le s_a(t) \tag{11.1.6}$$

which on integration yields

$$E(x,t) = \frac{en}{\epsilon_0}[x - s_a(t)] \tag{11.1.7}$$

The boundary condition is $E \approx 0$ at $x = s_a$ because E is continuous across the plasma–sheath interface (no surface charge) and the electric field is small in the plasma. The displacement current flowing through sheath *a* into the plasma is

$$I_{ap}(t) = \epsilon_0 A \frac{\partial E}{\partial t} \tag{11.1.8}$$

Substituting (11.1.7) in (11.1.8), we obtain

$$I_{ap}(t) = -en A \frac{ds_a}{dt} \tag{11.1.9}$$

From (11.1.9), the sheath boundary s_a oscillates linearly with the applied current. Setting $I_{ap}(t) = I_{rf}(t)$, where $I_{rf} = I_1 \cos \omega t$, we integrate (11.1.9) to obtain

$$s_a = \bar{s} - s_0 \sin \omega t \tag{11.1.10}$$

where

$$s_0 = \frac{I_1}{en\omega A} \tag{11.1.11}$$

is the sinusoidal oscillation amplitude about the dc value $\bar{s}$. The voltage across the sheath is given by

$$V_{ap}(t) = \int_0^{s_a} E\, dx = -\frac{en}{\epsilon_0} \frac{s_a^2}{2} \tag{11.1.12}$$

From (11.1.12), the sheath voltage is a nonlinear function of s_a and therefore of the applied current. Substituting (11.1.10) in (11.1.12), we obtain

$$V_{ap} = -\frac{en}{2\epsilon_0} \left(\bar{s}^2 + \frac{1}{2}s_0^2 - 2\bar{s}s_0 \sin \omega t - \frac{1}{2}s_0^2 \cos 2\omega t \right) \tag{11.1.13}$$

We see that the nonlinearity leads to second-harmonic voltage generation and a constant average value.

Similarly for sheath b we obtain

$$I_{bp} = -en A \frac{ds_b}{dt} \qquad (11.1.14)$$

and the voltage across this sheath is

$$V_{bp} = -\frac{en}{\epsilon_0} \frac{s_b^2}{2} \qquad (11.1.15)$$

By continuity of current, $I_{bp} = -I_{ap}$, so that adding (11.1.9) and (11.1.14) we find

$$\frac{d}{dt}(s_a + s_b) = 0$$

Integrating, we obtain

$$s_a + s_b = 2\bar{s}, \text{ a constant} \qquad (11.1.16)$$

so that $d = l - 2\bar{s} = \text{const}$, as previously stated. For sheath b,

$$s_b = \bar{s} + s_0 \sin \omega t \qquad (11.1.17)$$

with the nonlinear voltage response, using (11.1.15),

$$V_{bp} = -\frac{en}{2\epsilon_0}\left(\bar{s}^2 + \frac{1}{2}s_0^2 + 2\bar{s}s_0 \sin \omega t - \frac{1}{2}s_0^2 \cos 2\omega t\right) \qquad (11.1.18)$$

Although V_{ap} and V_{bp} are nonlinear, the combined voltage $V_{ab} = V_{ap} - V_{bp}$ across both sheaths, obtained by subtracting (11.1.18) from (11.1.13), is

$$V_{ab} = \frac{en\bar{s}}{\epsilon_0}(s_b - s_a)$$

Substituting for s_b and s_a from (11.1.10) and (11.1.17) we find

$$V_{ab} = \frac{2en\bar{s}s_0}{\epsilon_0} \sin \omega t \qquad (11.1.19)$$

which is a linear voltage response. We obtain the surprising result that although each sheath is nonlinear, the combined effect of both sheaths is linear. This is true only for the simplified model assumptions of a symmetric, homogeneous (constant ion density) discharge. The total voltage $\tilde{V}_{rf}$ across the discharge is the sum of $\tilde{V}_{ab}$ and $\tilde{V}_p$. However, for typical discharge conditions, we usually have $|\tilde{V}_p| \ll |\tilde{V}_{ab}|$, and we often approximate $\tilde{V}_{rf} \approx \tilde{V}_{ab}$.

(b) Conduction Current   Although the conduction current in each sheath is small, the average sheath thickness $\bar{s}$ is determined by the balance between ion and electron conduction currents. By assumption (a), there is a steady flow of ions from the plasma through sheath a, carrying a steady current

$$\bar{I}_i = enu_B A \tag{11.1.20}$$

where the loss velocity is taken to be the Bohm velocity u_B.

By symmetry, the time-average conduction current flowing to plate a is zero. There is a steady flow of ions to the plate. For the basic model, the electron density is assumed zero in the sheath. The sheath thickness $s_a(t)$ must therefore collapse to zero at some time during the rf cycle in order to transfer electrons from the plasma to the plate. It follows from (11.1.10) and (11.1.11) that

$$\bar{s} = s_0 = \frac{I_1}{en\omega A} \tag{11.1.21}$$

and from (11.1.13) that

$$V_{pa} = \frac{en}{2\epsilon_0} s_0^2 (1 - \sin\omega t)^2 \tag{11.1.22}$$

Since the sheath voltage collapses to zero at the time that the electrons are transferred to the plate, this acts like an ideal diode across the sheath whose preferred direction of current flow is into the plasma. A similar result holds for sheath b.

We can define a linear sheath capacitance C_s because the voltage (11.1.19) across both sheaths is sinusoidal. Differentiating (11.1.19) and substituting for I_{rf} using (11.1.21) we obtain the simple result

$$I_{rf} = C_s \frac{dV_{ab}}{dt}$$

where

$$C_s = \frac{\epsilon_0 A}{2s_0} \tag{11.1.23}$$

is a linear capacitance. Physically, this capacitance is the series combination of the two nonlinear capacitances $C_a = \epsilon_0 A / s_a(t)$ and $C_b = \epsilon_0 A / s_b(t)$.

The voltages $V_{ap}(t)$, $V_{pb}(t)$ and their sum $V_{ab}(t)$ are plotted versus t in Fig. 11.2. The manner in which the sum of the two nonsinusoidal voltages yields the V_{ab} sinusoid is clearly seen. The time-averaged value $\bar{V}$ for V_{pb} is also shown as the horizontal dashed line. The spatial variation of the total potential at various times within the rf cycle is shown (solid lines) in Fig. 11.3. It is assumed that the right-hand electrode is grounded (held at $V = 0$ at all times). The dashed curve shows the spatial variation of the time average potential.

Particle and Energy Balance

To complete the analysis we need to evaluate expressions for particle and energy balance as developed in Chapter 10. Particle balance per unit area, for a uniform plasma, is straightforwardly given by

$$nK_{iz}n_g d = 2nu_B \qquad (11.1.24)$$

as in (10.2.7), with $d_{eff} = d/2$. If the sheaths are thin, such that $d \approx l$, we can evaluate the temperature from (11.1.24) alone.

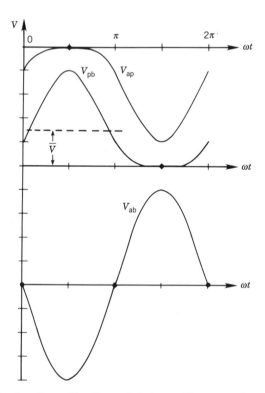

FIGURE 11.2. Sheath voltages V_{ap}, V_{bp}, and their sum V_{ab} versus time; the time-average value $\bar{V}$ of V_{bp} is also shown.

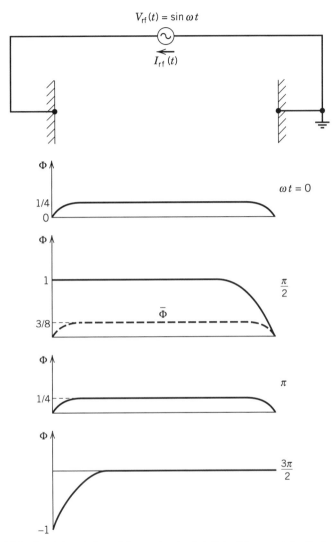

FIGURE 11.3. Spatial variation of the total potential Φ (solid curves) for the homogeneous model of Section 11.1, at four different times during the rf cycle. The dashed curve shows the spatial variation of the time-average potential $\bar{\Phi}$.

To calculate the plasma density, we must evaluate the time-average power per unit area absorbed by the electrons, S_e, which involves the rf currents and voltages, and the sheath oscillations.

(a) Ohmic Heating   The time-average power per unit area deposited by ohmic heating in the bulk plasma, $\bar{S}_{\text{ohm}}$, is due to collisional momentum transfer between the oscillating electrons and the neutrals. Integrating (4.2.30) over the bulk plasma

length d, we obtain

$$\bar{S}_{\text{ohm}} = \frac{1}{2}J_1^2 \frac{d}{\sigma_{\text{dc}}} \tag{11.1.25}$$

where $J_1 = I_1/A$ and σ_{dc} is the dc plasma conductivity. Substituting (4.2.22) for σ_{dc} into (11.1.25), we find

$$\bar{S}_{\text{ohm}} = \frac{1}{2}J_1^2 \frac{m\nu_m d}{e^2 n} \tag{11.1.26}$$

(b) Stochastic Heating   Electrons reflecting from the large decelerating fields of a moving high-voltage sheath can be approximated by assuming the reflected velocity is that which occurs in an elastic collision of a ball with a moving wall

$$u_r = -u + 2u_{\text{es}} \tag{11.1.27}$$

where u and u_r are the incident and reflected electron velocities parallel to the time-varying electron sheath velocity u_{es}. If the parallel electron velocity distribution at the sheath edge is $f_{\text{es}}(u, t)$, then in a time interval dt and for a speed interval du, the number of electrons per unit area that collide with the sheath is given by $(u - u_{\text{es}})f_{\text{es}}(u, t)dudt$. This results in a power transfer per unit area,

$$dS_{\text{stoc}} = \frac{1}{2}m(u_r^2 - u^2)(u - u_{\text{es}})f_{\text{es}}(u, t)\, du \tag{11.1.28}$$

Using $u_r = -u + 2u_{\text{es}}$ and integrating over all incident velocities, we obtain

$$S_{\text{stoc}} = -2m \int_{u_{\text{es}}}^{\infty} u_{\text{es}}(u - u_{\text{es}})^2 f_{\text{es}}(u, t)\, du \tag{11.1.29}$$

In the physical problem f_{es} varies with time, as the sheath oscillates, and the problem becomes quite complicated. For our uniform density model we note that

$$\int_{-\infty}^{\infty} f_{\text{es}}(u, t)\, du = n_{\text{es}}(t) = n, \text{ a constant} \tag{11.1.30}$$

Furthermore, for the purpose of understanding the heating mechanism we make the simplifying approximations that $f_{\text{es}}(u, t)$ can be approximated by a Maxwellian, ignoring the plasma drift, and that $u_{\text{es}} \ll \bar{v}_e$, the mean electron speed. These approximations simplify the calculation. We give the results using more realistic approximations in Section 11.2. Consistent with our approximation, we can set the lower limit in (11.1.29) to zero. Before performing the average over the distribution function, we substitute

$$u_{es} = u_0 \cos \omega t \qquad (11.1.31)$$

in (11.1.29) and average over time. Only the term in $\sin^2 \omega t$ survives giving

$$\bar{S}_{stoc} = 2mu_0^2 \int_0^\infty uf_{es}(u)\,du \qquad (11.1.32)$$

Now, consistent with our approximation that f_{es} is Maxwellian, we note that the integral gives the usual random flux $\Gamma_e = \frac{1}{4}n\bar{v}_e$, and (11.1.32) becomes

$$\bar{S}_{stoc} = \frac{1}{2}mu_0^2 n\bar{v}_e \qquad (11.1.33)$$

Inside the plasma the rf current is almost entirely conduction current, such that

$$I_1 = J_1 A = -enu_0 A \qquad (11.1.34)$$

Substituting (11.1.34) into (11.1.33) yields the stochastic electron power in terms of the (assumed) known current. Since we are calculating the power per unit area, we use the current density, to obtain, for a single sheath,

$$\bar{S}_{stoc} = \frac{1}{2}\frac{m\bar{v}_e}{e^2 n}J_1^2 \qquad (11.1.35)$$

Discharge Parameters

Adding (11.1.35) (for two sheaths) and (11.1.26), the total time-average electron power per unit area is

$$S_e = \frac{1}{2}\frac{m}{e^2 n}(\nu_m d + 2\bar{v}_e)J_1^2 \qquad (11.1.36)$$

Assuming J_{rf}, ω, n_g, A, and l are the specified control parameters, we determine n by setting (11.1.36) equal to the electron energy lost, as in (10.2.2),

$$S_e = 2enu_B(\mathcal{E}_c + \mathcal{E}_e) \qquad (11.1.37)$$

($\mathcal{E}_e = 2T_e$ for a Maxwellian) and obtaining

$$n = \frac{1}{2}\left[\frac{m(\nu_m d + 2\bar{v}_e)}{e^3 u_B(\mathcal{E}_c + \mathcal{E}_e)}\right]^{1/2} J_1 \qquad (11.1.38)$$

With the temperature assumed known from (11.1.24) and if we again let $d \approx l$, the density can be calculated. With n known the sheath thickness is calculated from (11.1.21). If $2\bar{s}$ is a significant fraction of l, then we determine $d \approx l - 2\bar{s}$, and the equations can be iterated to determine more accurate values for T_e, n, and $d = l - 2\bar{s}$. However, this iteration compromises the simplicity of the model.

Finally, to obtain the total power dissipated, we must calculate the power lost by the ions. To do this we need the average voltage across each sheath, which is found by time averaging (11.1.13):

$$\bar{V} \equiv \bar{V}_{pa} = \frac{3}{4} \frac{en}{\epsilon_0} s_0^2 = \frac{3}{4} \frac{J_1^2}{e \epsilon_0 n \omega^2} \tag{11.1.39}$$

where the second equality comes from substituting for s_0 from (11.1.21). Using (10.2.13), the power per unit area lost by the ions is

$$S_i = 2 e n u_B \left(\bar{V} + \frac{T_e}{2} \right) \approx \frac{3}{2} u_B \frac{J_1^2}{\epsilon_0 \omega^2} \tag{11.1.40}$$

where the factor of two is for two sheaths. The latter approximate equality holds for the usual case with $\bar{V} \gg T_e$. The total power absorbed per unit area, S_{abs}, is found by adding (11.1.36) and (11.1.40).

The stochastic heating $\bar{S}_{stoc}$ leads to equivalent sheath resistances R_a and R_b defined by $\bar{S}_{stoc} = (1/2)J_1^2 A R_{a,b}$. These resistances are in series with the sheath capacitances, as shown in Fig. 11.4. The ion heating S_i can be modeled as equivalent dc current sources $\bar{I}_i = \bar{J}_i A$, as shown in the figure. Because this dc current flows across a dc sheath voltage $\bar{V}$, it represents a power dissipation within the sheath. Note that R_a and $\bar{I}_i$ are not constants, but are functions of the rf voltage. For typical discharges, the inductive impedance of the bulk plasma is small compared to the capacitive impedance of the sheaths, such that almost all of the applied rf voltage appears across the two sheath capacitors. This situation was described in Chapter 4 (see Fig. 4.4, along with (4.2.25) and accompanying discussion in the text). Although the voltage drops across the resistors are generally small for an electropositive plasma, the power dissipation due to the flow of current through these resistors is important, as we have described. At very high frequencies (typically much exceeding 13.56 MHz), the bulk plasma inductance and the sheath capacitance can resonate, leading to a *resonant discharge* regime in which the overall rf voltage across the discharge drops to a very low value (of order a few volts). In an electronegative plasma with $n_e \ll n_i$, the voltage drops across the resistors can be comparable to, or even exceed, the voltage drops across the sheath capacitors, and the discharge can enter a resistive regime.

In real devices, the control parameter is usually V_{rf} or S_{abs}, rather than J_{rf}. This also would make the above calculations more cumbersome. We address this issue in the next section, where we make more quantitatively correct calculations and give examples of calculating the parameters in real discharges.

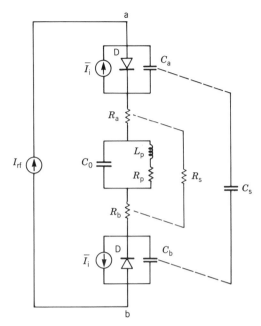

FIGURE 11.4. Nonlinear circuit model of the homogeneous rf plasma discharge. The dashed lines indicate that the series connection of the nonlinear elements C_a and C_b, and R_a and R_b, yield the corresponding linear elements C_s and R_s, respectively.

11.2 INHOMOGENEOUS MODEL

In this section we describe a realistic inhomogeneous model for a discharge and give the set of equations that are required for a quantitative calculation of the discharge parameters. For the inhomogeneous model we retain approximations (a)–(d) in Section 11.1, but allow the plasma and the sheath to be inhomogeneous. The inhomogeneity in the plasma is not critical, taking different forms depending on the pressure, as discussed in detail in Chapters 5 and 10. The inhomogeneous sheath, however, strongly modifies the results, and the consequences of this are the main subject of this section. The basic processes are the following. The decreasing ion density within the sheath between the plasma–sheath edge and the collecting boundary leads to a Child law variation of the density and an increased sheath width compared to the matrix sheath width in Section 11.1. It also leads to an increase of the sheath velocity in the regions of decreasing ion density. This follows because the rf current must be continuous, while the electron density is decreasing to preserve charge neutrality. The result is a substantial increase in the stochastic sheath heating. Due to the partial shielding of the ion space charge by the oscillating electrons, the Child law (6.3.12) for the ions is also modified. The increase in sheath width decreases the total sheath capacitance. A self-consistent analysis must consider all of these effects together. The analysis for a collisionless sheath is given somewhat briefly in the first part of this section; a more detailed calculation can be found in Lieberman (1988). The results required to make a quantitative calculation of the discharge parameters are summarized in (11.2.32)–(11.2.38), and their use is illustrated in several following examples. The reader who wishes to calculate parameters for a given discharge can skip to these equations without following the preceding analysis.

At higher pressures where the ion mean free path $\lambda_i < s_m$, the sheath width, collisional models similar to those described in Section 6.5 must be used to describe the self-consistent sheath dynamics. We summarize the results for these models in this section. We also briefly describe nonideal effects for the self-consistent sheath, including low to moderate rf driving voltages, ohmic heating in the sheaths, and fast sheath heating.

Collisionless Sheath Dynamics

The structure of the rf sheath is shown in Fig. 11.5. Ions crossing the ion sheath boundary at $x = 0$ accelerate within the sheath and strike the electrode at $x = s_m$ with high energies. Since the ion flux $n_i u_i$ is conserved and u_i increases as ions transit the sheath, n_i drops. This is sketched as the heavy solid line in Fig. 11.5. The ion particle and energy conservation equations are respectively

$$n_i u_i = n_s u_B \tag{11.2.1}$$

$$\frac{1}{2} M u_i^2 = \frac{1}{2} M u_B^2 - e\bar{\Phi} \tag{11.2.2}$$

where n_s is the plasma density at the plasma sheath edge at $x = 0$ and $\bar{\Phi}$ is the time-average potential within the sheath; $\bar{\Phi}$, n_i, and u_i are functions of x. The Poisson equation for the instantaneous electric field $E(x, t)$ within the sheath is

$$\begin{aligned}
\frac{\partial E}{\partial x} &= \frac{e}{\epsilon_0} n_i(x) & s(t) < x \\
&= 0 & s(t) > x
\end{aligned} \tag{11.2.3}$$

Here, $s(t)$ is the distance from the ion sheath boundary at $x = 0$ to the electron sheath edge. Time averaging (11.2.3) over an rf cycle, we obtain the equations for

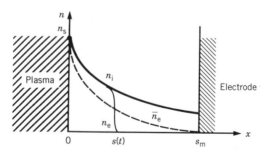

FIGURE 11.5.   Schematic plot of the densities in a high-voltage, capacitive rf sheath.

the time-average electric field $\bar{E}(x)$:

$$\frac{d\bar{E}}{dx} = \frac{e}{\epsilon_0}(n_i(x) - \bar{n}_e(x)) \tag{11.2.4}$$

$$\frac{d\bar{\Phi}}{dx} = -\bar{E} \tag{11.2.5}$$

where $\bar{n}_e(x)$ is the time-average electron density within the sheath. We determine $\bar{E}$ and $\bar{n}_e$ from $s(t)$ as follows. We note that $n_e(x, t) = 0$ during the part of the rf cycle where $s(t) < x$; otherwise, $n_e(x, t) = n_i(x)$. We therefore have

$$\bar{n}_e(x) = \left(1 - \frac{2\phi}{2\pi}\right) n_i(x) \tag{11.2.6}$$

where $2\phi(x) = 2\omega t$ is the phase interval during which $s(t) < x$. Qualitatively, we sketch $\bar{n}_e$ as the dashed line in Fig. 11.5. For x near zero, $s(t) < x$ during only a small part of the rf cycle; therefore, $2\phi \approx 0$ and $\bar{n}_e \approx n_i(x)$. For x near s_m, $s(t) < x$ during most of the rf cycle; therefore, $2\phi \approx 2\pi$ and $\bar{n}_e \approx 0$. To determine the time averages quantitatively, we assume that a sinusoidal rf current density passes through the sheath, which, equated to the conduction current at the electron sheath boundary, gives the equation for the electron sheath motion:

$$-en_i(s)\frac{ds}{dt} = -J_1 \sin \omega t \tag{11.2.7}$$

The solutions to these equations are rather involved, and we present only a few results. Combining (11.2.1)–(11.2.7), we obtain (see Lieberman (1988) for details)

$$\frac{x}{s_0} = (1 - \cos\phi) + \frac{H}{8}\left(\frac{3}{2}\sin\phi + \frac{11}{18}\sin 3\phi - 3\phi\cos\phi - \frac{1}{3}\phi\cos 3\phi\right) \tag{11.2.8}$$

for $0 \leq \phi \leq \pi$, as sketched in Fig. 11.6; and at the electron sheath edge the ion density is determined to be

$$\frac{n_i(x)}{n_s} = \left[1 - H\left(\frac{3}{8}\sin 2\phi - \frac{1}{4}\phi\cos 2\phi - \frac{1}{2}\phi\right)\right]^{-1} \tag{11.2.9}$$

Here

$$s_0 = \frac{J_1}{e\omega n_s} \tag{11.2.10}$$

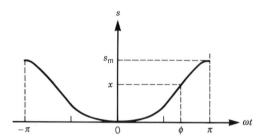

FIGURE 11.6. Sketch of the electron sheath thickness s versus ωt, showing the definition of the phase $\phi(x)$.

is an effective oscillation amplitude, and

$$H = \frac{J_1^2}{\pi e \epsilon_0 T_e \omega^2 n_s} = \frac{1}{\pi} \frac{s_0^2}{\lambda_{De}^2} \tag{11.2.11}$$

with $\lambda_{De} = (\epsilon_0 T_e / e n_s)^{1/2}$ the electron Debye length at the ion sheath edge ($x = 0$, $n_i = n_s$). The ion density and average electron density are as sketched in Fig. 11.5 in the usual regime of a high-voltage sheath with $V_{rf} \gg T_e$.

Child Law

The Child law for the self-consistent ion sheath is obtained by integrating (11.2.4) with $n_i(x)$ and $\bar{n}_e(x)$ given by (11.2.9) and (11.2.6). Performing the integrations we find

$$\frac{\bar{\Phi}}{T_e} = \frac{1}{2} - \frac{1}{2} \left[1 - H \left(\frac{3}{8} \sin 2\phi - \frac{1}{4} \phi \cos 2\phi - \frac{1}{2} \phi \right) \right]^2 \tag{11.2.12}$$

The ion sheath voltage $\bar{V}$ is then found by putting $\phi = \pi$ at $\bar{\Phi} = -\bar{V}$ in (11.2.12) to obtain, for $H \gg 1$,

$$\frac{\bar{V}}{T_e} = \frac{9\pi^2 H^2}{32} \tag{11.2.13}$$

Similarly putting $\phi = \pi$ at $x = s_m$ in (11.2.8), we obtain

$$\frac{s_m}{s_0} = \frac{5\pi H}{12} \tag{11.2.14}$$

The ion current is obtained from the Bohm flux at the plasma edge where $n_i = n_s$. Substituting for H from (11.2.11), we use (11.2.13) and (11.2.14) to construct the

Bohm flux, finding

$$\bar{J}_i = e n_s u_B = K_i \epsilon_0 \left(\frac{2e}{M}\right)^{1/2} \frac{\bar{V}^{3/2}}{s_m^2} \tag{11.2.15}$$

where $K_i = 200/243 \approx 0.82$. This has the same scaling with $\bar{V}$ and s_m as the normal Child law (6.3.12) without electron shielding, which has $K_i = 4/9 \approx 0.44$. For a fixed current density and sheath voltage, the self-consistent rf ion sheath thickness s_m is larger than the Child law sheath thickness by the factor $\sqrt{50/27} \approx 1.36$. This increase is produced by the reduction in space charge within the sheath due to the nonzero, time-average electron density.

Sheath Capacitance

To obtain a complete self-consistent model we need a relationship between the rf voltage and rf current, which involves the total capacitance of both sheaths. Unlike the uniform model in Section 11.1, the sum of the two sheath capacitances is no longer a constant, producing harmonics at the rf driving frequency. In the model the current has been taken to be sinusoidal; hence we Fourier decompose the voltage to obtain a capacitance associated with the fundamental component of the voltage

$$I_{ab}(t) \equiv C_{ab} \frac{d}{dt} V_{ab1}(t) \tag{11.2.16}$$

Using (11.2.8) and integrating Poisson's equation twice, to obtain the time-varying total voltage, we find

$$V_{ab} = -\frac{\pi}{4} H T_e \left\{ 8 \cos \omega t + H \left[\frac{10}{3} \pi \cos \omega t - \frac{5}{9} \sin 2\omega t \right. \right.$$
$$\left. \left. - \frac{25}{288} \sin 4\omega t + (2\omega t - \pi) \left(\frac{3}{8} + \frac{1}{3} \cos 2\omega t + \frac{1}{48} \cos 4\omega t \right) \right] \right\} \tag{11.2.17}$$

for $0 \le \omega t \le \pi$. The peak-to-peak value of V_{ab} is $2V(0)$, with $V(0)$ given by

$$V(0) = \frac{\pi}{4} H T_e \left[8 + H \left(\frac{125\pi}{48} \right) \right] \tag{11.2.18}$$

The amplitude of the fundamental voltage harmonic is

$$V_{ab1} = -\frac{\pi}{4} H T_e \left[8 + H \left(\frac{10\pi}{3} - \frac{4096}{675\pi} \right) \right] \tag{11.2.19}$$

Evaluating (11.2.19) and substituting in (11.2.16) we find

$$C_{ab} \approx \frac{0.613\,\epsilon_0\,A}{s_m} \tag{11.2.20}$$

There is no second harmonic, and the third harmonic of the voltage is only 4.2% of the fundamental. Hence, to a good approximation, a sinusoidal sheath current produces a sinusoidal voltage across the sum of the two sheaths in a symmetric rf discharge.

From (11.2.16) and (11.2.20), we obtain

$$J_1 \approx 1.23\frac{\omega\epsilon_0}{s_m}V_1 \tag{11.2.21}$$

where $V_1 = V_{ab1}/2$ is the fundamental rf voltage amplitude across a single sheath. From (11.2.13) and (11.2.19) with $H \gg 1$, we also find

$$\bar{V} \approx 0.83\,V_1 \tag{11.2.22}$$

The relation between V_1, J_1, and n_s is found by eliminating $\bar{V}$ and s_m from (11.2.15), (11.2.21), and (11.2.22) to obtain

$$\frac{J_1^2}{n_s} \approx 1.73\,e\epsilon_0\omega^2\mathrm{T}_e^{1/2}V_1^{1/2} \tag{11.2.23}$$

Ohmic Heating

The ohmic heating is obtained straightforwardly as in Section 11.1, except that the density and therefore the resistivity is a function of position. The time-average ohmic power per unit area can therefore be written

$$\bar{S}_{ohm} \approx \frac{1}{2}J_1^2 \int_{-l/2+s_m}^{l/2-s_m} \frac{m\nu_m}{e^2 n(x)}\,\mathrm{d}x \tag{11.2.24}$$

where $n(x)$ is the only function of position, depending on the equilibrium solution as calculated in Section 10.2, and the approximate equality is due to the approximation of the integration limits. At low pressures, $\lambda_i/d > \mathrm{T}_i/\mathrm{T}_e$, for which the density profile is rather flat, the central density can be substituted for $n(x)$, without significant error. At low pressures the ohmic heating is small compared to the stochastic heating, such that the errors are negligible. At higher pressures, $\lambda_i/d < \mathrm{T}_i/\mathrm{T}_e$, most of the ohmic heating occurs at the plasma edge and the mean free path of the energetic (ionizing) electrons is generally less than the discharge length. As pointed out in Section 10.2, this leads to a flattening of the cosine solution $n = n_0 \cos \beta x$ of (10.2.18). However, we ignore this effect here and use (10.2.18) to integrate $1/n(x)$ to incorporate the density variation. Thus we have

$$\bar{S}_{ohm} = \frac{1}{2} J_1^2 \frac{m\nu_m}{e^2 n_0} d \qquad\qquad \lambda_i > \left(\frac{T_i}{T_e}\right) d \quad (11.2.25a)$$

$$\bar{S}_{ohm} = \frac{1}{2} J_1^2 \frac{m\nu_m}{e^2 n_0} \frac{2}{\beta} \ln \tan\left(\frac{\pi}{4} + \frac{\beta d}{4}\right) \qquad \lambda_i < \left(\frac{T_i}{T_e}\right) d \quad (11.2.25b)$$

where $d \approx l - 2s_m$ is the plasma length and $\cos(\beta d/2) = n_s/n_0$.

Stochastic Heating

The power transferred to the electrons by the sheath is found from (11.1.29) as in Section 11.1, but now f_{es} is not a fixed Maxwellian, but is a time-varying function with a time-varying density $n_{es}(t)$ at the electron sheath edge $s(t)$. To determine f_{es}, we first note that the sheath is oscillating because the electrons in the bulk plasma are oscillating in response to a time-varying electric field. If the velocity distribution function within the plasma at the ion sheath edge $x = 0$ in the absence of the electric field is a Maxwellian $f_m(u)$ having density n_s, then the distribution within the plasma at the ion sheath edge is $f_s(u, t) = f_m(u - u_s)$, where $u_s(t) = -u_0 \sin \omega t$ is the time-varying oscillation velocity of the plasma electrons. At the moving electron sheath edge, because $n_{es} < n_s$, not all electrons having $u > 0$ at $x = 0$ collide with the sheath at s. Many electrons are reflected within the region $0 < x < s$ where the ion density drops from n_s to n_{es}. This reflection is produced by an ambipolar electric field whose value maintains quasineutrality $n_e \approx n_i$ at all times. The transformation of f_s across this region to obtain f_{es} is complicated. However, the essential features to determine the stochastic heating are seen if we approximate

$$f_{es} = \frac{n_{es}}{n_s} f_m(u - u_s), \qquad u > 0 \qquad (11.2.26)$$

Inserting (11.2.26) into (11.1.29) and transforming to a new variable $u' = u - u_s$, we obtain

$$S_{stoc}(t) = -\frac{2m}{n_s} \int_{u_{es}-u_s}^{\infty} u_{es} n_{es}[u'^2 - 2u'(u_{es} - u_s) + (u_{es} - u_s)^2] f_m(u') \, du' \quad (11.2.27)$$

From (11.2.7) we note that

$$n_{es} u_{es} = u_0 n_s \sin \phi \qquad (11.2.28)$$

and differentiating (11.2.8), we obtain

$$v_R(\phi) \equiv u_{es} - u_s = \pm \frac{u_0 H}{8} \left(-\frac{3}{2} \cos \phi + 3\phi \sin \phi + \frac{3}{2} \cos 3\phi + \phi \sin 3\phi\right)$$

$$(11.2.29)$$

where the plus sign is used for the integration from 0 to π and the minus sign for the integration from $-\pi$ to 0. Substituting (11.2.28) and (11.2.29) into (11.2.27), we find the average stochastic power for a single sheath to be

$$\bar{S}_{\text{stoc}} = -\frac{mu_0}{\pi} \int_{-\pi}^{\pi} \sin\phi \, d\phi \int_{v_R(\phi)}^{\infty} u'^2 f_m(u') \, du'$$
$$+ \frac{2mu_0}{\pi} \int_{-\pi}^{\pi} v_R(\phi) \sin\phi \, d\phi \int_{v_R(\phi)}^{\infty} u' f_m(u') \, du' \qquad (11.2.30)$$
$$- \frac{mu_0}{\pi} \int_{-\pi}^{\pi} v_R^2(\phi) \sin\phi \, d\phi \int_{v_R(\phi)}^{\infty} f_m(u') \, du'$$

or, for notational convenience, $\bar{S}_{\text{stoc}} = \bar{S}_1 + \bar{S}_2 + \bar{S}_3$.

If the assumption is made that the sheath motion is much slower than the electron thermal velocity, as in Section 11.1, then $v_R(\phi)$ is small, and we can make the lower limit of the u' integrals equal to zero. Since $v_R(\phi)$ is an odd function, the $\bar{S}_1$ and $\bar{S}_3$ integrands integrate to zero, with the $\bar{S}_2$ integral yielding

$$\bar{S}_{\text{stoc}} = \frac{3\pi}{32} H m n_s \bar{v}_e u_0^2 \qquad (11.2.31)$$

Self Consistent Model Equations

We summarize the complete set of equations which can be used to calculate the parameters for an electropositive plasma, given a set of control parameters for a symmetric plane parallel geometry. In addition to $f = \omega/2\pi$, l, A and p, we have assumed that J_{rf} is known in deriving the self-consistent set. However, usually V_{rf} or the total absorbed power P_{abs} is the specified control parameter. The model includes assumptions that are only approximately satisfied, so we should not expect very close quantitative agreement with more detailed numerical simulations, or with actual experiments. In addition, for experiments it is very difficult to control the transverse uniformity of the plasma, as implied in the plane-parallel assumption. However, reasonably accurate scaling of plasma parameters with control parameters can still be determined. In this subsection we use the basic set of equations for sample calculations of plasma parameters. We then indicate the scaling that can be employed to estimate a wider set of plasma parameters, keeping in mind that the various regimes have different coefficients in the scaling, and sometimes different scalings. In Section 11.3 we shall compare analytic results to simulations and experiments, with the symmetric plane parallel assumption. Then, in Section 11.4, we model asymmetric discharges.

The approximate self-consistent model equations are summarized here. We assume $d \approx l - 2s_m$, with an initial estimate $s_m \approx 1$ cm for numerical computations, which is a nominal value for low-pressure capacitive discharges. We can iterate on this value if we believe it will improve overall accuracy. From particle conservation (10.2.7) at intermediate and low pressures, we have

$$\frac{K_{iz}}{u_s} = \frac{1}{n_g d_{eff}} = \frac{2}{n_g d}\frac{n_s}{n_0}, \qquad \lambda_i \gtrsim \left(\frac{T_i}{T_e}\right)d \qquad (11.2.32a)$$

where $u_s = u_B$ for a collisionless sheath and where n_s/n_0 is given by (10.2.3). At higher pressures, from (10.2.23), we have

$$\frac{(K_{mi}K_{iz})^{1/2}}{u_B} = \frac{\pi}{n_g d}, \qquad \lambda_i \lesssim \left(\frac{T_i}{T_e}\right)d \qquad (11.2.32b)$$

These equations determine T_e given n_g and d. Substituting (11.2.23) into (11.2.25), we obtain the electron ohmic heating power per unit area,

$$\bar{S}_{ohm} \approx 1.73\frac{m}{2e}\frac{n_s}{n_0}\epsilon_0\omega^2\nu_m T_e^{1/2}V_1^{1/2}d, \qquad \lambda_i \gtrsim \left(\frac{T_i}{T_e}\right)d \qquad (11.2.33a)$$

$$\bar{S}_{ohm} \approx 1.73\frac{m}{2e}\frac{n_s}{n_0}\epsilon_0\omega^2\nu_m T_e^{1/2}V_1^{1/2}\frac{2}{\beta}\ln\tan\left(\frac{\pi}{4} + \frac{\beta d}{4}\right), \qquad \lambda_i \lesssim \left(\frac{T_i}{T_e}\right)d \qquad (11.2.33b)$$

where $\cos(\beta d/2) = n_s/n_0$. Substituting (11.2.23) into (11.2.31) with $u_0 = J_1/en_s$ and using (11.2.10) and (11.2.11) for a single sheath in the slow sheath limit, we obtain

$$\bar{S}_{stoc} \approx 0.45\left(\frac{m}{e}\right)^{1/2}\epsilon_0\omega^2 T_e^{1/2}V_1, \qquad \omega s_m \lesssim \bar{v}_e \qquad (11.2.34)$$

We also have, from (11.2.22), that the ion kinetic energy per ion hitting the electrode is

$$\mathcal{E}_i = \bar{V} \approx 0.83\,V_1 \qquad (11.2.35)$$

The electron power balance equation is

$$S_e = \bar{S}_{ohm} + 2\bar{S}_{stoc} = 2en_s u_s(\mathcal{E}_c + 2T_e) \qquad (11.2.36)$$

Since $\bar{S}_{ohm}$ and $\bar{S}_{stoc}$ are both functions of V_1 alone, independent of n_s and J_1, (11.2.36) explicitly determines n_s if V_1 is the specified electrical control parameter. The total power absorbed per unit area is then found as

$$S_{abs} = 2en_s u_s(\bar{V} + \mathcal{E}_c + 2T_e) \qquad (11.2.37)$$

Eliminating n_s from these two equations and using (11.2.35) for $\bar{V}$, we obtain

$$S_{abs} \approx (\bar{S}_{ohm} + 2\bar{S}_{stoc})\left(1 + \frac{0.83V_1}{\mathcal{E}_c + 2T_e}\right) \qquad (11.2.38)$$

If S_{abs} is the specified control parameter, then (11.2.38) implicitly determines V_1 by substituting for $\bar{S}_{ohm}$ and $\bar{S}_{stoc}$ from (11.2.33) and (11.2.34). In this case, (11.2.36) or (11.2.37) can then be used to find n_s. The center density n_0 is then found using (10.2.3) or (10.2.5), and $\bar{V}$ is found from (11.2.38). To complete the summary, s_m and J_1 are found from (11.2.15) and (11.2.21), respectively.

Example 1   We take the following parameters:

- $p = 3$ mTorr argon at 300 K
- $l = 10$ cm
- $A = 1000$ cm^2
- $f = 13.56$ MHz ($\omega = 8.52 \times 10^7$ s^{-1})
- $V_{rf} = 500$ V

Starting with an estimate $s_m \approx 1$ cm, and using $\lambda_i = 1/n_g\sigma_i$ we find from (3.5.7) with $n_g = 1.0 \times 10^{20}$ m^{-3} at 300 K, that $\lambda_i = 1.0$ cm $= 0.01$ m. Thus with $d = l - 2s_m = 0.08$ m, $\lambda_i/d \approx 0.125$, which is in the intermediate mean free path regime, in which the plasma is relatively flat in the center. The ratio between the edge density and center density is given in (10.2.3), with d replacing l, as $n_s/n_0 = 0.325$. Then $n_g d_{eff} \approx 1.23 \times 10^{19}$ m^{-2} from (11.2.32a). Solving the particle balance using Fig. 10.1, or numerically, using (3.5.5), we find $T_e \approx 3.1$ V. This gives $u_B = (eT_e/M)^{1/2} \approx 2.7 \times 10^3$ m/s. From Fig. 3.17, $\mathcal{E}_c \approx 47$ V and $\mathcal{E}_c + 2T_e \approx 53$ V. Estimating $\nu_m \approx K_{el}n_g$ with K_{el} given by (3.5.6), we obtain $\nu_m \approx 1.0 \times 10^7$ s^{-1}. Then (11.2.33a) can be evaluated to obtain

$$\bar{S}_{ohm} \approx 0.145 \, V_1^{1/2} \text{ W/m}^2 \tag{11.2.39}$$

Similarly evaluating (11.2.34) yields

$$\bar{S}_{stoc} \approx 0.121 \, V_1 \text{ W/m}^2 \tag{11.2.40}$$

Neglecting the voltage drop across the bulk plasma, and letting $V_1 \approx V_{rf}/2 = 250$ V in (11.2.39) and (11.2.40), we find $\bar{S}_{ohm} \approx 2.29$ W/m^2 and $\bar{S}_{stoc} \approx 30.3$ W/m^2. We see for this example that $\bar{S}_{stoc}$ considerably exceeds $\bar{S}_{ohm}$. Using these values in the electron power balance (11.2.36), we obtain $n_s \approx 1.37 \times 10^{15}$ m^{-3}. Since $n_s/n_0 \approx 0.325$, we have $n_0 \approx 4.23 \times 10^{15}$ m^{-3}. From (11.2.35), we find $\bar{V} = \mathcal{E}_i \approx 208$ V; from the two equations in (11.2.15), $\bar{J}_i \approx 0.59$ A/m^2 and $s_m \approx 0.90 \times 10^{-2}$ m, and from (11.2.21), $J_1 \approx 25.8$ A/m^2. The total power absorbed per unit area is then obtained from (11.2.37) to be $S_{abs} \approx 308$ W/m^2. For $A = 0.1$ m^2, the discharge power is 30.8 W. Since s_m is reasonably close to our initial estimate, the plasma parameters are probably calculated within the accuracy of the calculation, and therefore an iteration is not useful.

Example 2   We take the following parameters, with the absorbed power as the specified electrical parameter:

- $p = 3$ mTorr argon at 300 K
- $l = 10$ cm
- $A = 1000$ cm^2
- $f = 13.56$ MHz ($\omega = 8.52 \times 10^7$ s^{-1})
- $P_{abs} = 200$ W

As in Example 1, $n_s/n_0 \approx 0.325$, $T_e \approx 3.1$ V, $u_B \approx 2.7 \times 10^3$ m/s, and $\mathcal{E}_c + 2T_e \approx 53$ V. Because n_g and T_e are the same as in Example 1, $\bar{S}_{ohm}$ is given by (11.2.39) and $\bar{S}_{stoc}$ is given by (11.2.40). Substituting these into (11.2.38) with $S_{abs} = P_{abs}/A = 2000$ W/m^2, we obtain

$$2000 = \left(0.145V_1^{1/2} + 0.242V_1\right)\left(1 + \frac{0.83V_1}{53}\right) \qquad (11.2.41)$$

Dropping the first (small) terms in each parenthesis yields an approximate solution $V_1 \approx 727$ V. A numerical solution of (11.2.41) gives a more exact result $V_1 = 687$ V. Then $V_{ab1} = 2V_1 \approx V_{rf} \approx 1374$ V, and (11.2.35) yields $\mathcal{E}_i \approx 570$ V. Using this in (11.2.37), we obtain $n_s \approx 3.72 \times 10^{15}$ m^{-3} and, with $n_s/n_0 \approx 0.325$, we find $n_0 \approx 1.14 \times 10^{16}$ m^{-3}. The ion current density and the sheath width are found from (11.2.15) to be $\bar{J}_i \approx 1.6$ A/m^2 and $s_m \approx 1.16 \times 10^{-2}$ m, and the rf current density is found from (11.2.21) to be $J_1 \approx 54.9$ A/m^2.

Scaling

We can use the basic equations to obtain the most important scalings of the plasma parameters with control parameters. These scalings can also be compared to the scalings obtained from simulations and experiments to investigate the validity of the various approximations. We assume that $d \approx l - 2s_m$ is essentially constant as the voltage and pressure are varied over reasonable ranges. We can then combine the model equations to obtain the scalings in various limiting cases. We assume that the pressure is sufficiently low that ohmic heating can be neglected. We leave the ohmic heating scalings to a problem. From (11.2.34), we have

$$\bar{S}_{stoc} \propto \omega^2 T_e^{1/2} V_{rf} \qquad (11.2.42)$$

Dropping the ohmic term in (11.2.36), such that $S_e = 2\bar{S}_{stoc}$, assuming $\mathcal{E}_c \gg T_e$, and substituting for $\bar{S}_{stoc}$ from (11.2.34), we obtain

$$n_s \propto \frac{\omega^2 V_{rf}}{\mathcal{E}_c} \qquad (11.2.43)$$

For low sheath voltages, taking $\mathcal{E}_i \ll \mathcal{E}_c$ in (11.2.37), we obtain

$$S_{abs} \propto \omega^2 T_e^{1/2} V_{rf} \qquad (11.2.44a)$$

For the more common situation of high sheath voltages, $\mathcal{E}_i \gg \mathcal{E}_c$, with $\bar{V} \propto V_{rf}$, we obtain

$$S_{abs} \propto \frac{\omega^2 T_e^{1/2} V_{rf}^2}{\mathcal{E}_c} \tag{11.2.44b}$$

The weak dependence of s_m is found by substituting n_s from (11.2.43) into (11.2.15) to obtain

$$s_m \propto \frac{V_{rf}^{1/4}}{\mathcal{E}_c^{1/2} T_e^{1/4}} \tag{11.2.45}$$

and, using this scaling in (11.2.21), we find

$$J_{rf} \propto \omega V_{rf}^{3/4} \mathcal{E}_c^{1/4} T_e^{1/4} \tag{11.2.46}$$

Note that in the low-pressure regime, where stochastic heating dominates, variations in the temperature only enter logarithmically through the change in pressure. The generally strong frequency dependences should be noted.

We can equally well consider the total absorbed power as the independent variable and solve for V_{rf}, n_s, s_m, and J_{rf}. Using the same approximations as above, we find, for high voltages, $\mathcal{E}_i \gg \mathcal{E}_c$, that

$$V_{rf} \propto S_{abs}^{1/2} \mathcal{E}_c^{1/2} / \omega T_e^{1/4} \tag{11.2.47}$$

$$n_s \propto S_{abs}^{1/2} \omega / \mathcal{E}_c^{1/2} T_e^{1/4} \tag{11.2.48}$$

$$s_m \propto S_{abs}^{1/8} / \omega^{1/4} T_e^{5/16} \mathcal{E}_c^{3/8} \tag{11.2.49}$$

$$J_{rf} \propto S_{abs}^{3/8} \omega^{1/4} T_e^{1/16} \mathcal{E}_c^{7/8} \tag{11.2.50}$$

The above scalings are independent of pressure, except implicitly through the weak dependence of T_e on pressure. These scalings can be easily compared to experimental results. Since T_e and therefore $\mathcal{E}_c$ only vary logarithmically with change in pressure, they can usually be held constant in comparing scalings.

Collisional Sheaths

If $\lambda_i \lesssim s_m$, then the ions suffer one or more collisions as they cross the sheath and the collisionless analysis is not valid. For argon with λ_i given by (3.5.7) and with $s_m \sim 1$ cm, we find $p \lesssim 3$ mTorr for a collisionless sheath, at the low end of typical processing discharges. At higher pressures a self-consistent analysis of the collisional sheath is required, which has been given by Lieberman (1989a) and Godyak and Sternberg (1990b). These authors assume $\lambda_i = $ const, independent of velocity. The basic ion dynamical equations (11.2.1) and (11.2.2) are then modified, as in (6.5.1) and (6.5.2), to

$$n_i u_i = n_s u_s \tag{11.2.51}$$

and

$$u_i = \frac{2e\lambda_i}{\pi M u_i}\bar{E} \tag{11.2.52}$$

where

$$u_s = u_B\left(1 + \frac{\pi\lambda_{De}}{2\lambda_i}\right)^{-1/2} \tag{11.2.53}$$

is the modified Bohm velocity given by (6.5.9), with $\lambda_{De} = (\epsilon_0 T_e/en_s)^{1/2}$ the Debye length at the plasma sheath edge. Carrying out the analysis as in the first part of this section, the dc ion current density is found to be

$$\bar{J}_i = en_s u_s \approx 1.68\,\epsilon_0\left(\frac{2e}{M}\right)^{1/2}\frac{\bar{V}^{3/2}\lambda_i^{1/2}}{s_m^{5/2}} \tag{11.2.54}$$

where the coefficient is 1.68 for the self-consistent calculation rather than 1.43 as given in (6.5.7). Note that (11.2.54) differs from the collisionless Child law (11.2.15) because $\bar{J}_i$ now scales with λ_i and scales differently with s_m. The collisional sheath capacitance is found to be $0.76\,\epsilon_0\,A/s_m$, leading to

$$J_1 \approx 1.52\frac{\omega\epsilon_0}{s_m}V_1 \tag{11.2.55}$$

in place of (11.2.21). We also find

$$\bar{V} \approx 0.78\,V_1 \tag{11.2.56}$$

in place of (11.2.22).

The average ion-bombarding energy $\mathcal{E}_{ic}$ is reduced below $\bar{V} \equiv V_s$ because ion energy is lost during charge transfer and elastic collisions in the sheath, creating fast neutrals there. The ion-bombarding energy is found to be

$$\mathcal{E}_{ic} = \frac{1}{2}Mu_i^2(s_m) \approx 0.62\frac{\lambda_i}{s_m}\bar{V} \tag{11.2.57}$$

Note, however, that the total kinetic energy lost per ion transiting the sheath is still $\bar{V}$, as for the collisionless sheath, and as used in (11.2.37). Thus, the effect of collisions in the sheath is to reduce the ion bombarding energy but to proportionally increase the total energetic particle flux (ions + fast neutrals) to the electrode.

The slow sheath stochastic heating is found to be

$$\bar{S}_{stoc} \approx 0.59 \left(\frac{2\lambda_i s_0}{\pi^2 \lambda_{De}^2} \right)^{1/2} \left(1 + \frac{\pi \lambda_{De}}{2\lambda_i} \right)^{1/2} m n_s \bar{v}_e u_0^2 \qquad (11.2.58)$$

in place of (11.2.31). Substituting (11.2.53)–(11.2.56) into (11.2.58), we obtain (see Problem 11.2)

$$\bar{S}_{stoc} \approx 0.61 \left(\frac{m}{e} \right)^{1/2} \epsilon_0 \omega^2 T_e^{1/2} V_1 \qquad (11.2.59)$$

in place of (11.2.34). We see that, except for the numerical coefficients and the somewhat different scaling between (11.2.15) and (11.2.54), the equations for the collisional and collisionless sheaths have the same form. The procedure for calculating the discharge parameters for the collisionless sheath can therefore be applied to the collisional sheath, with minor modifications.

Because $\lambda_{De} \ll s_m$ in a high-voltage collisional sheath, there are actually two collisionality regimes:

1. Intermediate pressures, $\lambda_{De} \lesssim \lambda_i \lesssim s_m$. In this regime (Lieberman, 1989a) we have $u_s \approx u_B$ in (11.2.51). Since $s_m \sim 20$–$50\,\lambda_{De}$, the range of pressures, in argon, extends from 3 mTorr up to 60–150 mTorr. In this regime, for a fixed n_s and $\bar{V}$ (and T_e), the sheath thickness s_m decreases with increasing n_g according to (11.2.54).

2. High pressures, $\lambda_i \lesssim \lambda_{De}$. This regime, for which $u_s \approx u_B (2\lambda_i/\pi\lambda_{De})^{1/2}$, has been treated by Godyak and Sternberg (1990b), who also present a unified picture, including numerical solutions, of the collisionless sheath, the two collisional sheath regimes, and the transitions among them. In this regime, s_m is independent of n_g. Because u_s is a function of n_s in this regime, the equations for the discharge equilibrium must be solved iteratively; e.g., T_e as determined from (11.2.32) is a function of n_s, although the dependence on n_s is weak. At high pressures, ohmic heating is usually the dominant heating mechanism, allowing considerable simplifications in the analysis.

Low and Moderate Voltages

Godyak and Sternberg (1990b) have treated the regimes from $V_{rf} \ll T_e$ to $V_{rf} \gg T_e$ in a unified manner. For $V_{rf} \ll T_e$, their results reduce to that of an undriven dc sheath, as in (6.2.17). At high voltages, $V_1/T_e \gtrsim 200$, their numerical results asymptotically approach the analytic results $\bar{V} \propto V_1$, but these voltages are at the upper end of typical processing discharge regimes. At more moderate voltages, $50 \lesssim V_1/T_e \lesssim 200$, $\bar{V}$ is seen to have a weaker scaling with V_1, such that $\bar{V} \propto V_1^\beta$, with $\beta \approx 2/3$–$3/4$. With the weaker scaling, s_m in (11.2.45) is found to be nearly independent of V_{rf}, and from (11.2.21), the J_{rf} versus V_{rf} discharge characteristics are nearly linear, as observed in many experiments (see Godyak and Sternberg (1990b) for further discussion).

Ohmic Heating in the Sheath

Ohmic heating due to collisions of the oscillating electrons with the neutral gas can be an important additional electron heating mechanism. For a Child law scaling, the density within the sheath is, from (6.3.8), of order

$$n \sim n_s \left(\frac{T_e}{V_{rf}} \right)^{1/2}$$

Because the ohmic power density p_{ohm} scales as J_{rf}^2/n and J_{rf} is not spatially varying, we see that p_{ohm} is a factor of $(V_{rf}/T_e)^{1/2}$ larger within the sheath than at the sheath edge. Hence for a uniform bulk plasma, the ohmic heating within the sheath exceeds the bulk heating when

$$s_m \left(\frac{V_{rf}}{T_e} \right)^{1/2} \gtrsim d. \tag{11.2.60}$$

This condition can be met in a high-voltage discharge. For a collisionless sheath, Misium et al. (1989) give the expression

$$\bar{S}_{ohm,sh} \approx 1.73 \frac{m}{2e} \epsilon_0 \omega^2 \nu_m s_m (T_e V_1)^{1/2} \left[0.235 \left(\frac{V_1}{T_e} \right)^{1/2} + 1.16 + 4.39 \left(\frac{T_e}{V_1} \right)^{1/2} \right] \tag{11.2.61}$$

This should be added to (11.2.33) for each sheath. Although $\bar{S}_{ohm,sh}$ depends on s_m, explicitly, a nominal value $s_m = 1$ cm can be assumed initially, and the equations can be iterated if greater accuracy is required.

Fast Sheath Heating

For $u_{es} \gtrsim \bar{v}_e$ in (11.2.30), the stochastic heating result (11.2.31) is not correct. In the limit $u_{es} \gg \bar{v}_e$, the lower limit of the u' integration is $\pm\infty$, where the minus sign applies for $-\pi \le \phi \le 0$ and the plus sign applies for $0 \le \phi \le \pi$. In this case $\bar{S}_2 \approx 0$, since its integrand is odd. Furthermore, we find $\bar{S}_3 \gg \bar{S}_1$. Performing the integration for $\bar{S}_3$ and averaging over the phase, we obtain

$$\bar{S}_{fast} \approx \left(\frac{5\pi}{56} - \frac{2048}{15435\pi} \right) H^2 m n_s u_0^3 \tag{11.2.62}$$

It is instructive to examine the ratio of the average powers under the fast and slow sheath assumptions:

$$\frac{\bar{S}_{fast}}{\bar{S}_{stoc}} \approx 0.809 \frac{u_0}{\bar{v}_e} H \tag{11.2.63}$$

Inserting (11.2.14) with $u_0 = \omega s_0$ into (11.2.61), we obtain

$$\frac{\bar{S}_{\text{fast}}}{\bar{S}_{\text{stoc}}} = 0.618 \frac{\omega s_m}{\bar{v}_e} \tag{11.2.64}$$

The ratio $\bar{S}_{\text{fast}}/\bar{S}_{\text{stoc}}$ is 0.45 at $\omega/2\pi = 13.56$ MHz, $s_m = 1$ cm, and $T_e = 3$ V, which are typical experimental values. We mainly use the slow sheath limit in calculations. A numerical calculation using the complete expression for the stochastic heating from (11.2.30) gives a somewhat larger plasma density and power dissipation at the higher voltages, than that obtained using the slow sheath approximation (Wood, 1991). However, the differences are probably within the overall accuracy of the calculation. We should note that the fast sheath analysis is not on firm theoretical ground because the fast sheath strongly perturbs the distribution of electrons within the sheath, such that the sheath calculation is no longer self-consistent. This will become apparent in the next section in which we present results from simulations. Nevertheless, the concept of a fast sheath is useful for understanding the general behavior.

11.3 EXPERIMENTS AND SIMULATIONS

Models are based on a particular set of assumptions that must be tested by experimentally determining if the observable consequences of those assumptions are in agreement with the experiment that the model is designed to represent. In recent years the tool of computer simulation, added to analytic modeling and experiments, has improved our understanding. The particular simulations described here for modeling low-pressure discharges, called particle-in-cell (PIC) simulations, follow large numbers of representative particles acted upon by the basic forces. Many of the assumptions of the analytic models need not be used. It is also possible to determine various microscopic quantities that are not observable experimentally. In these ways the simulations serve as an intermediary between the models and the experiments.

In this section we shall first give some experimental observations and relate the results to the model of Section 11.2. We then present the results of simulations to obtain further understanding of the plasma behavior. Finally, we will comment on some of the modeling improvements that might increase the accuracy of the predictions of plasma source operation. Throughout this section we restrict our attention to symmetric plane parallel geometry. Simulations can be performed in more complicated geometries, but the calculations become more involved and lengthy.

Experimental Results

An early experiment to investigate stochastic sheath heating is described in Godyak's (1986) review. In an apparatus designed to approximate a plane parallel discharge, an effective collision frequency ν_{eff} was measured versus pressure, using the relationship for the power absorbed per unit area,

$$S_{\text{abs}} = \frac{1}{2} \frac{|\tilde{J}_{\text{rf}}|^2}{e^2 n} m \nu_{\text{eff}} \tag{11.3.1}$$

S_{abs}, $\tilde{J}_{\text{rf}}$, and n being simultaneously measured. The measurements were done at relatively low voltages, such that $\bar{V} \lesssim \mathcal{E}_c$; consequently, $S_{\text{abs}} \sim S_e$, the power per unit area absorbed by the electrons. The result is shown in Fig. 11.7. Both the asymptotic leveling off of ν_{eff} at low pressure p, characteristic of stochastic heating which is independent of p, and the linear increase of ν_{eff} with p at high p, characteristic of ohmic heating, are clearly visible. The good agreement of the measurements with ν_{eff} calculated from the stochastic heating formula is somewhat fortuitous, however, as a uniform sheath rather than a self-consistent sheath was used in the calculation, and the ion power loss S_i was neglected in determining ν_{eff} from the measurements.

An experimental study of symmetric rf discharge characteristics in argon at 13.56 MHz has been performed by Godyak et al. (1991). The discharge length and diameter were 6.7 cm and 14.3 cm, respectively, approximating a uniform plane parallel configuration. Measurements were made of rf voltage, rf current, total power absorbed, dc bias voltage, the central plasma density n_0, mean electron energy $\langle \mathcal{E}_e \rangle$, and electron distribution function f_e. The time-average power was determined by averaging $V_{\text{rf}}(t)I_{\text{rf}}(t)$ over an rf cycle (see Section 11.6), and n_0, $\langle \mathcal{E}_e \rangle$ and f_e were determined using Langmuir probes (see Section 6.6). Measurements were performed

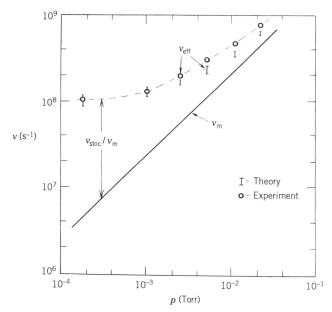

FIGURE 11.7. Effective collision frequency ν_{eff} versus pressure p, for a mercury discharge driven at 40.8 MHz. The solid line shows the collision frequency due to ohmic dissipation alone (after Popov and Godyak, 1985).

over a wide range of pressures from 3 mTorr to 3 Torr and for powers up to 100 W. The corresponding rf voltage amplitudes were up to 1500 V, and the rf current amplitudes were up to 2 A. Figures 11.8 and 11.9 show V_{rf}, P_{abs}, n_0, and $\langle \mathcal{E}_e \rangle$ versus I_{rf} at relatively low (10 mTorr) and relatively high (100 mTorr) pressures. At 10 mTorr, where ohmic heating is small, and at low to moderate voltages, the voltage scales roughly linearly with the current, with a transition to the scaling $V_{rf} \propto I_{rf}^{4/3}$ predicted from (11.2.46) at the higher voltages. The power scales as $P_{abs} \propto I_{rf} \propto V_{rf}$ at low voltages, with a transition to $P_{abs} \propto V_{rf}^2$ at higher voltages, in agreement with (11.2.44). The density, however, scales more strongly with the voltage than the linear scaling predicted by (11.2.42), and the mean electron energy $\langle \mathcal{E}_e \rangle$, which corresponds to $\frac{3}{2} T_e$ for a Maxwellian distribution, falls significantly at the higher voltages, contrary to the analytic model in which T_e depends only on the pressure and is independent of the voltage.

Generally, the experimental density is higher than the model predictions, indicating somewhat more efficient electron power absorption at a given applied voltage, which may be partly due to the effect of secondary electron emission. The discrepancy can also be partly understood by examining the measured electron energy probability function g_p (see (6.6.18)), which is plotted versus $\mathcal{E} = mv^2/2e$ in Fig. 11.10. We see a transition from a single Maxwellian for $V_{rf} \lesssim 100$ V to a two-temperature distribution at higher voltages, with most of the electrons in the lower energy class, which therefore determines $\langle \mathcal{E}_e \rangle$. The high-temperature tail maintains the ionization balance required by (11.2.32a), allowing $\langle \mathcal{E}_e \rangle$ to drop to low values. As we will see from simulations, a two-temperature distribution is characteristic of stochastic heating. A similar behavior is seen at 3 and 30 mTorr. At 100 mTorr, ohmic heating dominates the electron power absorption below approximately 300 V, leading to a single temperature Maxwellian, as seen in Fig. 11.10b with $\langle \mathcal{E}_e \rangle \approx$ 4–5 V. From Fig. 11.8b, we see a near-linear scaling of P_{abs} with V_{rf} at low voltages, with a transition to a steeper scaling of power with voltage at higher voltages. At higher V_{rf} there is a transition to a two temperature distribution, as seen in Fig. 11.10b, with $\langle \mathcal{E}_e \rangle$ falling to 1.5–2 V. These results indicate a transition from ohmic heating at low voltage to stochastic heating at high voltage.

In Figs. 11.11 and 11.12, the experimental results of Godyak et al. (1991) and Godyak and Piejak (1990b) for the density and total power are given as a function of the applied rf voltage and compared with the nonuniform density model results for a 3-mTorr argon discharge. At this low pressure, stochastic heating is the principal heating mechanism. For the modeling results, $\bar{S}_{stoc}$ was computed numerically from (11.2.30), and other quantities were computed using these numerical results. The self-consistent single Maxwellian temperature was 3.2 V. The total power in the experiments was then normalized to the electrode area. As can be seen from the figures, the model results are in qualitative agreement with the experiments, but there are also some significant disagreements. Generally, the experimental density is more steeply varying with rf voltage than the model predictions, which is at least in part related to the changing electron distribution function, which allows u_B, related to the lower temperature component, to decrease. As expected, this effect would not be prominent in the variation of power with rf voltage.

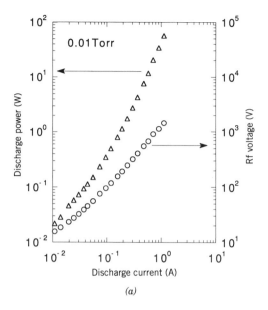

(a)

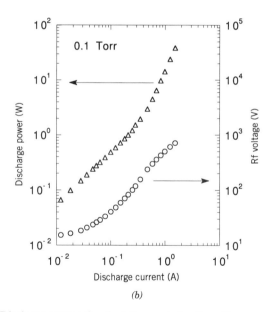

(b)

FIGURE 11.8. Discharge power absorbed P_{abs}, and rf voltage V_{rf} versus discharge current I_{rf} at (a) $p = 10$ mTorr and (b) $p = 100$ mTorr in argon (Godyak et al. 1991, © 1991 IEEE).

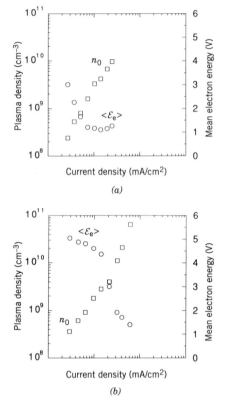

FIGURE 11.9. Central plasma density n_0 and mean electron energy $\langle \mathcal{E}_e \rangle$ versus discharge current density J_{rf} at (a) $p = 10$ mTorr and (b) $p = 100$ mTorr in argon gas (Godyak, 1990b).

Experimentally, the dc voltage $\bar{V}$ across a single sheath is found to track the rf voltage V_{ab} across both sheaths, with $\bar{V} \approx 0.4\,V_{ab} = 0.8\,V_1$ at high voltages, as predicted. For lower voltages, we find a weaker variation $\bar{V} \propto V_{ab}^\beta$, with $\beta \sim 2/3$–3/4, as described in Section 11.2. At 10 mTorr, we find that $V_{ab} \approx V_{rf}$ over the entire voltage range. However, at 100 mTorr, V_{ab} falls below V_{rf} at low V_{rf}, due to the additional rf voltage V_p dropped across the bulk plasma.

Some discrepancy exists for the sheath width, with the experimental widths being somewhat larger, but scaling more weakly with V_{rf}, than the high-voltage model. This may be a consequence of the somewhat weaker-than-linear scaling of $\bar{V}$ with V_{rf} at moderate discharge voltages. Despite these differences, the model can be considered to have reasonable predictive power. We shall discuss the discrepancies further after giving simulation results.

Particle-in-Cell Simulations

The symmetric measurements of Godyak and Piejak (1990a) have been compared to PIC simulations by Vahedi et al. (1994). The comparisons are in argon for a electrode

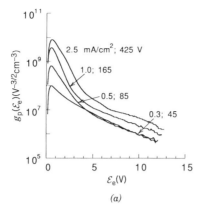

(a)

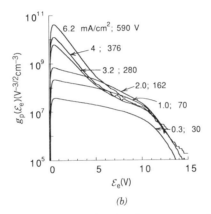

(b)

FIGURE 11.10. Electron energy probability function g_p versus $\mathcal{E}_e$ for various discharge currents for argon gas with $f = 13.56$ MHz and $l = 6.7$ cm: (a) $p = 10$ mTorr and (b) $p = 100$ mTorr (Godyak, 1990b).

diameter of 14.3 cm, a discharge length of 2 cm, and an external current source of 2.56 mA/cm² at 13.56 MHz. The gas pressure was varied between 70 and 500 mTorr to observe the transition from stochastically to ohmically dominated electron heating. Except for the normalization, the f_e's obtained from the simulations agree well with the measured f_e's, showing the transition from a two–temperature distribution at 70 mTorr to a single-temperature distribution at 500 mTorr. The simulation temperatures are in good agreement with the measured temperatures over the entire range of pressures. The plasma densities are compared in Fig. 11.13. Two sets of simulation results are shown, with and without secondary emission due to ion impact on the electrodes with $\gamma_{se} = 0.2$ (see discussion of secondaries below). The plasma density shows a better agreement with measurements when secondaries are included, but is lower than the measurements by roughly a factor of 1.5 at low gas pressures. Possible explanations include incomplete modeling of the atomic collision processes; e.g., neglect of energetic ion–neutral ionization processes within the sheaths, and neglect of metastable atom production, electron impact ionization, and quenching.

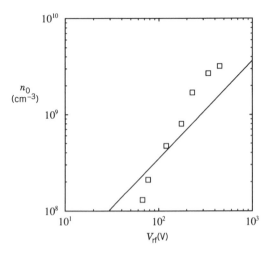

FIGURE 11.11. Central plasma density n_0 versus rf voltage V_{rf} from the simulations (solid line), compared to the data (squares) of Godyak et al. (1991), for a 3-mTorr argon discharge.

Another simulation of discharge behavior (Wood, 1991), was performed at $p = 3$ mTorr (argon) with a spacing of 10 cm between parallel plates, and over a range of rf voltages between 100 and 1000 V. A two-temperature distribution was found, as in the experiments, and the distribution varied in both space and time. It is clear that a deeper understanding of the discharge behavior involves the space and time variations of f_e. Figure 11.14 shows the one-dimensional electron distribution function $f_e(x, v_x, t)$ versus v_x at 15 positions near the sheath region ($x = 0$–3 cm) and at eight different times during the rf cycle. Each plot covers $\frac{1}{32}$ of a cycle temporally, and each line

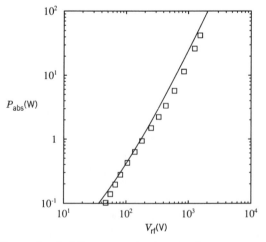

FIGURE 11.12. Power absorbed P_{abs} versus rf voltage V_{rf} from the simulations (solid line), compared to the data (squares) of Godyak et al. (1991), for a 3-mTorr argon discharge.

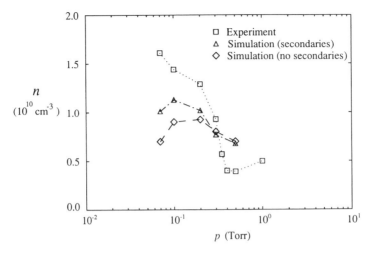

FIGURE 11.13. Central plasma density versus pressure in a 2-cm gap, current-driven argon rf discharge at 13.56 MHz. The squares are measurements by Godyak and Piejak (1990a), the diamonds are simulation results without secondary electrons, and the triangles are simulation results with secondary electron emission (Vahedi et al., 1994).

in a plot covers a 2-mm-thick region spatially. The units on the vertical axis are proportional to f_e. At time $\frac{0}{32}$, the sheath is fully expanded, and the two-temperature nature of the discharge near the sheath can be seen as the wide "base" and narrow "peak" of the distribution. As the rf cycle progresses to time $\frac{8}{32}$, the distributions in the sheath region at each position display a drift toward the electrode (negative velocity) that is approximately equal to the sheath velocity. By time $\frac{12}{32}$, fast electrons have arrived from the opposite electrode, moving at a velocity of about 4×10^6 m/s (small peak at extreme left of figure). At time $\frac{16}{32}$, the sheath is fully collapsed, the drift in the sheath has disappeared, and the fast electron group moving toward the electrode shows a lower velocity as slower electrons arrive from the opposite electrode. As the sheath begins to expand, as shown here at times $\frac{18}{32}$ and $\frac{20}{32}$, the electrons in the sheath region are strongly heated, and the beginning of an electron beam produced by this expansion can be seen moving away at a positive velocity. As the sheath continues to expand, the drift of the distribution in the sheath away from the electrode can be seen to initially match the sheath velocity (time $\frac{22}{32}$) but then decays (time $\frac{24}{32}$) to a velocity much slower than when the sheath was collapsing. One consequence of the complicated f_e near the sheath edge is that the average electron velocity at the moving sheath edge does not correspond to the sheath velocity during the entire rf cycle, as predicted from the model equations. This is shown in Fig. 11.15 where we see that the observed average electron drift velocity deviates from the predicted value at the sheath edge as the sheath edge oscillates.

The existence of more energetic electrons near the plasma edge due to stochastic heating increases the ionization there, tending to flatten the plasma profile. Furthermore, the ionization is not constant, but follows the density variations in space and time of the more energetic electrons. This is shown for a PIC simulation by Vender

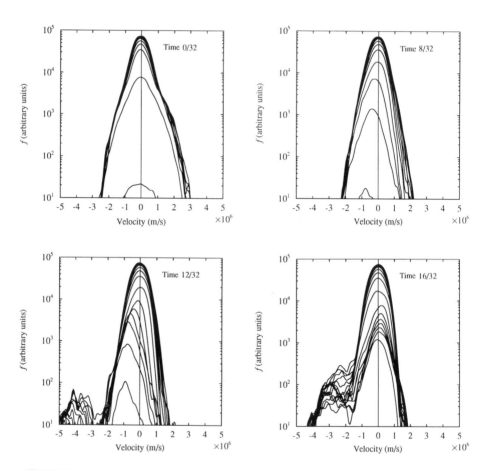

FIGURE 11.14. One-dimensional electron velocity distribution function $f_e(x, v_x, t)$ for a 10-cm electrode spacing; each plot covers a time window of $\frac{1}{32}$ of an rf cycle. Each line on a plot represents a spatial window of 2 mm (Wood, 1991).

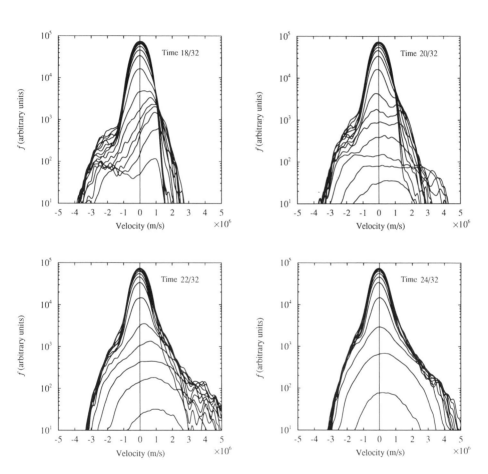

FIGURE 11.14 (*Continued*)

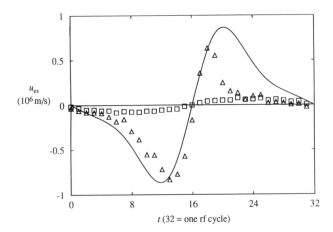

FIGURE 11.15. Drift of the electron velocity distribution versus time at the electron sheath edge $s(t)$ (triangles, simulation; solid curve, theory) and at s_m (squares, simulation) (Wood, 1991).

and Boswell (1990) in the plot of Fig. 11.16, in which the darkness of each square is proportional to the number of ionizing collisions within that square of position and time intervals. Most of the ionization is seen to occur along a path of fastest electrons that are reflected off of the sheath at the phase at which it is most rapidly expanding. There is also somewhat more ionization near the sheaths, an effect that becomes more pronounced at higher pressures where the ionization mean-free-path is shorter, which has been observed in various experiments.

In Fig. 11.17, the time-dependent ion and electron conduction currents are given as a function of time, with V_{rf} and V_{pb} also indicated. We see that $J_i(t)$ is nearly constant, as assumed, but also contains some ripple which is not important for the modeling. $J_e(t)$, on the other hand, is spread over a significant fraction of the rf cycle, when significant voltages exist between the plasma and the electrode. This is possible because of the distribution of electron energies. Because the time-average electron and ion fluxes must balance, the average potential of the plasma with respect to the electrode must decrease slightly.

To obtain a simple estimate of this effect, we assume a single-temperature Maxwellian and a sinusoidal voltage. The average electron flux $\bar{\Gamma}_e$ can be expressed as

$$\bar{\Gamma}_e = \frac{1}{4}\langle n_e(t)\rangle \bar{v}_e \tag{11.3.2}$$

where, from Boltzmann's law,

$$n_e(t) = n_s \exp\left(\frac{-\bar{V} + V_1 \sin \omega t}{T_e}\right) \tag{11.3.3}$$

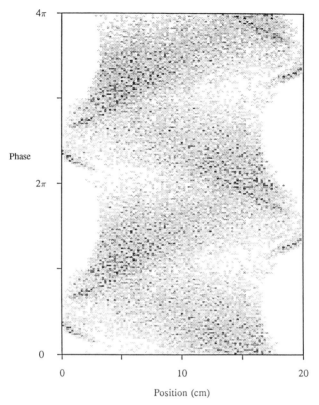

FIGURE 11.16. Spatiotemporal distribution of ionizing collisions collected over 20 rf cycles, for a 10-MHz, 1-kV, 20-mTorr hydrogen discharge (Vender and Boswell, 1990, © 1990 IEEE).

Substituting (11.3.3) in (11.3.2) we have

$$\bar{\Gamma}_e = \frac{1}{4} n_s \bar{v}_e \, e^{-\bar{V}/T_e} \frac{1}{2\pi} \int_0^{2\pi} \exp\left(\frac{V_1 \sin \omega t}{T_e}\right) d(\omega t) \qquad (11.3.4)$$

The averaged integral yields

$$I_0\left(\frac{V_1}{T_e}\right) \approx \left(\frac{T_e}{2\pi V_1}\right)^{1/2} e^{V_1/T_e}$$

where the approximate equality holds for $V_1 \gg T_e$. Using this and equating $\bar{\Gamma}_e$ to the Bohm ion flux, we have

$$e^{(-\bar{V}+V_1)/T_e} = \left(\frac{2\pi V_1}{T_e} \frac{2\pi m}{M}\right)^{1/2} \qquad (11.3.5)$$

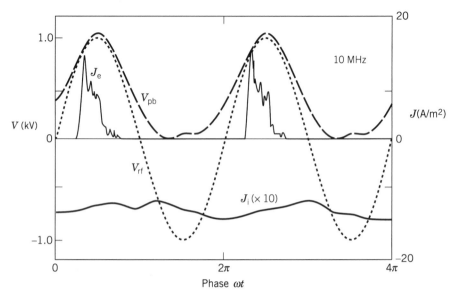

FIGURE 11.17. Central plasma potential V_{pb} (dashed), driving voltage V_{rf} (dotted), and electron (positive) and ion (negative) currents to the electrode. The ion current is plotted ten times enlarged to show modulation within the rf cycle (Vender and Boswell, 1990, © 1990 IEEE).

Taking the logarithm of both sides and solving for $\bar{V}$, we obtain

$$\bar{V} = V_1 + \frac{T_e}{2}\left(\ln\frac{M}{2\pi m} - \ln\frac{2\pi V_1}{T_e}\right) \tag{11.3.6}$$

We see that the zero-order value of $\bar{V} = V_1$ is increased by the usual thermal term but is decreased because the electrons reach the electrode over a finite time interval. For example, with $V_{rf} = 500$ V ($V_1 = 250$ V) and $T_e = 3.4$ V in argon gas, we find that the thermal enhancement to the voltage is the usual $4.8\, T_e$ while the finite electron loss effect reduces $\bar{V}$ by $3.2\, T_e$, leading to a net increase in $\bar{V}$ over V_1 of $1.6\, T_e = 5.4$ V. Recall, however, for our model in Sections 11.1 and 11.2 with sinusoidal current and nonsinusoidal single-sheath voltage, that the zero-order result is $\bar{V} = 0.83\, V_1$. In this case, for $V_1 \gg T_e$, the correction is not significant compared to other approximations in the model. The relation between $\bar{V}$ and V_1 for sinusoidal current drive has been obtained over the entire range of V_1/T_e by Godyak and Sternberg (1990b).

The effect of a group of fast beamlike particles, traversing the plasma from one sheath to the other, indicates that the discharge length may enter into the dynamics of stochastic heating in a more sensitive manner than in the model equations. This has been demonstrated by following a class of representative electrons, with energy greater than the ionization energy, over several rf cycles, for $l = 13$, 10, and 7.5 cm. One finds a tendency to have a resonant increase of energy near $l = 10$ cm, producing a higher-energy tail on the distribution, because for this length, electrons heated at one sheath arrive at the opposite sheath at a phase resulting in further heating. The average electron energy lost from the discharge also varies but in all cases is increased by a factor of 2–3 over the value of $2T_e$ for a single Maxwellian. Other interesting effects have been observed in simulations and sometimes confirmed in experiments. For example, high-harmonic components of the driving frequency have been observed, both in simulations and in experiments, that can be much larger than predicted from the sheath nonlinearity. This has been shown to result from a series resonance of the bulk plasma inductance L_p with the sheath capacitance C_s, occurring below the electron plasma frequency (Problem 11.3). The consequences on macroscopic plasma properties and on processing have not been explored.

Role of Secondaries

The steady flux Γ_i of ions striking the electrodes generates a flux $\Gamma_e = \gamma_{se}\Gamma_i$ of secondary electrons that are accelerated back into the plasma. Typically $\gamma_{se} \sim 0.1$–0.2 for metal electrodes. Depending on the phase of the rf voltage the secondaries gain various energies up to $\sim 2V_1$. The effect of secondaries is generally not important for argon at pressures below 100 mTorr at the usual operating frequency of 13.56 MHz. Unlike dc discharges, in which continuity of current requires secondaries, the rf current at 13.56 MHz can be sustained by the sheath capacitance. High-energy secondaries do produce some ionization. However, $\Gamma_e \sim (0.1$–$0.2)\,\Gamma_i$ is small, and at low pressures most of the secondary electrons are lost from the discharge before significant ionization occurs (Problem 11.6). Therefore, the effect is generally small compared to the beamlike electrons generated by stochastic heating.

Another effect of secondaries is to increase the power dissipation. Some of this additional energy loss goes into ionization and other collisional processes. Part of the energy is lost directly to surfaces at a lower potential than the emitting surface, and this acts as a power drain. Both effects of increased ionization and increased power loss can be included in a self-consistent model (Misium et al., 1989). With $\gamma_{se} = 0.1$ at p = 10 mTorr, there was little effect on most plasma parameters, except for an increase of up to 30% in the total power absorbed.

At higher pressures the effect of secondaries becomes greater, as more of their energy is captured by the plasma. In a still higher pressure regime, a transition can take place with increasing voltage to a different mode in which the plasma is sustained by ionization from secondaries. For example, if we take $\gamma_{se} = 0.1$ and assume that all secondary energy is absorbed in the plasma with $\mathcal{E}_c = 60$ V (energy per electron–ion pair created), then when the average secondary energy is $\bar{\mathcal{E}}_{se} = 600$ V, the ionization just balances the escaping ion flux, $\Gamma_{iz}/\Gamma_i = (600/60)\,\gamma_{se} = 1$. The bulk electron temperature falls to prevent additional ionization, and the plasma is sustained by secondaries alone. This transition to the γ-*mode* has been observed experimentally and predicted theoretically by Godyak and associates (see Godyak, 1986). The experiments were performed primarily at $p = 3$ Torr, where the secondaries are mostly absorbed. At these and higher pressures, secondary multiplication within the sheath can occur, leading to a mechanism that sustains the discharge similar to that which sustains a dc glow discharge (see Chapter 14).

As the rf frequency is reduced, I_{rf} decreases with ω at fixed V_{rf}, and secondaries again play a more important role in sustaining the plasma. We do not discuss this low-frequency regime in this chapter. In Chapter 14 we consider dc discharges in which the entire current in the sheath is sustained by secondaries and electrons that are created by ionization in the sheath. We have ignored this electron multiplication factor in our simple discussion here, but it is of essential importance in sustaining a dc discharge.

Implications for Modeling

What are the consequences of the comparison of model results with those obtained from simulations and experiments? Clearly the model assumptions are only a crude approximation on a detailed microscopic level. For example, the existence of a "two-temperature" electron distribution with a high-density cool population and a low-density warm population implies that the Bohm velocity is more characteristic of the cool temperature, and thus the density might be expected to be somewhat higher for a given absorbed power. The higher-energy electrons also interact more strongly and with relatively higher incident flux at the oscillating sheath, increasing the heating rate, and this also tends to increase the density. However, increased energy loss from the high-energy escaping electrons tends to decrease the plasma density. It is clear that a self-consistent model with a two-temperature distribution would be a significant improvement in the model. However, models employing additional assumptions lose their simplicity and may introduce internal inconsistencies of unknown consequence.

11.4 ASYMMETRIC DISCHARGES

Capacitive Voltage Divider

Most capacitive discharges are asymmetric, because more electrode surfaces are naturally grounded than driven. The dc voltage between the plasma and the driven electrode is then larger than the dc voltage between the plasma and the grounded electrode. This is easily seen from a model of the rf voltage drops across the two sheaths connecting the driven electrode to ground, as shown in Fig. 11.18, along with the linear relation (11.2.22) between the rf and dc voltages, which lead to

$$V_{ab1} = V_{a1} + V_{b1} = \frac{1}{0.83}(\bar{V}_a + \bar{V}_b) \tag{11.4.1}$$

where the dc voltages are taken between the plasma and the electrodes. An easily measurable dc *bias voltage* is set up at the driven electrode with respect to ground,

$$V_{bias} = -(\bar{V}_a - \bar{V}_b) \tag{11.4.2}$$

which is negative in the usual case that $\bar{V}_a > \bar{V}_b$. Since the voltage drops across the sheaths are inversely proportional to the sheath capacitances, the sheath with

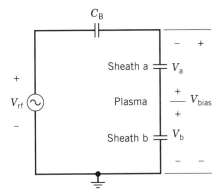

FIGURE 11.18. Capacitive voltage divider model of bias voltage formation in an asymmetric discharge.

the smaller area has a smaller capacitance and therefore a larger voltage drop. The situation is more complicated, because the sheath thickness also depends on the voltage across it, through Child's law, which must be solved self-consistently to obtain the voltage. This is relatively easy to do within various simplifying assumptions.

Consider arbitrary electrodes a and b having areas A_a and A_b and voltage drops $\bar{V}_a$ and $\bar{V}_b$, independent of the vector position $\mathbf{x}$ along the sheath. We then have the proportionalities

$$J_{a1}(\mathbf{x}) \propto \frac{\bar{V}_a}{s_a(\mathbf{x})} \tag{11.4.3}$$

for the capacitive sheath, and

$$n_a(\mathbf{x}) \propto \frac{\bar{V}_a^{3/2}}{s_a^2(\mathbf{x})} \tag{11.4.4}$$

for the collisionless Child law. In terms of the total rf current

$$I_{a1} = \int_{A_a} J_{a1}(\mathbf{x})\, d^2 x \tag{11.4.5}$$

we can eliminate s_a in favor of n_a by substituting (11.4.4) in (11.4.3) to get

$$I_{a1} \propto \bar{V}_a^{1/4} \int_{A_a} n_a^{1/2}(\mathbf{x})\, d^2 x$$

and similarly

$$I_{b1} \propto \bar{V}_b^{1/4} \int_{A_b} n_b^{1/2}(\mathbf{x})\, d^2 x$$

For rf current continuity we can equate $I_{a1} = I_{b1}$ to obtain

$$\frac{\bar{V}_a}{\bar{V}_b} = \left(\frac{\int_{A_b} n_b^{1/2} d^2 x}{\int_{A_a} n_a^{1/2} d^2 x} \right)^4 \qquad (11.4.6)$$

In the simplest plasma model we set $n_a = n_b$, independent of x, to find the scaling

$$\frac{\bar{V}_a}{\bar{V}_b} = \left(\frac{A_b}{A_a} \right)^4 \qquad (11.4.7)$$

This very strong scaling with area is not in accordance with most experimental observations which have typically found

$$\frac{\bar{V}_a}{\bar{V}_b} \approx \left(\frac{A_b}{A_a} \right)^q \qquad (11.4.8)$$

with $q \lesssim 2.5$. The experiments were mainly done at higher pressure, where the sheath dynamics do not follow the collisionless Child–Langmuir law. If, for example, we consider a collisional, constant λ_i sheath, as described in Section 6.4, the proportionality

$$n_a(x) \propto \frac{\bar{V}_a^{3/2}}{s_a^{5/2}(x)} \qquad (11.4.9)$$

leads to

$$\frac{\bar{V}_a}{\bar{V}_b} = \left[\frac{\int_{A_b} n_b^{2/5} d^2 x}{\int_{A_a} n_a^{2/5} d^2 x} \right]^{5/2} \qquad (11.4.10)$$

For the simple assumption of $n_a = n_b = \text{const}$, we obtain

$$\frac{\bar{V}_a}{\bar{V}_b} = \left(\frac{A_b}{A_a} \right)^{5/2} \qquad (11.4.11)$$

which is much closer to the experimental range. However, some experiments had sheaths that were more collisionless than collisional, so the sheath dynamics is not the only factor involved.

Spherical Shell Model

It is clear that geometric factors alone, even at low density, will make $n_a \neq n_b$. Additionally, for higher pressure, factors such as local ionization can further increase the density at the higher-voltage electrode. We now consider these effects using the

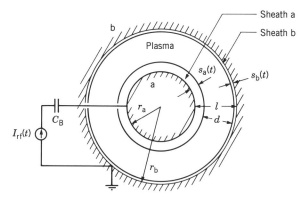

FIGURE 11.19. Spherical shell model of an asymmetric rf discharge (Lieberman, 1989b).

one-dimensional spherical shell model shown in Fig. 11.19 (Lieberman, 1989b). The powered electrode is the inner sphere a having radius r_a, and the grounded electrode is the outer sphere b having radius r_b. The electrode separation l, plasma thickness d, and sheath thicknesses s_a and s_b are defined in the figure. The discharge is driven by an rf current source through a blocking capacitor C_B having negligible impedance at the driving frequency. Since the system is spherically symmetric, the model is purely one-dimensional (along r). The freedom to choose not only the discharge length $l = r_b - r_a$ but also the powered-to-grounded electrode area ratio $A_a/A_b = r_a^2/r_b^2 < 1$ allows us to model an asymmetric discharge. We consider the intermediate mean free path regime, where the ion drift velocity is much greater than the ion thermal velocity, $u_i \gg v_{thi}$, and assume that the dominant ion collisional process is charge exchange of the ion with the parent neutral gas atom, such that the mean free path $\lambda_i = (n_g \sigma_i)^{-1}$ is nearly constant.

For this intermediate pressure regime we found the drift velocity in (5.3.5) to be

$$u_i^2 = \left| \frac{eT_e}{M} \frac{2}{\pi} \frac{\lambda_i}{n} \frac{dn}{dr} \right| \tag{11.4.12}$$

Substituting this into the continuity equation (5.3.4), in spherical coordinates, we obtain an equation for the density,

$$\frac{1}{r^2} \frac{d}{dr} \left[r^2 n \left| \frac{eT_e}{M} \frac{2}{\pi} \frac{\lambda_i}{n} \frac{dn}{dr} \right|^{1/2} \right] = \nu_{iz} n \tag{11.4.13}$$

Using various transformations to simplify (11.4.13), Lieberman (1989b) numerically solved this differential equation to obtain a simple expression for the ratio of densities at the grounded and powered electrodes. In the usual regime for processing discharges for which $(2\lambda_i \nu_{iz}/\pi u_B)(A_a/A_b)^{3/4} \ll 1$, this gives

$$\frac{n_b}{n_a} \approx \left(\frac{A_a}{A_b} \right)^{0.29} \tag{11.4.14}$$

With this geometric scaling of density we can recompute the voltage ratios from (11.4.6) and (11.4.9) to obtain, for a collisionless Child law sheath,

$$\frac{\bar{V}_a}{\bar{V}_b} = \left(\frac{A_b}{A_a}\right)^{3.42} \tag{11.4.15}$$

and, for a collisional (λ_i = const) sheath,

$$\frac{\bar{V}_a}{\bar{V}_b} = \left(\frac{A_b}{A_a}\right)^{2.21} \tag{11.4.16}$$

The above results do not exhaust the assumptions that can be made to describe the plasma glow region and the sheath region. Spatially uniform ionization by secondaries, and edge ionization by stochastically heated electrons can also be introduced as dominant ionization processes. For the sheath dynamics, a collisional constant mobility sheath law (6.5.8) or a homogeneous sheath can be introduced. The scaling results for all these cases (except the homogeneous sheath) are shown in Table 11.1. We note that considerably lower values of q can be found with the appropriate combination of glow and sheath dynamics, but we hasten to add that many of these combinations are mutually incompatible. Generally, the higher-pressure assumptions, to the right and down in the table, produce the lower values of q.

The results in the table do not give the whole story. For large area ratios, the voltage at the large area electrode saturates at its dc value given by (6.2.17) for an undriven sheath, and the scaling laws in the table must be modified. Geometries that would more closely resemble physical reactors, on which experimental measurements have been made, have also been investigated using more complicated models. One such study investigated various cylindrical and coaxial systems and compared the results with similar experimental configurations, obtaining reasonable agreement (Lieberman and Savas, 1990). For two-dimensional geometries, such as a cylinder, the voltage ratio does not simply scale as a power of the area ratio, but depends in a complicated way on the cylinder length and radius. For details the reader should consult the original paper.

TABLE 11.1. Scaling Exponent q for the Dependence of the Powered-to-Grounded Voltage Ratio $\bar{V}_a/\bar{V}_b$ on the Area Ratio A_b/A_a, Where q is Defined by $\bar{V}_a/\bar{V}_b = (A_b/A_a)^q$.

	Sheath Physics	Child's Law	Constant λ_i Law	Constant μ_i Law
Glow Physics	Scaling Law	$J \propto V^{3/2}/s^2$	$J \propto V^{3/2}/s^{5/2}$	$J \propto V^2/s^3$
Homogeneous	n = const	4.0	2.5	3.0
Thermal electron	$n \propto A^{-7/24}$	3.42	2.21	2.71
Secondary electron	$n \propto A^{-1/2}$	3.0	2.0	2.5
Local ionization	$n \propto V$	1.33	1.25	1.5

Source: After Lieberman (1989b).

11.5 MAGNETICALLY ENHANCED DISCHARGES

Capacitive discharges have been the most widely used source for low-pressure materials processing. However, they suffer from the disadvantages of high sheath voltages with consequent low ion density (ion flux) and high ion-bombarding energy at a given power level. The ion-bombarding energy also cannot be varied independently of the ion flux in these devices. To circumvent these disadvantages, other sources have been employed and various attempts have been made to improve the performance of the capacitive discharge. In the following two chapters we consider alternative methods of producing and heating plasmas, particularly to achieve lower sheath voltages, higher densities, and independent control of both. In this section we consider one modification of the capacitive discharge that can also achieve these goals, the *magnetically enhanced reactive ion etcher* (MERIE) or *rf magnetron*. In this discharge, a controllable, relatively weak (50–200 G) dc magnetic field is imposed parallel to the surface of the rf powered electrode.

The magnetic field introduces a number of effects that act to increase the density and reduce the sheath voltage at a fixed absorbed power. (1) As described in Section 5.4, the electron motion can be strongly inhibited across the field, leading to a reduced nonambipolar flux in the cross-field directions. If most of the collecting surface is across field lines then the power loss can be significantly reduced, resulting in higher density at a given power. (2) The magnetic field can confine the energetic (ionizing) electrons to a small volume near the electrode. This both reduces the overall effective loss area of the discharge and directs a greater fraction of the escaping plasma to the powered electrode, increasing the ion flux there. (3) A transverse field can increase the efficiency of stochastic heating due to multiple correlated collisions of electrons with the oscillating sheaths, increasing the density. (4) There is an increase in the efficiency of ohmic heating due to higher electric fields in the plasma. In a simple model presented below, we show that some of the experimental results in low-pressure discharges can be understood in terms of effects (3) and (4), the increased efficiency of stochastic heating and ohmic heating.

Although application of the magnetic field results in a reduced sheath voltage and increased plasma density, the plasma generated is strongly nonuniform both radially and azimuthally due to $\mathbf{E} \times \mathbf{B}$ drifts, where $\mathbf{E}$ and $\mathbf{B}$ are the local dc electric and magnetic fields, respectively. To increase process uniformity (at least azimuthally), the magnetic field can be rotated in the plane of the wafer at a low frequency (~ 0.5 Hz). While this is an improvement, MERIE systems do not have good uniformity, which may limit their performance. A strongly nonuniform plasma over the wafer can give rise to a lateral dc current within a film on the wafer that can damage the film.

In the following analysis we return to the homogeneous model of Section 11.1 for the plasma and the sheath. This is necessitated by the complexity of the dynamics due to the steady magnetic field. The approximation does not introduce significant error in the bulk plasma at low pressures, but, as described in Section 11.2, the sheath dynamics is not correctly treated. Effect 1, nonambipolar ion losses, could be included in the model if desired, but the effect may not be significant at the typical 50–200 G fields. Effect 2 may be quite important but cannot be treated in a homogeneous model.

Theory

The model is the same as given in Fig. 11.1, except that a uniform magnetic field B_0 is oriented parallel to the electrode surfaces in the x direction. As in Section 11.1, a uniform sinusoidal current density $J_x(t) = \mathrm{Re}\, J_1 e^{j\omega t}$ flows between the plates. In the plasma, the current density is related to the electric field vector through the dielectric tensor. Letting $J_\alpha(t) = \mathrm{Re}\, \tilde{J}_\alpha e^{j\omega t}$ and $E_\alpha(t) = \mathrm{Re}\, \tilde{E}_\alpha e^{j\omega t}$, where $\alpha = x$, y, or z, we have

$$
\begin{pmatrix} \tilde{J}_x \\ \tilde{J}_y \\ \tilde{J}_z \end{pmatrix} = j\omega\epsilon_0 \begin{pmatrix} \kappa_\perp & -\kappa_\times & 0 \\ \kappa_\times & \kappa_\perp & 0 \\ 0 & 0 & \kappa_\parallel \end{pmatrix} \begin{pmatrix} \tilde{E}_x \\ \tilde{E}_y \\ \tilde{E}_z \end{pmatrix}
\tag{11.5.1}
$$

where the tensor elements are given in Section 4.4. Since $\tilde{J}_x = J_1$, and $\tilde{J}_y = \tilde{J}_z = 0$, we can solve (11.5.1) to obtain $\tilde{E}_z = 0$,

$$
\tilde{E}_y = -\frac{\tilde{E}_x \kappa_\times}{\kappa_\perp}
\tag{11.5.2}
$$

and

$$
\tilde{E}_x = \left[j\omega\epsilon_0 \left(\kappa_\perp + \frac{\kappa_\times^2}{\kappa_\perp} \right) \right]^{-1} J_1
\tag{11.5.3}
$$

In the sheath region a (see Fig. 11.1), the x component of the electric field is found by integrating Poisson's equation to obtain

$$
E_{xa}(x, t) = en\frac{x - s_a(t)}{\epsilon_0} + E_x(t)
\tag{11.5.4}
$$

where $E_x(t)$ is the field in the plasma, and we have chosen $E_{xa} = E_x$ at the instantaneous position of the sheath edge $x = s_a$. Although the usual assumption is that $|E_x| \ll |E_{xa}|$, the field in the plasma can be significant for sufficiently large magnetic fields. The effect of finite $E_x(t)$ is discussed in Lieberman et al. (1991). For 50- to 200-G fields, it suffices to set $E_x(t) \equiv 0$ in (11.5.4), as in (11.1.7). The analysis of the sheath then proceeds as in Section 11.1, with the total rf voltage drop across both sheaths $V_{ab}(t)$ in this symmetric discharge given by (11.1.19) with $\bar{s} = s_0$. The complex amplitude of $V_{ab}(t)$ is

$$
\tilde{V}_{ab} = -\frac{2jens_0^2}{\epsilon_0}
\tag{11.5.5}
$$

where s_0 is given by (11.1.11). Adding to this the voltage drop $\tilde{E}_x d$ across the plasma, where $d = l - 2s_0$ is the bulk plasma thickness, we obtain the complex amplitude of the discharge voltage

$$\tilde{V}_{rf} = -\frac{2jens_0^2}{\epsilon_0} + \tilde{E}_x d \qquad (11.5.6)$$

The dc voltage across a single sheath is given by (11.1.39).

The dynamics of the sheath heating is profoundly changed by the addition of magnetic fields. For weak magnetic fields, we assume that the sheath motion remains unchanged, but that the particle interaction is modified due to multiple correlated collisions of electrons with the moving sheath. A gyrating electron that collides once with the moving sheath collides again in a time interval of approximately half a gyroperiod. The electron trajectory can be coherent over many such sheath collisions, leading to large energy gains. The mechanism is illustrated in Fig. 11.20. The coherent motion is destroyed on the timescale for electron collisions with neutral gas atoms.

To determine the heating quantitatively, we start with the basic sheath heating equation (11.2.22). Because the sheaths are expected to be thin at the usual applied frequencies, we expect only slow sheath heating. For the homogeneous model this simplifies to

$$\bar{S}_{stoc} = 2m\Gamma_e\langle\Delta u(\Delta u - u_{es})\rangle_\phi \qquad (11.5.7)$$

where Δu is the change in electron velocity for a set of multiple collisions. To determine Δu for the multiple sheath collisions, we let $2u_{es}(\omega t)$ be the change in electron velocity for a single collision with the sheath at time t. For a slowly moving sheath, as illustrated in Fig. 11.20, successive collisions take place at time intervals of $\Delta t = \pi/\omega_{ce}$, where $\omega_{ce} = eB_0/m$ is the electron gyration frequency. These collisions result in coherent energy gain. However, the coherent energy gain is terminated by

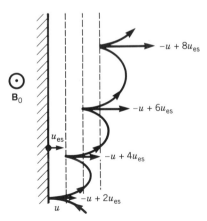

FIGURE 11.20. Electron trajectory and energy gain due to collisions with the moving sheath and with applied transverse B_0 (Lieberman et al. 1991, © 1991 IEEE).

electron collisions with neutral gas atoms. Hence we write

$$\Delta u = \sum_{i=0}^{\infty} u_{\mathrm{es}}(\omega t + i\omega \Delta t)\, e^{-i\nu_{\mathrm{el}}\Delta t} \tag{11.5.8}$$

where ν_{el} is the electron–neutral elastic scattering frequency. The exponential factor in (11.5.8) gives the fraction of electrons that have not collided with neutral gas atoms after a time $i\Delta t$. We are interested in the regime for which ω, $\nu_{\mathrm{el}} \ll \omega_{\mathrm{ce}}$, so we can convert the sum to an integral,

$$\Delta u = \frac{\omega_{\mathrm{ce}}}{\pi\omega} \int_0^{\infty} u_{\mathrm{es}}(\phi + \phi')\, e^{-\nu_{\mathrm{el}}\phi'/\omega}\, d\phi' \tag{11.5.9}$$

where $\phi = \omega t$ and $\phi' = i\omega \Delta t = i\pi\omega/\omega_{\mathrm{ce}}$. Substituting $u_{\mathrm{es}} = \mathrm{Re}\,\tilde{u}_{\mathrm{es}} e^{j\omega t}$ into (11.5.9) and integrating, we find

$$\Delta u(\phi) = \mathrm{Re}\,\tilde{u}_{\mathrm{es}} \left(\frac{\omega_{\mathrm{ce}}}{\pi(\nu_{\mathrm{el}} - j\omega)} \right) e^{j\phi} \tag{11.5.10}$$

Substituting this into (11.5.7) and averaging, we obtain the time-average power per unit area delivered to the electrons by the oscillating sheath,

$$\bar{S}_{\mathrm{stoc}} = \frac{1}{4} m n \bar{v}_e |\tilde{u}_{\mathrm{es}}|^2 \frac{\omega_{\mathrm{ce}}}{\pi(\nu_{\mathrm{el}}^2 + \omega^2)} \left(\nu_{\mathrm{el}} + \frac{\omega_{\mathrm{ce}}}{\pi} \right) \tag{11.5.11}$$

We note again that the derivation of (11.5.11) is valid when $\omega_{\mathrm{ce}} \gg \omega$, ν_{el}. To complete the model we add the usual equilibrium conditions of flux balance

$$2n u_{\mathrm{B}} = n n_{\mathrm{g}} K_{\mathrm{iz}} d \tag{11.5.12}$$

and electron power balance

$$\bar{S}_{\mathrm{ohm}} + 2\bar{S}_{\mathrm{stoc}} = 2 e n u_{\mathrm{B}}(\mathcal{E}_{\mathrm{c}} + 2T_{\mathrm{e}}) \tag{11.5.13}$$

where $\bar{S}_{\mathrm{ohm}} = \frac{1}{2}\mathrm{Re}\,\tilde{E}_x \tilde{J}_1 d$ and $\bar{S}_{\mathrm{stoc}}$ is given by (11.5.11). The total power per unit area absorbed by the discharge is

$$S_{\mathrm{abs}} = 2 e n u_{\mathrm{B}}(\mathcal{E}_{\mathrm{c}} + 2T_{\mathrm{e}} + \bar{V}) \tag{11.5.14}$$

where $\bar{V}$ is given by (11.1.39). After we eliminate $\tilde{E}_x$ from $\bar{S}_{\mathrm{ohm}}$ and $|\tilde{u}_{\mathrm{es}}|^2$ from $\bar{S}_{\mathrm{stoc}}$ in terms of J_1, then the left-hand side of (11.5.13) is proportional to J_1^2. Similarly eliminating s_0 in favor of J_1 in (11.1.39) yields $\bar{V} \propto J_1^2$. Hence by solving (11.5.13) for J_1^2 as a function of n and inserting this into (11.5.14), we obtain a single equation that can be solved numerically to determine n for a given S_{abs}. Equation (11.5.13) then yields J_1; the amplitudes $\tilde{E}_x$, $\tilde{E}_y$, $\tilde{V}_{rf}$, and the ion-bombarding energy $\bar{V}$ then follow.

The magnetic field has two main effects on the discharge equilibrium. (1) The stochastic heating increases with increasing B_0, provided most of the rf voltage appears across the sheaths; and (2) a significant fraction of the total rf discharge voltage can be dropped across the bulk plasma at high magnetic fields. If the bulk plasma voltage is small, then we can estimate the scaling of the discharge equilibrium with S_{abs} and B_0, in various regimes, as follows. We first note by current continuity that $J_{rf} \propto ns_0$. Since the sheaths are capacitive, $J_{rf} \propto V_{rf}/s_0$. Hence it follows that $n \propto V_{rf}/s_0^2$. Using this result in (11.5.11), we obtain the scaling of the stochastic heating power $\bar{S}_{stoc} \propto B_0^2 V_{rf}$, for $\omega_{ce} \gg \omega$, ν_{el}. Similarly, scaling the ohmic power yields $\bar{S}_{ohm} \propto V_{rf}^{1/2}$. The power balance equations can then be evaluated in various limiting cases, depending on whether stochastic heating or ohmic heating is the dominant heating mechanism and on whether ion energy losses or electron energy losses are the dominant loss mechanism. Consider the high-voltage case for which ion losses are dominant. Then at low pressures where stochastic heating dominates, we obtain the scaling

$$V_{rf} \propto \frac{S_{abs}^{1/2}}{B_0}$$

$$n \propto S_{abs}^{1/2} B_0 \qquad (11.5.15)$$

$$s_0 \propto B_0^{-1}$$

$$\bar{S}_{stoc} \propto S_{abs}^{1/2} B_0$$

At high pressures where ohmic heating dominates, we find the scaling

$$V_{rf} \propto S_{abs}^{2/3}$$

$$n \propto S_{abs}^{1/3} \qquad (11.5.16)$$

$$s_0 \propto S_{abs}^{1/6}$$

$$\bar{S}_{ohm} \propto S_{abs}^{1/3}$$

We leave the details to Problem 11.7. At very high B_0 the bulk plasma voltage drop can dominate the sheath drop, in which case the scaling can become very different, but this is not the usual regime for rf magnetrons.

Experimental Results

Measurements have been made in a commercial etch chamber in argon at 13.56 MHz to compare with the model. In the experiment, the dc bias voltage $V_{bias} = -(\bar{V}_{pa} - \bar{V}_{pb})$ was measured. Since the discharge is strongly asymmetric the measured value of $-V_{bias}$ is compared with the model result for $\bar{V}_{pa}$. The plasma density n was measured with a Langmuir probe approximately 3 cm in front of the 200 cm^2 powered electrode. In Fig. 11.21, we compare the experimental measurements of $-V_{bias}$ with model results of $\bar{V}_{pa}$, and in Fig. 11.22 we compare experiment and model results for the density, for three magnetic fields of 10, 30 and 100 G, and three power densities of 0.25, 0.5 and 1.0 W/cm², with $p = 10$ mTorr. The experimental magnetic field dependence is somewhat weaker than predicted, at the higher magnetic fields where stochastic heating is expected to dominate. However, the general trends of theory and experiment are similar for all cases. The qualitative agreement between model and experiment is evidence that the basic modeling approach includes much of the essential physics.

11.6 MATCHING NETWORKS AND POWER MEASUREMENTS

Although this text is mainly concerned with the internal dynamics of the plasma, some knowledge of the external circuit is necessary. If the discharge is driven directly by an rf power source, then generally power is not transferred efficiently from the

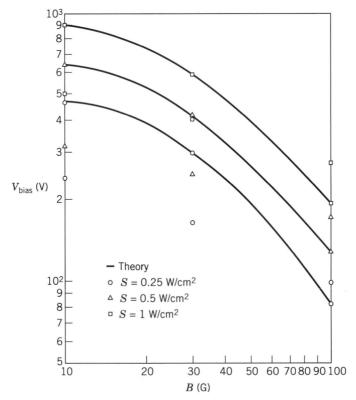

FIGURE 11.21. Measured bias voltage V_{bias} versus magnetic field B_0 at power densities of 0.25, 0.5, and 1.0 W/cm², along with model results for $\bar{V}_{pa}$ (solid curves) (after Lieberman et al., 1991).

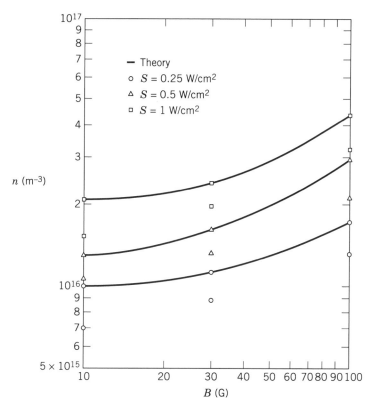

FIGURE 11.22. Measured plasma density n versus magnetic field B_0 at power densities of 0.25, 0.5, and 1.0 W/cm^2, along with model results (solid curves) (after Lieberman et al., 1991).

source to the discharge. To understand this, consider a discharge modeled as a load having impedance $Z_D = R_D + jX_D$, where R_D is the discharge resistance and X_D is the discharge reactance. The power source connected to Z_D is modeled by its Thevenin-equivalent circuit, consisting of a voltage source with complex amplitude $\tilde{V}_T$ in series with a source resistance R_T. The time average power flowing into the discharge is

$$\bar{P} = \frac{1}{2}\mathrm{Re}\,(\tilde{V}_{rf}\tilde{I}_{rf}^*) \tag{11.6.1}$$

where $\tilde{V}_{rf}$ is the complex voltage across Z_D. Solving for $\tilde{I}_{rf}$ and $\tilde{V}_{rf}$ for these series elements, we obtain

$$\tilde{I}_{rf} = \frac{\tilde{V}_T}{R_T + R_D + jX_D} \tag{11.6.2}$$

$$\tilde{V}_{rf} = \tilde{I}_{rf}(R_D + jX_D) \tag{11.6.3}$$

Substituting (11.6.2) and (11.6.3) into (11.6.1), we obtain

$$\bar{P} = \frac{1}{2}|\tilde{V}_T|^2 \frac{R_D}{(R_T + R_D)^2 + X_D^2} \tag{11.6.4}$$

For fixed source parameters $\tilde{V}_T$ and R_T, maximum power transfer is obtained by setting $\partial\bar{P}/\partial X_D = 0$ and $\partial\bar{P}/\partial R_D = 0$, which gives $X_D = 0$ and $R_D = R_T$. The maximum power supplied by the source to the load is then

$$\bar{P}_{max} = \frac{1}{4}\frac{|\tilde{V}_T|^2}{R_T} \tag{11.6.5}$$

If maximum power transfer is obtained, then we say that the source and load are matched.

Since X_D is not zero, and, typically, $R_D \ll R_T$, the power $\bar{P}$ is generally much less than $\bar{P}_{max}$. To increase $\bar{P}$ to $\bar{P}_{max}$, thus matching the source to the load, a loss-less *matching network* can be placed between them. Because R_D and X_D are two independent components of Z_D, the simplest matching network consists of two independent components. The most common configuration, called an "L-network," is shown inserted between the source and the load in Fig. 11.23. It consists of a shunt capacitor having susceptance $B_M = \omega C_M$ and a series inductor having a reactance $X_M = \omega L_M$.

To determine X_M and B_M, we write the admittance looking toward the right at location 2 in Fig 11.23 as the inverse of the impedance:

$$Y_2 \equiv G_2 + jB_2 \equiv Z_2^{-1} = (R_D + jX_2)^{-1} \tag{11.6.6}$$

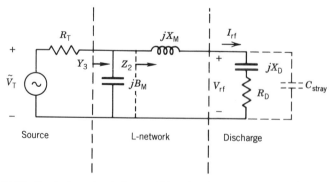

FIGURE 11.23. Equivalent circuit for matching the rf power source to the discharge using an L-network.

where $X_2 = X_M + X_D$. Separating real and imaginary parts and solving for G_2 and B_2, we obtain

$$G_2 = \frac{R_D}{R_D^2 + X_2^2} \tag{11.6.7}$$

$$B_2 = -\frac{X_D}{R_D^2 + X_2^2} \tag{11.6.8}$$

Next we note from Fig. 11.23 that $Y_3 = G_3 + jB_3$, with $G_3 = G_2$ and $B_3 = B_2 + B_M$. The matched condition of maximum power transfer is $G_3 = 1/R_T$ and $B_3 = 0$. Setting $G_3 = 1/R_T$ in (11.6.7), we solve for X_2 to obtain

$$X_2 = (R_D R_T - R_D^2)^{1/2} \tag{11.6.9}$$

Since $X_2 = X_M + X_D$, the required X_M is

$$X_M = (R_D R_T - R_D^2)^{1/2} - X_D \tag{11.6.10}$$

Since X_D is negative, X_M must be positive; i.e., a matching inductor $L_M = X_M/\omega$ must be used. Using (11.6.9) in (11.6.8) and setting $B_3 = 0$ ($B_M = -B_2$), we obtain

$$B_M = \left(\frac{1}{R_T R_D} - \frac{1}{R_T^2} \right)^{1/2} \tag{11.6.11}$$

Since B_M is positive, a matching capacitor $C_M = B_M/\omega$ is required. Because there must be real solutions for B_M and X_M, we see from (11.6.10) or (11.6.11) that R_D must be less than R_T for a match to be achieved with an L-network. This is the usual regime at the higher power levels used for typical processing discharges. For low powers, $R_D > R_T$ and a different form of matching network must be used (Problem 11.8). A three-element (T or Π) network can be used to match any discharge; hence such networks are commonly used to provide added flexibility. The three elements are not uniquely determined by the maximum power condition, but the inductive element is usually a fixed value, and the two capacitors can be varied to achieve the match.

Because R_D and X_D are actually functions of the discharge voltage or the absorbed power, we must specify these to determine the matched condition. For a specified voltage or absorbed power, we can determine $\tilde{I} \equiv I_1$ and s_m as in example 1 or 2 of Section 11.2. Then R_D and X_D are determined from

$$P_{\text{abs}} = \frac{1}{2} I_1^2 R_D \tag{11.6.12}$$

and

$$X_D = -\frac{1}{\omega C_{ab}} \tag{11.6.13}$$

where C_{ab} is given by (11.2.20).

Because typically $X_D \gg R_D$, the voltage and current across the discharge are nearly 90° out of phase. Setting the current and voltage across the discharge to be

$$I_{rf}(t) = I_1 \cos \omega t \tag{11.6.14}$$

$$V_{rf}(t) = V_{rf} \cos \left(\omega t + \frac{\pi}{2} + \psi \right) \tag{11.6.15}$$

we find that

$$V_{rf} = (R_D^2 + X_D^2)^{1/2} I_1 \approx X_D I_1 \tag{11.6.16}$$

$$\psi = \tan^{-1} \frac{R_D}{X_D} \approx \frac{R_D}{X_D} \tag{11.6.17}$$

The time-average power absorbed by the discharge is

$$P_{abs} = \frac{1}{\tau} \int_0^\tau V_{rf}(t) I_{rf}(t) \, dt \tag{11.6.18}$$

$$= \frac{1}{2} I_1 V_{rf} \sin \psi \tag{11.6.19}$$

where $\tau = 2\pi/\omega$. Under matched conditions, the voltage and current at the source are in phase with each other, $\tilde{V}_T = R_T \tilde{I}_T$, with the power supplied by the source,

$$P_T = \frac{\frac{1}{2} |\tilde{V}_T|^2}{R_T} \tag{11.6.20}$$

For a lossless matching network, $P_{abs} = P_T$. Equating (11.6.19) and (11.6.20) and solving for $|\tilde{V}_T|$, we obtain

$$|\tilde{V}_T| = \frac{(R_D R_T)^{1/2}}{X_D} V_{rf} \tag{11.6.21}$$

For the usual discharge conditions, $|\tilde{V}_T| \ll V_{rf}$.

Power Measurements

An rf wattmeter placed between the source and the matching network is conventionally used to measure the time-average power P_T supplied by the source. This instrument is often an integral part of the rf power supply. For sinusoidal voltages and currents, the time-average powers P_f and P_b flowing in the forward and backward directions are then measured, with

$$P_T = P_f - P_b \qquad (11.6.22)$$

For the voltage and current nearly in phase at the measurement location, we have $P_b \ll P_f$, and the measurements accurately determine P_T. Under strongly out-of-phase conditions, $P_b \approx P_f \gg P_T$, such that subtracting P_b from P_f does not determine P_T accurately. Hence rf wattmeters cannot be placed between the matching network and the discharge to determine the power P_{abs} absorbed by the discharge. For a lossless matching network, $P_{abs} = P_T$, but the nonideal matching networks used in typical processing systems often absorb a considerable fraction of the source power, such that $P_{abs} < P_T$. The usual source of loss is the finite resistance of the wire with which the matching inductor is wound. This nonideal inductor can be modeled as an ideal inductor L_M in series with a resistor R_M.

Equation (11.6.19) can be used to determine P_{abs} if I_1, V_{rf}, and ψ can be accurately measured. The discharge current is conventionally measured with a miniature current transformer. If the time-varying current and voltage are displayed on an oscilloscope or measured using to a dual channel vector voltmeter, than V_{rf}, I_1, and ψ can be determined. However, because ψ can be as small as 2–3°, phase shifts between the measured voltage and current signals due to nonideal instrumental and cabling effects can render the measurement meaningless. Accurate calibration of the phase shift for known calibration loads is essential. This can be a difficult measurement to make at 13.56 MHz in a practical processing discharge, where, for example, a 4-cm length of coaxial cable has a phase shift of approximately 1°. If the measured voltage and current waveforms are not approximately sinusoidal, than the power must be determined by direct averaging of the $I_{rf}V_{rf}$ product using (11.6.18).

In most discharges there is generally a large stray capacitance C_{stray} in parallel with the discharge impedance $R_D + jX_D$. C_{stray} represents the capacitance to ground of the powered electrode and center conductor of its coaxial cable feed; typically $C_{stray} \sim$ 100–200 pF in processing discharges. Then an effective procedure to determine the power absorbed by the discharge alone if the voltage and current waveforms are reasonably sinusoidal in shape is as follows:

1. Measure the voltage V_{rf} across the discharge and the source power P_T. This is the power absorbed by the lossy matching inductor and by the discharge.

2. Extinguish the discharge by raising the pressure in the chamber to 1 atm or by reducing the pressure to a very low value. The voltage measured will in general change.

3. Readjust the source voltage V_T so that V_{rf} is the same as that measured in (1), and measure the source power $P_T^{(0)}$. The matching capacitor can be retuned if desired, but a perfect match is not necessary. This measurement yields the power absorbed by the lossy inductor only.

Because V_{rf} is the same with and without the plasma present, and the impedance is approximately the same, determined by C_{stray}, the same current passes through the inductor and stray capacitor to ground. Hence the same power is lost in the inductor with and without the plasma present. Therefore, the power absorbed by the plasma is

$$P_{abs} = P_T - P_T^{(0)} \qquad (11.6.23)$$

For further discussion of power measurements in rf discharges, the reader is referred to Godyak and Piejak (1990b).

PROBLEMS

11.1. Plasma Admittance   Derive expression (11.1.3) for the bulk plasma admittance Y_p and show, using assumption (b), that the displacement current that flows through C_0 is much smaller than the conduction current that flows through L_p and R_p.

11.2. Stochastic Heating   Derive the stochastic heating expression (11.2.59), showing that $\bar{S}_{stoc}$ is proportional to the rf voltage, independent of the rf discharge current and plasma density.

11.3. Resonant Discharge   Consider a uniform density bulk plasma of thickness d with two vacuum sheaths, each of thickness s_m, such that $l = d + 2s_m$. Find the frequency ω_{res} of the series resonance between the bulk plasma inductance L_p (see (11.1.3)) and the overall sheath capacitance C_s, in terms of ω_{pe}, s_m, and d. Evaluate ω_{res} for $d = 8$ cm, $s_m = 1$ cm, and $n_0 = 2 \times 10^{10}$ cm^{-3}, and compare ω_{res} to the driving frequency ω for a 13.56-MHz discharge.

11.4. Low-Pressure Rf Discharge Equilibrium With Voltage Specified   Consider a symmetric, capacitively coupled rf discharge in argon gas having a total applied rf voltage of amplitude 800 V at a frequency $f = \omega/2\pi = 13.56$ MHz. The plate separation is $l = 0.1$ m, and the gas pressure is 5 mTorr. Use a low pressure discharge model, as in Example 1 of Section 11.2.

(a) Determine the electron temperature T_e in the discharge.

(b) Determine the plasma density n_0 and the ion flux.

(c) Determine the dc potential of the plasma with respect to the plates and the ion-bombarding energy (in volts).

(d) Determine the sheath thickness s_m.

(e) Determine the total rf power per unit area required to sustain the discharge.

(f) Determine the rf current amplitude drawn by the discharge.

(g) Use the scaling formulas to check your answers against Example 1.

11.5. Design of a Matching Network

(a) For the discharge of Example 1, find the effective resistance R_D (in ohms) and capacitance C_D (in farads) for a series RC model of the discharge.

(b) Design an L-type matching network to match the discharge to a 50-Ω rf generator.

11.6. Secondary Electrons in a Low-Pressure Discharge   Consider a symmetric, capacitively coupled rf discharge in argon gas having a total applied rf voltage of amplitude 800 V at a frequency $f = \omega/2\pi = 13.56$ MHz. The plate separation is $l = 0.1$ m, and the gas pressure is 5 mTorr. Use a low-pressure discharge model, as in Example 1 of Section 11.2.

(a) Sketch the total (rf + dc) potential $\Phi(x, t)$ inside the discharge versus the distance x between the plates, at four times $\omega t = 0, \pi/2, \pi$, and $3\pi/2$.

(b) Because energetic ions bombard the plates, secondary electrons are released which can accelerate to high energies through the sheaths back into the discharge, where they may become electrostatically trapped. Assume that these "hot electrons" have a temperature $T_h \sim 100$ V and are weakly collisional ($\lambda_e \gg l$). Estimate the range of times within an rf cycle $\omega t = 2\pi$ over which secondary electrons will be accelerated to high energies ($T_h \gg T_e$) and become trapped. Estimate the timescale for these hot electrons to be lost from the discharge. To do this, consider the time-varying behavior of the trapping potential $\Phi(x, t)$, as shown by your sketches in (a).

(c) From your answers in (b), and the hot-electron particle conservation law, show that the steady-state hot-electron density can be estimated as $n_h \approx \gamma_{se} n_0 u_B/(\omega l)$, where u_B is the Bohm velocity and γ_{se} is the secondary emission coefficient.

(d) These hot electrons will ionize argon atoms, producing electron–ion pairs. For $\gamma_{se} = 0.1$, over what range of discharge frequencies ω will this mechanism be important compared to ionization by the thermal ($T_e \sim 3$ V) electrons in the discharge? (You will need to use the data shown in Fig. 3.16).

11.7. Magnetically Enhanced Rf Discharge   Derive the scaling results quoted in (11.5.15) and (11.5.16) for an rf magnetron discharge.

11.8. Low-Power Matching Networks   Design an L-type matching network to match a discharge impedance $Z_D = R_D + jX_D$ to an rf generator having characteristic impedance R_T, for the low-power case $R_D > R_T$.

11.9. Low-Pressure Rf Discharge Equilibrium with Power Specified   Verify the results of Example 2.

11.10. Intermediate Pressure Rf Discharge Equilibrium With Voltage Specified   Consider a symmetric, capacitively coupled rf discharge in argon gas having a total applied rf voltage of amplitude 800 V at a frequency $f = \omega/2\pi = 13.56$ MHz. The plate separation is $l = 0.1$ m, and the gas pressure is 30 mTorr. Use a low-pressure discharge model, as in Example 1 of Section 11.2, but use the collisional sheath results (11.2.54)–(11.2.59) instead of the collisionless sheath results.

(a) Determine the electron temperature T_e in the discharge.

(b) Determine the plasma density n_0 and the ion flux.

(c) Determine the dc potential of the plasma with respect to the plates and the ion bombarding energy (in volts).

(d) Determine the sheath thickness s_m.

(e) Determine the total rf power per unit area required to sustain the discharge.

(f) Determine the rf current amplitude drawn by the discharge.

(g) For each plate of the discharge having a cross-sectional area of 1000 cm^2, find, using your results above, the effective resistance R_D (in ohms) and capacitance C_D (in farads) for a series RC model of the discharge.

(h) Design a matching network to match the discharge to a 50-Ω rf generator.

11.11. Fast Sheath Heating   Using the result in (11.2.60), derive the scalings of plasma parameters versus V_{rf} for fast sheath heating for the case in which sheath heating dominates electron heating and the collisionless Child law applies. Compare the results to those for slow sheath heating given in (11.2.42)–(11.2.46), and comment considering also the experimental and simulation results of Section 11.3.

CHAPTER 12

INDUCTIVE DISCHARGES

The limitations of capacitive rf discharges and their magnetically enhanced variants have led to the development of various low-pressure, high-density plasma discharges. The distinction between low- and high-density discharges is described in Sections 10.1 and 10.2. A few examples are shown schematically in Fig. 1.14, and typical parameters are given in Table 1.1. In addition to high density and low pressure, a common feature is that the rf or microwave power is coupled to the plasma across a dielectric window or wall, rather than by direct connection to an electrode in the plasma, as for a capacitive discharge. This noncapacitive power transfer is the key to achieving low voltages across all plasma sheaths at electrode and wall surfaces. The dc plasma potential, and hence the ion acceleration energy, is then typically 20–40 V at all surfaces. To control the ion energy, the electrode on which the substrate is placed can be independently driven by a capacitively coupled rf source. Hence independent control of the ion/radical fluxes (through the source power) and the ion-bombarding energy (through the substrate electrode power) is possible. The relation between substrate electrode power and ion-bombarding energy at the substrate is described in Chapter 11. High-density inductive rf discharges are described in this chapter, and high-density wave heated discharges are described in Chapter 13. Nonresonant inductive discharges operated at high densities and low pressures, which are driven at frequencies below the self-resonant frequency of the exciting coil, are described in Section 12.1. Other operating regimes and power transfer considerations are described in Section 12.2. The planar coil configuration, which is commonly used for materials processing, is described in Section 12.3, with emphasis on experimental measurements. Excitation at the self-resonant frequency leads to the *helical resonator* configuration, described in Section 12.4. Some other aspects of high-density discharges, including issues of plasma transport and substrate damage, are described in a review article by Lieberman and Gottscho (1994), from which some of the material in Chapters 12 and 13 is drawn.

12.1 HIGH-DENSITY, LOW-PRESSURE DISCHARGES

Inductive discharges are nearly as old as the invention of electric power, with the first report of an "electrodeless ring discharge" by Hittorf in 1884. He wrapped a coil around an evacuated tube and observed a discharge when the coil was excited with a Leyden jar. A subsequent 50-year controversy developed as to whether these discharges were capacitively driven by plasma coupling to the low- and high-voltage ends of the cylindrical coil, as in a capacitive discharge (see Chapter 11), or were driven by the induced electric field inside the coil. This issue was resolved with the recognition that the discharge was capacitively driven at low plasma densities, with a transition to an inductive mode of operation at high densities. Succeeding developments, which focused on pressures exceeding 20 mTorr in a cylindrical coil geometry, are described in a review article by Eckert (1986). The high-pressure regime was intensively developed in the 1970s with the invention of the open air induction torch and its use for spectroscopy. In the late 1980s, the planar coil configuration was developed, renewing interest in the use of high-density inductive discharges for materials processing at low pressures (< 50 mTorr) and in low aspect ratio geometries ($l/R \lesssim 1$ for a cylindrical discharge). Such discharges can be driven with either planar or cylindrical coils. It is this regime that is the primary focus here.

Plasma in an inductive discharge is created by application of rf power to a nonresonant inductive coil. Inductive sources have potential advantages over high-density wave-heated sources, including simplicity of concept, no requirement for dc magnetic fields (as required for ECRs and helicons, discussed in Chapter 13), and rf rather than microwave source power. The nonresonant operation may be either an advantage or a disadvantage, depending on use; a resonant version, the *helical resonator,* is considered in Section 12.4.

In contrast to ECRs and helicons, which can be configured to achieve densities $n_0 \gtrsim 10^{13}$ cm^{-3}, as we will see in Chapter 13, inductive discharges may have natural density limits, $n_0 \lesssim 10^{13}$ cm^{-3}, for efficient power transfer to the plasma. However, the density regime $10^{11} \lesssim n_0 \lesssim 10^{12}$ cm^{-3} for efficient inductive discharge operation, typically a factor of 10 times higher than for capacitive discharges, is of considerable interest for low-pressure processing. Inductive discharges for materials processing are sometimes referred to as TCPs (transformer-coupled plasmas), ICPs (inductively coupled plasmas), or RFI plasmas (rf inductive plasmas).

Inductive Source Configurations

The two coil configurations, cylindrical and planar, are shown in Fig. 12.1 for a low aspect ratio discharge. The planar coil is a flat helix wound from near the axis to near the outer radius of the discharge chamber ("electric stovetop" coil shape).

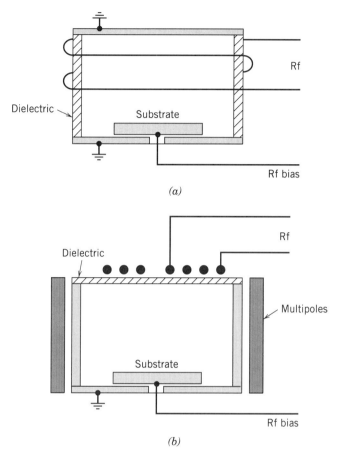

FIGURE 12.1. Schematic of inductively driven sources in (a) cylindrical and (b) planar geometries.

Multipole permanent magnets (see Section 5.6) can be used around the process chamber circumference, as shown in Fig. 12.1b, to increase radial plasma uniformity. The planar coil can also be moved close to the wafer surface, resulting in a close-coupled or near-planar source geometry ($l < R$) having good uniformity properties even in the absence of multipole confinement. In the close-coupled configuration, the coil can be wound nonuniformly to control the radial plasma uniformity.

Inductive coils are commonly driven at 13.56 MHz or below, using a 50-Ω rf supply through a capacitive matching network, which we describe later in this section. The coil can also be driven push–pull using a balanced transformer, which places a virtual ground in the middle of the coil and reduces the maximum coil-to-plasma voltage by a factor of two. This reduces the undesired capacitively coupled rf current flowing from coil to plasma by a factor of two. An electrostatic shield placed between the coil and the plasma further reduces the capacitive coupling if desired, while allowing the inductive field to couple unhindered to the plasma. For the dc plasma potential to be clamped to a low value, 20–40 V, in the presence of stray capacitive coupling from the exciting coil and from the capacitively driven substrate holder, it is essential that the plasma be in contact with a grounded metal surface of substantial area (see Fig. 12.1 and Section 11.4).

Power Absorption and Operating Regimes

In an inductively coupled plasma, power is transferred from the electric fields to the plasma electrons within a skin depth layer of thickness δ near the plasma surface by collisional (ohmic) dissipation and by a collisionless heating process in which bulk plasma electrons "collide" with the oscillating inductive electric fields within the layer. In the latter situation, electrons are accelerated and subsequently thermalized much like stochastic heating in capacitive rf sheaths, which we discussed in Section 11.1. We first consider the ohmic heating process.

The spatial decay constant α within a plasma for an electromagnetic wave normally incident on the boundary of a uniform density plasma is (Problem 12.1)

$$\alpha = \frac{\omega}{c} \operatorname{Im} \kappa_p^{1/2} \equiv \delta^{-1} \tag{12.1.1}$$

From (4.2.18), the relative plasma dielectric constant is

$$\kappa_p = 1 - \frac{\omega_{pe}^2}{\omega(\omega - j\nu_m)} \approx -\frac{\omega_{pe}^2}{\omega^2(1 - j\nu_m/\omega)} \tag{12.1.2}$$

with ω_{pe} the plasma frequency near the boundary, and ν_m the electron–neutral momentum transfer frequency. There are two collisionality regimes.

(a) For $\nu_m \ll \omega$, we drop ν_m/ω in (12.1.2) to obtain

$$\alpha = \frac{\omega_{pe}}{c} \equiv \frac{1}{\delta_p} \tag{12.1.3}$$

where δ_p is the collisionless skin depth. Substituting for ω_{pe} in (12.1.3), we find

$$\delta_p = \left(\frac{m}{e^2 \mu_0 n_s}\right)^{1/2} \tag{12.1.4}$$

(b) For $\nu_m \gg \omega$, we drop the 1 in the parentheses of (12.1.2) and substituting the imaginary part of $\kappa_p^{1/2}$ into (12.1.1), we obtain

$$\alpha = \frac{1}{\sqrt{2}} \frac{\omega_{pe}}{c} \left(\frac{\omega}{\nu_m}\right)^{1/2} \equiv \frac{1}{\delta_c} \tag{12.1.5}$$

where δ_c is the collisional skin depth. Substituting for the dc conductivity $\sigma_{dc} = e^2 n_s / m\nu_m$ from (4.2.22), δ_c can be written in the more usual form,

$$\delta_c = \left(\frac{2}{\omega\mu_0\sigma_{dc}}\right)^{1/2} \tag{12.1.6}$$

(c) There is a third situation (Turner, 1993) for which electrons incident on a skin layer of thickness δ_e satisfy the condition

$$\frac{\bar{v}_e}{2\delta_e} \gg \omega, \nu_m \tag{12.1.7}$$

where δ_e is determined below. In this case the interaction time of the electrons with the skin layer is short compared to the rf period or the collision time. In analogy to collisionless heating at a capacitive sheath, a stochastic collision frequency can be defined (see Appendix C) by

$$\nu_{stoc} = \frac{C_e \bar{v}_e}{\delta_e} \tag{12.1.8}$$

where C_e is a quantity of order unity that depends weakly on $\bar{v}_e$, δ_e, and ω, provided the ordering (12.1.7) is satisfied. A somewhat simplified calculation of C_e is given in Appendix C. We then substitute ν_{stoc} for ν_m in (12.1.2) and expand for $\nu_{stoc} \gg \omega$ as in (b), to obtain

$$\delta_e = \frac{c}{\omega_{pe}}\left(\frac{2C_e\bar{v}_e}{\omega\delta_e}\right)^{1/2}$$

Solving for δ_e, we find

$$\delta_e = \left(\frac{2C_e c^2 \bar{v}_e}{\omega\omega_{pe}^2}\right)^{1/3} = \left(\frac{2C_e\bar{v}_e}{\omega\delta_p}\right)^{1/3}\delta_p \tag{12.1.9}$$

where δ_e is the *anomalous skin depth* (see Alexandrov et al, 1984).

At 13.56 MHz in argon, we find $\nu_m = \omega$ for $p^* \approx 25$ mTorr. We are interested primarily in the low-pressure regimes with $p \ll p^*$, which we consider first. For each pressure regime, we also distinguish two density regimes:

(a) High density, $\delta \ll R, l$
(b) Low density, $\delta \gtrsim R, l$

For typical low-pressure processing discharges, we are generally in the regime for which the frequency ordering is $\omega \sim \bar{v}_e/2\delta \gtrsim \nu_m$ and in which the skin depth is approximately δ_p. For typical plasma dimensions $R, l \sim 10$ cm, we are in the high density regime. We shall discuss the high-pressure ($\nu_m \gg \omega$) and low-density ($\delta \gg R, l$) regimes later when we consider the minimum current and power necessary to generate an inductively coupled plasma.

Discharge Operation and Coupling

Although many systems are operated with planar coils (see Fig. 12.1b), finite geometry effects make these configurations difficult to analyze. To illustrate the general principles of inductive source operation, we examine a uniform density cylindrical discharge (Fig. 12.1a) in the geometry $l \gtrsim R$. We take the coil to have $\mathcal{N}$ turns at radius $b > R$. Integrating the power flow into the plasma, with $\delta_p \ll R$, we have (Problem 12.2)

$$P_{abs} = \frac{1}{2} \frac{J_\theta^2}{\sigma_{eff}} 2\pi R l \delta_p \qquad (12.1.10)$$

where J_θ is the amplitude of the induced rf azimuthal current density at the plasma edge near $r = R$ (opposite in direction to the applied azimuthal current in the coil). In analogy to the conductivity in (4.2.22),

$$\sigma_{eff} = \frac{e^2 n_s}{m \nu_{eff}} \qquad (12.1.11)$$

with $\nu_{eff} = \nu_m + \nu_{stoc}$, a sum of collisional and stochastic heating. In the regime of interest, ν_{stoc} is given by (C.17). Equation (12.1.10) is equivalently the power loss of a uniform sinusoidal current density flowing within a skin thickness δ_p. Letting $I_p = J_\theta / \delta_p$ be the total induced rf current amplitude and defining the plasma resistance through $P_{abs} = \frac{1}{2} I_p^2 R_p$, we obtain

$$R_p = \frac{2\pi R}{\sigma_{eff} l \delta_p} \qquad (12.1.12)$$

The plasma inductance L_p is found using $\Phi = L_p I_p$, where Φ is the total magnetic flux linked by the skin current. Using $\Phi = \mu_0 \pi R^2 H_z$, where $H_z = J_\theta \delta_p$ is the magnetic field produced by the skin current, we obtain

$$L_p = \frac{\mu_0 \pi R^2}{l} \qquad (12.1.13)$$

Letting the coil have $\mathcal{N}$ turns at a radius $b \gtrsim R$, where $b - R$ is the "thickness" of the dielectric interface separating coil and plasma, then we can model the source as the transformer shown in Fig. 12.2. Evaluating the inductance matrix for this transformer,

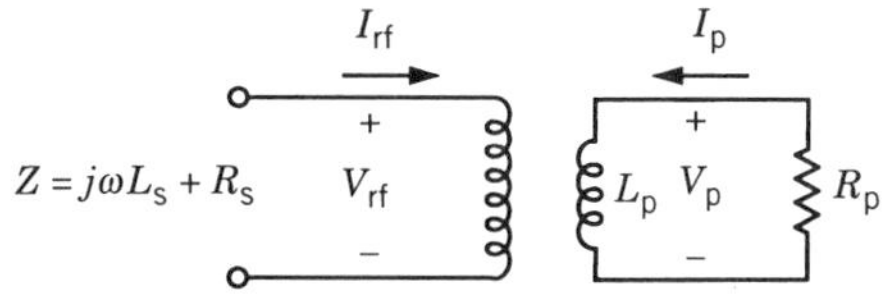

FIGURE 12.2. Equivalent transformer coupled circuit model of an inductive discharge. (From "Design of High-Density Sources for Materials Processing" from the work "Physics of Thin Films," Vol.18, by Academic Press, Inc., Publisher in Press)

defined through (Schwarz and Oldham, 1984)

$$\tilde{V}_{rf} = j\omega L_{11}\tilde{I}_{rf} + j\omega L_{12}\tilde{I}_p \tag{12.1.14}$$

$$\tilde{V}_p = j\omega L_{21}\tilde{I}_{rf} + j\omega L_{22}\tilde{I}_p \tag{12.1.15}$$

where the tildes denote the complex amplitudes, e.g., $V_{rf}(t) = \text{Re}\,\tilde{V}_{rf}\,e^{j\omega t}$, we obtain (Problem 12.3)

$$L_{11} = \frac{\mu_0 \pi b^2 N^2}{l} \tag{12.1.16}$$

$$L_{12} = L_{21} = \frac{\mu_0 \pi R^2 N}{l} \tag{12.1.17}$$

$$L_{22} = L_p = \frac{\mu_0 \pi R^2}{l} \tag{12.1.18}$$

Using $\tilde{V}_p = -\tilde{I}_p R_p$ (see Fig. 12.2) in (12.1.15) and inserting into (12.1.14), we can solve for the impedance seen at the coil terminals:

$$Z_s = \frac{\tilde{V}_{rf}}{\tilde{I}_{rf}} = j\omega L_{11} + \frac{\omega^2 L_{12}^2}{R_p + j\omega L_p} \tag{12.1.19}$$

For $\delta_p \ll R$, it can easily be seen from (12.1.12) and (12.1.13) that $R_p^2 \ll \omega^2 L_p^2$. Hence expanding the denominator in (12.1.19), we obtain

$$L_s \approx \frac{\mu_0 \pi R^2 N^2}{l}\left(\frac{b^2}{R^2} - 1\right) \tag{12.1.20}$$

$$R_s \approx N^2 \frac{2\pi R}{\sigma_{eff} l \delta_p} \tag{12.1.21}$$

where $Z_s = R_s + j\omega L_s$. The power balance,

$$P_{abs} = \frac{1}{2}|\tilde{I}_{rf}|^2 R_s \tag{12.1.22}$$

then yields the required rf source current, and the rf voltage is determined from

$$\tilde{V}_{rf} = \tilde{I}_{rf}|Z_s| \tag{12.1.23}$$

Example   We let $R = 10$ cm, $b = 15$ cm, $l = 20$ cm, $\mathcal{N} = 3$ turns, $n_g = 1.7 \times 10^{14}$ cm^{-3} (5 mTorr argon at 298 K), $\omega = 8.5 \times 10^7$ s^{-1} (13.56 MHz), and $P_{abs} = 600$ W. At 5 mTorr, $\lambda_i \approx 0.6$ cm. Then from (10.2.3) and (10.2.4) $h_l \approx 0.20$, $h_R \approx 0.18$, and from (10.2.8), $d_{eff} \approx 17.9$ cm. For argon we then obtain from Fig. 10.1 that $T_e \approx 2.6$ V, and from Fig. 3.17, that $\mathcal{E}_c \approx 58$ V. From (10.2.1) and using (10.2.9) to obtain $\mathcal{E}_i + 2T_e \approx 19$ V, we find $\mathcal{E}_T \approx 77$ V. The Bohm velocity is $u_B \approx 2.5 \times 10^5$ cm/s, and from (10.2.15), $A_{eff} \approx 350$ cm^2. Then from (10.2.14), we obtain $n_0 \approx 5.6 \times 10^{11}$ cm^{-3} and $n_s = h_R n_0 \approx 1.0 \times 10^{11}$ cm^{-3}. Estimating ν_m for argon from Fig. 3.16, we find $\nu_m \approx 1.4 \times 10^7$ s^{-1}. Using (12.1.4), we find $\delta_p \approx 1.7$ cm. Evaluating ν_{stoc} from (C.17) with $\bar{v}_e \approx 1.1 \times 10^8$ cm/s and $\delta = \delta_p$, we obtain $\nu_{stoc} \approx 1.1 \times 10^7$ s^{-1}, such that $\nu_{eff} \approx 2.5 \times 10^7$ s^{-1}. Using this in (12.1.11), we find $\sigma_{eff} \approx 113$ mho/m. Evaluating (12.1.21) and (12.1.20), we find $R_s \approx 14.7\Omega$ and $L_s \approx 2.2\ \mu$H, such that $\omega L_s \approx 190\Omega$. Equations (12.1.22) and (12.1.23) then yield $I_{rf} \approx 9.0$ A and $V_{rf} \approx 1720$ V.

We note that $\omega > \nu_{eff}$ for this example, such that $\delta \approx \delta_p$, the collisionless skin depth, verifying our assumed ordering. We also note that $\nu_{stoc} \sim \nu_m$, such that stochastic and collisional heating contribute roughly equally to the overall heating. At lower pressures, e.g., 1 mTorr, stochastic heating dominates over collisional heating. For some choices of parameters, e.g., a lower driving frequency, ν_{stoc} can be comparable to or exceed both ω and ν_m. In these cases, ν_{stoc} is determined by (C.16) with $C_e \sim 1$ and $\delta = \delta_e$, the anomalous skin depth.

Godyak et al. (1993) have measured ν_{eff} in an inductive discharge, finding that ν_{eff} is independent of pressure at low pressures, indicating the dominance of stochastic over collisional heating. The measured ν_{eff} was $\sim 2\bar{v}_e/\delta_e$, equivalent to (C.16) with $C_e \sim 2$. However, for this experiment, $\delta_e \sim \delta_p \sim \delta_c$, so the scale length dependence could not be distinguished. The theory of collisionless heating in inductive discharges is not well established. The calculation presented in Appendix C is in the spirit of similar calculations for capacitive discharges (see Section 11.2) and ECR discharges (see Section 13.1). Appendix C should not be regarded as definitive, but as illustrative of the physical principles that determine the heating.

Under some conditions, the electron drift velocity associated with the induced rf plasma current within the skin depth layer can be larger than the electron thermal velocity. This drifting Maxwellian distribution with large mean energy can produce an increased ionization, leading to a lowering of T_e and significant changes in the density profile over that found for the global (constant T_e) ionization model (10.2.7) (see Problem 12.4(b)).

Capacitive Coupling

At this point the reader might ask: since the voltage $V_{rf} \approx 1720$ volts at the high-voltage end of the coil, comparable to capacitive discharges, why is the discharge considerably more efficient? The answer is that only a small fraction of that voltage appears across the sheath, such that the ion energy loss is considerably lower. To estimate the rf voltage across the sheath, $\tilde{V}_{sh}$, at the high-voltage end of the coil, we note that the sheath capacitance per unit area is $\sim \epsilon_0/s_m$ and the capacitance per unit area of the dielectric cylinder is $\sim \epsilon_0/(b - R)$. Assuming that the plasma is at ground potential, then the voltage across the sheath is found from the capacitive voltage divider formula,

$$\tilde{V}_{sh} = V_{rf}\frac{s_m}{b - R + s_m} \tag{12.1.24}$$

Using the modified Child law (11.2.15), we calculate the sheath thickness from

$$en_s u_B = 0.82\,\epsilon_0 \left(\frac{2e}{M}\right)^{1/2} V_{rf}^{3/2} \left(\frac{s_m}{b - R + s_m}\right)^{3/2} \frac{1}{s_m^2} \tag{12.1.25}$$

which is a cubic equation in s_m. However, for the usual case for which $s_m \ll b - R$, (12.1.25) simplifies to

$$s_m \approx \left(\frac{0.82\,\epsilon_0}{en_s u_B}\right)^2 \left(\frac{2e}{M}\right) \frac{V_{rf}^3}{(b - R)^3} \tag{12.1.26}$$

The right-hand side is generally small for the usual voltages of inductive discharges, so that s_m is much smaller than in a capacitive discharge. In our example, we find $s_m \approx 6.4 \times 10^{-4}$ cm, so that, from (12.1.24), $\tilde{V}_{sh} \approx 0.22$ V. Actually, for a sheath this thin, the high-voltage sheath relation (12.1.25) is not valid. From (2.4.23), the Debye length is $\lambda_{De} \approx 3.8 \times 10^{-3}$ cm. The sheath is a few Debye lengths thick. Using calculations for capacitive discharges (see Godyak and Sternberg, 1990b), we estimate $s_m \sim 2 \times 10^{-2}$ cm, such that (12.1.24) yields $\tilde{V}_{sh} \sim 9$ V, which contributes only a small correction to the dc sheath voltage.

From the relations (12.1.20)–(12.1.23), we can see why the designer wants to keep the number of turns of the exciting coil small. From (12.1.20) and (12.1.21), we see that $Z_s \propto N^2$, and at fixed P_{abs}, from (12.1.22), we find $\tilde{I}_{rf} \propto 1/N$. Then (12.1.23) gives $\tilde{V}_{rf} \propto N$. From (12.1.26), we see that $s_m \propto N^3$ at a fixed density, such that doubling N would increase s_m by almost an order of magnitude. The increased ion energy loss across this larger sheath (increased $\mathcal{E}_T$) leads to lower density and generally less favorable discharge parameters.

Matching Network

The high inductive voltage required for this three-turn coil can be supplied from a 50-Ω rf power source through a capacitive matching network, as shown in Fig. 12.3. The admittance looking to the right at the terminals A–A' is

$$Y_A \equiv G_A + jB_A = \frac{1}{R_s + j(X_1 + X_s)} \tag{12.1.27}$$

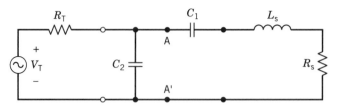

FIGURE 12.3.   Equivalent circuit for matching an inductive discharge to a power source.

where the conductance is

$$G_A = \frac{R_s}{R_s^2 + (X_1 + X_s)^2} \tag{12.1.28}$$

and the susceptance is

$$B_A = -\frac{X_1 + X_s}{R_s^2 + (X_1 + X_s)^2} \tag{12.1.29}$$

and where $X_1 = -(\omega C_1)^{-1}$. As described in Section 11.6, we must choose G_A to be equal to $1/R_T$, for maximum power transfer, where $R_T = 50\ \Omega$ is the Thevenin-equivalent source resistance. For $R_s \approx 14.7\ \Omega$ and $X_s \approx 190\ \Omega$, we obtain from (12.1.28) that $X_1 \approx -167\ \Omega$. Hence $C_1 \approx 70$ pF. Evaluating B_A for this value of X_1, we obtain $B_A \approx -0.031\ \Omega^{-1}$. We must choose C_2 to cancel this susceptance, i.e., $B_2 = \omega C_2 = -B_A$, which determines $C_2 = 364$ pF to achieve the matched condition. In practice, C_1 and C_2 are variable capacitors that are tuned to achieve the match. The power absorption, $P_{abs} = \frac{1}{2}I_T^2 R_T$, then determines $I_T \approx 4.9$ A and $V_T = 2I_T R_T \approx 490$ V.

12.2   OTHER OPERATING REGIMES

Low Density Operation

Since the effective conductivity $\sigma_{eff} \propto n_0$ and $\delta_p \propto n_0^{-1/2}$, it is apparent from (12.1.21) and (12.1.22) that at fixed driving current I_{rf}, we have the scaling at high densities:

$$P_{abs} \propto n_0^{-1/2} I_{rf}^2 \tag{12.2.1}$$

However, at low densities, such that $\delta_p \gg R$, the conductivity is low and the fields fully penetrate the plasma. In this case, applying Faraday's law to determine the induced electric field E_θ within the coil, we obtain

$$E_\theta(r) = \frac{\frac{1}{2} j \omega r \mu_0 N I_{rf}}{l} \tag{12.2.2}$$

and, writing $J_\theta = j\omega\epsilon_0\kappa_p E_\theta$ for $\nu_m \ll \omega$, we have $J_\theta \propto n_0 r I_{rf}$. Evaluating the power absorbed for this case, we have

$$
\begin{aligned}
P_{abs} &= \frac{1}{2}\int_0^R \frac{J_\theta^2(r)}{\sigma_{eff}} 2\pi r l \, dr \\
&= \frac{1}{2}I_{rf}^2 \frac{\pi e^2 n_0 \nu_{eff}\mu_0^2 \mathcal{N}^2 R^4}{8ml}
\end{aligned}
\tag{12.2.3}
$$

such that

$$
P_{abs} \propto n_0 I_{rf}^2
\tag{12.2.4}
$$

Comparing (12.2.1) with (12.2.4), we see that holding I_{rf} fixed, P_{abs} versus n_0 has a maximum near $\delta_p \sim R$. This corresponds to a variation with density as sketched in Fig. 12.4 for several different values of I_{rf}. Now consider the power balance requirement (10.2.14), which is plotted as a straight line in the figure. The intersection of the line with the curves defines the equilibrium point for discharge operation. We see that inductive source operation is impossible if the source current I_{rf} is below some minimum value I_{min}. In this regime, any discharge must be capacitively driven.

Power Transfer Efficiency

Let us note that the driving coil (primary of the transformer shown in Fig. 12.2) has some resistance R_{coil}. Hence, even if the discharge is extinguished ($n_0 = 0$),

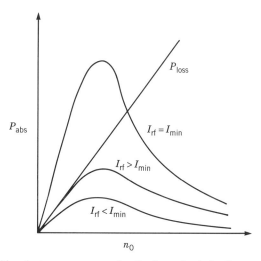

FIGURE 12.4.   Absorbed power versus density from the inductive source characteristics (curves) and power lost versus density (straight line); the curves are drawn for different values of the driving current I_{rf}.(From "Design of High-Density Sources for Materials Processing" from the work "Physics of Thin Films," Vol.18, by Academic Press, Inc., Publisher in Press)

there is a minimum power $P_{Tmin} = \frac{1}{2}I_{min}^2 R_{coil}$ supplied by the source before the inductive discharge can form. Because $P_{abs} \propto n_0^{-1/2}$ at high densities, we see from Fig. 12.4 that the power transfer efficiency P_{abs}/P_T falls continually as n_0 is increased, hence limiting source operation at high densities because of power supply limitations. Although $\delta_p \sim R$ is the preferred operating regime for maximum power efficiency, other considerations often indicate operation at higher density. The poor power transfer to the plasma at very low and at very high densities is analogous to the well-known property of an ordinary transformer with an open and a shorted secondary winding. In both cases no power is dissipated in the load (here the plasma), but in both cases there is power dissipated in the primary winding (here the coil) due to its inherent resistance. Piejak et al. (1992) have given a complete analysis of an inductive discharge in terms of measurable source voltages and currents, based on this analogy.

At very high densities, the electron–ion collision frequency may be larger than the electron–neutral collision frequency. In this collisional regime, ν_{90} in (3.3.7) replaces ν_m in determining σ_{dc}. Since $\nu_{90} \propto n_0$ (the Spitzer conductivity is independent of n_0), the scaling (12.1.1) is replaced by

$$P_{abs} \propto I_{rf}^2$$

independent of n_0 in this regime. However, low-pressure inductive discharges for materials processing are rarely operated at such high densities.

High-Pressure Discharges

In the regime $\nu_m \gg \omega$, the penetration of the rf fields into the discharge is governed by the collisional skin depth (12.1.6). At high densities $\delta_c \ll R$, with a transition to the solution (12.2.2) and (12.2.3) at very low densities. Thomson (1927) determined the solution that spans the range of densities for a uniform density cylindrical plasma with $l \gg R$, and we briefly describe his treatment here. In the limit that the displacement current $j\omega\epsilon_0\tilde{E}$ is small compared to the conduction current, Maxwell's equations (2.2.1) and (2.2.2) for the $\tilde{E}_\theta$ and $\tilde{H}_z$ field components become

$$\frac{d}{dr}(r\tilde{E}_\theta) = -j\omega\mu_0 r\tilde{H}_z \tag{12.2.5}$$

$$r\frac{d\tilde{H}_z}{dr} = -\sigma_{dc}(r\tilde{E}_\theta) \tag{12.2.6}$$

Eliminating $r\tilde{E}_\theta$ from these equations, we obtain

$$\frac{d^2\tilde{H}_z}{dr^2} + \frac{1}{r}\frac{d\tilde{H}_z}{dr} - j\omega\mu_0\sigma_{dc}\tilde{H}_z = 0 \tag{12.2.7}$$

which is Bessel's equation. With the boundary condition that $\tilde{H}_z(R) = H_{z0}$, the solution is

$$\tilde{H}_z = H_{z0}\frac{J_0(-j^{1/2}\beta r)}{J_0(-j^{1/2}\beta R)} \tag{12.2.8}$$

where

$$\beta = (\omega\mu_0\sigma_{dc})^{1/2} = \frac{\sqrt{2}}{\delta_c} \tag{12.2.9}$$

We see that the Bessel functions have complex argument. Using (12.2.8) to evaluate the left-hand side of (12.2.6), and solving for $\tilde{E}_\theta$, we obtain

$$\tilde{E}_\theta = -H_{z0}\frac{j^{1/2}\beta}{\sigma_{dc}}\frac{J_1(-j^{1/2}\beta r)}{J_0(-j^{1/2}\beta R)} \tag{12.2.10}$$

The time-average power flowing into the discharge can be written in terms of the field amplitudes at the plasma surface $r = R$ using the complex Poynting theorem (Ramo et al., 1984),

$$P_{abs} = 2\pi Rl S_{abs} = 2\pi Rl \operatorname{Re}\left(\frac{1}{2}\tilde{E}_{\theta 0}H_{z0}\right) \tag{12.2.11}$$

Using (12.2.8)–(12.2.10) in (12.2.11), we obtain

$$S_{abs} = H_{z0}^2 \frac{\pi\omega\mu_0 R}{\beta R} \operatorname{Re}\left[\frac{j^{1/2}J_1(-j^{1/2}\beta R)}{J_0(-j^{1/2}\beta R)}\right] \tag{12.2.12}$$

As σ_{dc} is increased from zero at fixed ω, R, and H_{z0} (equivalent to holding the coil current fixed), then P_{abs} rises from zero to a maximum at $\beta R \approx 2.5$ ($\delta_c \approx 0.57R$), and then falls to zero. The variation is similar to that shown in Fig. 12.4. Hence, the maximum power efficiency in the high-pressure regime for a uniform density cylindrical source occurs for $\delta_c \approx 0.57R$.

Other issues of inductive discharge operation include finite geometry effects ($l \sim R$), planar coil source operation, startup, and self-resonant coil effects due to stray coil capacitances. We address some of these issues in the following sections, and refer the reader to the literature (Piejak et al., 1992; Eckert, 1986; Hopwood et al., 1993a,b) for further information.

12.3 PLANAR COIL CONFIGURATION

The planar coil discharge shown in Fig. 12.1b, with or without multipole magnets, is a commonly used configuration for materials processing, typically generating relatively uniform low aspect ratio plasmas with densities between 10^{11} and 10^{12} cm^{-3} over substrate diameters of 20 cm or more. In axisymmetric geometry, the coil generates an inductive field having magnetic components $\tilde{H}_r(r, z)$ and $\tilde{H}_z(r, z)$, and an electric component $\tilde{E}_\theta(r, z)$. As shown in Fig. 12.5a, the rf magnetic field lines in the absence of a plasma encircle the coil and are symmetric with respect to the plane of the coil. If a plasma is formed below the coil, as shown in Fig. 12.5b, then from Faraday's law (2.2.1), an azimuthal electric field $\tilde{E}_\theta$ and an associated current density $\tilde{J}_\theta$ are induced within the plasma. The plasma current, opposite in direction to the coil current, is confined to a layer near the surface having a thickness of order the skin depth δ. The total magnetic field, which is the sum of the fields due to the $\mathcal{N}$ turn exciting coil current and the "single-turn" induced plasma current, is shown in Fig. 12.5b. The dominant magnetic field components within the plasma are $\tilde{H}_z$ near the axis and $\tilde{H}_r$ away from the axis, as shown. Near the axis, Faraday's law implies that both $\tilde{E}_\theta$ and $\tilde{J}_\theta$ vanish as $\tilde{E}_\theta$, $\tilde{J}_\theta \propto r$. This implies that the absorbed power density,

$$p_{abs} = \frac{1}{2} \operatorname{Re} \tilde{J}_\theta \tilde{E}_\theta^* \tag{12.3.1}$$

vanishes on axis, leading to a ring shaped profile for the absorbed power.

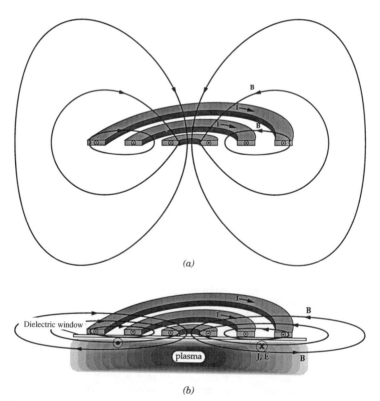

(a)

(b)

FIGURE 12.5. Schematic of the rf magnetic field lines near a planar inductive coil (a) without nearby plasma and (b) with nearby plasma (after Wendt, 1993).

The rf magnetic fields within the plasma have been measured by Hopwood et al. (1993a) for an inductive discharge excited by a planar square coil, which was separated from a rectangular aluminum plasma chamber 27 cm on a side and 13 cm high by a 2.54-cm-thick quartz window. Although this system is not axisymmetric, the general structure of the fields and the absorbed power profile are similar to those in an axisymmetric system. We use Hopwood's results to illustrate the general features observed in planar inductive discharges.

Figure 12.6 shows the measured variation of $B_r \equiv |\tilde{B}_r|$ with z at $r = 6.3$ cm in a 5-mTorr oxygen discharge. The field decreases exponentially with distance from the window, with a maximum of 2.7–5.1 G, depending on the incident power P_{inc}, and with a skin depth δ (characteristic length for the exponential decay) varying from 2.1 to 2.7 cm, and scaling roughly as $P_{inc}^{-1/2}$, in agreement with (12.1.4) or (12.1.6) with $n_0 \propto P_{inc}$. In general, the skin depth lies between the values δ_p and δ_c given by (12.1.4) and (12.1.6), and agrees with δ_c to within 25%. Figure 12.7 shows the measured variation of B_r with r (along the diagonal of the chamber) at three different positions below the window in a 5-mTorr, 500-W, argon discharge. We see that B_r falls to zero on the axis and has a maximum at approximately 9.5 cm off the axis.

The rf electric field $\tilde{E}_\theta$ can be related to $\tilde{B}_r \equiv \mu_0 \tilde{H}_r$ by using the r component of Faraday's law (2.2.1),

$$-\frac{\partial \tilde{E}_\theta}{\partial z} = j\omega\mu_0\tilde{H}_r \tag{12.3.2}$$

Assuming that $\tilde{E}_\theta = \tilde{E}_{\theta 0}\, e^{-z/\delta}$ and inserting this into (12.3.2), we obtain

$$\tilde{E}_\theta = j\omega\mu_0\delta\tilde{H}_r \tag{12.3.3}$$

Hence $\tilde{E}_\theta$ has the same axial and radial variation as $\tilde{B}_r$ given in Figs. 12.6 and 12.7.

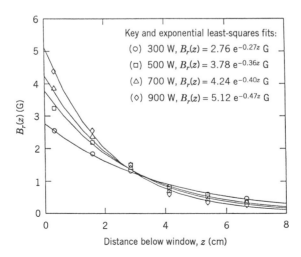

FIGURE 12.6. Rf magnetic induction amplitude $|\tilde{B}_r|$ versus z in a 5-mTorr oxygen discharge. The solid lines are a least-squares fit to the data (Hopwood et al., 1993a).

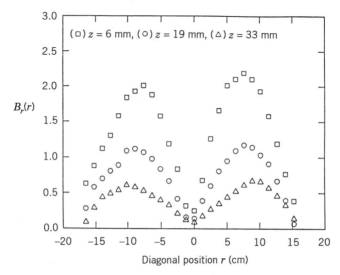

FIGURE 12.7. Rf magnetic induction amplitude $|\tilde{B}_r|$ versus diagonal radius r at three different distances below the window as measured in a 5-mTorr, 500-W argon discharge (Hopwood et al., 1993a).

In addition to the field measurements, Langmuir probes (see Section 6.6) were used to determine the ion density n_i, electron temperature T_e, and plasma potential V_s (Hopwood et al., 1993b). The ion density measurement was confirmed by comparison to a 35-GHz microwave interferometer measurement (see Section 4.6). Figure 12.8 shows n_i versus incident power P_{inc} at a location on-axis and 5.7 cm ($\sim$ 3 skin depths) below the window, for pressures between 0.5 and 15 mTorr in argon. We see that n_i varies linearly with P_{inc}, but that $n_i \approx 0$ (on a 10^{11}-cm^{-3} scale) at approximately 100 W. Below this incident power, an inductive discharge cannot be sustained (see

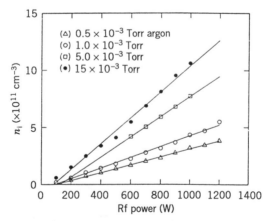

FIGURE 12.8. Ion density versus rf power and argon pressure (Hopwood et al., 1993b).

Section 12.2), and a low-density plasma is sustained by capacitive coupling between the coil and the plasma.

Figure 12.9 shows the measured variation of n_i, T_e, and V_s with argon pressure for $P_{inc} = 500$ W. We see that T_e falls slowly as p increases, as determined from the ion particle balance relation (10.2.7), plotted in Fig. 10.1. The ion density is seen to increase with increasing pressure. This can be shown to be consistent with the power balance relation, which indicates that the density varies inversely with the effective plasma area, as follows. For this discharge, with $l = 13$ cm and $R \gg l$, we can estimate that A_{eff} in (10.2.15) scales as

$$A_{eff} \propto h_l \propto \left(3 + \frac{l}{2\lambda_i}\right)^{-1/2} \qquad (12.3.4)$$

Using (3.5.7) to determine the ion–neutral mean free path λ_i in argon, we find $\lambda_i \approx 3$ cm at 1 mTorr and $\lambda_i \approx 0.15$ cm at 20 mTorr. Hence from (10.2.14), the predicted density ratio is

$$\frac{n_i(20 \text{ mTorr})}{n_i(1 \text{ mTorr})} \approx \frac{A_{eff}(1 \text{ mTorr})}{A_{eff}(20 \text{ mTorr})} \approx \frac{0.44}{0.15} \approx 3.0$$

which is in reasonable agreement with the measured ratio of ~ 3.3 obtained from Fig. 12.9. The plasma potential V_s is seen to lie between 12 and 30 V, roughly consistent with (10.2.9). The potential increases as the pressure decreases, in qualitative agreement with the scaling predicted from (10.2.9).

All preceding measurements were performed with multipole magnets placed along the four 27-cm × 13-cm sidewall areas (see Section 5.4). In Fig. 12.10, the normalized ion saturation current (proportional to the density) is plotted along a diagonal within

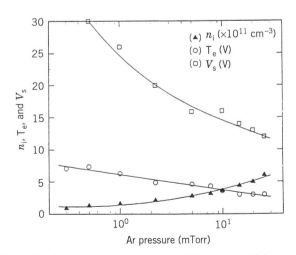

FIGURE 12.9. Ion density, electron temperature, and plasma potential versus argon pressure in a 500-W discharge with magnetic multipole confinement (Hopwood et al., 1993b).

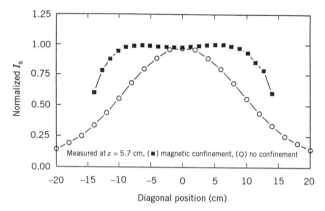

FIGURE 12.10. Normalized ion saturation current measured across the diagonal of the plasma chamber with and without magnetic multipole confinement (Hopwood et al., 1993b).

the chamber with and without the multipole magnets in place, for a 5-mTorr oxygen discharge. We see that the multipole magnets greatly increase the uniformity of the density. The ratio of the standard deviation to the average density across the central 20 cm of the discharge with multipole magnets was measured to be 2.5%.

12.4 HELICAL RESONATOR DISCHARGES

Helices have long been used to propagate electromagnetic waves with phase velocity $v_{ph} \ll c$, the velocity of light. This property allows a helix to resonate in the MHz range such that it can be used for efficient plasma generation at low pressures. Low-pressure operation makes it particularly useful for etching, and it has also been useful for deposition of silicon dioxide and silicon nitride films (Cook et al., 1990). Helical resonator plasmas operate conveniently at radio frequencies (3–30 MHz) with simple hardware, do not require a dc magnetic field (as do ECRs and helicons; see Chapter 13), exhibit high Q (600–1500 typically without the plasma present) and high characteristic impedance (Z_0), and can be operated without a matching network. As shown in Fig. 12.11, the source consists of a coil surrounded by a grounded coaxial cylinder. The composite structure becomes resonant when an integral number of quarter waves of the rf field fit between the two ends. When this condition is satisfied, the electromagnetic fields within the helix can sustain a plasma with low matching loss at low gas pressure.

We have seen for inductive discharges at low density that the ohmic heating is not sufficient to sustain the discharge, which then operates similarly to a capacitive discharge at much lower density and higher sheath voltage (larger sheaths). We expect the same general behavior to occur in the resonant helical discharge, because of the voltage divider action between the rf voltage across the sheath and that which appears across the fixed spacing between the helix and the vacuum chamber. Because the helix is a resonant device, large voltages necessarily appear between the open end of the helix and the plasma, such that it is more difficult to operate in a purely inductive regime. To force operation in this regime, an electrostatic shield can be added between the helix and the plasma column to reduce the capacitive coupling to a negligible value. The shield is typically a metal cylinder slotted along z that allows the inductive field $\tilde{E}_\theta$ to penetrate into the plasma, while shorting out the capacitive $\tilde{E}_r$ and $\tilde{E}_z$ fields. Using this configuration, discharges as large as 25 cm in diameter have been produced experimentally, driven by up to 5 kW of rf power at 13.56 MHz and producing plasmas with densities (in argon) exceeding 2×10^{12} cm^{-3}. However, little in the way of plasma parameter measurements and comparison with theory has been done in this configuration.

A more thorough analysis, with comparison to experiments, has been performed on an unshielded helical discharge at lower powers, where both stochastic heating and ohmic heating are important (Niazi et al., 1994). We describe some important features of this discharge here. Some, but not all, of the results are relevant to shielded discharges at higher powers.

The basic design parameters for a helical resonator discharge consist of pressure, rf power, source length, plasma radius, helix radius, outer cylinder radius, winding pitch angle, and excitation frequency. This is a complicated system that is not fully understood. A first step is to determine the helical slow-wave modes and their inter-action with the plasma. For an unshielded plasma column, the wave dispersion, k_z versus ω, and the relation among the field quantities, have been determined in the approximation of a uniform, collisionless ($\omega \gg \nu_m$) plasma having relative dielectric constant $\kappa_p = 1 - \omega_{pe}^2/\omega^2$ (see (4.2.24)) by using a *developed sheath helix model*, in which the rf current in the helical wires is replaced by a continuous current sheet ("sheath") and the cylindrical (r, θ, z) geometry is unfolded into a rectangular (x, y, z) geometry ("developed"). This is a standard analytical technique for treating helical systems that retains most essential physics. In the absence of the plasma, it can be shown that there is a single mode that tends to propagate along the helical wire, as expected for this two-conductor transmission line system.[*] For source parameters, $R = 3$ cm, $b = 5$ cm, $d = 10$ cm, $l = 30$ cm, and $\Psi = 0.1$ radians, where the parameters are defined in Fig. 12.11, k_z is shown versus $f = \omega/2\pi$ in Fig. 12.12, at low plasma density $n_0 = 10^9$ cm^{-3} (triangles and circles), at high density $n_0 = 10^{11}$ cm^{-3} (crosses and squares), and without plasma $n_0 = 0$ (pluses). For comparison, the upper line shows a wave following the geometrical helix pitch,

$$k_{zh} = \frac{\omega}{c \tan \Psi} \tag{12.4.1}$$

and the lower line shows a light wave $k_{z0} = \omega/c$. Without a plasma, there is only one mode of propagation, with k_z somewhat smaller than k_{zh} (the wave velocity ω/k_z is somewhat larger than $c \tan \Psi$); i.e., the wave tends to propagate along the helix wire above a ground plane. As n_0 increases, the wave speeds up, and as $n_0 \to \infty$,

---

[*]This is analogous to wave propagation in a coaxial line, with the helical coil being the inner conductor and the grounded cylinder being the outer conductor.

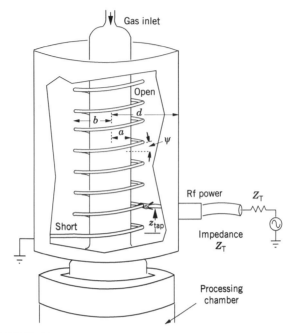

FIGURE 12.11.   Schematic of a helical resonator plasma source.

$\omega/k_z \to c$. For this "coax" mode, at large n_0, both the helix and the plasma are at a high voltage with respect to the outer cylinder.

A second "helix" mode appears when n_0 is such that $\omega_{pe} > \omega$, a condition that is always met for typical discharge operation. Hence, both modes can coexist. The wave velocity for the second mode is always smaller than the helix velocity $c \tan \Psi$. The mode appears as a resonance $k_z \to \infty$ at n_0 such that $\omega_{pe} = \omega$, and the wave speeds up as n_0 increases. For the helix mode at large n_0, the plasma and outer cylinder are at nearly the same voltage, and the helix is at a high voltage with respect to them both. In the high-density limit the plasma acts like a conducting cylinder, in which the helix is nearly transparent to the coax mode, while the helix mode is similar to the mode that exists without a plasma.

The axial wave numbers for the two modes are very different. For example, at $f = 25$ MHz and at $n_0 = 10^{11}$ cm^{-3}, $k_z(\text{coax}) \approx 0.5$ m^{-1} and $k_z(\text{helix}) \approx 5.5$ m^{-1}. Since the source length l is chosen to be roughly a quarter wavelength at the helix geometrical pitch, $k_{zh}l \approx \pi/2$, the coax mode is not resonantly excited [$k_z(\text{coax}) \ll k_{zh}$]. However, this mode does play a role in source operation at startup. During typical source operation, only the helix mode is resonant, and it dominates the source operation. A simple estimate of the resonant frequency from (12.4.1) gives $f \approx 25$ MHz at $l = 30$ cm ($k_z = 5.2$ m^{-1}). Greater accuracy can be obtained by analyzing the helix–plasma configuration in cylindrical geometry. End effects can change the resonant frequency due to additional capacitive coupling.

Once the resonant frequency for quarter wavelength operation is determined, then the fields within all regions inside the helical resonator can be found. From the fields in

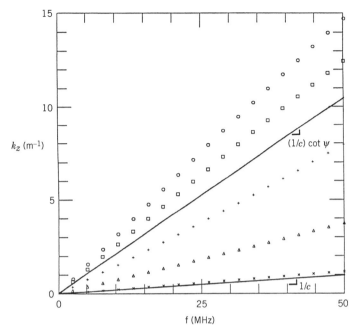

FIGURE 12.12. Axial wave number k_z versus frequency f for the coax and helix modes, with density n_0 as a parameter; circles, helix mode, $n_0 = 10^9$ cm^{-3}; squares, helix mode, $n_0 = 10^{11}$ cm^{-3}; plusses, coax mode, $n_0 = 0$; triangles, coax mode, $n_0 = 10^9$ cm^{-3}; crosses, coax mode, $n_0 = 10^{11}$ cm^{-3}. (From "Design of High-Density Sources for Materials Processing" from the work "Physics of Thin Films," Vol.18, by Academic Press, Inc., Publisher in Press)

the plasma, the absorbed power can be found. This has been done for a quasistatic field approximation in the capacitively coupled regime where both ohmic and stochastic heating (see Sections 11.1 and 11.2) contribute to the power absorbed by the electrons. The calculation has not been performed for the inductive configuration (with an electrostatic shield), where only $\tilde{E}_\theta$ contributes to the absorbed power, but should be similar to that used for conventional inductive sources (see Section 12.1 and below).

For the cylindrical sheath-current helix, within the quasistatic approximation, the helix voltage $\tilde{V}_{max}$ at the high-voltage end of the coil is determined in terms of the azimuthal component of sheath current per unit length K_θ at the low-voltage end of the coil to be

$$\tilde{V}_{max} = \zeta \left(\frac{\mu_0}{\epsilon_0} \right)^{1/2} bK_\theta \qquad (12.4.2)$$

where ζ is a geometric constant of order unity. For a helix with $\mathcal{N}$ turns the helix current $\tilde{I}_{max}$ at the low-voltage end of the coil is obtained from $\mathcal{N}\tilde{I}_{max} = lK_\theta$. Combining this with (12.4.2), we obtain the helix impedance

$$Z_0 = \frac{\tilde{V}_{max}}{\tilde{I}_{max}} = \zeta \left(\frac{\mu_0}{\epsilon_0} \right)^{1/2} \frac{\mathcal{N}b}{l}. \qquad (12.4.3)$$

For the example given above, $\mathcal{N} = 48$ and $\zeta = 0.33$, giving $Z_0 = 995\ \Omega$. For a constant l and a constant pitch angle of the helix, $\mathcal{N}b/l$ remains essentially constant with varying helix diameter, and since ζ changes relatively slowly, the impedance $Z_0 \sim 1\ \mathrm{k\Omega}$ over a range of aspect ratios. This high helix impedance allows easy matching to transmission lines.

From (12.4.3), we see that the impedance, and also the voltage $\tilde{V}_{\mathrm{max}}$, increases as the number of helix turns $\mathcal{N}$ increases. We also found that to be true for inductive discharges in Section 12.1. For the geometry considered here, with $\mathcal{N}$ large, the result is to also have large sheath fields and accompanying large ion energy losses. The discharge is then operating in a mainly capacitive regime. The geometry of the helical resonator can also be changed to have a larger diameter with fewer turns. The approximate resonance condition, $k_{\mathrm{zh}}l = \pi/4$, can also be written, using (12.4.1) with $\tan \Psi = l/2\pi b\mathcal{N}$, as

$$2\pi b\mathcal{N} = \frac{\lambda}{4} \tag{12.4.4}$$

where λ is the free space wavelength. A much larger radius b with only five turns has, in fact, been experimentally investigated as a prototype reactor. In this configuration, the main difference between a five-turn inductive discharge of the type considered in Section12.1 and the five-turn helical resonator considered here is that the latter is resonant, thus allowing a simple coupling, described in the following paragraph. However, a variable frequency (expensive) power supply is required to maintain the resonance condition. Hence helical resonators are usually driven using a fixed frequency supply in conjunction with a matching network.

To determine the power coupled from an external generator to the resonator, and the condition for a match (maximum power transfer), we consider the system shown schematically in Fig. 12.11. The rf generator and its transmission line have characteristic impedances Z_{T}, with one side of the transmission line connected to the helix at the tap position z_{tap} and the other side connected to the outer shield. Since the helix characteristic impedance Z_0 given in (12.4.3) is typically large compared to Z_{T}, we expect a match to occur with the tap made near the shorted end of the helical resonator, where the voltage is small and the current is large. This is, in fact what is found from a transmission line calculation and also in the experiment. The shorted end of the helix acts as part of the matching system, allowing a good match over a wide range of parameters, without a matching network, provided the operating frequency can be varied over a modest range.

The analysis described above can and has been improved. The cold plasma wave dispersion for the sheath helix can be solved directly in cylindrical coordinates to improve the accuracy of the calculation. In addition to the stochastic heating at the sheath edge and ohmic heating in the bulk plasma, ohmic heating in the sheath region and ionization in the sheath can also be included. Including these effects, a comparison has been made between theory and experiment, for the device shown schematically in Fig. 12.11, for argon and nitrogen at 2- and 20-mTorr pressures. The results for argon at 20 mTorr are shown in Fig. 12.13. In Fig. 12.13a, the density is plotted versus absorbed power indicating, approximately, a $n \propto P_{\mathrm{abs}}^{1/2}$ scaling, which is found

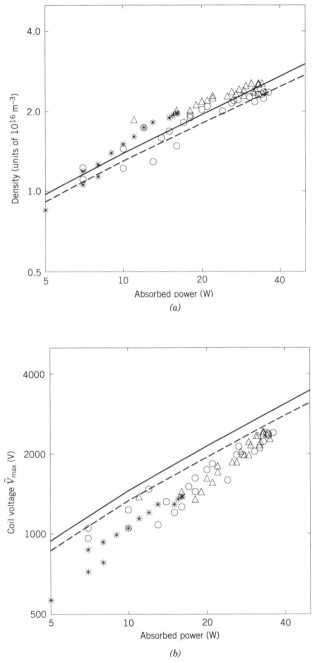

FIGURE 12.13. Measurements of (a) density and (b) coil voltage versus absorbed power in a helical resonator. The symbols give the measured values for various runs; the solid and dashed lines give the analytic model results without and with the additional power required to account for sheath losses, respectively (Niazi et al., 1994).

in capacitive discharges when stochastic heating is the primary heating and when most of the energy loss is due to ions accelerated through a high-potential sheath. The dashed curve shows the additional power required to account for sheath losses. Figure 12.13b shows the variation of the maximum coil voltage $\tilde{V}_{max}$ with power. As we have seen in Chapter 11, this variation is more complicated than the density variation, with an approximate scaling $\tilde{V}_{max} \propto P_{abs}^{1/2}$ in the regime for which stochastic heating is most important and ion energy losses exceed electron losses. (Note that because the sheath thickness s_m varies weakly with P_{abs}, then $\tilde{V}_{sh} \propto \tilde{V}_{max}$.) However, in the same range of sheath voltage and pressure, ohmic heating is more important in the helical resonator than in a capacitive rf discharge, due to an additional (here, axial) component of electric field.

PROBLEMS

12.1. Skin Depth   Consider a uniform electric field,

$$E_z(x, t) = \operatorname{Re} \tilde{E}_z(x)\, e^{j\omega t}$$

at the surface of a half-space $x > 0$ of plasma having dielectric constant $\epsilon_p = \epsilon_0 \kappa_p$ given by (12.1.2).

(a) Using Maxwell's equations (2.2.1) and (2.2.2) with $\mathbf{J} = j\omega\epsilon_p \mathbf{E}$ and with $\mathbf{E} = \hat{z} E_z$ in the form given above, show that

$$\frac{d^2 \tilde{E}_z}{dx^2} = -\frac{\omega^2}{c^2} \kappa_p \tilde{E}_z.$$

(b) Obtain the solution for $\tilde{E}_z(x)$ with the boundary conditions that $\tilde{E}_z = E_0$ at $x = 0$ and that $\tilde{E}_z$ is noninfinite as $x \to \infty$, and show that the electric field magnitude $|\tilde{E}_z(x)|$ decays exponentially into the plasma with a decay constant

$$\alpha = \frac{\omega}{c} \operatorname{Im} \kappa_p^{1/2}$$

(c) Evaluate α in the two limits $\nu_m \ll \omega$ and $\nu_m \gg \omega$, thus verifying (12.1.3) and (12.1.5).

12.2. Power Dissipation   Starting from the basic expression for power dissipation,

$$P_{abs} = \frac{1}{2} \int \mathbf{J} \cdot \mathbf{E}^* \, d\mathcal{V}$$

and with $|\mathbf{E}|$ decaying exponentially into the plasma with a decay constant α given by (12.1.3), with skin depth $\delta_p \ll R$, and with an effective collision frequency ν_{eff}, then obtain expression (12.1.10) for the absorbed power P_{abs}.

12.3. Self- and Mutual Inductance of Concentric Solenoids   Consider two concentric solenoids of length l. The outer solenoid has $\mathcal{N}_1$ turns at radius b, and the inner solenoid has $\mathcal{N}_2$ turns at radius R. The elements of the inductance matrix are defined as

$$\Phi_1 = L_{11}I_1 + L_{12}I_2$$
$$\Phi_2 = L_{21}I_1 + L_{22}I_2$$

where Φ_i is the total magnetic flux linking the $\mathcal{N}_i$ turns of solenoid i and I_i is the feed current. The magnetic induction inside a solenoid having $\mathcal{N}_i$ turns each carrying a current I_i is uniform and given by $B_{zi} = \mu_0 \mathcal{N}_i I_i / l$. Using this and the above definition, for $\mathcal{N}_1 = \mathcal{N}$ and $\mathcal{N}_2 = 1$, obtain (12.1.16)–(12.1.18) for the elements of the inductance matrix.

12.4. Inductive Discharge Equilibrium

(a) Verify all calculations for the example of inductive discharge equilibrium given in Section 12.1.

(b) Estimate the electron drift velocity v_e within the skin depth layer, compare $\mathcal{E}_e = \frac{1}{2}mv_e^2$ to T_e, and comment on the validity of the global ionization model (10.2.7) for these discharge parameters.

12.5. Discharge Equilibrium at High Pressure   For the same $R, b, l, \mathcal{N}, f, P_{abs}$, as in the example, but with a higher pressure $p = 50$ mTorr, find all the equilibrium discharge parameters.

12.6. Discharge Equilibrium With Anomalous Skin Depth   For the same R, b, l, $\mathcal{N}, p$, and P_{abs} as in the example, but with a lower frequency $f = 2$ MHz, find all the equilibrium discharge parameters. Assume that $\bar{v}_e / 2\delta_e \gg \omega, \nu_m$.

12.7. Discharge Equilibrium and Matching Network

(a) Verify all calculations for the values of the matching network capacitors C_1 and C_2 given at the end of Section 12.1.

(b) Suppose P_{abs} is increased from 600 to 1200 W in the example given in Section 12.1, with $R, b, l, \mathcal{N}, p$, and f remaining the same. Find all the equilibrium discharge parameters.

(c) For part (b), determine values of C_1 and C_2 to match the discharge to a 50-Ω rf power source, using the procedure given at the end of Section 12.1.

12.8. Minimum Current for an Inductive Discharge   For the same $R, b, l, \mathcal{N}, p$, and f as given in the example of Section 12.1, use (12.2.3) to determine the minimum rf current amplitude I_{min} to sustain an inductive discharge.

CHAPTER 13

WAVE-HEATED DISCHARGES

Waves generated near a plasma surface can propagate into the plasma or along the surface where they can be subsequently absorbed, leading to heating of plasma electrons and excitation of a discharge. For electron cyclotron resonance (ECR) discharges, described in Section 13.1, a right circularly polarized wave propagates along the dc magnetic field lines to a resonance zone, where the wave energy is absorbed by a collisionless heating mechanism. ECR discharges are generally excited at microwave frequencies (e.g., 2450 MHz), and the wave absorption requires application of a strong dc magnetic field (875 G at resonance). The aspect ratio of these discharges, l/R for a plasma cylinder, can range from $l/R \ll 1$ to $l/R \gg 1$. For helicon discharges, described in Section 13.2, a *whistler wave* launched by an antenna propagates along a plasma column and is subsequently absorbed by a collisional or collisionless mechanism, resulting in heating of the bulk plasma electrons. Helicon discharges are usually excited at rf frequencies (e.g., 13.56 MHz), and a weak magnetic field (20–200 G) is required for wave propagation and absorption. The aspect ratio ranges from $l/R \sim 1$ to $l/R \gg 1$. For surface wave discharges, described in Section 13.3, a wave launched along the surface of the plasma propagates and is absorbed by collisional heating of the plasma electrons near the surface. The heated electrons subsequently diffuse into the bulk plasma. Surface wave discharges can be excited by either rf or microwave sources and do not require dc magnetic fields, but generally a long propagation distance is needed for efficient wave absorption, leading to discharges with high aspect ratios, $l/R \gg 1$. In contrast to capacitive rf discharges (see Chapter 11), wave-heated discharges share with inductive discharges (see Chapter 12) the characteristic that the potential of the plasma with respect to all wall surfaces is low, of order $5\,T_e$.

As was shown in Section 10.1, this leads to high-density plasmas at reasonable absorbed power levels. A brief description of some of the characteristics of wave-heated discharges was given in Chapter 1.

13.1 ELECTRON CYCLOTRON RESONANCE DISCHARGES

Characteristics and Configurations

Microwave generation of plasmas has been employed since the invention of high-power microwave sources in World War II. At low plasma densities, the high electric fields obtainable in a resonant microwave cavity can break down a low-pressure gas and sustain a discharge. For good field penetration in the absence of a magnetic field, $\omega_{pe} \lesssim \omega$, which sets a critical density limit $n_c \lesssim \omega^2 \epsilon_0 m / e^2$, or, in practical units, $n_c (\mathrm{m}^{-3}) \lesssim 0.012 f^2$, with f in Hz. More restrictively, for the high fields required, the cavity Q must be high, further limiting the range of operation.

The introduction of a steady magnetic field **B**, in which there is a resonance between the applied frequency ω and the electron cyclotron frequency $\omega_{ce} = eB/m$ somewhere within the discharge, allows operation at high density and without a cavity resonance. Because of the cyclotron resonance, the gyrating electrons rotate in phase with the right-hand circularly polarized (RHP) wave, seeing a steady electric field over many gyro-orbits. Thus the high field of the cavity resonance, acting over a short time, is replaced by a much lower field, but acting over a much longer time. The net result is to produce sufficient energy gain of the electrons to allow ionization of the background gas. Furthermore, the injection of the microwaves along the magnetic field, with $\omega_{ce} > \omega$ at the entry into the discharge region, allows wave propagation to the absorption zone $\omega_{ce} \approx \omega$, even in a dense plasma with $\omega_{pe} > \omega$ ($n_0 > n_c$).

Figure 13.1a shows a typical high aspect ratio, i.e., $l > R$, ECR system, with the microwave power injected along the magnetic field lines. The power at frequency $f = \omega/2\pi$ is coupled through a vacuum end window into a cylindrical metal source chamber, which is often lined with a dielectric to minimize metal contamination resulting from wall sputtering. One or several magnetic field coils are used to generate a nonuniform, axial magnetic field $B(z)$ within the chamber. The magnetic field strength is chosen to achieve the ECR condition, $\omega_{ce}(z_{res}) \approx \omega$, where z_{res} is the axial resonance position. When a low-pressure gas is introduced, the gas breaks down and a discharge forms inside the chamber. The plasma streams or diffuses along the magnetic field lines into a process chamber toward a wafer holder. Energetic ions and free radicals generated within the entire discharge region (source and process chambers) impinge on the wafer. A magnetic field coil at the wafer holder is often used to modify the uniformity of the etch or deposition process.

Typical parameters for ECR discharges used for semiconductor materials processing are shown in the last column of Table 1.1. The electron cyclotron frequency $f_{ce}(\mathrm{MHz}) \approx 2.8B$, with B in gauss. For $f_{ce} = f = 2450$ MHz, we obtain a resonant magnetic field $B_{res} \approx 875$ G. A typical source diameter is 15 cm.

In some cases, there are multiple resonance positions, as shown by the heavy dashed line in Fig. 13.1b. A uniform profile can be used only for a low aspect ratio

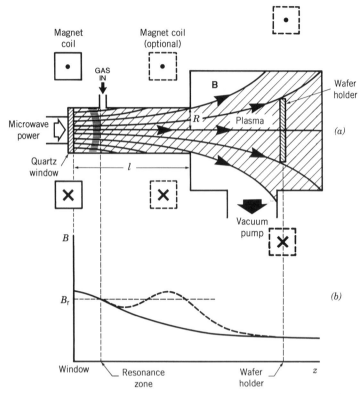

FIGURE 13.1. A typical high-profile ECR system: (a) geometric configuration; (b) axial magnetic field variation, showing one or more resonance zones. (From "Design of High-Density Sources for Materials Processing" from the work "Physics of Thin Films," Vol.18, by Academic Press, Inc., Publisher in Press)

system ($l \lesssim R$), where the substrate is located near the point of microwave power injection, because of the difficulty of maintaining exact resonance and the possibility of overheating the electrons. The monotonically decreasing profile $dB/dz < 0$ shown as the solid line in Fig. 13.1b, with one resonant zone near the window, is often used. The mirror profile shown as the heavy dashed line in Fig. 13.1b has one resonant zone near the window and two additional zones under the second magnet. This profile can yield higher ionization efficiencies, due to enhanced confinement of hot (superthermal) electrons that are magnetically trapped between the two mirror (high-field) positions. However, the longer length of a two-mirror system leads to enhanced radial diffusion at high pressures and consequently may reduce the plasma density at the substrate.

A typical microwave power system is shown in Fig. 13.2. A dc power supply drives a magnetron or klystron source coupled to the discharge by means of a TE_{10} waveguide transmission system. This consists of a circulator, to divert reflected power to a water-cooled, matched load; a directional coupler, to monitor the transmitted and reflected power; a multiscrew tuner, to match the source to the load through the

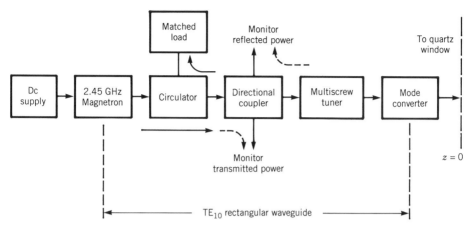

FIGURE 13.2. Typical ECR microwave system. (From "Design of High-Density Sources for Materials Processing" from the work "Physics of Thin Films," Vol.18, by Academic Press, Inc., Publisher in Press)

dielectric window, achieving a condition of low reflected power; and, often, a mode converter, to convert the TE_{10} linear polarized, rectangular waveguide mode to an appropriate mode in the cylindrical source chamber.

The simplest mode converter (Fig. 13.3a) is from TE_{10} rectangular to TE_{11} circular mode. At 2450 MHz, the minimum source chamber diameter for TE_{11} mode propagation (in vacuum) is 7.18 cm. However, the electric field profile and corresponding power flux are peaked on axis and are not azimuthally symmetric for this mode, leading to possible nonaxisymmetric processing profiles on the wafer. A common converter to an axisymmetric mode configuration (Fig. 13.3b) is from TE_{10} rectangular to TM_{01} circular mode, having a minimum diameter for mode propagation of 9.38 cm at 2450 MHz. The profile is ringlike, with a vanishing on-axis power flux. The electric field for both modes is linearly polarized, consisting of equal admixtures of RHP and LHP waves. The basic power-absorption mechanism is the absorption of the RHP wave on a *magnetic beach,* where the wave propagates from higher to lower magnetic field to the resonance $\omega_{ce}(B) \approx \omega$. The fate of the LHP wave is unclear, but it is probably inefficiently converted to a RHP wave due to multiple reflections from waveguide feed or source surfaces, or, more efficiently, from a critical density layer in the source (Musil and Zacek, 1970, 1971). An efficient scheme uses a microwave polarizer to convert from TE_{10} rectangular to a TE_{11} circular mode structure that rotates in the right-hand sense at frequency ω. This yields a time-averaged azimuthally symmetric power profile peaked on axis and having an on-axis electric field that is right-hand polarized. Hence, most of the power can be delivered to the plasma in the form of the RHP wave alone.

There are a variety of ECR processing discharges, with somewhat different coupling of the microwave power to the resonance zone. Three categories are (1) traveling wave propagation mainly along **B** (wave vector $\mathbf{k} \| \mathbf{B}$), (2) propagation mainly across **B** ($\mathbf{k} \perp \mathbf{B}$), and (3) standing wave excitation (mainly cavity coupled). While these

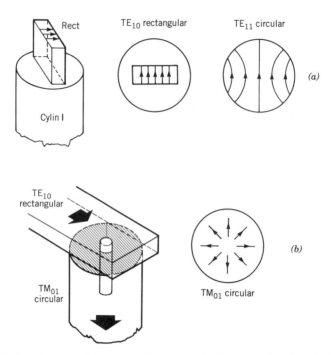

FIGURE 13.3. Microwave field patterns for ECR excitation. (From "Design of High-Density Sources for Materials Processing" from the work "Physics of Thin Films," Vol.18, by Academic Press, Inc., Publisher in Press)

distinctions are significant, most of these ECR sources rely on the magnetic beach absorption of the RHP wave. Additionally, the sources are not neatly broken into these categories; e.g., wave propagation is at an angle to **B**, and absorption can involve standing waves.

Various ECR configurations are shown in Fig. 13.4. A high aspect ratio system with the source plasma far from the wafer and with microwave injection along **B** is shown in Fig. 13.4a. The resonance (heating) zone can be ring- or disk-shaped (the latter is shown) and may be as much as 50 cm from the wafer. Expansion of the plasma from the resonance zone to the wafer reduces the ion flux and increases the ion impact energy at the wafer. Hence high aspect ratio systems have given way to low aspect ratio systems, as shown in Fig. 13.4b, where only a single high-field magnet is used and where the resonance zone is placed within the process chamber and may be only 10–20 cm from the wafer. Uniformity is controlled at least in part by shaping the axial magnetic field. Uniformity can be further improved and density increased by adding 6–12 linear multipole permanent magnets around the circumference of the process chamber, as shown in Fig. 13.4c. Multipole magnetic confinement is described in Section 5.6. As another variation, a strong (rare earth) permanent magnet that generates a diverging axial magnetic field can also replace the source coil. Another approach to achieving adequate uniformity and density is to combine the source and process chambers and place the resonance zone close

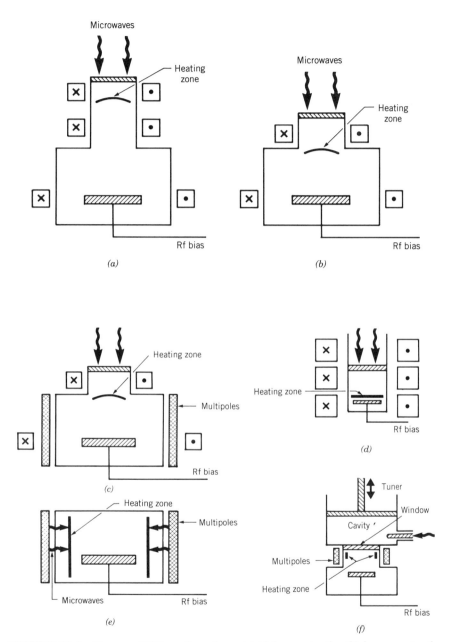

FIGURE 13.4. Common ECR configurations: (a) high aspect ratio; (b) low aspect ratio; (c) low aspect ratio with multipoles; (d) close-coupled; (e) distributed (DECR); (f) microwave cavity excited. (From "Design of High-Density Sources for Materials Processing" from the work "Physics of Thin Films," Vol.18, by Academic Press, Inc., Publisher in Press)

to the wafer, leading to a low aspect ratio *close-coupled* configuration, shown in Fig. 13.4d. Uniformity requirements can be met by using a relatively flat, radially uniform resonance zone.

The multipole, distributed ECR (DECR) system shown in Fig. 13.4e is powered by microwave injection perpendicular to the strong, permanent magnet, multipole magnetic fields. Typically, four or more microwave applicators are arranged around the circumference to achieve adequate uniformity. Each applicator creates an approximately linear resonance zone near the process chamber wall, as shown.

A microwave cavity source is shown in Fig. 13.4f. The coaxial feed is tuned using a sliding short on top and a stub tuner from the side. In earlier, lower-density versions, a grid was used below the plasma generation region providing microwave containment while allowing the plasma to diffuse out. The linear resonance zones, similar to those in the DECR (Fig. 13.4e), are generated by a set of 8–12 strong permanent magnets arranged around the circumference of the source chamber as shown.

Electron Heating

The basic principle of ECR heating is illustrated in Fig. 13.5. A linearly polarized microwave field launched into the source chamber can be decomposed into the sum of two counterrotating circularly polarized waves. Assuming a sinusoidal steady state with the incident wave polarized along $\hat{x}$,

$$\mathbf{E}(\mathbf{r}, t) = \mathrm{Re}\,\hat{x}E_x(\mathbf{r})e^{j\omega t} \qquad (13.1.1)$$

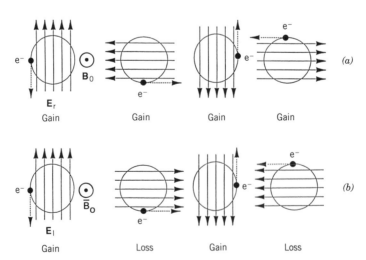

FIGURE 13.5. Basic principle of ECR heating: (a) continuous energy gain for right-hand polarization; (b) oscillating energy for left-hand polarization. (From "Design of High-Density Sources for Materials Processing" from the work "Physics of Thin Films," Vol. 18, by Academic Press, Inc., Publisher in Press)

where the complex amplitude E_x is here taken to be pure real, we have

$$\hat{x}E_x = (\hat{x} - j\hat{y})E_r + (\hat{x} + j\hat{y})E_l \qquad (13.1.2)$$

where $\hat{x}$ and $\hat{y}$ are unit vectors along x and y and where E_r and E_l are the amplitudes of the RHP and LHP waves, with $E_r = E_l = E_x/2$. The electric field vector of the RHP wave rotates in the right-hand sense around the magnetic field at frequency ω while an electron in a uniform magnetic field B_0 also gyrates in a right-hand sense at frequency ω_{ce}. Consequently, as shown in Fig. 13.5a, for $\omega_{ce} = \omega$, the force $-eE$ accelerates the electron along its circular orbit, resulting in a continuous transverse energy gain. In contrast, as shown in Fig. 13.5b, the LHP wave field produces an oscillating force whose time average is zero, resulting in no energy gain.

(a) Collisionless Heating Calculation   To determine the overall heating power, the nonuniformity in the magnetic field profile $B(z)$ must be considered. For $\omega_{ce} \neq \omega$, an electron does not continuously gain energy, but rather its energy oscillates at the difference frequency $\omega_{ce} - \omega$. As an electron moving along z passes through resonance, its energy oscillates as shown in Fig. 13.6, leading to a transverse energy gained (or lost) in one pass. For low power absorption, where the electric field at the resonance zone is known, the heating can be estimated as follows. We expand the magnetic field near resonance as

$$\omega_{ce}(z') = \omega(1 + \alpha z') \qquad (13.1.3)$$

where $z' = z - z_{res}$ is the distance from exact resonance, $\alpha = \partial\omega_{ce}/\partial z'$ is proportional to the gradient in $B(z)$ near the resonant zone, and we approximate $z'(t) \approx v_{res}t$, where v_{res} is the parallel speed at resonance.

The complex force equation for the right-hand component of the transverse velocity, $v_r = v_x + jv_y$, can be written in the form

$$\frac{dv_r}{dt} - j\omega_{ce}(z)v_r = -\frac{e}{m}E_r e^{j\omega t} \qquad (13.1.4)$$

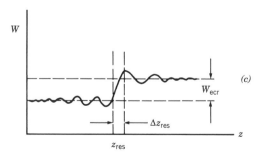

FIGURE 13.6. Energy change in one pass through a resonance zone. (From "Design of High-Density Sources for Materials Processing" from the work "Physics of Thin Films," Vol.18, by Academic Press, Inc., Publisher in Press)

where E_r is the amplitude of the RHP wave with

$$\mathbf{E} = \text{Re}\left[(\hat{x} - j\hat{y})E_r e^{j\omega t}\right] \qquad (13.1.5)$$

Using (13.1.3) and substituting $v_r = \tilde{v}_r \exp(j\omega t)$ into (13.1.4), we obtain

$$\frac{d\tilde{v}_r}{dt} - j\omega \alpha v_{res} t \tilde{v}_r = -\frac{e}{m}E_r \qquad (13.1.6)$$

Multiplying by the integrating factor $e^{-j\theta(t)}$ and integrating (13.1.6) from $t = -T$ to $t = T$, we obtain

$$\tilde{v}_r(T)e^{-j\theta(T)} = \tilde{v}_r(-T)e^{-j\theta(-T)} - \frac{eE_r}{m}\int_{-T}^{T} dt' \, e^{-j\theta(t')} \qquad (13.1.7)$$

where

$$\theta(t) = \omega \alpha v_{res} t^2 / 2 \qquad (13.1.8)$$

In the limit $T \gg (2\pi/\omega|\alpha|v_{res})^{1/2}$, the integral in (13.1.7) is the integral of a Gaussian of complex argument, which has the standard form

$$\int_{-T}^{T} dt' \, e^{-j\theta(t')} = (1 - j)\left(\frac{\pi}{\omega|\alpha|v_{res}}\right)^{1/2} \qquad (13.1.9)$$

Substituting (13.1.9) into (13.1.7), multiplying (13.1.7) by its complex conjugate, and averaging over the initial "random" phase $\theta(-T)$, we obtain

$$|\tilde{v}_r(T)|^2 = |\tilde{v}_r(-T)|^2 + \left(\frac{eE_r}{m}\right)^2 \left(\frac{2\pi}{\omega|\alpha|v_{res}}\right) \qquad (13.1.10)$$

The average energy gain per pass is thus

$$W_{ecr} = \frac{\pi e^2 E_r^2}{m\omega|\alpha|v_{res}} \qquad (13.1.11)$$

This can also be written as

$$W_{ecr} = \frac{1}{2}m(\Delta v)^2 \qquad (13.1.12)$$

where $\Delta v = (eE_r/m)\Delta t_{res}$, and

$$\Delta t_{res} = \left(\frac{2\pi}{\omega|\alpha|v_{res}}\right)^{1/2} \qquad (13.1.13)$$

is the effective time in resonance. The effective resonance zone width (see Fig. 13.6) is

$$\Delta z_{res} \equiv v_{res} \Delta t_{res} = \left(\frac{2\pi v_{res}}{\omega |\alpha|} \right)^{1/2} \tag{13.1.14}$$

which, for typical ECR parameters, gives $\Delta z_{res} \sim 0.5$ cm.

The absorbed power per unit area, or energy flux, is found by integrating (13.1.11) over the flux $n v_{res}$ of electrons incident on the zone, yielding

$$S_{ecr} = \frac{\pi n e^2 E_r^2}{m \omega |\alpha|} \tag{13.1.15}$$

We can understand the form of Δt_{res} as follows. An electron passing through the zone coherently gains energy for a time Δt_{res} such that

$$\left[\omega - \omega_{ce}(v_{res} \Delta t_{res}) \right] \Delta t_{res} \approx 2\pi \tag{13.1.16}$$

Inserting (13.1.3) into (13.1.16) and solving for Δt_{res}, we obtain (13.1.13). A more careful derivation of the absorbed power (13.1.15), including the effect of non-constant v_{res} during passage through resonance, is presented by Jaeger et al. (1972), giving similar results. For both calculations, S_{ecr} is proportional to the density and the square of the RHP electric field amplitude at the resonance. The calculation here also gives S_{ecr} proportional to α^{-1} and independent of the axial electron velocity, which are not true for nonconstant v_{res}.

(b) Collisional Heating Calculation   The fact that (13.1.15) is independent of v_{res} suggests that we can examine the $v_{res} \rightarrow 0$ limit in considering the effects of electron collisions. Adding collisional (friction) terms $-\nu_m \tilde{v}_x$ and $-\nu_m \tilde{v}_y$ to the force equations (4.4.1*a*) and (4.4.1*b*), respectively, and solving for the transverse velocity amplitudes, we obtain

$$\tilde{v}_x + j\tilde{v}_y = -\frac{2eE_r}{m} \frac{1}{\nu_m + j(\omega - \omega_{ce})} \tag{13.1.17}$$

where ν_m is the electron momentum transfer frequency. The time-average power absorbed per electron is

$$\bar{p}_{ecr} = \frac{1}{2} \text{Re} \left(-e\tilde{E}_x \tilde{v}_x^* - e\tilde{E}_y \tilde{v}_y^* \right) \tag{13.1.18}$$

Substituting (13.1.17) into (13.1.18) with $\tilde{E}_x = E_r, \tilde{E}_y = -jE_r$, we find

$$\bar{p}_{ecr} = m \left(\frac{eE_r}{m} \right)^2 \frac{\nu_m}{\nu_m^2 + (\omega - \omega_{ce})^2} \tag{13.1.19}$$

For $\omega_{ce} \rightarrow \omega$, we see that

$$\bar{p}_{ecr} \rightarrow \frac{e^2 E_r^2}{m \nu_m}$$

This implies a singular behavior $\bar{p}_{ecr} \rightarrow \infty$ as $\nu_m \rightarrow 0$. However, this behavior is found only at exact resonance. To obtain the total heating power, we average (13.1.19) over the distribution of electrons near the resonance zone. Substituting the linear expansion (13.1.3) in (13.1.19), we obtain

$$\bar{p}_{ecr} = \frac{e^2 E_r^2}{m} \frac{\nu_m}{\nu_m^2 + \omega^2 \alpha^2 z'^2} \tag{13.1.20}$$

Multiplying (13.1.20) by $n \, dz$ and integrating from $z = -z_0$ to $z = z_0$, we obtain

$$\bar{S}_{ecr} = \frac{2 e^2 E_r^2 n}{m \omega |\alpha|} \tan^{-1} \left(\frac{\omega |\alpha| z_0}{\nu_m} \right) \tag{13.1.21}$$

The total power absorbed is obtained by letting $z_0 \rightarrow \infty$ such that $\tan^{-1} \rightarrow \pi/2$ and (13.1.21) reduces to (13.1.15). We see that the power absorbed is independent of ν_m for constant electric field, and the nonlinear collisionless and the collisional power absorption calculations correspond. If we insert $z_0 \equiv \Delta z_{res}$, from (13.1.14) into (13.1.21), we find that, since $\nu_m \ll \omega |\alpha| \Delta z_{res}$, almost all of the power is absorbed by collisionless heating within the resonance zone. This is the usual regime for ECR processing discharges.

Resonant Wave Absorption

A serious limitation on the result (13.1.15) is that it assumes that the electric field within the resonance zone is constant and known from the input power. That this cannot be true in the case of strong absorption is clear, since the absorbed power cannot exceed the incident power. The resolution of this difficulty lies in the attenuation of the wave in the resonance zone, so that the resonant value of E_r is in fact much smaller than the value of the incident E_r.

The propagation and absorption of microwave power in ECR sources is an active area of research and is not fully understood. For excitation at an end window (Figs. 13.4a–d), the waves in a cylindrical magnetized plasma are neither exactly RHP nor propagating exactly along **B**. The waves are not simple plane waves and the mode structure in a magnetized plasma of finite dimension must be considered. Nevertheless, the essence of the wave coupling, and transformation and absorption at the resonance zone, can be understood by considering the one-dimensional problem of an RHP plane wave propagating strictly along **B** in a plasma that varies spatially only along z. For right-hand polarization (13.1.5), where now $E_r(z)$ is the spatially varying electric field amplitude, the wave equation for plane waves propagating along

B parallel to z can be written as (Problem 13.1)

$$\frac{d^2 E_r}{dz^2} + k_0^2 \kappa_r E_r = 0 \tag{13.1.22}$$

Far from resonance such that $\omega - \omega_{ce} \gg \nu_m$, we have the relative dielectric constant (4.5.5a),

$$\kappa_r = 1 - \frac{\omega_{pe}^2(z)}{\omega\left[\omega - \omega_{ce}(z)\right]} \tag{13.1.23}$$

with $k_0 = \omega/c$ and c the velocity of light. κ_r varies with z due to the dependence of ω_{pe}^2 on the density $n(z)$ and of ω_{ce} on the magnetic field $B(z)$. If the variation of κ_r with z is weak,

$$\frac{d\lambda}{dz} \ll 1 \tag{13.1.24}$$

with $\lambda = 2\pi/k$, and

$$k(z) = k_0 \kappa_r^{1/2}(z) \tag{13.1.25}$$

then a Wentzel–Kramers–Brillouin (WKB) wave expansion can be made (Stix, 1992):

$$E_r(z) = E_{r1}(z) \exp\left[-j \int^z k(z')\, dz'\right] \tag{13.1.26}$$

where $E_{r1}(z) = E_{r0} k_0^{1/2}/k^{1/2}(z)$ is the spatially varying amplitude of the wave. The WKB wave propagates without absorption for $\kappa_r > 0$, where k is real, and the wave is evanescent for $\kappa_r < 0$, where k is imaginary. The WKB result can be understood from a calculation of the time averaged power per unit area carried by the wave,

$$\bar{S}_r = \frac{1}{2} Z_0^{-1} \kappa_r^{1/2} E_{r1}^2 = \text{const} \tag{13.1.27}$$

where $Z_0 = (\mu_0/\epsilon_0)^{1/2} \approx 377\,\Omega$ is the impedance of free space. This indicates that the propagation is without reflection. The result is characteristic of slowly varying solutions called *adiabatic*. We note that for propagation close to cyclotron resonance, from (13.1.23), κ_r becomes large, as does k, and from (13.1.26), E_{r1} becomes small. However, as resonance is approached from the propagating side ($\omega_{ce} > \omega$), the condition of slow spatial variation (13.1.24) is no longer satisfied, and the WKB approximation breaks down.

What happens as a wave propagates through the resonance into an evanescent region where $\omega_{ce} < \omega$? The answer was obtained analytically by Budden (1966), for the approximation of constant density and linear magnetic field. Reintroducing collisions into the plasma dielectric constant, (13.1.22) becomes

$$\frac{d^2 E_r}{dz^2} + k_0^2 \left\{ 1 - \frac{\omega_{pe}^2(z)}{\omega \left[\omega - \omega_{ce}(z) - j\nu_m(z) \right]} \right\} E_r = 0 \tag{13.1.28}$$

Taking ω_{pe} and ν_m to be constants independent of z, and linearizing ω_{ce} about the resonance point, (13.1.28) reduces to

$$\frac{d^2 E_r}{ds^2} + \left(1 + \frac{\eta}{s + j\gamma} \right) E_r = 0 \tag{13.1.29}$$

where we have normalized z by $s = k_0(z - z_{res})$, ω_{pe}^2 by $\eta = \omega_{pe}^2/(\omega c |\alpha|)$, and ν_m by $\gamma = \nu_m/(c|\alpha|)$. The dielectric function has both a pole and a zero, with the pole, in the absence of collisions, occurring at $s = 0$ ($z = z_{res}$) and the zero at $s = -\eta$. In this approximation, Budden has obtained a solution to (13.1.29) in the limit of $\gamma \to 0$. For a wave traveling into a decreasing magnetic field (the magnetic beach), he obtained

$$\frac{S_{abs}}{S_{inc}} = 1 - e^{-\pi\eta} \tag{13.1.30}$$

$$\frac{S_{trans}}{S_{inc}} = e^{-\pi\eta} \tag{13.1.31}$$

$$\frac{S_{refl}}{S_{inc}} = 0 \tag{13.1.32}$$

Hence some of the wave power is absorbed at the resonance while some tunnels through to the other side, but no power is reflected. Taking a typical case for which $\alpha = 0.1$ cm^{-1} and $k_0 = 0.5$ cm^{-1}, we find that $\eta > 1$ corresponds to $\omega_{pe}^2/\omega^2 > 0.2$. Thus at 2450 MHz we expect that most of the incident power will be absorbed for a density $n_0 \gtrsim 1.5 \times 10^{10}$ cm^{-3}.

The Budden result can be qualitatively understood in terms of the dispersion diagrams of k versus ω_{ce}/ω, as shown in Fig. 13.7, with $\omega_{pe}^2 = $ const. In Fig. 13.7a, for low density ($\omega_{pe} < \omega$), the wave is evanescent downstream of the resonance in the region

$$1 - \frac{\omega_{pe}^2}{\omega^2} < \frac{\omega_{ce}}{\omega} < 1$$

and is propagating otherwise. For $\omega_{pe} \ll \omega$, the region of evanescence is thin (in z), and the wave can tunnel through this region to propagate again further downstream. In Fig. 13.7b, for high density ($\omega_{pe} > \omega$), the wave is always evanescent downstream such that the tunneling fields fall off exponentially. Within Budden's approximation, we have the nonintuitive result that there is no reflected power for a wave incident on the resonance zone from the high-field side.

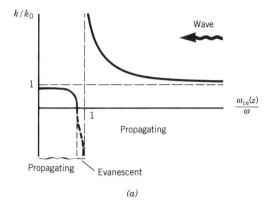

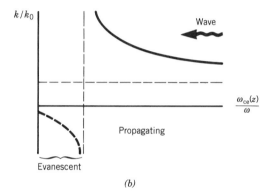

FIGURE 13.7. k/k_0 versus ω_{ce}/ω for (a) low density $\omega_{pe}/\omega \ll 1$ and (b) high density $\omega_{pe}/\omega \gg 1$. The heavy dashed curves denote imaginary values for k.

Since from (10.2.14) and (10.2.15), with the pressure dependence of h_l and h_R determined from (10.2.3) and (10.2.4), the bulk density scales as $n_0 \propto S_{abs}$ at low pressures and as $n_0 \propto p^{1/2}S_{abs}$ at high pressures, we obtain from (13.1.30) the region of good power absorption $\eta \gtrsim 1$, as sketched in Fig. 13.8. For parameters well within this region, the incident microwave power is efficiently absorbed over the entire cross section of the resonance zone. For operation outside this region, considerable microwave power can impinge on the substrate.

The minimum S_{inc} to sustain an ECR discharge can similarly be found. Expanding (13.1.30) for small η (n_0 small) yields $S_{abs} = \pi\eta S_{inc}$. Substituting this into (10.2.14), we obtain the minimum value of S_{inc} to sustain the discharge. At a given pressure, this minimum is found to be a factor of two below the $\eta = 1$ condition for good power absorption, as illustrated in Fig. 13.8. The situation is analogous to the case for an inductive discharge (see Fig. 12.4 and accompanying discussion), in which there is a tangency between power absorbed and power lost versus density, at low density. As in the inductive discharge, the plasma does not turn off, but has a transition to a considerably lower density state (sometimes referred to as a "low mode"). This can

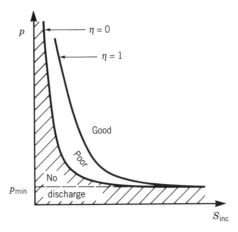

FIGURE 13.8. Parameters for good ECR source operation: pressure p versus incident power P_{inc}. (From "Design of High-Density Sources for Materials Processing" from the work "Physics of Thin Films," Vol.18, by Academic Press, Inc., Publisher in Press)

be qualitatively understood in terms of cavity resonance effects (Williamson et al., 1992). We should also note that the discharge cannot be sustained if the pressure drops below some minimum value p_{min}, because the particle balance equation (10.2.7) has no solution for T_e. This limit is also illustrated in Fig. 13.8.

In a number of respects, the Budden theory is rather idealized for direct application to a physical system. The reflections in a plasma chamber generate interference of waves that can significantly affect the absorption. The variation of axial density causes initial upstream power reflection. The collisionality is locally enhanced by nonlinear absorption of power in the resonance zone. These effects can be taken into account in a one-dimensional model by numerical integration of the fundamental equation (13.1.28) with boundary conditions imposed at each end of the region of interest (Williamson et al., 1992). The results indicate that the Budden theory holds reasonably well for strong absorption.

The size, shape, and location of the resonant zone are set by the magnet coil configuration and the magnet currents. The zone shape and location are also modified by the Doppler effect for electrons incident on the zone. The actual resonance position is determined by the Doppler-shifted frequency,

$$\omega + k v_{res} = \omega_{ce}(z_{res}) \tag{13.1.33}$$

At high densities, from (13.1.25), k can be large near the zone, leading to a large Doppler shift. For example, for $k = 6.3 \text{ cm}^{-1}$ ($\lambda = 1$ cm), a typical value at the edge of the resonance zone, and $v_{res} = 10^8$ cm/s (a 3 V electron), we obtain $k v_{res}/\omega \approx 0.094$. Hence the resonant magnetic field is 910 G for this electron and not 875 G. For $\alpha = 0.1 \text{ cm}^{-1}$, this leads to a broadening of the zone of ± 0.4 cm. By using a coaxial electrostatic probe to sample the microwave field in an ECR discharge and beating that signal against a reference signal from the incident microwaves, Stevens et al.

(1992) have measured the microwave field amplitude as a function of position and verified that the resonant zone is Doppler broadened, with the absorption beginning at ~ 975 G for their case.

Radial density and magnetic field variations can lead to wave refraction effects that are significant. The radial gradients are generally much larger than the axial gradients in high aspect ratio ($l/R \gg 1$) sources. A radial density profile that is peaked on axis leads to a dielectric constant κ_r that is peaked on axis. This in turn can lead to a self-focusing effect that can increase the sharpness of the microwave power profile as the wave propagates to the zone, adversely affecting uniformity. The mechanism is analogous to the use of a graded dielectric constant optical fiber to guide an optical wave. However, the ECR refraction problem is much more complicated because the density profile is not known a priori and the magnetized plasma medium cannot be represented as an isotropic dielectric. A simplified picture of the refraction is obtained in the geometrical optics limit by examining the trajectories of optical rays as they propagate. The ray dynamics are derivable from the dispersion equation and have a Hamiltonian form (Born and Wolf, 1980, Appendix II), with $(k_\perp, r)$ and (k_z, z) canonically conjugate variable pairs and with $\omega(k_\perp, k_z, r, z)$ the Hamiltonian. For high densities and magnetic fields ($\omega_{pe}, \omega_{ce} \gg \omega$) and propagation at an angle to the magnetic field, the dispersion equation reduces to that of whistler waves (see Problem 4.12), with

$$\omega = \frac{k_0^2 \omega_{pe}^2}{k k_z \omega_{ce}} \tag{13.1.34}$$

where $k = (k_\perp^2 + k_z^2)^{1/2}$ is the wave-vector magnitude and $k_\perp$ and k_z are the radial and axial components. Choosing $\omega_{pe}^2/\omega_{ce}$ to have radial variation only, independent of z, Hamilton's equations are

$$\frac{dk_\perp}{dt} = -\frac{\partial \omega}{\partial r}$$

$$\frac{dr}{dt} = \frac{\partial \omega}{\partial k_\perp}$$

$$\frac{dk_z}{dt} = -\frac{\partial \omega}{\partial z} \equiv 0$$

$$\frac{dz}{dt} = \frac{\partial \omega}{\partial k_z}$$

They show that k_z is conserved along the path of a ray. If $\omega_{pe}^2/\omega_{ce}$ is a decreasing function of r, then (13.1.34) shows that $k_\perp$ decreases with increasing r, implying that the ray bends toward the axis, a focusing action. On the other hand, for some parameter choices, e.g., $\omega_{pe} \sim \omega \sim \omega_{ce}$, a refraction of the wave away from the axis has been found by numerical integration of the ray equations, leading, for this particular case, to an increased uniformity of the power flux profile (Stevens et al., 1992).

For some source concepts (e.g., DECR in Fig. 13.4e), the microwave power is injected perpendicular to the magnetic field, and not parallel to the field. In this case, the feed structure excites the extraordinary (x) wave (see Section 4.5), which in the WKB limit has a resonance at the upper hybrid frequency (4.5.15), $\omega_{UH} = (\omega_{pe}^2 + \omega_{ce}^2)^{1/2}$, where the wave power is absorbed. Since ω_{UH} depends on both ω_{pe} and ω_{ce}, we see that the shape and location of the resonance zone depends on the density as well as the magnetic configuration. Furthermore, the x wave is evanescent for frequencies such that $\omega_{UH} < \omega < \omega_R$, where ω_R is given by (4.5.11). For a fixed driving frequency $\omega > \omega_{ce}$, the x wave must tunnel through this evanescent layer on its journey from the feed structure to the zone. For $\omega_{pe} > \omega$, the tunneling becomes rapidly small with increasing density and the wave cannot propagate to the zone. This can limit the density obtainable in these sources to of order 2×10^{12} cm^{-3} at 2450 MHz. However, the limitation is not severe for typical processing applications. Microwave cavity sources (Fig. 13.4f) can suffer from similar limitations. On the other hand, densities as high as 3×10^{13} cm^{-3} have been generated using 2450-MHz RHP wave injection along **B**.

Model and Simulations

The spatially averaged model of the discharge equilibrium described in Section 10.2 can be applied to determine the plasma parameters for a given geometry, magnetic field, pressure, and absorbed power. For a cylindrical plasma of radius R and length l with a strong axial magnetic field, particle balance (10.2.7) can be used to determine T_e, and power balance (10.2.14) then determines n_0. The procedure was described in Example 2 of Section 10.2 for the case appropriate to ECR discharges where particle loss to the walls is inhibited by the strong field. In this case, d_{eff} in (10.2.8) is $l/2h_l$, A_{eff} in (10.2.15) is $2\pi R^2 h_l$, and the ion-bombarding energy is $\mathcal{E}_i \approx V_s + \frac{1}{2}T_e$, with V_s given by (10.2.9).

Wu et al. (1994) have compared results from a two-dimensional simulation of plasma transport in a low aspect ratio ($l/R \lesssim 1$) ECR source with predictions from the spatially averaged model. A two-dimensional hybrid simulation was used. The electrons were treated as a fluid, with the particle, momentum, and energy conservation equations (2.3.7), (2.3.9), and (2.3.21) coupled by the electric field in two dimensions (r, z) to the motion of the ions. These were treated as a collection of particles acted on by the Lorentz force (2.2.12), along with Monte Carlo collisions against the background neutral gas. Argon gas with a simplified set of cross sections, similar to that described in Chapter 3, was used. The source geometry and magnetic field lines are shown in Fig. 13.9. At $P_{abs} = 850$ W, the spatially averaged electron temperature, ion-bombarding energy at the endwall $z = 21.5$ cm, and density are shown versus the pressure p for the simulation and the model in Figs. 13.10a, b, and c, respectively. The model and simulation agree to within about 10% over the pressure range 0.5–10 mTorr and power range 850–1500 W. The model provides insight into the discharge behavior and scaling with control parameters, while the simulation provides spatial profile information of the plasma density and the ion-bombarding energy and flux at the substrate surface.

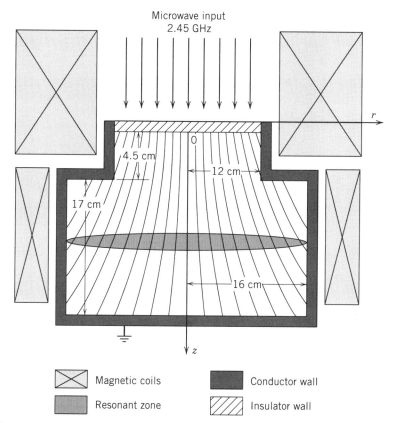

FIGURE 13.9. Schematic of ECR configuration used to compare model with hybrid simulation.

Plasma Expansion

In a high aspect ratio system, $l \gg R$, where the plasma flows from a small-diameter source chamber into a larger-diameter process chamber along the diverging magnetic field lines, the plasma density n_0 in the source chamber can considerably exceed the density n_s in the process chamber where the substrate is located. In this case, illustrated in Fig. 13.11, a *distributed potential* V_d exists between the source and process chambers, with V_d related to n_0 and n_s by the Boltzmann factor,

$$n_s = n_0 \, e^{-V_d/T_e} \tag{13.1.35}$$

This dc potential acts to accelerate ions from the source exit to the plasma–sheath edge near the substrate. As the ions cross the sheath, they are further accelerated by the wall sheath potential V_s. The drop in density in a high aspect ratio system is due to the expansion in the area of the plasma as it flows along field lines, to the increase in ion velocity at a fixed flux, and to particle loss by radial diffusion to the walls.

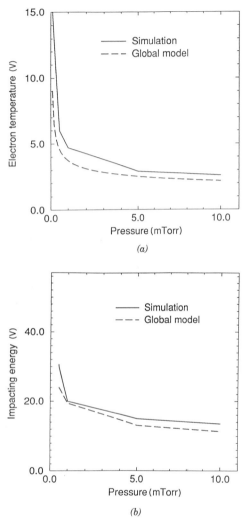

FIGURE 13.10. Comparison between spatially averaged model and hybrid simulation predictions of (a) electron temperature and (b) ion impact energy for $P_{abs} = 850$ W (Wu et al., 1994).

An estimate of the potentials can be made in the collisionless (very low pressure) limit, ignoring radial diffusion, using the model in Fig. 13.11. The assumptions are that ions are generated only within the source chamber and flow out of the source with a characteristic velocity $u_i \sim T_e \ln(n_{max}/n_0) \sim u_B$, the Bohm velocity, where the magnetic field is B_0 and the cross-sectional area is A_0. The ions flow along the magnetic field lines as the magnetic field decreases, such that the cross-sectional area expands. By conservation of magnetic flux, we have

$$A_s = A_0 \frac{B_0}{B_s} \tag{13.1.36}$$

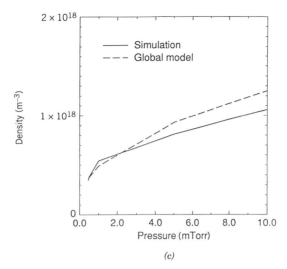

FIGURE 13.10. *(continued)* Comparison between spatially averaged model and hybrid simulation predictions of (c) plasma density versus neutral gas pressure, for $P_{abs} = 850$ W (Wu et al., 1994).

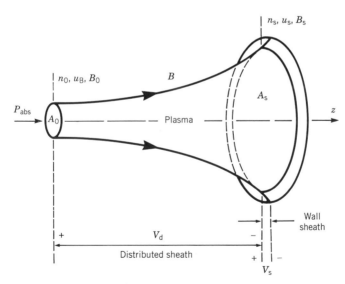

FIGURE 13.11. Model used to calculate the distributed potential V_d and the sheath potential V_s in a diverging field ECR system.

The ion particle and energy balance equations are

$$n_s u_s A_s = n_0 u_B A_0 \tag{13.1.37}$$

$$\frac{1}{2} M u_s^2 = \frac{1}{2} M u_B^2 + e V_d \tag{13.1.38}$$

Equations (13.1.35)–(13.1.38) can be solved numerically to determine V_d/T_e, u_s/u_B, and n_s/n_0 as functions of the area expansion ratio A_s/A_0. The results for V_d/T_e are shown in Fig. 13.12. The further potential drop across the wall sheath can then be found by equating ion and electron fluxes at the wall. With

$$\Gamma_i = n_s u_s \tag{13.1.39}$$

$$\Gamma_e = \frac{1}{4} n_s \bar{v}_e \, e^{-V_s/T_e} \tag{13.1.40}$$

and setting $\Gamma_i = \Gamma_e$, we obtain V_s/T_e as shown in Fig. 13.12. The total potential drop $V_T = V_d + V_s$ from the source to the wall is also shown in the figure.

The ion-bombarding energy for a collisionless ion flow from the source to the wall is

$$\mathcal{E}_i = V_d + V_s + \frac{1}{2} T_e \tag{13.1.41}$$

where the last term is the initial ion energy at the source exit. Figure 13.12 shows that $\mathcal{E}_i \approx 5$–$8\, T_e$ over a wide range of area expansion ratios. However, the ion flow across the distributed potential is collisionless only at very low pressures, such that

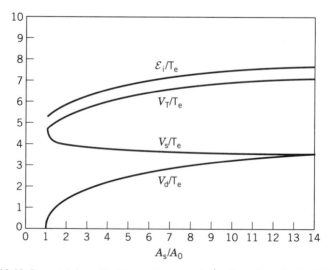

FIGURE 13.12 Potential drops V_T, V_s, and V_d versus A_s/A_0 for a diverging field ECR system.

$\lambda_i \gtrsim l_d$, where l_d is the length of the distributed potential region (see Fig. 13.11). For $\lambda_i < l_d$, the more usual pressure regime, the ion energy is modified to

$$\mathcal{E}_i \approx \frac{\lambda_i}{l_d} V_d + V_s + \frac{1}{2} T_e \tag{13.1.42}$$

The first term in (13.1.42) is reduced below V_d because ion energy is lost during charge transfer and elastic collisions in the expansion region. The situation is similar to that described in Section 11.2 for the ion-bombarding energy of the collisionless and collisional sheaths in a capacitive rf discharge. For $\lambda_i \ll l_d$, V_s reduces to the usual sheath voltage (10.2.9), and the first term in (13.1.42) is negligible. Hussein and Emmert (1990) have given a more complete description of the potential drops and ion-bombarding energies in a diverging magnetic field ECR system.

Measurements

Most measurements of ion energy distributions in high-density sources have been done for diverging field systems. The work of Matsuoka and Ono (1988) is typical. Microwaves are launched from a cavity into a high magnetic field region so that the RHP wave propagates and then is absorbed, heating electrons in the process. Because the magnetic field continues to decrease, the plasma expands, the plasma density decreases, and an ambipolar field is created that accelerates ions along the magnetic field gradient. At some point downstream, ions are sampled through a 50-μm pinhole and energy analyzed using two grids and a collector. Although the relatively large orifice diameter and the use of arbitrary units for spatial distance makes the work of mostly qualitative value, the trends are notable and are borne out in many other experiments.

Matsuoka and Ono focused primarily on the effects of magnetic field configuration and pressure. By varying the dc current i_m in an electromagnet located near the sampling plane, they modified the divergence of the magnetic field, varying the field from a mirror at high current to a cusp at negative current ($i_c = -i_m$). Figure 13.13 shows their ion energy distributions for different currents i_m. As the field is made uniform (large i_m), the parallel ion temperature T_i (spread in $N(\mathcal{E}_i)$) decreases and the average energy $\bar{\mathcal{E}}_i$ shifts to lower values. At the same time, the ion current density increases, the plasma potential (deduced from Langmuir probe current–voltage characteristics) decreases, and the plasma potential gradient or electric field decreases. These effects are all consistent with reduced plasma expansion. By contrast, the largest $\bar{\mathcal{E}}_i$ and T_i are obtained when the subcoil magnet is used to produce a cusp (highly diverging field) before the sampling orifice. Under these conditions, the plasma expansion is largest as the magnetic field decreases to zero and then reverses on the other side of the cusp. Note that ions and electrons do not follow field lines through a cusp since the field decreases to zero. Regardless of the magnetic field configuration, both $\bar{\mathcal{E}}_i$ and T_i decrease as the pressure is increased and charge exchange cools the ions.

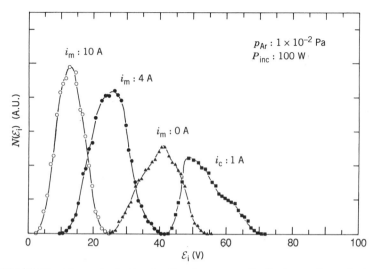

FIGURE 13.13. Change in the bombarding ion energy distribution as the wafer-level coil current i_m is varied (Matsuoka and Ono, 1988).

13.2   HELICON DISCHARGES

Helicon generation of plasmas was first employed by Boswell (1970), following a ten-year history of helicon propagation studies, first in solid-state and then in gaseous plasmas. A detailed theory of helicon propagation and absorption is given by Chen (1991). Helicons are propagating whistler wave modes in a finite diameter, axially magnetized plasma column, with dispersion as given in (13.1.34) (see also Problem 4.12). The electric and magnetic fields of the modes have radial, axial, and, usually, azimuthal variation, and they propagate in a low-frequency, low magnetic field, high-density regime characterized by

$$\omega_{LH} \ll \omega \ll \omega_{ce} \qquad (13.2.1)$$

$$\omega_{pe}^2 \gg \omega\omega_{ce} \qquad (13.2.2)$$

where ω_{LH} is the lower hybrid frequency given by (4.5.20), with ω_{pi} and ω_{ci} the ion plasma frequency and ion gyrofrequency, respectively. The driving frequency is typically 1–50 MHz, with 13.56 MHz commonly used for processing discharges. The magnetic fields vary from 20 to 200 G for processing discharges, while fields up to 1000 G have been employed for some fundamental plasma studies. Plasma densities range from 10^{11}–10^{14} cm^{-3}, with 10^{11}–10^{12} cm^{-3} typical for processing.

Helicons are excited by an rf-driven antenna that couples to the transverse mode structure across an insulating chamber wall. The mode then propagates along the column, and the mode energy is absorbed by plasma electrons due to collisional or collisionless damping. All helicon applications to materials processing to date have utilized a process chamber downstream from the source. A typical helicon system was shown in Fig. 1.14b. The plasma potential in helicon discharges is typically low, of order 15–20 V, as in ECRs. However, the magnetic field is much lower than the 875 G required for ECRs, and the helicon power is supplied by rf rather than microwave sources. The smaller magnetic field, in particular, may lead to lower cost for helicon sources when compared to the ECR sources. However, as we will see, the resonant coupling of the helicon mode to the antenna can lead to nonsmooth variation of density with source parameters, known as "mode jumps," restricting the operating regime for a given source design.

The rf power system driving the helicon antenna can be of conventional design (as for rf capacitive discharges; see Section 11.6). A 500- to 5000-W, 50-Ω, 13.56-MHz supply can be used to drive the antenna through a matching network to minimize the reflected power seen by the supply. The matching network can be an L-design with two variable capacitors, as for inductive discharges. The antenna can also be driven through a balanced transformer so that the antenna coil is isolated from ground. This reduces the maximum antenna–plasma voltage by a factor of two, thus also reducing the undesired capacitive current coupled to the plasma by a factor of two. Since low aspect ratio geometries have not been developed for helicons, as they have for ECRs (see Fig. 13.4d), the transport and diffusion of the source plasma into the process chamber may be a significant limitation. The process chamber can have multipole confinement magnets to increase uniformity (see Section 5.5) or can have a wafer-level magnet coil (e.g., as in Fig. 13.1a) to keep the source plasma more tightly focused, thus increasing the etch rate but with some reduction in uniformity.

Helicon Modes

Helicon modes are a superposition of low-frequency *whistler waves* propagating at a common (fixed) angle to $\mathbf{B}_0$. Hence, although helicons have a complex transverse mode structure, they have the same dispersion equation as whistler waves, which is, from (13.1.34),

$$\frac{kk_z}{k_0^2} = \frac{\omega_{\mathrm{pe}}^2}{\omega\omega_{\mathrm{ce}}} \tag{13.2.3}$$

where

$$k = (k_\perp^2 + k_z^2)^{1/2} \tag{13.2.4}$$

is the wave-vector magnitude, $k_\perp$ and k_z are the radial and axial components, and $k_0 = \omega/c$. The helicon modes are mixtures of electromagnetic ($\nabla \cdot \mathbf{E} \approx 0$) and quasistatic ($\nabla \times \mathbf{E} \approx 0$) fields having the form

$$\mathbf{E}, \mathbf{H} \sim \exp j(\omega t - k_z z - m\theta)$$

where the integer m specifies the azimuthal mode. For an insulating (or conducting) wall at $r = R$ and assuming a uniform plasma density, the boundary condition on the total radial current density amplitude $\tilde{J}_r = 0$ (or $\tilde{E}_\theta = 0$) leads to (Chen, 1991)

$$mkJ_m(k_\perp R) + k_z J_m'(k_\perp R) = 0 \qquad (13.2.5)$$

where the prime denotes a derivative of the Bessel function, J_m, with respect to its argument. For a given frequency ω, density n_0, and magnetic field B_0, (13.2.3)–(13.2.5) can be solved to obtain $k_\perp$, k_z, and k.

Helicon sources based on excitation of the $m = 0$ mode and the $m = 1$ mode have been developed. Since the $m = 0$ mode is axisymmetric and the $m = 1$ mode has a helical variation, both modes generate time-averaged, axisymmetric field intensities. The transverse electric field patterns and the way these propagate along z are shown in Fig. 13.14a for the $m = 0$ mode and in Fig. 13.14b for the $m = 1$ mode. Undamped helicon modes have $\tilde{E}_z = 0$ (i.e., the quasistatic and electromagnetic components of $\tilde{E}_z$ exactly cancel). The antenna couples to the transverse electric or magnetic fields to excite the modes.

Equation (13.2.5) can be solved for $k_\perp R$ as a function of k_z/k. There are an infinite number of solutions corresponding to different radial field variations, and in any real system a mixture of modes is very likely excited. For simplicity, let us consider the first radial mode, shown in Fig. 13.14. For $m = 0$, from (13.2.5), $J_0'(k_\perp R) = 0$, which gives

$$k_\perp R = 3.83$$

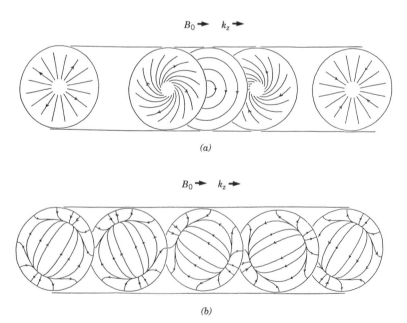

(a)

(b)

FIGURE 13.14. Transverse electric fields of helicon modes at five different axial positions: (a) $m = 0$; (b) $m = 1$ (Chen, 1991).

for any k_z/k. For $m = 1$, we solve (13.2.5) numerically to obtain the graph shown in Fig. 13.15, with the limiting values

$$k_\perp R = 3.83 \qquad (k_z \ll k_\perp)$$
$$k_\perp R = 2.41 \qquad (k_z \gg k_\perp)$$

To design an antenna for efficient power coupling, we must solve (13.2.3)–(13.2.5) and determine $k_\perp$ and k_z. Rewriting (13.2.3) in more physical terms,

$$kk_z = \frac{e\mu_0 n_0 \omega}{B_0} \tag{13.2.6}$$

we see that the $k_z \ll k_\perp$ limit corresponds to low density, and the $k_z \gg k_\perp$ limit corresponds to high density. These two limits can be treated analytically. We distinguish them by setting the condition $n = n_0^*$ for which $k_z = k_\perp$ for the $m = 1$ mode. We have $k = \sqrt{2}k_z$ and, from Fig. 13.15, $k_z = k_\perp \approx 2.5/R$. Choosing typical source parameters of $R = 5$ cm, $f = 13.56$ MHz, and $B_0 = 200$ G, we obtain $n_0^* \approx 4.0 \times 10^{12}$ cm^{-3}. Hence for this source with $n_0 \ll n_0^*$, we have $k_\perp \gg k_z$ and, from (13.2.4), $k \approx k_\perp$. For this case, (13.2.6) yields the axial wavelength of the helicon mode for low-density operation:

$$\lambda_z = \frac{2\pi}{k_z} = \frac{3.83}{R} \frac{B_0}{e\mu_0 n_0 f} \tag{13.2.7}$$

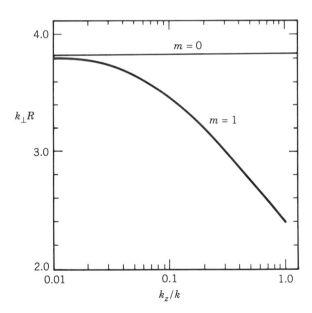

FIGURE 13.15. $k_\perp R$ versus k_z/k for helicon modes. (From "Design of High-Density Sources for Materials Processing" from the work "Physics of Thin Films," Vol.18, by Academic Press, Inc., Publisher in Press)

This regime is of limited interest for materials processing because, setting the antenna length $l_a \sim \lambda_z$ (see the next subsection), the condition on $k_\perp R$ requires $R \ll l_a < l$ for a cylindrical discharge of radius R and length l. Hence, the source would be long and thin, and uniformity over a large area would be compromised. However, (13.2.7) is useful in understanding source operation, as described below.

For $n_0 \gg n_0^*$, we have $k_z \gg k_\perp$ and $k \approx k_z$. In this high-density regime, we find

$$\lambda_z = \left(\frac{2\pi B_0}{e\mu_0 n_0 f} \right)^{1/2} \tag{13.2.8}$$

This regime is also of limited interest because it requires the antenna length $l_a \sim \lambda_z \ll R$, which leads to inefficient coupling of power from the antenna to the plasma. For a given current, only a small axial voltage is induced, leading to a small axial charge separation to drive the helicon mode. The regime of most interest for materials processing sources is $n_0 \sim n_0^*$, for which $k_z \sim k_\perp$; hence we have $R \sim l_a \sim l$, yielding an aspect ratio of order unity. This regime is not easy to analyze. For $m = 1$, the solution must be found numerically. One usually chooses $k_\perp$ somewhat larger than k_z; hence we can use (13.2.7) for simple estimates of source operation. Komori et al. (1991) have measured the helicon wave magnetic field using a magnetic pick-up coil. The dependence of λ_z on B_0/n_0 was found to roughly follow (13.2.7) at densities below n_0^*.

Recall from power balance (10.2.14) that the bulk density n_0 is determined by the absorbed power P_{abs} and the pressure p. Once B_0, f, and R (for low density) are chosen, then (13.2.7) or (13.2.8) determine λ_z. Ideally, the antenna must be designed to excite modes having that particular λ_z. At first sight, this seems to limit source operation to one particular density unless B_0 or f can be conveniently varied. Fortunately, antennas excite a range of λ_z's, thus allowing source operation over a range of n_0's.

Antenna Coupling

A typical antenna used to excite the $m = 1$ mode is shown in Fig. 13.16. Other antennas are described by Chen (1992). Looking at the x–y transverse coordinates shown in the figure, we see that this antenna generates a $\tilde{B}_x$ field over an axial antenna length l_a, which can couple to the transverse magnetic field of the helicon mode. The antenna also induces a current within the plasma column just beneath each horizontal wire, in a direction opposite to the currents shown. This current produces charge of opposite signs at the two ends of the antenna, which in turn generates a transverse quasistatic field $\tilde{E}_y$, which can couple to the transverse quasistatic fields of the helicon mode (see Fig. 13.14b). The conditions for which each form of coupling dominates are not well understood.

To illustrate the wavelength matching condition for helicon excitation, we consider an ideal antenna field for quasistatic coupling:

$$\tilde{E}_y(z) \sim \tilde{E}_{y1} \Delta z \left[\delta \left(z + \frac{l_a}{2} \right) - \delta \left(z - \frac{l_a}{2} \right) \right] \tag{13.2.9}$$

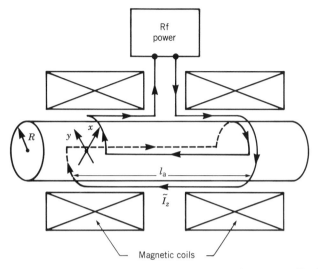

FIGURE 13.16. The antenna for $m = 1$ helicon mode excitation. (From "Design of High-Density Sources for Materials Processing" from the work "Physics of Thin Films," Vol.18, by Academic Press, Inc., Publisher in Press)

where δ is the Dirac delta function. This ideal field is sharply peaked in Δz near the two antenna ends, as shown schematically in Fig. 13.17a. Taking the Fourier transform,

$$E_y(k_z) = \int_{-\infty}^{\infty} dz\, \tilde{E}_y(z) \exp(-jk_z z)$$

and squaring this to obtain the spatial power spectrum of the antenna, we obtain

$$E_y^2(k_z) = 4\tilde{E}_{y1}^2 (\Delta z)^2 \sin^2 \frac{k_z l_a}{2} \qquad (13.2.10)$$

which is plotted in Fig. 13.17b. We see from (13.2.10) that the antenna couples well to the helicon mode ($E_y^2(k_z)$ is a maximum) for $k_z \approx \pi/l_a$, $3\pi/l_a$, etc., corresponding to $\lambda_z \approx 2l_a$, $2l_a/3$, etc. The coupling is poor ($E_y^2(k_z) \approx 0$) for $k_z \approx 0$, $2\pi/l_a$, $4\pi/l_a$, etc., corresponding to $\lambda_z \to \infty$, $\lambda_z \approx l_a$, $\lambda_z \approx l_a/2$, etc.

Figure 13.18 shows the effect of the antenna coupling on the density n_0 as the power P_{inc} supplied to the antenna is increased, using a 36-GHz microwave interferometer to measure n_0 (see Section 4.6 for details of the measurement technique). For $P_{inc} <$ 350 W, n_0 determined from the power balance (10.2.14) is low, leading to $k_z \ll \pi/l_a$ and, from (13.2.10), poor coupling to the helicon mode. The discharge in this regime is probably capacitively driven, with a relatively high antenna voltage (~ 2 kV) and plasma potential (> 30 V). The transition to helicon mode operation with $k_z \approx \pi/l_a \approx$ $0.4k_\perp$ for $P_{inc} \approx 400$ W and $n_0 \approx 1.4 \times 10^{11}$ cm^{-3} is clearly seen. A further increase in power is not reflected in a proportional density increase, as the antenna coupling

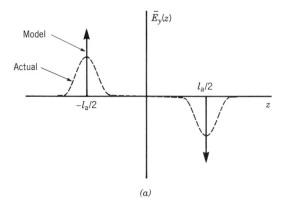

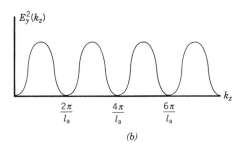

FIGURE 13.17. The quasistatic antenna coupling field $\tilde{E}_y$: (a) ideal and actual field; (b) spatial power spectrum of a typical field. (From "Design of High-Density Sources for Materials Processing" from the work "Physics of Thin Films," Vol.18, by Academic Press, Inc., Publisher in Press)

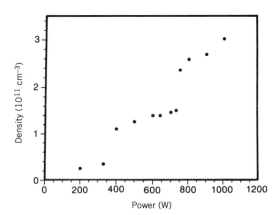

FIGURE 13.18. Measured density as a function of input power for $B_0 = 80$ G at 5 mTorr in argon (Perry et al., 1991).

becomes increasingly inefficient. A second transition is seen to $k_z \approx 3\pi/l_a \approx k_\perp$ with $n_0 \approx 2.7 \times 10^{11}$ cm^{-3}. Standing helicon wave effects may also play a role in this transition, as described in the next subsection. Figure 13.19 shows the roughly linear scaling of n_0 with B_0 predicted from (13.2.7) or (13.2.8), for a different source than that of Fig. 13.18. Again we see the density steps imposed by the antenna coupling condition. Depending on the specific experimental configuration, for example, the distance between the antenna and the outer surface of the source dielectric cylinder, the density steps are not always as evident as shown in these data. They may also be produced by large relaxation oscillations as the discharge "hunts" between helicon and inductive excitation modes. The antenna can also be designed to couple efficiently to a wide range of k_z's, reducing the importance of mode jumps in the density range of interest. Similar effects are seen for $m = 0$ mode helicons. This mode is excited by an antenna consisting of two circular coils of radius R, separated by a length l_a, carrying oppositely directed currents.

Helicon Mode Absorption

The helicon mode energy is believed to be transferred to the plasma electrons as the mode propagates along the column by collisional or collisionless (Landau) damping. The former mechanism transfers the energy to the thermal (bulk) electron population, while the latter mechanism can act to preferentially heat a nonthermal electron population to energies greatly exceeding the bulk electron temperature. There is considerable evidence that collisional absorption is too weak to account for energy deposition at low pressures (< 10 mTorr argon), although this mechanism may dominate at higher pressures. Landau damping is a process by which a wave transfers energy to electrons having velocities near the phase velocity $v_{\rm ph} = \omega/k_z$ of the wave. (See, for example, Chen (1984) for an exposition of the phenomenon.) Chen (1991) has estimated the effective collision frequency $\nu_{\rm LD}$ for Landau damping of the helicon

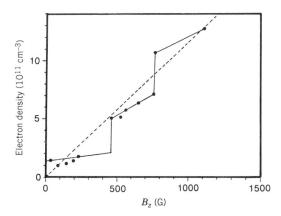

FIGURE 13.19. Measured density as a function of magnetic field at a fixed input power. The dashed line represents the resonance condition imposed by the antenna (Perry et al., 1991).

mode as

$$\nu_{LD} \approx 2\sqrt{\pi}\omega\zeta^3 \exp(-\zeta^2) \qquad \zeta \gg 1 \qquad (13.2.11)$$

$$\nu_{LD}(\text{max}) \approx 1.45\,\omega \qquad \zeta \approx 1.2 \qquad (13.2.12)$$

where $\zeta = \omega/(k_z\sqrt{2}v_{th})$, with $v_{th} = (eT_e/m)^{1/2}$ (here m is the electron mass). From (13.2.7) or (13.2.8) we see that ζ decreases with increasing density. Thus for $\zeta \gg 1$, ν_{LD} increases with increasing electron density at constant magnetic field. However, in typical helicon sources where ζ may be less than or of order unity, ν_{LD} can decrease with increasing n. The total effective collision frequency can be written as

$$\nu_T = \nu_c + \nu_{LD}$$

where here ν_c is the sum of the electron–neutral and electron–ion collision rates. The axial decay length α_z^{-1} for helicon mode damping is (see Problem 13.2)

$$\alpha_z^{-1} \approx \frac{\omega_{ce}}{k_\perp \nu_T} \qquad (13.2.13)$$

for low density ($k_z \ll k_\perp$); and

$$\alpha_z^{-1} \approx \frac{2\omega_{ce}}{k_z \nu_T} \qquad (13.2.14)$$

for high density ($k_z \gg k_\perp$). For efficient power transfer to the plasma electrons, we require that $\alpha_z^{-1} \lesssim l$, where l is the helicon chamber source length.

By choosing the antenna length l_a such that $k_z \approx \pi/l_a$, it is possible to heat electrons, by Landau damping, whose energies are near that corresponding to the wave phase velocity

$$e\mathcal{E} = \frac{1}{2}m\left(\frac{\omega}{k_z}\right)^2 \qquad (13.2.15)$$

If $\mathcal{E}$ is chosen near the peak of the ionization cross section (~ 50 V in argon), then the collisional energy $\mathcal{E}_c$ lost per electron–ion pair created can be reduced to a low value, of order of twice the ionization energy $\mathcal{E}_{iz}$. It follows from (10.2.14) that this can lead to a significant increase in density for the same absorbed power. However, the effective collision frequency ν_{LD} falls precipitously for $\omega/k_z \gg v_{th}$, leading to a low spatial decay rate which is not compatible with materials processing sources having $l \sim R$. Experimental evidence of Landau damping has been reported, but other absorption mechanisms, such as nonlinear excitation of plasma instabilities, or electron acceleration due to the axially nonuniform helicon mode amplitude, may also play a role in helicon mode energy transfer.

Example   As an example of helicon design, let $R = 5$ cm, $l = 20$ cm, $B_0 = 200$ G, $n_g = 3.3 \times 10^{13}$ cm^{-3} (1 mTorr), $\omega = 85 \times 10^6$ s^{-1} (13.56 MHz), and $P_{abs} = 3000$ W. At 1 mTorr, $\lambda_i \approx 3$ cm. Then from (10.2.3) and (10.2.4), $h_l \approx h_R \approx 0.33$, and from (10.2.8), $d_{eff} \approx 6.1$ cm. For argon we then obtain from Fig. 10.1 that $T_e \approx 5.2$ V, and from Fig. 3.17, that $\mathcal{E}_c \approx 39$ V. Using (10.2.1), we find $\mathcal{E}_T \approx 76$ V. The Bohm velocity is $u_B \approx 3.5 \times 10^5$ cm/s, and from (10.2.15), $A_{eff} \approx 260$ cm^2. Then from (10.2.14), we obtain $n_0 \approx 2.7 \times 10^{12}$ cm^{-3}. Using the value of $n_0^* \approx 4.0 \times 10^{12}$ cm^{-3} at $B_0 = 200$ G from our previous discussion, we see that $n_0 < n_0^*$. From (13.2.7), we find $\lambda_z = 20.8$ cm, and hence we choose an antenna length $l_a = \lambda_z/2 = 10.4$ cm to optimize power coupling. We note that $\omega/k_z = 2.8 \times 10^8$ m/s, compared with the electron thermal velocity $v_{th} = 1.0 \times 10^8$ cm/s. Hence, $\zeta \approx 2$, not too far from the peak of the Landau damping rate for thermal electrons.

13.3 SURFACE WAVE DISCHARGES

Electromagnetic surface waves that propagate along a cylindrical plasma column can be efficiently absorbed by the plasma, hence sustaining a discharge. Surface waves, which are propagating modes having strong fields only near the plasma surface, were described by Smullin and Chorney (1958) and Trivelpiece and Gould (1959). Moisan and his group at the Université de Montréal have extensively analyzed the concept and developed high-power wave-launching systems over a wide frequency range (1 MHz–10 GHz). Cylindrical surface wave sources have been reviewed by Moisan and Zakrzewski (1991). Discharges having diameters as large as 15 cm have been operated, although diameters of 3–10 cm are more commonly used. The simplest sources operate without an imposed axial magnetic field. At the high densities of interest here, the sources must be driven at microwave frequencies in the range of 1–10 GHz. Although there have been some applications to materials processing, the absorption length α_z^{-1} for the surface modes tends to be long, such that $l \gg R$ for these discharges. Hence, the cylindrical configuration cannot be operated as a low aspect ratio source. However, planar (rectangular) configurations have been developed (Komachi, 1993) that may be suitable for large area processing applications.

Planar Surface Waves

Two types of configurations can support electromagnetic surface waves at an interface between a dielectric and a plasma. (1) At an interface between a semi-infinite plasma and a dielectric, a solution can be found for which the wave amplitude decays in both directions away from the plasma–dielectric interface. Maxwell's equations admit solutions of the form ($x > 0$ in the plasma)

$$\tilde{H}_{yd} = H_{y0}\, e^{\alpha_d x - jk_z z} \tag{13.3.1}$$

$$\tilde{H}_{yp} = H_{y0}\, e^{-\alpha_p x - jk_z z} \tag{13.3.2}$$

where we have assumed that H_y is continuous across the interface at $x = 0$. From the wave equation, the transverse decay constants are related to the propagation constant k_z by

$$-\alpha_d^2 + k_z^2 = \kappa_d \frac{\omega^2}{c^2} \tag{13.3.3}$$

and

$$-\alpha_p^2 + k_z^2 = \kappa_p \frac{\omega^2}{c^2} \tag{13.3.4}$$

where κ_p, given by (4.2.24), is the lossless plasma relative dielectric constant. From Maxwell's equations (2.2.1), we obtain the electric field components (e.g., see Ramo et al., 1984, Chapter 8)

$$\tilde{E}_{zd} = H_{y0} \frac{\alpha_d}{j\omega\epsilon_0 \kappa_d} e^{\alpha_d x - jk_z z} \tag{13.3.5}$$

and

$$\tilde{E}_{zp} = -H_{y0} \frac{\alpha_p}{j\omega\epsilon_0 \kappa_p} e^{-\alpha_p x - jk_z z} \tag{13.3.6}$$

Using continuity of $\tilde{E}_z$ at the interface $x = 0$, we can eliminate the arbitrary constant H_{y0} by equating (13.3.5) to (13.3.6) to obtain

$$\frac{\alpha_p}{\kappa_p} = -\frac{\alpha_d}{\kappa_d} \tag{13.3.7}$$

Substituting (13.3.3) and (13.3.4) into (13.3.7), we obtain

$$\kappa_d^2 \left(k_z^2 - \kappa_p \frac{\omega^2}{c^2} \right) = \kappa_p^2 \left(k_z^2 - \kappa_d \frac{\omega^2}{c^2} \right) \tag{13.3.8}$$

which can be solved for k_z to determine the wave dispersion,

$$k_z = \kappa_d^{1/2} \frac{\omega}{c} \left[\frac{\omega_{pe}^2 - \omega^2}{\omega_{pe}^2 - (1 + \kappa_d)\omega^2} \right]^{1/2} \tag{13.3.9}$$

Figure 13.20 shows k_z versus ω for the lossless case. We see that k_z is real for $\omega \leq \omega_{res}$, where $\omega_{res} = \omega_{pe}/(1 + \kappa_d)^{1/2}$ gives the resonance $k_z \to \infty$ of the surface wave. For $\omega \ll \omega_{res}$, we see that $k_z \approx \kappa_d^{1/2}\omega/c$. The region of interest for surface wave sources

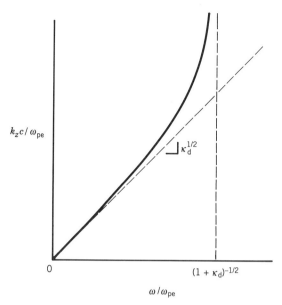

FIGURE 13.20. Surface wave dispersion k_z versus ω. (From "Design of High-Density Sources for Materials Processing" from the work "Physics of Thin Films," Vol.18, by Academic Press, Inc., Publisher in Press)

is ω near but just below ω_{res}. Hence for high-density sources, the frequencies of interest are above 1 GHz; i.e., microwave frequencies. Fixing ω for the source, we introduce the resonance value of the density $n_{\text{res}} = \epsilon_0 m \omega^2 (1 + \kappa_d)/e^2$ (here, m is the electron mass). Then the surface wave propagates for densities $n_0 \geq n_{\text{res}}$.

(2) A configuration, in which the plasma is separated from a conducting plane by a dielectric slab of thickness d, also admits a wave that decays into the plasma region. However, this wave does not decay into the dielectric, but is confined within the dielectric by the conducting plane. This type of surface wave, similar to that used for optical wave guiding, has also been used for surface wave discharges (Komachi, 1993). We will not consider this configuration here.

Cylindrical Surface Waves

A surface wave also propagates on a nonmagnetized plasma column of radius R confined by a thick dielectric tube of outer radius b. In analogy to the wave in slab geometry, assuming $b \gg R$, so that interaction with external surfaces can be neglected, the azimuthally symmetric mode has $\tilde{H}_z = 0$ and

$$\tilde{E}_{zp} = \tilde{E}_{z1} \frac{I_0(\alpha_p r)}{I_0(\alpha_p R)} \exp j(\omega t - k_z z) \qquad r < R \qquad (13.3.10)$$

$$\tilde{E}_{zd} = \tilde{E}_{z1} \frac{K_0(\alpha_d r)}{K_0(\alpha_d R)} \exp j(\omega t - k_z z) \qquad r > R \qquad (13.3.11)$$

where α_d and α_p are related to k_z by (13.3.3) and (13.3.4), and I_0 and K_0 are the modified Bessel functions of the first and second kind. We note from the form of the Bessel functions that the fields decay away from the surface of the plasma in both directions. The transverse fields are obtained from $\tilde{E}_z$ using Maxwell's equations. In particular, we find

$$\tilde{H}_\theta = -\frac{j\omega\epsilon_0\kappa}{\alpha^2}\frac{\partial\tilde{E}_z}{\partial r}$$

in the two regions. The continuity of the tangential magnetic field $\tilde{H}_\theta$ then yields the dispersion equation

$$\frac{\kappa_p}{\alpha_pR}\frac{I_0'(\alpha_pR)}{I_0(\alpha_pR)} = \frac{\kappa_d}{\alpha_dR}\frac{K_0'(\alpha_dR)}{K_0(\alpha_dR)} \tag{13.3.12}$$

From (13.3.10) and (13.3.11), if $\alpha R \gg 1$, then the surface modes decay rapidly, which greatly simplifies the analysis. Using the asymptotic expansions of the Bessel functions $I_0'/I_0 = 1$ and $K_0'/K_0 = -1$, we then obtain the result (13.3.7), i.e., the cylinder looks like a plane in this approximation. The dispersion is the same as in (13.3.9) and as illustrated in Fig. 13.20. However, in the cylinder, at low frequencies, the ordering $\alpha R \gg 1$ is not valid, and the complete dispersion equation (13.3.12) must be solved numerically. The result is similar to that shown in Fig. 13.20.

Power Balance

We treat the power balance in the geometrically simple case of a long, thin source, $l \gg R$, using the general principles described in Section 10.2. In particular, the local power balance along z determines the density n_0 for a given absorbed power P_{abs}' per unit length along the column, as in the derivation leading to (10.2.14). Letting P_w be the power carried by the wave along the column at the position z, at which the density is n_0, then

$$P_{abs}'(n_0) = 2\alpha_z(n_0)P_w \tag{13.3.13}$$

where α_z is the axial attenuation constant of the wave fields at the density n_0. Equating P_{abs}' to the power loss per unit length,

$$P_{loss}'(n_0) = en_0u_BA_{eff}'\mathcal{E}_T \tag{13.3.14}$$

where $A_{eff}' = 2\pi Rh_R$ is the effective (radial) loss area per unit length, we obtain $n_0(z)$ for a given wave power $P_w(z)$.

The mode attenuates as it propagates along z due to a nonzero electron–neutral momentum transfer frequency ν_m. Letting $\nu_m \ll \omega$ in (4.2.18), substituting this into (13.3.9), solving for the complex propagation constant k_z, and taking the imaginary part, we obtain the attenuation constant $\alpha_z(n_0) = -\operatorname{Im} k_z$ at a fixed ω. The expression is complicated and we give only the scaling for n_0 greater than, but not too near, resonance:

$$\alpha_z \propto \frac{n_0 \nu_m}{(n_0 - n_{res})^{3/2}} \tag{13.3.15}$$

At resonance, there is a finite α_z, while for $n_0 < n_{res}$, the wave does not propagate. For the variation of α_z in (13.3.15), P'_{abs} given by (13.3.13) is plotted versus n_0 for several different values of P_w in Fig. 13.21. The linear variation of P'_{loss} given by (13.3.14) is also plotted on the figure. The intersection of P'_{abs} with P'_{loss} determines the equilibrium density along the column. It can be seen that there is a minimum value $P_{wmin}(z)$, below which a discharge at that value of z cannot be sustained. A discharge forms near the position of surface wave excitation $z = 0$ for $P_{wmax} > P_{wmin}$. As the wave propagates, P_w attenuates along z due to wave absorption. A discharge cannot be sustained when P_w falls below P_{wmin} at $z = z_{max}$. Hence the discharge exists as a finite length plasma column over $0 < z < z_{max}$. Typical plasma column variations of n_0 and P_w are shown in Fig. 13.22. We note in Fig. 13.21 that there are generally two intersections of $P'_{abs}(n_0)$ with $P'_{loss}(n_0)$. The lower-density intersection is an unstable equilibrium because a fluctuation that decreases n_0 leads to $P'_{abs} < P'_{loss}$, thus further decreasing n_0. The higher-density intersection is stable by similar reasoning.

PROBLEMS

13.1. Wave Equation for Right Circularly Polarized Wave Starting from Maxwell's equations (2.2.1)–(2.2.4), with variation in z only, with J given by (4.4.3) and following, derive the wave equation (13.1.22), with κ_r given by (13.1.23).

13.2. Helicon Mode Decay Constants Starting from (13.2.3), perform the following:

   (a) Rederive the right-hand side to include collisions.

   (b) Introducing $k_z = \beta_z - j\alpha_z$ on the right-hand side, expand in the two limits $k_z \ll k_\perp$ and $k_z \gg k_\perp$ to obtain (13.2.13) and (13.2.14), respectively.

13.3. Helicon Discharge Equilibrium

   (a) Taking the example in Section 13.2, obtain all of the values given there by making the appropriate calculations.

   (b) Repeat for $p = 10$ mTorr and $P_{abs} = 600$ W. Discuss the modification of the results.

13.4. ECR Discharge Equilibrium An ECR discharge is excited in argon gas at a pressure of 1 mTorr by a 2.45 GHz, right circularly polarized wave carrying $P_{inc} = 1000$ W of incident microwave power through a quartz window at one

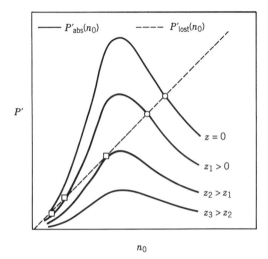

FIGURE 13.21. Determination of the equilibrium density in a surface wave discharge. The high-density intersection of $P'_{\rm abs}$ and $P'_{\rm loss}$ gives the equilibrium density (after Moisan and Zakrzewski, 1991).

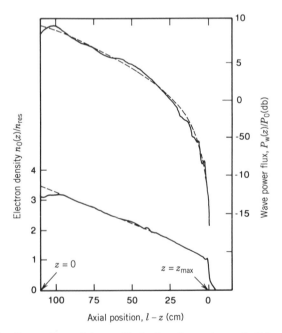

FIGURE 13.22. Comparison of theory (dashed) and experiment (solid) of density n_0 and wave power $P_{\rm w}$ versus z for a typical surface wave source (after Moisan and Zakrzewski, 1991).

end of a cylindrical discharge chamber of radius $R = 10$ cm and length $l = 50$ cm. The magnetic field monotonically decreases from the window into the chamber, and the logarithmic field gradient at the resonance zone is $\alpha = 4\text{m}^{-1}$.

(a) Assuming that all of the incident power is uniformly absorbed over the cross sectional area of the plasma and that there is axial loss, but no radial loss of plasma, find the electron temperature T_e and the central plasma density n_0.

(b) Using (13.1.30), verify that essentially all of the incident power is absorbed by the plasma.

(c) Find the minimum incident power that will sustain the discharge.

(d) Using your results in (a), (b), and (c), sketch n_0 versus P_{inc} (linear scales) for $0 < P_{inc} < 1000$ W.

CHAPTER 14

DC DISCHARGES

14.1 QUALITATIVE CHARACTERISTICS OF GLOW DISCHARGES

The dc glow discharge has been historically important, both in applications of weakly ionized plasmas and in studying the properties of the plasma medium. A dc discharge has one obvious feature, its macroscopic time independence, that is simpler than rf discharges. However, the need for the current, which provides the power for the discharge, to be continuous through the dc sheath provides an additional complication to the operation. This complication is not present in rf or microwave discharges where displacement current provides current continuity through the sheath. For studying the glow discharge we consider the usual configuration of a long glass cylinder with the positive anode at one end and a negative cathode at the other. Although not necessarily the configuration used in processing applications, it has the advantage of symmetry and has been well studied. The usual pressure range of operation is between 10 mTorr and 10 Torr. Typically, a few hundred volts between cathode and anode is required to maintain the discharge. The approximate characteristics of the discharge are shown in Fig. 14.1. It is clear from the many light and dark regions identified in Fig. 14.1a that the behavior is quite complicated. The length of the positive column region can be varied by changing the distance between electrodes at a constant pressure and approximately constant voltage drop, while the other regions maintain their lengths. It is therefore apparent that the positive column can be analyzed per unit length, while the other features must be analyzed in their entirety. All of the regions are gas,

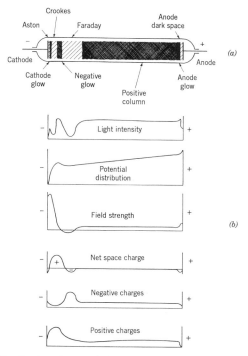

FIGURE 14.1.   Qualitative characteristics of a dc glow discharge (after Brown, 1959).

pressure, and voltage dependent in their size and intensity, with some of the smaller features being essentially absent over various parameter ranges.

We now describe qualitatively the essential operation of the various regions in maintaining the discharge. The treatment follows most closely that in Cobine (1958) where additional material and references can be found. Planar magnetron discharges are described in Section 14.5.

Positive Column

The axially uniform plasma is maintained by the $\mathbf{J} \cdot \mathbf{E}$ power integrated over the cross section, which balances the loss of energy per electron–ion pair created, which, in the axially uniform model, is assumed to be radial. The dynamics are very similar to that of the bulk rf discharge, with the power lost per electron–ion pair created going to excitation (the glow), ionization, electron–neutral elastic scattering energy losses, and kinetic energy of the electrons and ions striking the walls. The normal glow discharge tends to have a negative voltage–current characteristic (negative differential resistance (dV/dI) which is stabilized by an external resistor. The resistor is varied to adjust the current to the desired value. The power balance determines the (weak) E field required to maintain the positive column. Once E is known, the drift velocity of the electrons along the column can be found using the dc electron mobility and then, from J, the density can be determined. We use this prescription in the next section for calculating the characteristics of the positive column.

Cathode Sheath

This region, known also as the *cathode fall* or *Crookes dark space*, is the region over which most of the voltage drop occurs. The electrons, which carry most of the current in the positive column, are, of course, prevented from reaching the cathode. The massive ions, however, are incapable of carrying the full current. The discharge is maintained by secondary electrons produced at the cathode by the impact of the energetic ions. This process, which is incidental (although often important) in rf discharges, is essential for the operation of the dc discharge. The current is built up by ionization within the sheath, which is generated by the secondary electrons accelerating in the large electric fields of this region. The electron density and flux grow exponentially from the cathode, with the exponent known as the *first Townsend coefficient*. This mechanism is important, not only for the steady-state discharge, but also for understanding the *breakdown* that initiates the discharge. In breakdown the entire region between the cathode and the anode participates in the process, which requires a much higher voltage and therefore leads to hysteresis in the voltage–current characteristic. We analyze this dynamics in Section 14.3.

Negative Glow and Faraday Dark Space

The exponentially increasing density of high-velocity electrons near the cathode leads rapidly to a bright cathode glow in which intense ionization and excitation occurs. The electric field must decrease rapidly at the end of this region, where the transition to the positive column occurs. However, the high electron velocities must be dissipated by elastic and inelastic collisions before the equilibrium conditions of the positive column can be established. This is done in a rather complicated process in which the electrons first lose almost all of their energy and then are reaccelerated in a weak field over approximately a mean free path (the Faraday dark space). We give a simple approximate analysis of this behavior at the end of Section 14.3.

Anode Fall

The drift velocity of the electrons in the weak electric field of the positive column is typically less than their thermal velocity. This requires a retarding electric field in the neighborhood of the anode to prevent the full thermal electron current from reaching the anode. However, the anode itself must clearly be positive with respect to the positive column to maintain the current. The result is a *double layer*, which is also seen in various other types of discharges, for essentially the same reason. Since the total voltage drop in this region is small and plays little role in the overall dynamics, we will not analyze it quantitatively.

The various other regions indicated in Fig. 14.1 are not of particular significance for an overall understanding of the discharge behavior. In addition to the axial variations there are, of course, radial variations. In a long cylindrical discharge, we shall obtain the usual radial variation as part of our solution for the positive column given in Section 14.2. We may assume qualitatively similar radial variations of density in other regions, but quantitative calculations are very difficult. Additional radial features exist, such as an incomplete coverage of the cathode surface by the discharge, as we discuss in Section 14.3.

In the previous discussion we have considered the typical characteristics in the normal glow, which occurs over a range of current densities, typically between 10^{-5} and 10^{-3} A/cm^2. Considering current density as the controlling variable, the voltage–current characteristic of a dc discharge is shown in Fig. 14.2. The flat region with slightly negative slope dV/dI is that of the normal glow. From low currents, the region below I_A is called a *dark* or *Townsend discharge*. The glow gradually builds up until a transition is reached, with hysteresis, entering the *normal glow* at a voltage V_S. The voltage remains constant as the current increases until I_B, at which point there is an increasing voltage–current characteristic called the *abnormal glow*. A further increase in current results in a rather abrupt transition at I_C, again characterized by hysteresis, to a considerably lower voltage discharge known as an arc discharge. The voltage continues to decrease with increasing current, approaching an asymptote. For a typical pressure (say 1 Torr) and a typical discharge tube of a few centimeters cross section, the transitions might occur at $I_A \approx 10^{-6}$ A, $I_B \approx 10^{-2}$ A, and $I_C \approx 10^{-1}$ A, but these currents depend on various other factors such as gas and electrode surfaces. There are applications in these various regions that we say little about here.

In some pressure and voltage ranges there are also interesting time-varying phenomena, such as moving transverse striations and longitudinal filaments. At high pressures, arc spots can form at the cathode, which correspond to an entirely different range of operation, not considered here, in which the secondary emission process is thermionic. For further study the interested reader is referred to the literature (Cobine, 1958; Franklin, 1976; Raizer, 1991).

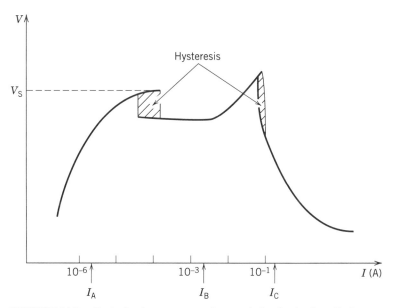

FIGURE 14.2. Typical voltage–current characteristic of a dc glow discharge.

A phenomenon that is not part of the discharge dynamics, but is important both for applications and in limiting the use of glow discharges, is cathode sputtering. The potential drop in a cathode sheath is typically several hundred volts. These ion-bombarding voltages lead to severe sputtering of the cathode surface and consequently deposit material on other surfaces. We describe physical sputtering in Section 9.3 and its application to the deposition of thin films in Section 16.3. Since there is little control over the large voltage drop in the cathode sheath, the existence of the sputtering is important in defining appropriate applications, and low aspect ratio dc discharges have been used for sputtering. To enhance sputtering at the cathode, a nonuniform dc magnetic field can be applied, leading to the *dc planar magnetron discharge,* described in Section 14.5. These discharges are commonly used as sputtering sources for deposition of metallic thin films on substrates.

14.2   ANALYSIS OF THE POSITIVE COLUMN

As in the analysis of rf and microwave discharges, there are various pressure regimes for which different dynamics apply. We will assume the following: (1) The pressure is sufficiently high, $\lambda_i \lesssim (T_i/T_e)R$, that a diffusion equation with a constant diffusion coefficient D_a applies. The low-pressure (collisionless) limit with freely falling ions, $\lambda_i \gtrsim R$, was described very early by Tonks and Langmuir (1929); and the intermediate pressure regime, $R \gtrsim \lambda_i \gtrsim (T_i/T_e)R$, is discussed in Godyak (1986). In fact, as discussed in Section 5.3, the radial distributions in the low and intermediate regimes tend to look quite similar. Franklin (1976) describes these various solutions and relations between them. (2) As discussed in Section 14.1 it is often adequate to assume only radial variation, which we do here.

Calculation of T_e

The calculation of T_e proceeds as for rf discharges (see also Chapter 10). Ion particle balance is obtained from the diffusion equation

$$-\nabla \cdot D_a \nabla n = \nu_{iz} n \tag{14.2.1}$$

where $n = n_e = n_i$ is the plasma density, D_a is the ambipolar diffusion coefficient, and $\nu_{iz} = K_{iz} n_g$ is the ionization rate as defined in (3.5.1). In cylindrical coordinates (14.2.1) becomes

$$\frac{d^2 n}{dr^2} + \frac{1}{r}\frac{dn}{dr} + \frac{\nu_{iz}}{D_a} n = 0 \tag{14.2.2}$$

Equation (14.2.2) is Bessel's equation with solution

$$n = n_0 J_0(\beta r) \tag{14.2.3}$$

where $\beta = (\nu_{iz}/D_a)^{1/2}$ and J_0 is the usual zero-order Bessel function. If the ion mean free path λ_i and the sheath thickness s ($s \approx$ few λ_{De}) are both small compared to the column radius R, then the solution is approximately

$$\beta = \left(\frac{\nu_{iz}}{D_a}\right)^{1/2} = \frac{\chi_{01}}{R} \tag{14.2.4}$$

where $\chi_{01} \approx 2.405$ is the first zero of the zero-order Bessel function. Although (14.2.4) does not give a completely self-consistent solution, since the finite ion flux at the wall implies infinite velocity at zero density (see Section 5.2), it can give a reasonably accurate value of T_e. The reason is that ν_{iz} is a very sensitive function of T_e of the form (see Chapter 3)

$$\nu_{iz} \propto p \exp\left(-\frac{\mathcal{E}_{iz}}{T_e}\right) \tag{14.2.5}$$

with p the pressure and with the ionization voltage $\mathcal{E}_{iz} \gg T_e$. Thus T_e depends only weakly on all parameters except for $\mathcal{E}_{iz}$. A more accurate solution is obtained by setting the flux Γ equal to $n_s u_s$, where, as previously, n_s is the density at the sheath edge and u_s is given by (6.5.9). For this case, since $\Gamma = -D_a dn/dr$, we can take a derivative of (14.2.3) to obtain a transcendental equation for the electron and ion flux to the wall (see also Section 10.2):

$$-(D_a \nu_{iz})^{1/2} J_1(\beta R) = J_0(\beta R) u_s \tag{14.2.6}$$

Because $\lambda_i \ll R$ for this constant D_a solution, (14.2.6) essentially reduces to (14.2.4).

In the intermediate- and low-pressure regimes, $\lambda_i \gtrsim (T_i/T_e)R$, the radial profile becomes relatively uniform, and the estimate for particle balance (3.3.14) applies,

$$\nu_{iz} \approx 2.2 \frac{u_B}{R} \left(4 + \frac{R}{\lambda_i}\right)^{-1/2} \tag{14.2.7}$$

A serious problem at low pressures is the deviation of the electron distribution from a Maxwellian. In using (14.2.5) we have assumed a Maxwellian, thus ignoring the electron drift motion u_e. This motion can readily be included (see Franklin, 1976); with $u_e \ll (eT_e/m)^{1/2}$ this does not appreciably change the results. More important at low densities is the truncation of the distribution at high velocities due to the greater ease with which the higher velocity electrons escape through the retarding sheath potential at the radial wall. This effect can significantly raise the average energy of the electrons in a positive column at low pressures.

Calculation of E and n_0

The electric field E along the z axis of the discharge is calculated by equating the input power absorbed to the power lost. In the rf discharge this was used to determine the density. Here the density cancels, leaving an expression for the electric field. However, once the field is known, a subsidiary condition immediately gives the density. Equating the ohmic power absorbed

$$P_{abs} = 2\pi \int_0^R \mathbf{J} \cdot \mathbf{E}\, r\, dr \tag{14.2.8}$$

to the power lost

$$P_{loss} = 2\pi R \Gamma_r e \mathcal{E}_T \tag{14.2.9}$$

where Γ_r is the radial particle flux and $e\mathcal{E}_T$ is the total energy carried out per electron–ion pair created, and substituting our radial density solutions, we have

$$en_0 \mu_e E^2 2\pi \int_0^R J_0(\beta r) r\, dr = 2\pi R (D_a \nu_{iz})^{1/2} n_0 J_1(\beta R) e \mathcal{E}_T \tag{14.2.10}$$

where we have assumed a constant mobility μ_e, substituted for the current density J along z using

$$J = en_0 \mu_e E \tag{14.2.11}$$

and have taken E out of the integral by assuming that it is a constant in the long thin approximation. We see that n_0 cancels from (14.2.10) giving an equation for E alone. Performing the integration we find that J_1 cancels, and we can solve directly for E to obtain

$$E = \left(\frac{\nu_{iz} \mathcal{E}_T}{\mu_e} \right)^{1/2} \tag{14.2.12}$$

Substituting $\mu_e = e/m\nu_m$, from (5.1.4), then (14.2.12) can also be written in the form

$$E = \left(\frac{m}{e} \nu_{iz} \nu_m \mathcal{E}_T \right)^{1/2} \tag{14.2.13}$$

We note that ν_{iz} and ν_m are both linearly dependent on pressure, and that the only other dependence on the right hand side is T_e. Although (14.2.12) gives E as a function of p and as an exponentially sensitive function of T_e through its dependence on ν_{iz}, we can eliminate ν_{iz} using (14.2.4) to obtain

$$E = \frac{\chi_{01}}{R} \left(\frac{D_a \mathcal{E}_T}{\mu_e} \right)^{1/2} = \frac{\chi_{01}}{R} \left(\frac{mK_{me}}{MK_{mi}} T_e \mathcal{E}_T \right)^{1/2} \tag{14.2.14}$$

which shows that E depends only on T_e, independent of p. Integrating (14.2.11) over the discharge cross section yields

$$I = en_0 \left(\frac{R^2}{\chi_{01}}\right) J_1(\chi_{01})\mu_e E \tag{14.2.15}$$

which can be solved to determine n_0 for a given discharge current I, with E given by (14.2.14).

14.3 ANALYSIS OF THE CATHODE SHEATH

Considering the analysis of the previous section, we take as an example an argon glow discharge at $p = 100$ mTorr and $T_e = 4$ V. The current density carried by the electrons in the glow is calculated from

$$J(r) = en(r)\mu_e E$$

with $\mu_e \approx 10^3$ m^2/V-s and $E = 60$ V/m. Continuity of current requires the same current at the edge of the cathode sheath region, where the current is carried only by the ions. This can be approximated by

$$J_i(r) = en_s(r)u_B$$

where for argon at $T_e = 4$ V we calculate $u_B = (eT_e/M)^{1/2} \approx 3 \times 10^3$ m/s. This is considerably less than the electron drift velocity $v_e = \mu_e E = 6 \times 10^4$ m/s, and thus, even ignoring the difference between n_s and n, it is not possible for the ions to carry the current in the cathode sheath. The resolution of this contradiction is that secondary electrons, created by ion impact at the cathode, are required to sustain the discharge. The process is similar to that involved in vacuum breakdown, and was first analyzed in that context. We first consider the more straightforward case of vacuum breakdown and then discuss the modifications required for treating the cathode sheath.

Vacuum Breakdown

Consider electrons emitted from a cathode at $z = 0$ being accelerated by an electric field and ionizing a neutral background. For a flux Γ_e in the z direction (the direction of the field) a differential equation for the increase in flux can be written

$$d\Gamma_e = \alpha(z)\Gamma_e \, dz \tag{14.3.1}$$

with the solution

$$\Gamma_e(z) = \Gamma_e(0) \exp\left[\int_0^z \alpha(z') \, dz'\right] \tag{14.3.2}$$

where $\alpha(z) \equiv 1/\lambda_{iz}(z)$ is the inverse of an "ionization" mean free path, analogous to the collisional mean free path defined in a similar way in Chapter 3. By continuity of total charge (creation of equal numbers of electron–ion pairs) the electron flux leaving the sheath edge at $z = d$, minus the electron flux emitted at $z = 0$, must be equal to the ion flux striking the cathode at $z = 0$, minus the ion flux that enters at $z = d$:

$$\Gamma_i(0) - \Gamma_i(d) = \Gamma_e(0) \left\{ \exp \left[\int_0^d \alpha(z') \, dz' \right] - 1 \right\} \tag{14.3.3}$$

where we have substituted for $\Gamma_e(d)$ from (14.3.2). For breakdown, the discharge must be self-sustaining. That is, setting $\Gamma_e(0) = \gamma_{se} \Gamma_i(0)$ where γ_{se} is the secondary electron emission coefficient at the cathode $z = 0$, then (14.3.3) must be satisfied with $\Gamma_i(d) = 0$. Solving for the exponential, we obtain

$$\exp \left(\int_0^d \alpha(z') \, dz' \right) = 1 + \frac{1}{\gamma_{se}} \tag{14.3.4}$$

as the self-sustaining condition. For a vacuum region, E is a constant and the electron drift velocity $v_e(z) = \mu_e E = \text{const}$. Hence the electron energy is a constant, allowing us to set $\alpha = \text{const}$ in (14.3.4). Taking the logarithm of both sides, we have

$$\alpha d = \ln \left(1 + \frac{1}{\gamma_{se}} \right) \tag{14.3.5}$$

the usual form for the breakdown condition of a dc discharge. The quantity α is known as the *first Townsend coefficient*. As might be expected from our knowledge of cross sections, α is a complicated function of the pressure and the accelerating field, which is very difficult to calculate. However, we might expect α to be expressed in the form

$$\alpha = \frac{\text{const}}{\lambda_e} \exp \left(-\frac{\mathcal{E}_{iz}}{E\lambda_e} \right) \tag{14.3.6}$$

where λ_e is the mean free path for electron scattering off neutrals, $E\lambda_e$ is a typical electron energy gain in the field, and $\mathcal{E}_{iz}$ is an energy for ionization. Here $E\lambda_e$ plays the role that T_e plays in (14.2.5). This form overestimates $\mathcal{E}_{iz}$, but is not used directly in calculating α. Rather, recognizing that $\lambda_e \propto p^{-1}$, then (14.3.6) can be written in the form

$$\frac{\alpha}{p} = A \exp \left(-\frac{Bp}{E} \right) \tag{14.3.7}$$

where A and B are determined experimentally and found to be roughly constant over a range of fields and pressures for any given gas. Some experimental values of α/p

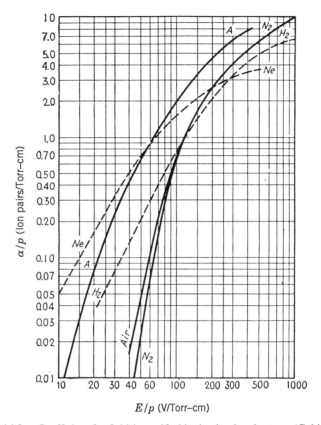

FIGURE 14.3.   Coefficient for field-intensified ionization by electrons (Cobine, 1958).

versus E/p are shown in Fig. 14.3. From curves such as these the coefficients in Table 14.1 are constructed.

Combining (14.3.7) with (14.3.5), and setting the breakdown voltage $V_b = Ed$, we have the relation

$$Apd\exp\left(-\frac{Bpd}{V_b}\right) = \ln\left(1 + \frac{1}{\gamma_{se}}\right) \qquad (14.3.8)$$

Solving (14.3.8) for V_b, we obtain

$$V_b = \frac{Bpd}{\ln Apd - \ln\left[\ln\left(1 + 1/\gamma_{se}\right)\right]} \qquad (14.3.9)$$

We see that the breakdown voltage is a function of the product pd. For large values of pd, V_b increases essentially linearly with pd. For small pd there is a minimum value of $Apd = \ln(1 + 1/\gamma_{se})$ below which breakdown cannot occur. The breakdown voltage is a minimum V_{min} at some intermediate value of pd. The curve $V_b(pd)$ is

TABLE 14.1. Constants of the Equation $\alpha/p = A \exp(-Bp/E)$

Gas	A	B	Range of E/p (V/cm-Torr)
Air	14.6	365	150–600
Ar	13.6	235	100–600
CO_2	20.0	466	500–1000
H_2	5.0	130	150–400
H_2O	12.9	289	150–1000
He	2.8	34	20–150

Source. Cobine (1958).

called the *Paschen curve,* and is a function of the gas and weakly a function of the electrode material. Typical breakdown curves for plane-parallel electrodes are shown in Fig. 14.4a, with the curve for air extended to high values of pd in Fig. 14.4b. As we shall see, the values of V_{min} and $(pd)_{min}$ play an important role in the more complicated problem of the cathode sheath.

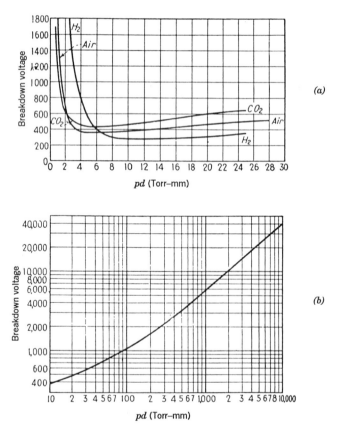

FIGURE 14.4. Breakdown voltage for plane parallel electrodes at 20°C: (a) low pressures; (b) high pressures for air (Cobine, 1958).

Cathode Sheath

We now consider the cathode sheath region of a discharge for which the electric field, and consequently α, is not a constant with position. Taking the logarithm of (14.3.4) we have

$$\int_0^d \alpha(z)\,dz = \ln\left(1 + \frac{1}{\gamma_{se}}\right) \tag{14.3.10}$$

An exact solution for $\alpha(z)$ would involve an integral equation for the field and be very difficult to solve. A simpler alternative is to measure the electric field distribution, which then becomes a known variation in determining $\alpha(z)$. Somewhat surprisingly (Cobine, 1958), it is found that the matrix sheath (constant ion space charge) well approximates the region, giving a linear field variation

$$E \approx E_0\left(1 - \frac{z}{d}\right) \tag{14.3.11}$$

with $z = 0$ at the cathode and $z = d$ at the sheath edge. Substituting (14.3.11) in (14.3.7) we have

$$\frac{\alpha}{p} = A\exp\left[-\frac{Bp}{E_0(1 - z/d)}\right] \tag{14.3.12}$$

and substituting (14.3.12) in (14.3.10) we obtain

$$\int_0^d Ap\exp\left[-\frac{Bp}{E_0(1 - z/d)}\right]dz = \ln\left(1 + \frac{1}{\gamma_{se}}\right) \tag{14.3.13}$$

which can be evaluated to give E_0 as a function of d. Integrating E in (14.3.11) from 0 to d, we can express E_0 in terms of the cathode sheath (cathode fall) voltage V_c as $E_0 = 2V_c/d$, which when substituted in (14.3.13) gives

$$\frac{AB(pd)^2}{2V_c}S\left(\frac{2V_c}{Bpd}\right) = \ln\left(1 + \frac{1}{\gamma_{se}}\right) \tag{14.3.14}$$

where

$$S(\zeta) = \int_0^\zeta e^{-1/y}\,dy \tag{14.3.15}$$

is a known tabulated integral. If one plots $V_c(pd)$ for a given gas (given A and B) and given electrode material (given γ_{se}) we find, as expected, curves that have a minimum $V_c = V_{cmin}$ at some $(pd)_{min}$. We might expect the discharge to adjust itself to this stable value of d, and this is indeed the case in the normal glow. Some values of

TABLE 14.2a. Normal Cathode Fall in Volts

Cathode	Air	Ar	H$_2$	He	Hg	N$_2$	Ne	O$_2$
Al	229	100	170	140	245	180	120	311
Ag	280	130	216	162	318	233	150	
C			240		475			
Cu	370	130	214	177	447	208	220	
Fe	269	165	250	150	298	215	150	290
Hg				142	340	226		
K	180	64	94	59		170	68	
Mg	224	119	153	125		188	94	310
Na	200		185	80		178	75	
Ni	226	131	211	158	275	197	140	
Pb	207	124	223	177		210	172	
Pt	277	131	276	165	340	216	152	364
Zn	277	119	184	143		216		354

Source. After Cobine (1958).

the cathode fall voltage are given in Table 14.2a, and some corresponding normal glow cathode fall thicknesses are given in Table 14.2b. These values are similar to the values for breakdown.

We have not quite reached the end of the story. It is also possible to eliminate d in favor of the current density and gain both new insight into the operation of the normal glow region and also understand the abnormal glow operation. The total current density at the cathode is given by

$$J(0) = en_i(0)v_i(0)(1 + \gamma_{se}) \qquad (14.3.16)$$

where n_i is the ion density, v_i is the ion velocity, and γ_{se} gives the fraction of the current due to secondary electrons. Using Poisson's equation with the assumption of constant charge density, we can write en_i in terms of the cathode fall potential $en_i(0) = \epsilon_0 2V_c/d^2$. Similarly, assuming a collisional sheath, we have $v_i(0) = \mu_i 2V_c/d$, where

TABLE 14.2b. Normal Cathode Fall Thickness pd in Torr-cm

Cathode	Air	Ar	H$_2$	He	Hg	N$_2$	Ne	O$_2$
Al	0.25	0.29	0.72	1.32	0.33	0.31	0.64	0.24
C			0.9		0.69			
Cu	0.23		0.8		0.6			
Fe	0.52	0.33	0.9	1.30	0.34	0.42	0.72	0.31
Hg			0.9					
Mg			0.61	1.45		0.35		0.25
Ni			0.9		0.4			
Pb			0.84					
Pt			1.0					

Source. After Cobine (1958).

μ_i is the ion mobility. Substituting these values in (14.3.16) we obtain

$$J(0) = \frac{4\epsilon_0 \mu_i V_c^2 (1 + \gamma_{se})}{d^3} \qquad (14.3.17)$$

from which we can eliminate d in favor of $J(0)$. Hence we can determine a Paschen-type curve of V_c versus $J(0)$. This is shown in Fig. 14.5 in terms of normalized parameters.

It is clear that with a fixed external voltage source V_T and resistance R_T, the dashed curve is unstable, such that if $J = I/A < J_{min}$, where A is the effective cathode area; that is, if

$$\frac{V_T - V_{cmin}}{R_T A} < J_{min}$$

then the cathode fall area will constrict to a smaller value. This is the normal glow region. On the other hand, for

$$\frac{V_T - V_{cmin}}{R_T A} > J_{min}$$

the solution is stable, and V_c will increase with increasing current density. It is this region that is called the *abnormal glow*, but as we can see, it is just as normal as the normal glow.

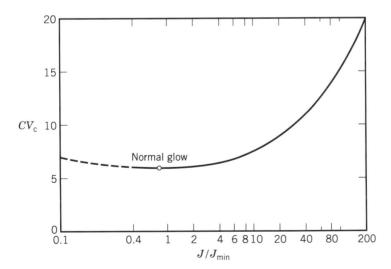

FIGURE 14.5. Cathode voltage drop versus discharge current, illustrating the normal and abnormal glow; $C = 2A/B \ln[1 + (1/\gamma_{se})]$ (Cobine, 1958).

14.4   THE NEGATIVE GLOW AND FARADAY DARK SPACE

As discussed qualitatively in Section 14.1, when the electrons have multiplied sufficiently that they can carry the current in the cathode sheath, the high electric field must decrease to create plasmalike, rather than sheathlike conditions. However, the majority of electrons have been accelerated by a high field and are thus far from equilibrium. A local region of high ionization and excitation must therefore exist while the electrons are slowing down. This is characterized by a mean free path for the combination of scattering and energy loss processes, sometimes referred to as the range of the energetic electrons. In fact, the electric field can actually reverse in this region to keep the electron current in balance, contributing to the slowing down process. Overall, the visual region of the negative glow has been correlated with the measured range of fast electrons in various gases, obtaining good agreement.

The reversal of the field tends to exclude ions from a region of the column, as shown in Fig. 14.1. This region of low ion density prevents the negative glow from joining directly onto the essentially neutral positive column, requiring one more transition region. Although an exact analysis is difficult, a simple calculation produces the correct scaling and surprisingly good quantitative agreement with experiment. If electrons are assumed to start from rest and accelerate through a mean free path, the kinetic energy gained is

$$\frac{1}{2}mv_{res}^2 = eE\lambda_e = eV_{res} \tag{14.4.1}$$

where V_{res} is known as the resonance voltage. Setting the current density as

$$J = env_{res} = en\left(\frac{2eE\lambda_e}{m}\right)^{1/2} \tag{14.4.2}$$

where we have substituted for v_{res} from (14.4.1), then n can be used in Poisson's equation to obtain

$$\frac{dE}{dz} = \frac{J}{\epsilon_0}\left(\frac{m}{2eE\lambda_e}\right)^{1/2} \tag{14.4.3}$$

Integrating (14.4.3) and substituting $E = V_{res}/\lambda_e$ from (14.4.1), we obtain

$$V_{res} = \left(\frac{2e}{m}\right)^{-1/3}\left(\frac{3}{2}\frac{J\lambda_e}{\epsilon_0}z\right)^{2/3} \tag{14.4.4}$$

Assuming that V_{res} is a constant then we find that the length of the Faraday dark space scales as

$$z \propto \frac{1}{\lambda_e J} \propto \frac{p}{J} \tag{14.4.5}$$

which is found to hold experimentally provided the pressure is sufficiently high that the Faraday dark space is collisional.

14.5   PLANAR MAGNETRON DISCHARGES

Dc planar magnetron discharges are widely used for sputter deposition of metallic thin films such as aluminum, tungsten, gold, and various alloys; e.g., Al/2%Cu and Ti/W. When powered by an rf source, these discharges are also used for sputter deposition of insulating films such as oxides, nitrides, and ceramics. Physical sputtering is described in Section 9.3, and its application to thin-film deposition is described in Section 16.3. In this section we first present the limitations of glow discharges as sputtering sources. We then describe the planar magnetron configuration and present a simple equilibrium model of these discharges that can be used to estimate discharge parameters and sputtering efficiency. For a more thorough understanding, the review articles by Thornton and Penfold (1978) and Waits (1978) and more recent studies by Wendt and Lieberman (1990) and Sheridan et al. (1991) should be consulted.

Limitations of Glow Discharge Sputtering Source

Low aspect ratio ($l/R < 1$ for a cylindrical plasma) dc glow discharges have long been used as sputtering sources for metallic materials and are still used as sputtering sources in some specialized applications. These are illustrated in Fig. 14.6a for a planar discharge in argon gas driven by a constant current dc source. The upper aluminum electrode is the cathode, which serves as the target for ion impact sputtering of aluminum atoms. The substrates, on which the sputtered atoms are deposited, are placed on the lower electrode, which is the anode. The cathode–anode gap is typically $l \sim 5$ cm. Almost all of the anode–cathode voltage appears across the cathode sheath (dark space or cathode fall). The negative glow extends almost to the anode, and the positive column is absent in these short discharges. High ion current densities, $J_{dc} \gtrsim 1$ mA/cm^2, are required in order to achieve, at best, commercially viable deposition rates of ≈ 350 Å/min. Hence the discharge is operated in the abnormal glow regime with a high discharge voltage, $V_{dc} \sim 2$–5 kV. The sputtering power efficiency (sputtered atoms/ion-volt) is relatively low at these high energies and decreases with increasing energy (see Section 9.3).

As described in Section 14.3, the discharge is maintained in the usual manner by secondary electron emission from the cathode, with the energetic secondary electrons providing the ionization required to maintain the discharge. However, operating pressures must be high enough,

$$p \gtrsim 30 \text{ mTorr} \tag{14.5.1}$$

so that secondary electrons are not lost to the anode or side walls. These pressures are higher than optimum for deposition of sputtered atoms onto the substrates due to scattering of sputtered atoms by argon atoms. This results in sputtered atom

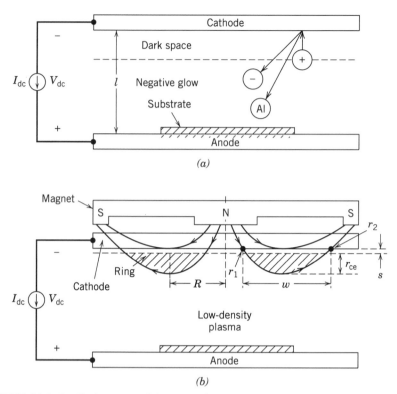

FIGURE 14.6. Dc discharges used for sputtering: (a) low aspect ratio dc glow discharge; (b) planar magnetron discharge.

redeposition on the cathode, deposition on the side walls, and, in some cases, poor adhesion for the sputtered film (see Section 16.3). For a neutral–neutral scattering cross section $\sigma \sim 2 \times 10^{-16}$ cm^2, setting the mean free path $\lambda = (n_g \sigma_{el})^{-1} \lesssim l$, we obtain

$$p \lesssim 30 \,\text{mTorr} \tag{14.5.2}$$

for acceptably low sputtered atom scattering. Equations (14.5.1) and (14.5.2) indicate that there is a narrow pressure range around 30 mTorr for dc glow discharge sputtering.

Magnetron Configuration

It is clearly desirable to operate a sputtering discharge at higher current densities, lower voltages, and lower pressures than can be obtained in a conventional glow discharge. This has led to the use of a dc magnetic field at the cathode to confine the secondary electrons. An axisymmetric dc magnetron configuration is shown in Fig. 14.6b. The permanent magnet placed at the back of the cathode target generates magnetic field lines that enter and leave through the cathode plate as shown. A discharge is formed when a negative voltage of 200 V or more is applied to the cathode. The discharge appears in the form of a high-density brightly glowing circular plasma ring of width w and mean radius R that hovers below the cathode, as illustrated in the figure, with sputtering occurring in a corresponding track on the cathode. The ring is embedded in a lower density bulk plasma. The plasma shields the electric field through most of the chamber, and a cathode sheath of the order of $s = 1$ mm develops, which sustains most of the externally applied voltage. Argon ions in the plasma, unconfined by the magnetic field, are accelerated toward the cathode and strike it at high energy. In addition to sputtering target material, the impact of the ions produces secondary electron emission. These electrons are accelerated back into the plasma and are confined near the cathode by the magnetic field. The electrons undergo a sufficient number of ionizing collisions to maintain the discharge before being lost to a grounded surface. Typical planar magnetron characteristics are

- $B_0 \sim 200$ G
- $p \sim 2$–5 mTorr argon
- $\bar{J}_i \sim 20$ mA/cm^2
- $V_{dc} \sim 800$ V
- Deposition rate ~ 2000 Å/min

Here B_0 is the magnetic field strength at the radius R where the magnetic field line is tangential to the cathode surface, and $\bar{J}_i$ is the average ion current density over the ring area.

Discharge Model

Because the magnetic field and discharge structure are highly nonuniform, a complete quantitative model of the discharge has not been developed. We present a qualitative model to indicate some issues that arise when determining the equilibrium properties of the discharge. The given discharge control parameters are I_{dc}, p, B_0, and R.

(a) Voltage V_{dc}   Almost all of the applied voltage is dropped across the cathode sheath. The secondary emission coefficient for argon ions on aluminum is $\gamma_{se} \sim 0.1$ for 200- to 1000-V argon ions. If $\mathcal{N}$ is the number of electron–ion pairs created by each secondary electron that is trapped within the ring, then an estimate for $\mathcal{N}$ is

$$\mathcal{N} \approx \frac{V_{dc}}{\mathcal{E}_c} \qquad (14.5.3)$$

where $\mathcal{E}_c$ is the energy lost per electron–ion pair created by secondary electrons. For 200- to 1000-V secondaries, we take $\mathcal{E}_c \approx 30$ V (see Fig. 3.17). Because of the tangential magnetic field, not all secondary electrons emitted at the cathode are trapped in the ring. Some electrons execute one or more gyro orbits and are reabsorbed at the cathode. This leads to an effective secondary emission coefficient γ_{eff} that is less than γ_{se}. Thornton and Penfold (1978) have estimated

$$\gamma_{eff} \approx \frac{1}{2} \gamma_{se} \tag{14.5.4}$$

In steady state, for ion particle balance, we require

$$\gamma_{eff} \mathcal{N} = 1 \tag{14.5.5}$$

Inserting (14.5.3) and (14.5.4) into (14.5.5), we obtain

$$V_{dc} \approx \frac{2\mathcal{E}_c}{\gamma_{se}} \tag{14.5.6}$$

For $\mathcal{E}_c = 30$ V and $\gamma_{se} = 0.1$, we find $V_{dc} \approx 600$ V.

The energetic electron gyroradius is found from (4.1.8) to be

$$r_{ce} = \frac{v_e}{\omega_{ce}} = \frac{1}{B_0} \left(\frac{2mV_{dc}}{e} \right)^{1/2} \tag{14.5.7}$$

where $v_e = (2eV_{dc}/m)^{1/2}$. For $B_0 = 200$ G and $V_{dc} = 600$ V, we obtain $r_{ce} \approx 0.5$ cm. The energetic ion gyroradius is similarly found from

$$r_{ci} = \frac{1}{B_0} \left(\frac{2MV_{dc}}{e} \right)^{1/2} \tag{14.5.8}$$

to give $r_{ci} \approx 1.3$ m for argon ions. Therefore, the energetic ions are not magnetized by the weak magnetic field in this discharge.

(b) Ring Width w   Referring to Fig. 14.7, we estimate that the ring has mean height (from the cathode) equal to the gyration radius r_{ce}. We assume that the sheath width $s \ll r_{ce}$, and will show this below. Hence energetic secondary electrons are trapped on a magnetic field line and can oscillate back and forth between radii r_1 and r_2. The main force that reflects the electrons at r_1 and r_2 is the electrostatic sheath potential; there can also be some mirroring, due to the nonuniform magnetic field, which results

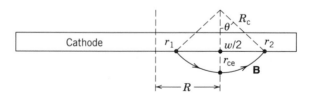

FIGURE 14.7.   Calculation of planar magnetron ring width.

in a parallel force as given by (4.3.9). For the field line having a radius of curvature R_c and height r_{ce} from the cathode, given $w \approx r_2 - r_1$, then, as shown in Fig. 14.7,

$$\frac{w/2}{R_c} = \sin\theta \tag{14.5.9}$$

and

$$r_{ce} + R_c \cos\theta = R_c \tag{14.5.10}$$

Eliminating θ from these two equations yields w as a function of r_{ce} and R_c. For simplicity, assuming that $w/2 \ll R_c$, then (14.5.9) and (14.5.10) become

$$\frac{w}{2R_c} \approx \theta$$

$$\frac{2r_{ce}}{R_c} \approx \theta^2$$

Eliminating θ, we obtain an estimate of the ring width:

$$w \approx 2(2r_{ce}R_c)^{1/2} \tag{14.5.11}$$

For $r_{ce} \approx 0.5$ cm and choosing $R_c = 4$ cm, we obtain $w \approx 4$ cm. This is not fully consistent with the simplifying assumption $w/2 \ll R_c$, but a more accurate result can be found by solving (14.5.9) and (14.5.10) (see Problem 14.5).

(c) Ion Current Density $\bar{J}_i$ and Sheath Thickness s   The ions are unmagnetized and the gas pressure is low; therefore the collisionless Child law (6.3.12) can be used to describe the flow of ions from the surface of the ring to the cathode:

$$\bar{J}_i = \frac{4}{9}\epsilon_0 \left(\frac{2e}{M}\right)^{1/2} \frac{V_{dc}^{3/2}}{s^2} \tag{14.5.12}$$

Assuming for simplicity that the ring is thin, $w \ll R$, then we find

$$\bar{J}_i = \frac{I_{dc}}{2\pi Rw} \tag{14.5.13}$$

Taking the typical parameters $I_{dc} = 5$ A, $R = 5$ cm, $w = 4$ cm, and $V_{dc} = 600$ V, then we obtain $\bar{J}_i \approx 40$ mA/cm^2 from (14.5.13) and $s \approx 0.56$ mm from (14.5.12).

(d) Plasma Density n_i   We use the Bohm flux at the edge of a collisionless plasma to estimate n_i in the ring from

$$0.61\, en_i u_B = \bar{J}_i \tag{14.5.14}$$

where the electron temperature T_e enters only weakly. Using a typical value for low-pressure discharges of $T_e \approx 3$ V, then from (14.5.14) with $\bar{J}_i \approx 40$ mA/cm^2 we obtain $n_i \approx 1.5 \times 10^{12}$ cm^{-3}. We note that the density n_{es} of energetic electrons within the ring is generally much smaller than n_i (see Problem 14.4).

(e) Sputtering Rate R_{sput} Letting γ_{sput} be the yield of sputtered atoms per incident ion, we have the sputtering rate

$$R_{sput} = \gamma_{sput} \frac{\bar{J}_i}{e} \frac{1}{n_{Al}} \text{ cm/s} \qquad (14.5.15)$$

Taking $n_{Al} \approx 6 \times 10^{22}$ cm^{-3} to be the atomic density of the aluminum target and evaluating (14.5.15) for $\gamma_{sput} \sim 1$ and $\bar{J}_i \approx 40$ mA/cm^2, we obtain $R_{sput} \approx 4.1 \times 10^{-6}$ cm/s. After 24 hours of operation, a target thickness of 3.6 mm has been sputtered. Thus the discharge digs an erosion track into the cathode material beneath the ring. The cathode must be replaced when the erosion track becomes comparable to the cathode thickness, which is usually an expensive proposition for the ultrapure materials commonly used as targets.

The discharge power is $P_{abs} = V_{dc}I_{dc}$. For our example with $I_{dc} = 5$ A and $V_{dc} = 600$ V, $P_{abs} = 3$ kW. Almost all of this power is absorbed at the cathode; hence the cathode must be cooled. Uniformity of the sputtered film deposited on the substrate is an important issue. For a ring-shaped source of sputtered atoms, the aspect ratio l/R controls the radial uniformity. As expected, for $l \ll R$, the radial distribution of deposited atoms has a maximum off the axis, while for $l \gg R$, the radial distribution is (mildly) peaked on-axis. Hence an optimum ratio of l/R exists that maximizes the deposition uniformity. Measurements and analysis show that $l/R \approx 4/3$ (Thornton and Penfold, 1978).

PROBLEMS

14.1. Positive Column of a Dc Glow Discharge

(a) A glow discharge in argon with $R = 2$ cm, $l = 25$ cm is operated at $p = 100$ mTorr. This gives $T_e \approx 4$ V, $\mathcal{E}_T \approx 40$ V, $\lambda_e \approx 0.4$ cm, and $D_a p \approx 10^4$ cm^2-Torr/s. Assuming the solution in Section 14.2, and using (14.2.14), find the electric field strength E. Assuming that most of the discharge is positive column, what is the voltage drop in the positive column?

(b) If the discharge current in part (a) is 10 mA, what is the plasma density?

14.2. Breakdown of a Dc Discharge   For the previous discharge, but at $p = 1$ Torr and a cathode secondary emission constant $\gamma_{se} = 0.12$, using values from Table 14.1, calculate the vacuum breakdown voltage for the discharge.

14.3. Operation of a Dc Discharge   For the parameters of Problem 14.2, take $T_e \approx 3$ V and $\mathcal{E}_T \approx 50$ V, and $D_a p \approx 10^4$ cm^2-Torr/s.

(a) Recompute the results for the positive column of Problem 14.1.

(b) Using Fig. 14.5, calculate the voltage drop across the cathode sheath.

(c) For an applied voltage $V_B = 1500$ V ($V_B > V_b$ in Problem 14.2) what resistance should be put in series with the applied voltage to supply the 10 mA required.

14.4. Planar Magnetron Discharge   An axially symmetric planar magnetron discharge in argon with an aluminum cathode has a magnetic field strength $B_0 = 200$ G at a radius $R = 10$ cm, where the field line is tangent to the cathode surface. The field line radius of curvature is $R_c = 3$ cm. The discharge current is $I_{dc} = 2$ A and the pressure is $p = 2$ mTorr.

(a) Assuming that the effective secondary emission coefficient $\gamma_{eff} \approx 0.05$ for Ar^+ ions on Al and that 20% of the secondary electrons are lost by diffusive transport to the anode before creating electron–ion pairs in the plasma ring, estimate the dc voltage V_{dc} across the discharge.

(b) Estimate the mean width w (in the r direction) of the ring (erosion track) and the ion current density $\bar{J}_i$ (mA/cm²) incident on the aluminum cathode over the erosion track area $2\pi Rw$.

(c) Assuming that the sputtering coefficient at the erosion track is unity (1 sputtered Al atom per incident Ar ion) and that sputtered atoms are deposited uniformly on the anode surface over an area of πR_a^2, where $R_a \approx 15$ cm, estimate the deposition rate (Å/min) for the aluminum film deposited on wafers located at the anode surface.

(d) Estimate the (low-temperature) plasma density n_i (cm⁻³) within the ring and the secondary electron density n_{se} within the ring. (Use the data given in Fig. 3.13 for the argon ionization cross section σ_{iz}(m²) for secondary electrons having energy eV_{dc}, and note that secondary electrons lose about 30 V per ionization. From this information, the secondary electron lifetime can be determined.) Note that you should find $n_{se} \ll n_i$.

14.5. Planar Magnetron Ring Width   Use the exact equations (14.5.9) and (14.5.10) to determine w for $r_{ce} \approx 0.5$ cm and $R_c = 4$ cm, and compare to the result $w \approx 4$ cm obtained in (14.5.11) for the simplified analysis.

CHAPTER 15

ETCHING

15.1 ETCH REQUIREMENTS AND PROCESSES

Plasma etching is a key process for removing material from surfaces. The process can be chemically selective, removing one type of material while leaving other materials unaffected, and can be anisotropic, removing material at the bottom of a trench while leaving the same material on the sidewalls unaffected. Plasma etching is the only commercially viable technology for anisotropic removal of material from surfaces. As such, it is an indispensable part of modern integrated circuit fabrication technology, as was described in Chapter 1. For a more complete description of this area, the reader should consult other sources; e.g., Manos and Flamm 1989, Chapters 1 and 2.

Although there are many other areas of application, nearly all modern developments in plasma etching have been driven by their potential for integrated circuit fabrication. In this chapter, we focus almost exclusively on this area, placing emphasis on the key concepts that determine etch rate, selectivity, and anisotropy in plasma etch processes. In this section, we introduce typical etch requirements and possible tradeoffs among them and describe the four types of plasma etch processes. In Section 15.2, some simple models of surface etching and discharge kinetics are described and a general chemical framework for plasma etching is introduced. In Section 15.3, the use of halogens to etch silicon is discussed. In particular, fluorine atom etching of silicon has been the most well-studied etch system, providing insight into other less well-characterized systems. In Section 15.4, some descriptions of silicon oxide and nitride etching, metal etching, and photoresist etching are given.

Plasma Etch Requirements

It is important to consider the entire set of processing requirements for a particular application. For pattern transfer by etching on a silicon wafer, these might include requirements on etch rate, anisotropy, selectivity, uniformity across the wafer, surface quality, and process reproducibility. Consider first the *etch rate* requirements for the typical set of films, shown in Fig. 15.1a, consisting of 1.5 μm (15000 Å) of resist, over 3000 Å of polysilicon, and over 300 Å of "gate" oxide on an epitaxial silicon wafer. For a single wafer process of commercial interest, each of these films must be etched in 2–3 minutes. This leads to minimum etch rate requirements E_{pr} = 5000 Å /min for the photoresist and E_{ox} = 100 Å /min for the gate oxide, with an intermediate requirement of E_{poly} = 1000 Å /min for the polysilicon.

Next, consider the *selectivity* requirements for the polysilicon etch. For etch of the 3000-Å polysilicon with the resist as a mask, a selectivity of

$$ s = \frac{E_{poly}}{E_{pr}} \gg \frac{0.3}{1.5} $$

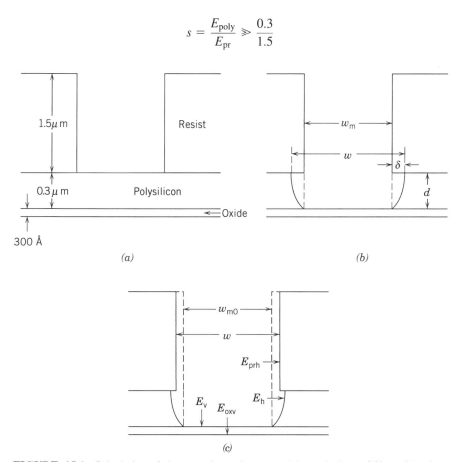

FIGURE 15.1. Calculation of plasma etch requirements: (a) a typical set of films; (b) anisotropy requirement for polysilicon etch; (c) uniformity requirement, including the effect of photoresist erosion.

is required in order to complete the polysilicon etch while not significantly eroding the resist. For this application, a selectivity of 2–3 might be acceptable. However, there is a second selectivity requirement. Due to a lack of *uniformity* across the wafer, it is necessary to overetch the polysilicon at some locations on the wafer in order to clear it from all unmasked regions. During overetch, the thin oxide is exposed to the etchant at some wafer locations. For an overetch of 100% (twice the etch time required to clear a perfectly uniform polysilicon film using a perfectly uniform process), a polysilicon to gate oxide selectivity of

$$s = \frac{E_{poly}}{E_{ox}} \gg \frac{3000}{300}$$

is required. Depending on the use, selectivities of 30–300 might be needed. Hence selectivity to the underlying layer can be a significant issue in film removal.

Consider now, as shown in Fig. 15.1b, the *anisotropy* required to etch a trench of width w into a film of thickness d, and let w_m be the minimum mask feature size that can be used (e.g., due to lithography limitations). The anisotropy for the film etch is defined as

$$a_h = \frac{E_v}{E_h} \tag{15.1.1}$$

where E_v and E_h are the vertical and horizontal etch rates, respectively. Assuming that the mask is not eroded, then after the etch, as shown in the figure, we find the relation

$$a_h = \frac{d}{\delta} \tag{15.1.2}$$

The maximum width of the trench is

$$w = w_m + 2\delta \tag{15.1.3}$$

Solving for δ and substituting this into (15.1.2), we obtain the anisotropy requirement

$$a_h \geq \frac{2d}{w - w_m} \tag{15.1.4}$$

As an example, for $w = 1\ \mu m$, $d = 2\ \mu m$, and $w_m = 0.5\ \mu m$, we obtain $a_h \geq 8$. Even for $w_m \equiv 0$ (physically unreasonable), an anisotropy of $a_h \geq 4$ is required. Evidently, the smallest feature size that can be made has a width

$$w \approx \frac{2d}{a_h} \tag{15.1.5}$$

For etching of deep trenches $(d/w \gg 1)$, the etch anisotropy requirements can be severe.

Consider now the impact of process *uniformity* on selectivity and anisotropy requirements. Refering to Fig. 15.1c, we now also consider erosion of the mask sidewalls due to a horizontal etch rate E_{prh} and erosion of the underlying gate oxide sublayer due to a vertical etch rate E_{oxv}. Introducing the selectivities

$$s_{pr} = \frac{E_v}{E_{prh}} \tag{15.1.6a}$$

$$s_{ox} = \frac{E_v}{E_{oxv}} \tag{15.1.6b}$$

where E_v is the polysilicon vertical etch rate, we let t_{max} be the time required to completely etch the polysilicon layer at all unmasked locations on the wafer. Then

$$t_{max} = f \frac{d}{E_v} \tag{15.1.7}$$

where f is an overetch nonuniformity factor due to film thickness variations,

$$d = d_0(1 \pm \alpha) \tag{15.1.8}$$

and etch rate variations across the wafer,

$$E_v = E_{v0}(1 \pm \beta) \tag{15.1.9}$$

Hence t_{max} is determined by the maximum thickness and minimum etch rate

$$f = \frac{1 + \alpha}{1 - \beta} \approx 1 + \alpha + \beta \tag{15.1.10}$$

where we have assumed that α, $\beta \ll 1$. Assuming also that $E_v \gg E_h$ in the polysilicon, the horizontal etch width δ can be estimated as the sum of two terms,

$$\delta \approx (E_h + E_{prh})t_{max} \tag{15.1.11}$$

where the horizontal etch rate in the resist mask leads to a horizontal etch of the polysilicon because the vertical etch in the polysilicon is fast compared to the horizontal erosion of the mask. Substituting (15.1.7) into (15.1.11), we find

$$\delta \approx fd \frac{E_h + E_{prh}}{E_v} \tag{15.1.12}$$

Substituting (15.1.12) in (15.1.3), we obtain

$$w \approx w_{m0} + 2fd\frac{E_h + E_{prh}}{E_v}$$

or

$$\frac{E_h + E_{prh}}{E_v} \approx \frac{w - w_{m0}}{2fd} \tag{15.1.13}$$

For example, letting $w = 1~\mu m$, $w_{m0} = 0.5~\mu m$, $d = 0.5~\mu m$, and $\alpha = \beta = 0.1$, we find $(E_h + E_{prh})/E_v \approx 0.42$. Hence, as shown in Fig. 15.2a, one can trade resist selectivity (E_{prh}) against etch anisotropy (E_h) within a triangular window in parameter space near the origin. For a deeper trench, e.g., $d = 2~\mu m$, we find $(E_h + E_{prh})/E_v \approx 0.10$, and the requirements on resist selectivity and etch anisotropy (dashed curve) become more severe.

Consider now the etch of the gate oxide sublayer. In a worst-case analysis, the etch begins at a time

$$t_{min} = \frac{d}{E_v}\frac{1 - \alpha}{1 + \beta} \tag{15.1.14}$$

The maximum sublayer thickness etched is then

$$\delta_{ox} = (t_{max} - t_{min})E_{oxv}$$

$$= \frac{d}{E_v}2(\alpha + \beta)E_{oxv} \tag{15.1.15}$$

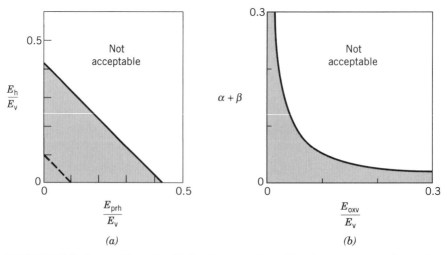

FIGURE 15.2. Acceptable trade-offs for plasma etching: (a) anisotropy versus photoresist selectivity; (b) uniformity versus oxide selectivity.

This can be rewritten as

$$(\alpha + \beta)\frac{E_{\mathrm{oxv}}}{E_{\mathrm{v}}} = \frac{\delta_{\mathrm{ox}}}{2d} \tag{15.1.16}$$

which shows that uniformity $(\alpha + \beta)$ can be traded against sublayer selectivity $(E_{\mathrm{oxv}}/E_{\mathrm{v}})$, within the hyperbolic region near the origin, as shown in Fig. 15.2b. As an example, for $d = 3000$ Å and $\alpha = \beta = 0.1$, a selectivity to achieve $\delta_{\mathrm{ox}} \leq 30$ Å of $s_{\mathrm{ox}} = E_{\mathrm{v}}/E_{\mathrm{oxv}} \geq 40$ is required. For a more detailed estimation of the trade-offs among anisotropy, selectivity, and uniformity, the review by Flamm and Herb (1989) should be consulted.

Etch Processes

There are four basic low-pressure plasma processes commonly used to remove material from surfaces: sputtering, pure chemical etching, ion energy driven etching, and ion inhibitor etching. Sputtering is the ejection of atoms from surfaces due to energetic ion bombardment. This process was described in Section 9.3 and is illustrated in Fig. 15.3a. The discharge supplies energetic ions to the surface, with the ions typically having energies above a few hundred volts. Sputtering is an unselective process since, from (9.3.14), the sputtering yield γ_{sput} roughly depends only on the surface binding energy $\mathcal{E}_t$ and the masses of the targets and projectiles. Typically, these do not vary by more than a factor of 2–3 among different materials. Hence the sputtering

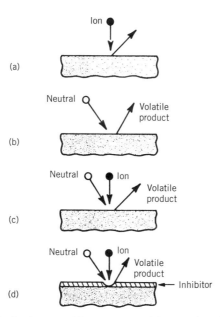

FIGURE 15.3. Four basic plasma etching processes: (a) sputtering; (b) pure chemical etching; (c) ion energy-driven etching; (d) ion-enhanced inhibitor etching (Flamm and Herb, 1989).

rates of different materials are roughly the same (see Table 9.2). Sputtering rates are generally low because the yield is typically of order one atom per incident ion, and ion fluxes incident on surfaces in discharges are often small compared to commercially significant rates for materials removal. Sputtering is, however, an anisotropic process, strongly sensitive to the angle of incidence of the ion. Fig. 15.4 shows typical angular dependences of the sputtering yield for aluminum and photoresist films. In both cases the yield rises from its normal (0°) incidence value to some maximum value γ_{max} at θ_{max} and then falls to zero at grazing incidence (90°). Therefore, there is essentially no sidewall removal of material for ions normally incident on a substrate. However, because the sputtering yield peaks at $\theta_{max} \neq 0$, topographical patterns might not be faithfully transferred during sputter etching. Figure 15.5 shows ions at normal incidence on a step (a) before and (b) after sputtering. A facet has developed after sputtering due to the peaking of the yield at θ_{max}. Sputtering is the only one of the four etch processes that can remove involatile products from a surface. This is important for removing low fraction involatile components during film etching using other processes, e.g., the sputter removal of copper during etching of Al–2%Cu films. It is also important for sputter deposition processes, which are described in Chapter 16. For these applications, the mean free path of the sputtered atoms must be large enough to prevent redeposition on the substrate or target. Consequently, these processes are generally carried out at low pressure.

A second etch process is pure chemical etching, in which the discharge supplies gas-phase etchant atoms or molecules that chemically react with the surface to form gas-phase products. This process can be highly chemically selective. Some examples

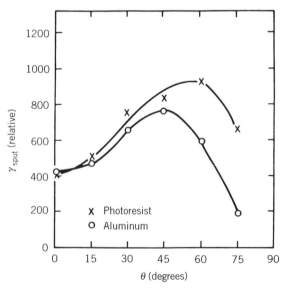

FIGURE 15.4. Relative sputtering yields for photoresist and aluminum versus angle of incidence θ (after Flamm and Herb, 1989).

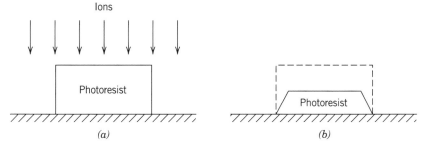

FIGURE 15.5. The development of facets due to sputtering of photoresist: (a) before sputtering; (b) after sputtering.

are

$$Si(s) + 4F \longrightarrow SiF_4(g)$$
$$photoresist + O(g) \longrightarrow CO_2(g) + H_2O(g)$$

Pure chemical etching, illustrated in Fig. 15.3b, is almost invariably isotropic, since the gas-phase etchants arrive at the substrate with a near uniform angular distribution. Therefore, unless the reaction is with a crystal having a rate depending on crystallographic orientation, one may expect a relatively isotropic etch rate. As described in Section 1.1, the etch products must be volatile. The etch rate can be quite large because the flux of etchants to the substrate can be high in processing discharges. However, etch rates are generally not limited by the rate of arrival of etchant atoms, but by one of a complex set of reactions at the surface leading to formation of etch products. For example, for F-atom etching of silicon, there is considerable evidence that the rate-limiting step involves the reaction of an F^- ion generated at the surface with the fluorinated surface layer. We consider some of these issues in Section 15.3.

A third etch process, illustrated in Fig. 15.3c, is *ion-enhanced energy-driven etching,* in which the discharge supplies both etchants (e.g., F atoms) and energetic ions to the surface. The combined effect of both etchant atoms and energetic ions in producing etch products can be much larger than that produced by either pure chemical etching or by sputtering alone, as is shown in Fig. 1.4. For etching of silicon with a high incident flux of F atoms, for example, a single 1-kV argon ion can cause the removal of as many as 25 silicon atoms (and 100 fluorine atoms) from the surface. Experiments suggest that the etching is chemical in nature, but with a reaction rate determined by the energetic ion bombardment. The etch rate generally increases with increasing ion energy above a threshold energy of a few volts. The etch product must be volatile, as for pure chemical etching. Because the energetic ions have a highly directional angular distribution when they strike the substrate, the etching can be highly anisotropic. However, ion energy-driven etching may have poor selectivity compared to pure chemical etching. The trade-off between anisotropy and selectivity is important in designing etch processes, as was shown earlier. The detailed mechanism for etch product formation and the rate of formation are not well understood. An empirical model is given in Section 15.2, and some proposed mechanisms are described in Section 15.3.

A fourth type of etch process, *ion-enhanced inhibitor etching*, illustrated in Fig. 15.3d, involves the use of an inhibitor species (see Section 7.5). The discharge supplies etchants, energetic ions, and inhibitor precursor molecules that adsorb or deposit on the substrate to form a protective layer or polymer film. The etchant is chosen to produce a high chemical etch rate of the substrate in the absence of either ion bombardment or the inhibitor. The ion-bombarding flux prevents the inhibitor layer from forming or clears it as it forms, exposing the surface to the chemical etchant. Where the ion flux does not fall, the inhibitor protects the surface from the etchant. Inhibitor precursor molecules include CF_2, CF_3, CCl_2, and CCl_3 molecules, which can deposit on the substrate to form fluoro- or chloro-carbon polymer films. A classic example of an ion inhibitor plasma etch, described in more detail in Section 15.5, is the anisotropic etching of aluminum trenches or holes using CCl_4/Cl_2 or $CHCl_3/Cl_2$ discharges. Both Cl and Cl_2 rapidly etch aluminum, but the resulting etch is isotropic. The addition of carbon to the feedgas mix results in formation of a protective chlorocarbon film on the surface. Ion bombardment clears the film from the trench bottom, allowing the etch process to proceed there. The same film on the sidewalls protects these from the etchant. With proper optimization, a highly anisotropic etch with vertical sidewalls can be formed. Ion inhibitor etching shares most other features of ion energy-driven etching. The process may not be as selective as pure chemical etching, and a volatile etch product must be formed. Contamination of the substrate and final removal of the protective inhibitor film are other issues that must be addressed for this etch process.

Although four etch processes have been distinguished, their use for a particular film etch often involves parallel or serial combinations of the processes, as has already been noted for Al–2%Cu etching. Consider, for example, the cutting of a vertical trench in a thick polysilicon layer that must stop with high selectivity at a silicon dioxide layer, as shown in Fig. 15.1. This might be accomplished by a two-step process. The first step might be a fast, highly anisotropic ion energy-driven etch. Pure chemical etching in parallel at the sidewalls might determine the anisotropy of this process. After 90% of the polysilicon has been removed, the final step might be a slow, highly selective, but relatively isotropic etch to remove the remaining polysilicon with minimum etching of the underlying oxide. The small undercut produced by this step might be acceptable if the polysilicon that remains after the first step is thin enough.

15.2 ETCHING KINETICS

With the exception of the physical sputtering of elemental materials, the detailed mechanisms for plasma etch processes are not well understood. Simple empirical models that incorporate some of the key observations can provide insight into the use of various processes. In this section, kinetic models for surface etch processes are introduced in which known neutral and ion fluxes at the surface are used to determine the etch rate and anisotropy. These fluxes, in turn, must be found using a discharge model that accounts for the generation of both etchant atoms and bombarding ions. Finally, a general framework for the chemistry of etch processes is introduced; this will be elaborated in subsequent sections.

Surface Kinetics

Consider first the example of an ion energy-driven process for O atom etching of a carbon substrate. We assume that the only reactions that occur are

$$O(g) + C(s) \xrightarrow{K_a} C : O \tag{15.2.1}$$

$$C : O \xrightarrow{K_d} CO(g) \tag{15.2.2}$$

$$\text{ion} + C : O \xrightarrow{Y_i K_i} CO(g) \tag{15.2.3}$$

Let θ be the fraction of surface sites (area density n_0') covered with $C : O$ bonds. We assume Langmuir kinetics, as shown in Fig. 15.6. All O atoms from the gas phase incident on the surface not covered with $C : O$ are assumed to react immediately to form $C : O$. The rate-limiting etch step is assumed to be desorption of $CO(g)$. The steady-state surface coverage is then found from

$$\frac{d\theta}{dt} = K_a n_{OS}(1 - \theta) - K_d \theta - Y_i K_i n_{is} \theta = 0 \tag{15.2.4}$$

where n_{OS} and n_{is} are the neutral and ion densities at the surface and the plasma-sheath edge, respectively,

$$K_a = \frac{1}{4} \frac{\bar{v}_O}{n_0'} = \frac{1}{4} \left(\frac{8kT_O}{\pi M_O}\right)^{1/2} \frac{1}{n_0'}$$

is the O-atom adsorption rate constant, K_d is the rate constant for thermal desorption of CO, Y_i is the yield of CO molecules desorbed per ion incident on a fully covered surface in the absence of other desorption mechanisms, and $K_i = u_B/n_0' = (eT_e/M_i)^{1/2}/n_0'$ is the rate constant for ions incident on the surface. For high ion

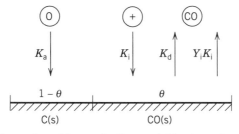

FIGURE 15.6. Surface etch model assuming Langmuir kinetics and rate limiting desorption.

energies, Y_i is typically much greater than unity. A crude model is that $Y_i \sim \eta \mathcal{E}_i / \mathcal{E}_b$ for ion energies $\mathcal{E}_i \gg \mathcal{E}_b$, where $\mathcal{E}_b$ is the energy that binds the molecule to the surface, and where $\eta \lesssim 1$ is a factor giving the efficiency of bond breaking by the incident ion. Solving (15.2.4) for θ, we obtain

$$\theta = \frac{K_a n_{OS}}{K_a n_{OS} + K_d + Y_i K_i n_{is}} \tag{15.2.5}$$

The flux of CO molecules leaving the surface is

$$\Gamma_{CO} = (K_d + Y_i K_i n_{is})\, \theta n_0' \tag{15.2.6}$$

The vertical etch rate is

$$E_v = \frac{\Gamma_{CO}}{n_C} \quad \text{(m/s)} \tag{15.2.7}$$

where n_C is the carbon atom density of the substrate. Inserting (15.2.5) and (15.2.6) into (15.2.7), we obtain

$$E_v = \frac{n_0'}{n_C} \frac{1}{\dfrac{1}{K_d + Y_i K_i n_{is}} + \dfrac{1}{K_a n_{OS}}} \tag{15.2.8}$$

Assuming that the ions strike the substrate surface at normal incidence, then the ion flux incident on a vertical trench sidewall is zero. In this limit, we obtain a purely chemical horizontal etch rate:

$$E_h = \frac{n_0'}{n_C} \frac{1}{\dfrac{1}{K_d} + \dfrac{1}{K_a n_{OS}}} \tag{15.2.9}$$

The normalized etch rates $(n_C/n_0')E_v/K_d$ and $(n_C/n_0')E_h/K_d$ are plotted versus the normalized neutral atom density $K_a n_{OS}/K_d$ in Fig. 15.7 in the regime $Y_i K_i n_{is} \gg K_d$, which is the usual regime for ion energy driven etching. For $K_a n_{OS} \ll K_d$, the surface is starved for etchant atoms and both E_h and E_v are determined by the rate of arrival of O atoms to the surface, with $\theta \ll 1$. As $K_a n_{OS}$ is increased beyond K_d, the normalized horizontal (sidewall) etch rate saturates at 1 and $\theta \to 1$, while the vertical etch rate continues to increase linearly with n_{OS}, with $\theta \ll 1$. This is the neutral flux limited regime of ion energy-driven etching. In turn, the normalized vertical etch rate saturates as $K_a n_{OS}$ is increased beyond $Y_i K_i n_{is}$. In this ion flux limited regime, both vertical and horizontal surfaces are flooded with O atoms ($\theta \to 1$ for both surfaces), and the vertical etch rate is determined by the rate of arrival of energetic ions to the surface.

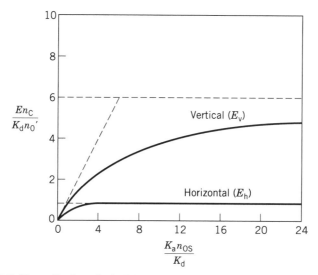

FIGURE 15.7. Normalized vertical (E_v) and horizontal (E_h) etch rates versus normalized gas-phase density n_{OS}, for $K_i Y_i n_{is}/K_d = 5$.

The etch anisotropy in the regime $K_a n_{OS}, Y_i K_i n_{is} \gg K_d$ is

$$a_h = \frac{E_v}{E_h} = \frac{Y_i K_i n_{is}}{K_d} \frac{1}{1 + \dfrac{Y_i K_i n_{is}}{K_a n_{OS}}} \tag{15.2.10}$$

and has its maximum value $Y_i K_i n_{is}/K_d$ when $\theta \to 1$ for both horizontal and vertical surfaces; i.e., $K_a n_{OS} \gg Y_i K_i n_{is} \gg K_d$. In this ion flux limited regime, high anisotropies can be achieved for high ion energies and fluxes (Y_i and n_{is} large) and low substrate temperatures (K_d small) provided n_{OS} is large enough. In the neutral flux limited regime $K_d \ll K_a n_{OS} \ll Y_i K_i n_{is}$, the anisotropy is

$$a_h = \frac{K_a n_{OS}}{K_d} \tag{15.2.11}$$

independent of ion energy and flux.

Additional chemistry and physics can be incorporated into such etch models, including sputtering of carbon,

$$\Gamma_C = \gamma_{sput} K_i n_{is} n_0'$$

associative and normal desorption of O atoms,

$$C:O \longrightarrow C + O(g)$$

$$2C:O \longrightarrow 2C + O_2(g)$$

ion energy driven desorption of O atoms

$$\text{ions} + C:O \longrightarrow C + O(g)$$

formation and desorption of CO_2 as an etch product; and the effect of non-zero ion angular bombardment of sidewall surfaces.

Consider now an etch system in which etchants are physisorbed and the rate-limiting step is formation and desorption of the etch product. As an example, suppose that the rate-limiting step for F-atom etching of silicon is reaction of a physisorbed F atom with a fluorinated silicon layer (SiF_x, $x \approx 3$) followed by desorption to form SiF_4:

$$F:S + SiF_3 \longrightarrow SiF_4(g).$$

For Langmuir kinetics, the adsorption–desorption steady state for F atoms is described by

$$\frac{d\theta}{dt} = K_a n_{FS}(1 - \theta) - K_d \theta = 0 \tag{15.2.12}$$

which yields the F-atom coverage

$$\theta = \frac{K_a n_{FS}}{K_a n_{FS} + K_d} \tag{15.2.13}$$

For physisorption of F on SiF_3, the activation energy for desorption is low; hence $K_d \gg K_a n_{FS}$ for the usual discharge conditions so that

$$\theta \approx \frac{K_a n_{FS}}{K_d} \ll 1 \tag{15.2.14}$$

The etch rate is

$$E_v = \frac{(K_r + Y_i K_i n_{is})\,\theta n_0'}{n_{SiF_3}} \tag{15.2.15}$$

where K_r is the reaction rate constant for pure chemical etching and $Y_i K_i n_{is}$ gives the enhancement in the rate due to ion bombardment. Substituting (15.2.13) into (15.2.15) yields

$$E_v = \frac{K_r + Y_i K_i n_{is}}{n_{SiF_3}} \frac{K_a n_{FS}}{K_d} \frac{n_0'}{1 + K_a n_{FS}/K_d} \tag{15.2.16}$$

which shows the ion enhancement. For $K_d \gg K_a n_{FS}$, E_v depends linearly on n_{FS}. E_h and a_h are found as in the previous example.

Although such ad hoc etch models can provide insight, they may not be faithful to the actual chemical physics for the etch process. Some of these issues will be addressed in Section 15.3, using the example of F atom etching of silicon, and a more complete model will be described.

Discharge Kinetics

A general framework for electropositive and electronegative discharge modeling was given in Chapter 10 and applied to various discharges in succeeding chapters. Given the feed gas, gas pressure, power absorbed, and discharge geometry, the self-consistent ion-bombarding fluxes and energies were estimated using these methods. However, little emphasis was placed on determining neutral etchant densities and fluxes. As has been shown, these are also crucial variables for etch processes. The complicated nature of the entire problem can be illustrated by considering the simplest discharge model, that of a one-dimensional, uniform, electropositive plasma slab in the low-pressure regime. For electrode separation l and area A, summarizing the analysis in Section 10.2, the ion particle balance ((10.2.6) with $R \gg l$) is

$$2n_{is}u_B A \approx K_{iz}n_g n_i l A \qquad (15.2.17)$$

where n_{is} and n_i are the ion (and electron) densities at the plasma-sheath edge and in the bulk, respectively, u_B is the Bohm velocity, and n_g is the neutral gas density. For low pressures, we write $n_{is} = h_l n_i$, where h_l is given by (10.2.3), and solve (15.2.17) to obtain

$$\frac{K_{iz}(T_e)}{u_B(T_e)} \approx \frac{2h_l}{n_g l} \qquad (15.2.18)$$

which determines T_e for a specified $n_g l$. The overall discharge power balance can be written

$$P_{abs} \approx 2Ae\mathcal{E}_T n_{is} u_B \qquad (15.2.19)$$

where P_{abs} is the power absorbed by the discharge and $\mathcal{E}_T$ is the total energy lost per electron-ion pair created, as defined in (10.2.1). The ion density at the sheath edge is determined from (15.2.19) as

$$n_{is} \approx \frac{P_{abs}}{2e\mathcal{E}_T u_B A} \qquad (15.2.20)$$

and the ion flux is

$$\Gamma_{is} \approx n_{is} u_B \qquad (15.2.21)$$

Consider now the production of etchant atoms by dissociation of the feedstock; e.g., for oxygen,

$$e + O_2 \xrightarrow{K_{diss}} 2O + e$$

where K_{diss} has an Arrhenius form:

$$K_{diss} = K_{diss0} \, e^{-\mathcal{E}_{diss}/T_e} \tag{15.2.22}$$

The ionization rate constant, as discussed in Chapter 3, can be fitted to a similar form:

$$K_{iz} = K_{iz0} \, e^{-\mathcal{E}_{iz}/T_e} \tag{15.2.23}$$

Raising (15.2.23) to the power $\mathcal{E}_{diss}/\mathcal{E}_{iz}$, we obtain

$$e^{-\mathcal{E}_{diss}/T_e} = \left(\frac{K_{iz}}{K_{iz0}} \right)^{\mathcal{E}_{diss}/\mathcal{E}_{iz}} \tag{15.2.24}$$

Substituting this into (15.2.22), we obtain

$$K_{diss} = C_0 K_{iz}^{\mathcal{E}_{diss}/\mathcal{E}_{iz}} \tag{15.2.25}$$

where $C_0 = K_{diss0}/K_{iz0}^{\mathcal{E}_{diss}/\mathcal{E}_{iz}}$. Substituting (15.2.18) into (15.2.25) to eliminate K_{iz}, we obtain

$$K_{diss} = C_0 \left(\frac{2h_l u_B}{n_g l} \right)^{\mathcal{E}_{diss}/\mathcal{E}_{iz}} \tag{15.2.26}$$

In this form, it can be seen that K_{diss} depends only weakly on the temperature T_e. Assume now that the electrodes are passivated (net flux of O atoms to the electrodes is zero) and that the only loss of O atoms is due to the vacuum pump. We will later discuss loading effects due to nonpassivated walls. Assuming no other significant sources or sinks of O atoms, then the steady-state rate equation is

$$Al \frac{dn_O}{dt} = Al K_{diss} n_g n_i - S_p n_{OS} = 0 \tag{15.2.27}$$

where S_p (m³/s) is the pumping speed, $n_{OS} \approx n_O$ is the O-atom density, and, for simplicity, we take $n_i \approx n_e$ (negligible negative ion density). Solving (15.2.27) for n_{OS},

$$n_{OS} = \frac{Al n_g}{S_p} K_{diss} n_i \tag{15.2.28}$$

and using $n_i \approx n_{is}/h_l$, with n_{is} given by (15.2.20), and K_{diss} given by (15.2.26), we obtain

$$n_{OS} = \frac{P_{abs}}{e\mathcal{E}_T S_p} C_0 \left(\frac{n_g l}{2h_l u_B} \right)^{1 - \mathcal{E}_{diss}/\mathcal{E}_{iz}} \qquad (15.2.29)$$

Typically, $\mathcal{E}_{diss}/\mathcal{E}_{iz} \approx 0.3\text{--}0.5$. The flux of O atoms incident on an electrode is then

$$\Gamma_{OS} = \frac{1}{4} n_{OS} \bar{v}_O \qquad (15.2.30)$$

where $\bar{v}_O = (8kT_O/\pi M_O)^{1/2}$. Equations (15.2.20) and (15.2.29) give the ion density at the plasma-sheath edge and the neutral etchant density at the surface, respectively, as required to determine the etch rates E_v and E_h using, for example, (15.2.8) and (15.2.9). The ion-bombarding energy depends on the type of discharge, as discussed in Chapters 11 and 12 (see also (10.2.9) and following discussion).

Consider, for example, the scaling behavior of the vertical etch rate E_v from (15.2.8) for a high-density source with $\mathcal{E}_i \approx 5T_e$. If P_{abs} is increased then both n_{is} and n_{OS} increase linearly with P_{abs}, and $\mathcal{E}_i$ is unchanged. In the usual ion-driven etch regime $Y_i K_i n_{is} \gg K_d$, we see that E_v increases linearly with P_{abs}. Since both n_{OS} and n_{is} increase, the etch regime (ion flux or neutral flux limited) is not altered. Now consider increasing n_g. From (15.2.18) T_e and therefore $\mathcal{E}_i$ fall slightly; n_{is} remains unchanged, and n_{OS} increases as $n_{OS} \propto n_g^{1 - \mathcal{E}_{diss}/\mathcal{E}_{iz}} \propto n_g^{0.5 - 0.7}$. From Fig. 15.7, we see that as n_{OS} increases, the system can enter the high etchant density regime where the etch rate is ion flux limited. Similar scaling laws can be determined for low-pressure capacitive rf discharges and for high-pressure discharges in which the ion transport is diffusion limited. Systems containing substrate holders that are independently rf biased can be treated similarly. We leave the estimation of scalings in these systems as exercises for the reader.

Consider now the effect on O-atom density of O-atom loss at the electrodes due to etch reactions with a set of i wafers each having area A_w. Assuming loss of all O atoms hitting the wafers, the O-atom rate equation (15.2.27) becomes

$$Al \frac{dn_O}{dt} = AlK_{diss} n_g n_i - S_p n_{OS} - iA_w \frac{1}{4} \bar{v}_O n_{OS} = 0 \qquad (15.2.31)$$

which can be solved to obtain

$$n_{OS} = \frac{Al n_g n_i K_{diss}}{S_p + iA_w \bar{v}_O/4} \qquad (15.2.32)$$

This can also be written in the form

$$\frac{1}{n_{OS}} = \frac{1}{n_{OS}^{(0)}} + \frac{iA_w \bar{v}_O}{4Al n_g n_i K_{diss}} \qquad (15.2.33)$$

where $n_{OS}^{(0)}$ is the etchant density in the absence of wafers ($i \equiv 0$). In the neutral flux limited regime where the etch rate is proportional to n_{OS}, we see that (15.2.33) leads to a reduction in the etch rate. This is called the *loading effect*. We see from (15.2.26) and (15.2.33) that to minimize the loading effect, $n_g n_i l$ should be large. In this calculation, we have assumed that O atoms do not recombine on the wall surfaces and that all O atoms incident on the substrate surfaces react (reaction probability $P_r = 1$). In fact, depending on the etchant (O, F, Cl, etc.) and the wall and substrate materials, there can be considerable recombination on walls and considerable emission of etchant atoms from substrates. For F atoms and most wall–substrate systems, it is generally the case that $P_{rec} \ll P_r$ and considerable loading effects are seen. For other etchants (e.g., Cl, Br, O), P_{rec} can be of the order of or can exceed P_r. For this case very weak or even negative loading effects are seen.

The neutral dynamics may not be as simple as a pure dissociation reaction. For example, consider the creation of F atom etchants in a CF_4 discharge by the processes

$$e + CF_4 \longrightarrow CF_3 + F + e \qquad \mathcal{E}_{diss} = 12.5 \text{ V}$$
$$e + CF_4 \longrightarrow CF_3 + F^- \qquad \mathcal{E}_{att} = 7.3 \text{ V}$$
$$e + F^- \longrightarrow F + 2e \qquad \mathcal{E}_{det} = 3.4 \text{ V}$$

A maximum estimate of the F^- density is found by neglecting positive–negative ion recombination,

$$\frac{dn_{F^-}}{dt} = K_{att} n_{CF_4} n_e - K_{det} n_{F^-} n_e = 0,$$

which yields

$$n_{F^-} = \frac{K_{att}}{K_{det}} n_{CF_4} \qquad (15.2.34)$$

Hence the generation rate for F atoms is

$$\frac{dn_F}{dt} = K_{diss} n_{CF_4} n_e + K_{det} n_{F^-} n_e$$
$$\equiv K_{eff} n_{CF_4} n_e \qquad (15.2.35)$$

where, using the maximum estimate of n_{F^-} from (15.2.34),

$$K_{eff} = K_{diss} + K_{att} \qquad (15.2.36)$$

is the effective dissociation rate constant. Although (15.2.35) has the form of simple dissociation, the rate constant K_{eff} has a complicated (non-Arrhenius) dependence on T_e. For this system, $K_{diss} \gtrsim K_{att}$ for $T_e \gtrsim 6$ V (Plumb and Ryan, 1986).

Chemical Framework

Feedgas mixes for plasma etching are usually complex because of the conflicting requirements on etch rate, selectivity to mask and underlayer, and anisotropy. This is especially true for ion inhibitor processes where a balance must be struck among etchant, inhibitor, and ion fluxes to the substrate. Furthermore, the plasma itself dissociates the feedgas into other, usually more reactive, species. The feedgas and its dissociated products may include chemical constituents such as (Flamm, 1989, Chapter 2):

- Saturates: CF_4, CCl_4, CF_3Cl, COF_2, SF_6, etc;
- Unsaturates: C_2F_6, C_3F_8, C_2F_4, CF_2, CF_3, CCl_3, etc;
- Etchants: F, Cl, Br, O (for resist), F_2, Cl_2, Br_2, etc;
- Oxidants: O, O_2, etc;
- Reductants: H, H_2, etc;
- Nonreactive Gases: N_2, Ar, He, etc.

These species react with each other in the gas phase and on the surface in reactions such as:

$$e + saturate \longrightarrow unsaturate + etchant + e,$$

$$etchant + substrate \longrightarrow volatile\ products,$$

$$unsaturate + substrate \longrightarrow films.$$

For some substrates (e.g., SiO_2), unsaturates can themselves be etchants:

$$unsaturate + substrate \longrightarrow volatile\ products.$$

At low pressures, three body reactions such as

$$etchant + unsaturate\ (+M) \longrightarrow saturate\ (+M)$$

are not important in the gas phase, but may be important at surfaces or at high gas pressures. If oxidants or reductants are added to the feedgas, commonly O_2 or H_2 respectively, or gases that contain these atoms, then additional reactions can occur:

$$oxidant + unsaturate \longrightarrow etchant + volatile\ product,$$

$$reductant + etchant \longrightarrow volatile\ products.$$

The ratio of etchant to unsaturate flux at the substrate is an important process parameter. As will be seen in the next section, a high ratio can lead to isotropic etching, while a low ratio can lead to film deposition. There can be an intermediate ratio in which inhibitor film can be deposited on sidewalls but cleared from trench bottoms by ion bombardment; this is the regime of anisotropic ion enhanced inhibitor etching.

Etchants (Cl_2, Br_2) can be added to the feedgas to increase the etchant/unsaturate ratio, and oxidants (O_2) can also be added to increase the ratio by burning unsaturates to produce etchants; e.g.,

$$CF_3 + O \longrightarrow COF_2 + F$$

Conversely, unsaturated feedgases (C_3F_8, C_2F_4) and H_2 can be added to reduce the etchant/unsaturate ratio, pushing the system toward increased sidewall protection if that is desired. Inert gas additives are sometimes used to control discharge electrical and substrate thermal properties, to dilute etchants, and to alter gas-phase chemistry through mechanisms such as Penning ionization and excitation (see Section 8.4). Other additives are sometimes used in etch processes to break through protective oxide layers (e.g., Al_2O_3 for aluminum etching) and to scavenge contaminants (e.g., H_2O) (Flamm, 1989, Chapter 2).

15.3   HALOGEN ATOM ETCHING OF SILICON

One of the most important applications of plasma etching is the selective, anisotropic removal of patterned silicon or polysilicon films. Halogen atom etchants (F, Cl, Br) are almost always used for this purpose. In fact, F-atom etching of silicon is experimentally the most well-characterized surface etch process and is often used as a paradigm for describing plasma etch processes, as we do here. In this section, we first give a summary of pure chemical and ion-enhanced surface etch processes for F-atom etching. We then describe the discharge chemistry, concentrating on the well-studied CF_4 feedstock system. Finally, we describe silicon etching using other halogen atoms. For more detailed descriptions of silicon etching, the reader should consult the reviews by Flamm (1989; 1990) and Winters and Coburn (1992).

Pure Chemical F-Atom Etching

F atoms are known to spontaneously attack silicon and silicon dioxide in the absence of ion bombardment. The etch rates have been measured to have roughly an Arrhenius form over a wide range of temperatures and to depend linearly on the gas-phase F-atom density near the surface up to densities as high as 5×10^{15} cm^{-3}. For undoped silicon and for thermally grown silicon dioxide, the etch rates are

$$E_{Si} \text{ (Å/min)} = 2.86 \times 10^{-12} n_{FS} T^{1/2} e^{-1248/T} \tag{15.3.1}$$

$$E_{SiO_2} \text{ (Å/min)} = 0.61 \times 10^{-12} n_{FS} T^{1/2} e^{-1892/T} \tag{15.3.2}$$

where n_{FS} (cm^{-3}) is the F-atom density near the surface and T (K) is the surface temperature. The silicon-to-silicon dioxide selectivity is then

$$s = 4.66\, e^{644/T}. \tag{15.3.3}$$

At room temperature (300 K), and for a typical F atom density of 3×10^{14} cm^{-3}, $E_{Si} \approx 230$ Å/min, $E_{SiO_2} \approx 5.9$ Å/min, and $s \approx 40$. There is also good selectivity over Si_3N_4 and reasonable selectivities over resists.

The mechanism for pure chemical F-atom etching of silicon has been studied for over fifteen years and is still not thoroughly understood. In the steady state, it is known that a fluorinated silicon SiF_x layer 2–5 monolayers thick forms at the surface. The fluorine to silicon ratio at the top of this layer is typically 3:1 (mostly SiF_3), and the ratio falls smoothly to zero at the SiF_x-Si interface. The layer thickness varies with etch conditions; typically the film is thin when the etch rate is high, and vice versa. The layer is stable at room temperature; i.e., if the incident flux of F atoms is terminated after the layer forms, then etching ceases. If the layer is then heated, it does not begin to decompose until temperatures of 300–400 °C have been reached. The decomposition products are $SiF_2(g)$ and $SiF_4(g)$, with the former being the most important. During etching, etch product measurements indicate that roughly 65% of the etch product at room temperature is $SiF_4(g)$, with $Si_2F_6(g)$ and $Si_3F_8(g)$ comprising the remaining product. As the temperature is raised, the $SiF_4(g)$ percentage slowly increases to 80–90 % of the total at 300 °C, with a corresponding reduction in Si_2F_6 and Si_3F_8. SiF_4 then begins to decrease and becomes a minor product above 600 °C. At high temperatures, the dominant etch product is $SiF_2(g)$, which increases from 5 to 10% of the total at 300 °C up to 40–50% at 600 °C.

The formation of a steady-state SiF_x layer whose thickness varies inversely with the etch rate is suggested by a model in which diffusion of F atoms (or, as will be seen below, F$^-$ ions) into the surface is balanced by an erosion of the surface due to the etching. Letting n_F be the volume density of diffusing F atoms in the solid and E_{Si} (m/s) be the etch rate, then in the frame $x' = x - E_{Si}t$ moving with the etched surface at $x' = 0$, the flux vanishes:

$$\Gamma'_F = -D\frac{dn_F}{dx'} - n_F E_{Si} = 0 \tag{15.3.4}$$

For a crude model in which the diffusion coefficient D is a constant, independent of n_F, this can be solved to obtain

$$n_F = n_{F0} \exp\left(-\frac{E_{Si}x'}{D}\right) \tag{15.3.5}$$

yielding a layer having characteristic thickness D/E_{Si}. This type of model is suggestive of the more complicated kinetics, including both diffusion and reactions within the solid, that lead to the formation of the SiF_x layer.

The measured linear dependence of the etch rate on n_{FS} up to densities as high as 5×10^{15} cm^{-3} is suggestive of weakly bound adsorbed F atoms as precursors to subsequent etch reactions. As pointed out in Section 15.2, the adsorption–desorption kinetics leads to a surface coverage for weakly adsorbed F atoms that is linear in the gas-phase F-atom density n_{FS} near the surface, as given by (15.2.14). The Arrhenius form of (15.3.1) at first sight suggests a single activated process for the etch reaction, but the current belief is that this is probably fortuitous. The etch product distributions and the decomposition properties of the SiF_x layer differ greatly at low and at high temperatures, implying that the etch mechanisms differ also. In addition, weakly and strongly non-Arrhenius etch rates are seen for etching of silicon with F_2 and with XeF_2, respectively. The existence of an activation energy of 1248 K ($\equiv 0.108$ V) for F-atom etching of silicon is not understood.

A significant feature of pure chemical etching using halogen atoms, known as the *doping effect,* is that the etch rate depends on the silicon doping levels, with n-type silicon etching faster than p-type. The dopants must be thermally activated in order to restore the crystalline structure. The effect is weak for F atoms, with a factor of two difference in etch rates, but is very strong for Cl atoms, where the etch rates can differ by many orders of magnitude.

The existence of a doping effect suggests that negative charge carriers play a role in etch reactions. The electron affinity of an F atom in free space is $\mathcal{E}_{aff} \approx 3.45$ V, but near the surface, the affinity is increased by the energy[*] of the electrostatic image force (see (9.3.2)). For an F atom a distance $a_{eff} = 1$ Å from the SiF_x surface, (9.3.2) yields a large affinity $\mathcal{E}_{aff} \approx 3.45 + 3.60 \approx 7.05$ V. SiF_3 similarly has a large affinity. Hence, negative ion formation at the surface is favored. Winters and Haarer (1987) suggest that the rate-limiting etching step involves reaction of adsorbed F atoms at negative charge centers on the SiF_x surface. The negative charge is supplied by electrons tunneling from the silicon substrate through the SiF_x layer. The variation of the charge density on the surface with doping level accounts for the etch rate variation in this model.

To understand the doping effect quantitatively, a brief description of the electronic properties of silicon is required. Intrinsic (undoped) silicon is a semiconductor with a room-temperature band gap energy $\mathcal{E}_C - \mathcal{E}_V \approx 1.1$ V, where, as shown in Fig. 15.8a, $\mathcal{E}_C$ and $\mathcal{E}_V$ are the electron energies at the bottom of the conduction band and the top of the valence band, respectively. At a temperature of absolute zero, the valence band is completely full, and the conduction band is completely empty, and the Fermi energy $\mathcal{E}_{FI}$ lies midway between the bands, $\mathcal{E}_{FI} = (\mathcal{E}_C + \mathcal{E}_V)/2$. At room temperature (300 K), this is still approximately true, but a small density n_e of electrons is thermally excited from the valence to the conduction band, leaving behind an equal density n_p of holes, with

$$n_e = n_p = n_I \approx 1.2 \times 10^{10} \text{ cm}^{-3} \tag{15.3.6}$$

The presence of activated dopants shifts the balance between electrons and holes. For example, taking the equilibrium between electron–hole recombination and thermal generation

$$n_e n_p = n_I^2 \tag{15.3.7}$$

---

[*]The energy for a surface having dielectric constant $\epsilon \gg 1$ is the same as that for a perfectly conducting surface (Ramo et al., 1984).

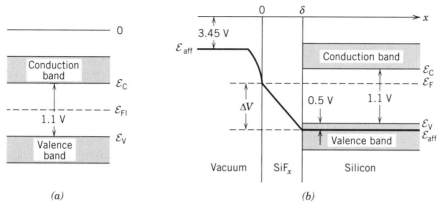

FIGURE 15.8. The mechanism for the dependence of the chemical etch rate on the silicon doping level: (a) intrinsic silicon showing the conduction and valence bands; (b) dielectric slab model in n-type silicon.

along with conservation of charge

$$n_e = n_p + n_D \tag{15.3.8}$$

with n_D the density of donor atoms, we can solve to determine n_e and n_p. For $n_D \gg n_I$, we obtain $n_e \approx n_D$ and $n_p \approx n_I^2/n_D$. The Fermi level shifts to reflect the new equilibrium in the presence of dopants, with

$$\mathcal{E}_F \approx \mathcal{E}_{FI} - T \ln \frac{n_D}{n_I} \tag{15.3.9}$$

for $10^{10} \ll n_D \lesssim 10^{19}$ cm^{-3} (the case for nondegenerate electrons).

Now consider, as shown in Fig. 15.8b, a simple dielectric slab model to determine the area density n'_- of negative charge centers on the SiF$_x$ surface at $x = 0$. For a surface charge density $-en'_-$ at $x = 0$ and a corresponding induced positive charge in the silicon near the SiF$_x$–Si interface at $x = \delta$, the voltage of the surface with respect to the interface is found from Poisson's equation (2.2.11) to be

$$\Delta V = -\frac{en'_- \delta}{\epsilon_{SiF_x}} \tag{15.3.10}$$

where $\epsilon_{SiF_x} \sim 3\epsilon_0$ is the dielectric constant. ΔV gives the difference between the affinity level at the interface and at the surface. As shown in Fig. 15.8b, the Fermi energy and the affinity level at the surface must coincide to insure detailed balancing for the affinity reaction and its opposing reaction:

$$e + SiF_x = SiF_x^-$$

Winters and Haarer also argue that a reasonable choice for the affinity level in the silicon is about 0.5 V below the top of the valence band. Hence from Fig. 15.8b,

$$\Delta V = \mathcal{E}_{FI} - \mathcal{E}_F + \frac{\mathcal{E}_C - \mathcal{E}_V}{2} + 0.5$$

$$= \mathcal{E}_{FI} - \mathcal{E}_F + 1.05 \text{ V} \qquad (15.3.11)$$

Inserting (15.3.9) into (15.3.11), we obtain

$$\Delta V = 1.05 + T \ln \frac{n_D}{n_I} \qquad (15.3.12)$$

Equating (15.3.12) to (15.3.10) and solving for n'_-, we obtain

$$n'_- = \frac{\epsilon_{SiF_x}}{e\delta} \left(1.05 + T \ln \frac{n_D}{n_I} \right) \qquad (15.3.13)$$

This result shows that n'_- increases for sufficiently large doping levels and that n'_- varies inversely with the layer thickness δ. The room-temperature etch rate in this model is proportional to n'_-, to n_{FS} through (15.2.14), and to the surface density of SiF_3:

$$E_{Si} = K_r n'_- n'_{SiF_3} n_{FS} \qquad (15.3.14)$$

where K_r is the rate constant. The linear dependence on n'_{SiF_3} arises because some fraction (roughly 1/2–1/3 for a fully fluorinated surface) of the adsorbed F atoms activated at the negative charge centers are presumed to attack the Si–SiF_3 bond holding an SiF_3 group to the surface, thus forming the SiF_4 etch product. Smaller concentrations of Si–SiF_2SiF_3 and Si–$SiF_2SiF_2SiF_3$ bonds are also attacked, yielding the observed Si_2F_6 and Si_3F_8 etch products in lesser concentrations. The remaining fraction (1/2–2/3) of the activated F atoms break Si–Si bonds within the layer, leading to growth of the layer. The activation mechanism is likely to be the formation of negative ions,

$$e + F : S \longrightarrow F^- : S$$

Such a negative ion sees a strong image force directed into the surface, promoting lattice penetration. Once inside the lattice, any F atom produced by neutralization of F^- is likely to attack an Si–Si bond. Winters and Coburn (1992) claim that the rate expression (15.3.14) is consistent with all experimental data on F-atom chemical etching of silicon, and is also consistent with data on F_2 and XeF_2 etching.

Ion Energy Driven F Atom Etching

Etch rates for a given F-atom flux can be increased by a factor of 5–10 for sufficiently high fluxes (and energies) of bombarding ions. A single 1-kV Ar^+ ion can cause the removal of as many as 25 silicon atoms and 100 F atoms from the surface. The total surface concentration of fluorine in the SiF_x layer is reduced by up to a factor of two in the presence of ion bombardment. Furthermore, the etch product distribution changes, and, notably, a significant fraction of $SiF_2(g)$ etch product is formed. Although the etch anisotropy can be as high as 5–10, this still implies a reasonably large pure chemical etch rate on trench sidewalls. Consequently, fluorine-based anisotropic silicon etches are not commonly used. Although inhibitor chemistries can be employed for sidewall protection and increased anisotropy, as we subsequently show, the contamination produced by the protective films is undesirable, and other halogen etch chemistries are generally used for strongly anisotropic etches.

Many mechanisms have been proposed to explain the enhanced etch rate due to ion energy driven F atom etching of silicon, including the following:

1. *Formation of a damaged region that is more reactive to subsequently arriving fluorine.* However, experiments show that the energetic ion bombardment influences the fluorine that is present within the SiF_x layer at the moment of impact. Hence, this is probably not the mechanism for F-atom etching of silicon. However, it is known that lattice damage is an important mechanism for some systems; e.g., ion-enhanced XeF_2 etching of tungsten.

2. *Temperature increase due to etch reactions or ion bombardment.* The temperature rise is not large enough.

3. *Chemically enhanced physical sputtering.* For this proposed mechanism, the binding energies $\mathcal{E}_t$ of $Si–SiF_2$ and $Si–SiF_4$ bonds at some locations on the SiF_x surface are supposed to be much smaller than those of a pure silicon surface, yielding significant physical sputtering rates for SiF_x in the presence of ion bombardment. However, the binding energies would have to be of order 0.3–0.5 V, much smaller than the usual Si–Si or Si–F bond energies, in order that the species remain on the surface without thermally desorbing and yet be easily sputtered (e.g., see (9.3.14)). Furthermore, the expected mass dependence given in (9.3.14) is not observed. In addition, time-resolved etch rate measurements using modulated beams of ions show that ion-enhanced etching is much slower than predicted by a physical sputtering model. Hence, the evidence suggests that chemically enhanced physical sputtering is not the major contributor to ion-enhanced etching at the neutral–ion flux ratios typically found in etching discharges. This conclusion is supported by molecular dynamics simulations (Barone and Graves, 1994).

4. *Chemical reaction and desorption due to ion bombardment.* In this mechanism, sometimes called chemical sputtering, the ion bombardment causes a chemical reaction to occur that produces an easily desorbed etch product. When an energetic ion collides with and penetrates the SiF_x layer, producing a collision cascade as for physical sputtering (see Section 9.3), then a large number of Si–Si and Si–F bonds can be broken and reformed, leading to molecules such as SiF_4 and SiF_2 that are weakly bound to the surface. These molecules can thermally desorb during or after the collision cascade. This mechanism on the

surface is similar to a reaction in the gas phase such as

$$\text{ion} + \text{Si}_2\text{F}_6 \longrightarrow \text{SiF}_4 + \text{SiF}_2 + \text{ion}$$

It is likely to be an important mechanism for ion energy-driven F-atom etching of silicon and is seen to be important in molecular dynamics simulations (Barone and Graves, 1994).

5. *Enhanced chemical etching.* In this mechanism, ion bombardment reduces the layer thickness, thus increasing the pure chemical etch rate given by (15.3.14), because n'_- increases. It is known experimentally that the pure chemical etch rate varies inversely with the layer thickness and that ion bombardment reduces this thickness. Hence, this mechanism can contribute to ion enhanced etching. However, the etch rate enhancement is unlikely to exceed a factor of two for typical plasma etch conditions.

For the simple phenomenological models for ion-assisted etching described in Section 15.2, the increased etch rate is due to ion-enhanced desorption of etch products or ion-enhanced reaction of etchants with the surface. Gray et. al. (1993) have developed a more complete model for Ar^+-enhanced F-atom etching of silicon (and silicon dioxide) that is consistent with high flux ion and atomic beam studies and with other data. The kinetics includes the following processes for silicon etching:

1. Physisorption and thermal desorption of F atoms

$$\text{F(g)} + \text{S} \underset{K_{d1}}{\overset{K_{a1}}{\rightleftharpoons}} \text{F} : \text{S}$$

2. Chemisorption of physisorbed F atoms at silicon dangling bond (Si^*) sites

$$2\text{F} : \text{S} + \text{Si}^* \overset{K_{a2}}{\longrightarrow} \text{SiF}_2 : \text{S}$$

3. Ion-induced desorption of SiF_2

$$\text{SiF}_2 : \text{S} \overset{Y_d K_i}{\longrightarrow} \text{SiF}_2(\text{g}) + 2\text{Si}^*$$

4. Creation of SiF_4 by ion beam mixing followed by ion induced desorption

$$2\text{F} : \text{S} + \text{SiF}_2 : \text{S} \overset{Y_{mix} K_i}{\longrightarrow} \text{SiF}_4(\text{g}) + 2\text{Si}^*$$

5. Physical sputtering of silicon

$$\text{Si}^* \overset{Y_{sput} K_i}{\longrightarrow} \text{Si(g)} + \text{Si}^*$$

6. Chemical etching of silicon

$$2F : S + SiF_2 : S \xrightarrow{K_r} SiF_4(g)$$

In this model, the ion bombardment increases the etch rate due to increases in the chemisorption site density n'_{Si^*} (mechanism 1), the rate of chemically enhanced physical sputtering of etch products (mechanism 3), and the rate of chemical reaction and thermal desorption of etch products (mechanism 4).

CF₄ Discharges

Because F_2 itself etches silicon, generally leaving a rough and pitted surface, it is not used as a feedstock. Common feedstock gases include CF_4, SF_6, and NF_3, along with unsaturate additions such as C_2F_6. The most well-studied system is CF_4, which we use to illustrate the effects of gas- and surface-phase chemistry in discharges used for F-atom etching of silicon. The overall etch reaction in a CF_4 discharge produces unsaturated CF products, e.g.,

$$4CF_4 + Si \longrightarrow 2C_2F_6 + SiF_4$$

Hence, the major effluent gases observed are CF_4, SiF_4, and C_2F_6.

CF_4 is a very stable tetrahedral molecule (symmetry group T_d with the carbon atom in the center) with an enthalpy of formation of -925 kJ/mol and a C–F bond distance of 1.3 Å. Its vibration frequencies ($\hbar\omega_{vib}/e$ in voltage units) are 0.054 V (doubly degenerate), 0.078 V (triply degenerate), 0.113 V (singly degenerate), and 0.159 V (triply degenerate), and its rotation constant is $B_{rot} = 2.4 \times 10^{-5}$ V (triply degenerate) for each of the three degrees of rotational freedom. The CF_3–F bond energy is 5.6 V. All excited states of CF_4 are repulsive; consequently, all electronic excitations of CF_4 are dissociative. In particular, the positive ion CF_4^+ is not stable.

The CF_3 radical is weakly bound and large, having pyramidal symmetry C_{3v}; the carbon atom is at the top of a flat pyramid with an equilateral triangular base of F atoms. The F–F bond distance is 4.1 Å and the C atom is 0.75 Å above the base. The CF_2–F bond energy is 3.8 V. CF_3 is electronegative with an electron affinity of 1.9 V, lower than the affinity 3.45 V of F atoms. The CF_3^- ion also has symmetry C_{3v}, with F–F bond distance and C atom height of 4.1 and 1.3 Å, respectively. The linear radical CF_2 has a CF–F bond energy of 5.8 V. The bond energy of the CF radical is 5.1 V. Some threshold energies and rate constants for important electron-CF_x reactions are given in Table 15.1.

Three-body neutral–neutral gas-phase recombination reactions among F atoms and CF_x radicals, as in (9.2.28), can be important because of the large size of the CF_x radicals and the large enthalpies of formation, in particular for C_2F_6 and C_2F_4 products. At the pressures of interest for etching, these reactions can be in the intermediate regime between low and high pressures. The effective two-body rate

TABLE 15.1. Selected Second-Order Reaction Rate Constants for Electron Impact Collisions in CF$_4$ Discharges

Number	Reaction	Rate Constant (cm^3/s)	Source
1	e + CF$_4$ → CF$_3$ + F$^-$	4.6E–9 T$_e^{-3/2}$ exp($-7/T_e$)	a
	→ CF$_3^-$ + F		
2	e + CF$_4$ → CF$_3$ + F + e	2E–9 exp($-13/T_e$)	b
3	e + CF$_4$ → CF$_2$ + 2F + e	5E–9 exp($-13/T_e$)	b
4	e + CF$_4$ → CF$_3^+$ + F + 2e	1.5E–8 exp($-16/T_e$)	a

Notes. T$_e$ between 3 and 6 V. The notation E–9 means 10^{-9}.
[a]Based on cross sections of Hayashi (1987).
[b]Based on data of Plumb and Ryan (1986).

constant K_{AB} in the intermediate regime is, from (9.2.33),

$$K_{AB} = \frac{K_3' n_M}{1 + K_3' n_M / K_2} \qquad (15.3.15)$$

where K_3' is defined by (9.2.36). Some values of K_2 and K_3' for important three-body association reactions are given in Table 15.2 (Plumb and Ryan, 1986).

Important electron collision reactions in CF$_4$ discharges are reaction 1 in Table 15.1 for F$^-$ and CF$_3^-$ creation, reaction 4 for CF$_3^+$ creation, and reactions 2 and 3 for CF$_3$, CF$_2$, and F creation. The recombination reactions 1 and 4 in Table 15.2 can be strong at moderate gas pressures. In view of these and the 1:2.5 branching ratio between CF$_3$/CF$_2$ production from electron collisions with CF$_4$, the CF$_2$ radical density can much exceed the CF$_3$ radical density. As for O$_2$ discharges, negative ions are lost by positive–negative ion recombination, and, possibly, by electron impact detachment or associative detachment in the volume, and positive ions are lost by this same recombination in the volume and by flow or diffusion to the walls.

Etchant atoms can be lost to a surface by adsorption followed by recombination, e.g.,

$$F:S + F:S \longrightarrow F_2(g) + 2S$$

TABLE 15.2. Selected Values of Rate Constants K_2 and K_3' for Association Reactions in CF$_4$ Discharges

Number	Reaction	K_2 (cm^3/s)	K_3' (cm^6/s)
1	F + CF$_3$ + M → CF$_4$ + M	2E–11	7.7E–27
2	F + CF$_2$ + M → CF$_3$ + M	1.3E–11	3.0E–29
3	CF + F + M → CF$_2$ + M	1E–11	3.2E–31
4	CF$_3$ + CF$_3$ + M → C$_2$F$_6$ + M	8.3E–12	2.8E–23
5	CF$_2$ + CF$_3$ + M → C$_2$F$_5$ + M	1E–12	2.3E–26

Notes. Here M represents CF$_4$. The notation E–11 means 10^{-11}.
Source. After Plumb and Ryan (1986).

or by reaction with the surface, e.g.,

$$F : S + SiF_x \longrightarrow SiF_4(g)$$

Atoms that are not lost are desorbed back into the discharge. For fluorine atoms, as shown in Problem 9.7, the probability P_{rec} that atoms adsorb and recombine on most surfaces is generally small at the substrate temperatures and atom densities characteristic of etching discharges. For example, F atoms incident on Al_2O_3, SiO_2, Pyrex, Teflon, stainless steel, Mo, Ni, and Al–0.1%Cu have recombination probabilities $P_{rec} \sim 10^{-4}$–10^{-3} at 300 K (Flamm, 1989). However, there are some exceptions: $P_{rec} \gtrsim 0.01, 0.05$, and 0.2 for F atoms on Cu, brass, and Zn, respectively. For chlorine and oxygen atoms, recombination probabilities can be larger than those for fluorine atoms, up to ~ 0.1 for many surfaces. The reaction probability P_r for F atoms is negligible on most surfaces, but can be significant for some surfaces, e.g., $P_r \approx 0.0017$ and 1 for Si and BN, respectively. In many cases, except by design (i.e., substrates to be etched), etchant atoms incident on surfaces in processing discharges are recycled back into the discharge as atoms.

The behavior of free radical molecules incident on surfaces is more complicated due to their possible dissociation. On nonactive surfaces, free radicals can be adsorbed without dissociation. Their subsequent probabilities for recombination or reaction with the surface are generally small, as for etchant atoms. An example is the adsorption of CF_3 on SiO_2, which is nondissociative. The recombination and reaction probabilities are small, so most CF_3 radicals incident on SiO_2 desorb as CF_3 radicals.

However, on active surfaces such as pure silicon, CF_3 and CF_2 radicals generally dissociatively adsorb, producing a C atom and three (or two) F atoms that each bond to the silicon. Although these radicals deliver etchant atoms to the surface, they also deliver C atoms, which can form a protective film on the surface that inhibits the etch reaction. Similarly, CF_3 and CF_2 can dissociatively or nondissociatively adsorb on an SiF_x layer, leading to a buildup of carbon or polymer film. It is unlikely that the film will be removed from the surface in the absence of ion bombardment except as $CF_4(g)$. Hence, the flux of CF_x radicals ($x < 4$) reduces the silicon etch rate. If Γ_{CF_x} is the *net* flux of CF_x adsorbed, Γ_F is the *net* flux of F atoms adsorbed, Γ_{SiF_4} is the flux of SiF_4 desorbed, and Γ_{CF_4} is the flux of CF_4 desorbed, then conservation of C atoms on the surface requires $\Gamma_{CF_4} = \Gamma_{CF_x}$, and conservation of F atoms on the surface requires

$$x\Gamma_{CF_x} + \Gamma_F = 4\Gamma_{CF_x} + 4\Gamma_{SiF_4} \tag{15.3.16}$$

Solving for the etch rate, we obtain

$$E_{Si} = \frac{\Gamma_{SiF_4}}{n_{Si}} = \frac{\Gamma_F - (4 - x)\Gamma_{CF_x}}{4n_{Si}} \tag{15.3.17}$$

For $\Gamma_F < (4 - x)\Gamma_{CF_x}$, there is net deposition of carbon and the etch rate is zero (Problem 15.6). For $x \approx 3$, we see that the condition for etching is $\Gamma_F > \Gamma_{CF_x}$.

Ion bombardment can shift the balance in (15.3.17) in one of two ways: (1) It can increase the ratio Γ_F/Γ_{CF_x} of net fluxes adsorbed. (2) It can lead to desorption of CF_y, $y < 4$, due to physical sputtering of CF_y polymer (and also of C) and due to ion energy driven etching of CF_y polymer, in the same manner that ion energy driven F atom etching of silicon leads to SiF_2 etch product release from the SiF_x surface. For desorption of CF_y, (15.3.17) is replaced by

$$E_{Si} = \frac{\Gamma_{SiF_4}}{n_{Si}} = \frac{\Gamma_F - (y - x)\Gamma_{CF_x}}{4n_{Si}} \qquad (15.3.18)$$

For $y \leq x$, there is always etching.

Evidently, the ratio F/C of fluorine atoms to CF_x radicals in the discharge is an important process parameter in determining whether etching or film deposition occurs. Figure 15.9 gives an illustrative picture (not quantitative) of the boundary between etching and deposition as the F/C feedstock gas ratio and the bias voltage (ion bombarding energy) are varied in typical fluorocarbon discharges. We describe methods for varying the F/C ratio below. The general trends indicated in this picture follow from (15.3.18). For F/C > 3, there is etching independent of bias voltage, and so both trench sidewalls and bottoms are etched. Although the horizontal (sidewall) rate is not ion assisted and can be small compared to the vertical rate, the sidewall is not protected by inhibitor film, and the etch anisotropy is not large. For 2 < F/C < 3, the sidewalls are protected by inhibitor film, but the ion bombardment exposes the trench bottoms to the etchants; this is the regime of highly anisotropic etching using fluorocarbon feedstocks. For F/C < 2, there is film deposition on both sidewall and bottom and etching ceases.

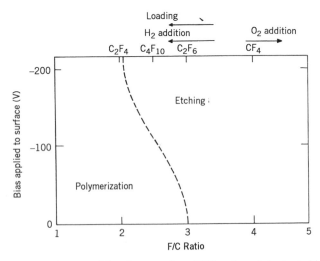

FIGURE 15.9. The influence of fluorine to carbon (F/C) ratio and electrode bias voltage on etching and polymerization processes in a fluorocarbon discharge (Coburn and Winters, 1979).

The loading effect was described in Section 15.2. For a fixed feedstock flow rate, the F-atom density in the discharge is depressed as the area of silicon being etched increases, due to the formation of SiF_4 etch product. Hence the F/C ratio decreases and the equilibrium for the system shifts toward the left, as indicated in Fig. 15.9. This can lead to polymer formation under heavy loading conditions.

O_2 and H_2 Feedstock Additions

Oxygen gas is often added to the feedstock. Figure 15.10 gives the variation of the silicon etch rate E_{Si} and the F atom concentration n_F versus $\%O_2$ for a CF_4/O_2 feedstock mix in a capacitive rf discharge reactor (Mogab et al., 1979). O_2, CO_2, CO, and COF_2 are now seen in addition to CF_4, SiF_4, and C_2F_6 effluents. Three different regimes are observed:

1. Up to roughly 16% O_2, E_{Si} and n_F increase with $\%O_2$.
2. Between 16 and 30% O_2, E_{Si} decreases with $\%O_2$, although n_F continues to increase.
3. Above roughly 30% O_2, both E_{Si} and n_F decrease.

It is generally agreed that the first regime of increasing E_{Si} and n_F with $\%O_2$ is due to O atom (and, possibly, O_2 molecule) "burning" of CF_x unsaturates either in the gas

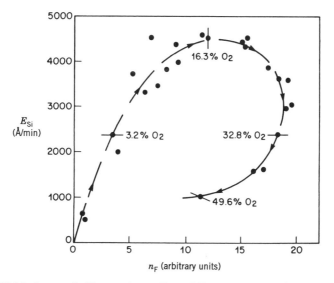

FIGURE 15.10. Locus of silicon etch rate E_{Si} and F-atom concentration n_F as the $\%O_2$ is varied in a CF_4/O_2 feedstock mix (Mogab et al., 1979).

phase or on the surface:

$$O + CF_3 \longrightarrow COF_2 + F$$
$$O + CF_2 \longrightarrow CO + 2F$$
$$\longrightarrow COF + F$$
$$\longrightarrow COF_2 \qquad\qquad (15.3.19)$$
$$O + COF \longrightarrow CO_2 + F$$
$$O + C \longrightarrow CO$$
$$\text{etc.}$$

In addition to destroying CF_x radicals, many of these reactions produce F atoms, thus increasing n_F. Furthermore, the *net* flux Γ_O of O atoms (or O_2 molecules) adsorbed on the surface modifies the etch rate by removing adsorbed carbon from the surface:

$$E_{Si} = \frac{\Gamma_F + \Gamma_O - (y - x)\Gamma_{CF_x}}{n_{Si}}, \qquad \Gamma_O < (y - x)\Gamma_{CF_x} \qquad (15.3.20)$$

Hence E_{Si} increases with $\%O_2$ because both Γ_F and Γ_O increase and because Γ_{CF_x} decreases. This shift in equilibrium to the right with O_2 addition is indicated in Fig. 15.9.

The second regime of increasing n_F and decreasing E_{Si} is believed to be due to the competition of O atoms for chemisorption sites on the SiF_x lattice. Hence in this regime, the surface layer becomes more "oxidelike", reducing the etch rate. In this regime, there is no C on the surface, and a crude model for the etch rate gives

$$E_{Si} = \frac{\Gamma_F}{n_{Si}(1 + \eta_O \Gamma_O / \Gamma_F)} \qquad (15.3.21)$$

where η_O gives the competitive efficiency for O atoms over F atoms to be adsorbed. The third regime in which both n_F and E_{Si} decrease is believed to be due to oxygen dilution effects; i.e., the flow of F atoms into the discharge is reduced by the $\%O_2$ in the feedstock.

The chemistry of CF_4/O_2 discharges is extremely complicated. In addition to the CF_4 reactions listed in Tables 15.1 and 15.2, and the neutral chemistry (15.3.19) in the gas phase and on the surface, electron dissociation of O_2, COF_2, and CO_2 is important, and some three-body gas-phase reactions, e.g.,

$$COF + F + M \longrightarrow COF_2 + M$$

might also be significant at high pressures. A fairly complete model of the gas-phase chemistry has been developed (Plumb and Ryan, 1986).

Hydrogen gas is sometimes added to the feedstock mix. The key additional reaction on the surface or in the gas phase is

$$H + F \longrightarrow HF$$

which reduces the F-atom concentration, thus shifting the equilibrium to the left in Fig. 15.9, toward increasing polymer formation.

Cl-Atom Etching

Chlorine atoms differ from fluorine atoms in two major respects for pure chemical silicon etching: (1) There are pronounced crystallographic effects, and (2) there is a large doping effect. Figure 15.11 shows the pure chemical etch rate E_{Si} at 400 K for doped silicon as a function of doping concentration n_D for various crystallographic conditions. The etch rates fit a generalized Arrhenius form (Ogryzlo et al., 1990):

$$E_{Si}(\text{Å/min}) = An_D^{\gamma}n_{ClS}T^{1/2}e^{-B/T} \qquad (15.3.22)$$

where the parameters A, B, and γ are given in Table 15.3. The activation energy $\mathcal{E}_a = kB/e \approx 0.19$ V is roughly independent of doping level and crystallographic orientation. The very strong dependence on n_D indicates that Cl^- ions formed on the surface must play a critical role in Cl-atom etching, as was found for F^- ions. However, the simple model leading to (15.3.13), while predicting a doping effect, does not give the correct order of magnitude and scaling with n_D. A model that accounts for depletion of mobile electrons in the silicon and yields $\gamma = 1/2$ is explored in

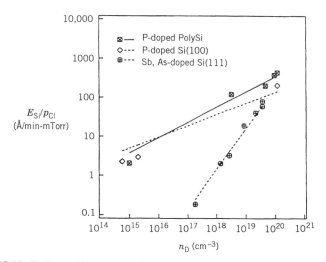

FIGURE 15.11. Etch rate E_{Si} versus doping level n_D and crystallographic orientation for Cl-atom etching of n-type silicon at 400 K; p_{Cl} is the partial pressure of Cl atoms (after Ogryzlo et al., 1990).

TABLE 15.3. Coefficients of the Modified Arrhenius Form for Cl Atom Etching of *n*-Type Silicon

Crystallographic Orientation	A ($Å$-cm$^{3+3\gamma}$/min-K$^{1/2}$)	B (K)	γ
Polysilicon	4×10^{-18}	2365	0.39
$\langle 100 \rangle$	1.1×10^{-17}	2139	0.29
$\langle 111 \rangle$	1.6×10^{-31}	2084	1.03

Source. Flamm (1990).

Problem 15.4. The dependence of the etch rate on crystallographic orientation can be ascribed to the different area densities of silicon atoms (or Si–Si bonds) at the surface. The 111 orientation has a higher density than the 100 orientation, which reduces Cl or Cl$^-$ penetration into the lattice for 111, leading to a lower etch rate.

Exposure of a pure silicon surface to Cl$_2$ leads to dissociative chemisorption which saturates at about one monolayer. Continued exposure can lead to a slow growing silicon chloride corrosion phase, but this regime is not of interest in typical etch applications. Etching rates are not significant for Cl$_2$ at room temperature. Hence, Cl$_2$ can and often does serve as a feedstock for Cl atom etching.

Exposure of a silicon surface to Cl atoms leads to formation of an SiCl$_x$ layer several monolayers thick, thinner than that formed using F atoms. Using (15.3.22), pure chemical etch rates are found to be very small for Cl atoms on undoped silicon, but can be substantial for heavily n-doped silicon. The etch products at room temperature are SiCl$_4$ and possibly Si$_2$Cl$_6$ and SiCl$_2$.

Almost all surface studies of ion-assisted etching have been with Cl$_2$ molecules, not Cl atoms. Etch products such as SiCl and SiCl$_2$ have been seen, in addition to SiCl$_4$ and Si$_2$Cl$_6$. There is general agreement that ion beam-induced mixing and recoil implantation of dissociated Cl$_2$ molecules on the surface lead to formation of an SiCl$_x$ layer more than one monolayer thick, similar to that formed for pure Cl-atom etching. The ion-assisted etch yields (silicon atoms removed per incident ion) with Cl$_2$ are comparable to those seen using F$_2$, but are a factor of 5–10 lower than those seen for ion-assisted F atom etching. For example, yields of 3–5 have been observed for 1-kV Ar$^+$ ions with an adequate flux of Cl$_2$ molecules. Yields for ion-assisted Cl-atom etching are not well known, but seem to be only somewhat larger than yields for ion-assisted Cl$_2$ etching. Hence Cl$_2$ can be an important etchant molecule for ion energy-driven etching, although it is unimportant for pure chemical etching. Br atoms are even less reactive than Cl atoms. No room-temperature pure chemical etching is observed, even for heavily n-doped silicon. At higher temperatures, etching is observed and a very large doping effect is seen.

15.4   OTHER ETCH SYSTEMS

In this section, we describe briefly some common etch systems. For a more thorough description, the review of Flamm (1989) should be consulted.

F and CF_x Etching of SiO_2

F atoms are known to etch SiO_2, although the pure chemical etch rate (15.3.2) is small and almost never significant in real etch systems. No more than a monolayer of fluorine is adsorbed on an SiO_2 surface. It is also known that CF_x radicals do not spontaneously etch SiO_2 and, furthermore, that these radicals do not dissociatively adsorb on SiO_2. Hence, there is essentially no pure chemical etching of SiO_2 in fluorocarbon plasmas, and all observed etching is ion energy driven.

Large ion-induced etch rates for SiO_2, $\gtrsim 2000$ Å/min, are seen for high ion-bombarding energies, $\gtrsim 500$ V, with both F atoms and CF_x radicals as the etchant species. The etching is anisotropic, and the etch rate correlates with the ion-bombarding energy and is independent of the substrate temperature. The loading effects are much smaller than those seen for F-atom etching of silicon. For F atoms, there is no selectivity for SiO_2 over silicon. Consequently, discharges rich in F atoms are generally not used to etch SiO_2 in the presence of silicon. High selectivity can be achieved for CF_x radical etchants that are produced using unsaturated fluorocarbon feedstocks or by adding hydrogen to saturated feedstocks, e.g., CF_4/H_2 mix. In both cases, the F-atom density is suppressed and a high density of CF_x unsaturates is created, as described in Section 15.3. Under these conditions, the etch products that are seen include SiF_4, SiF_2, $SiOF_2$, CO, CO_2, and COF_2. Figure 15.12 shows the variation of the gas phase densities n_F and n_{CF_2} and the SiO_2 etch rate E_{SiO_2} versus $\%H_2$ and $\%O_2$ added to a CF_4 parallel plate discharge (Flamm, 1989). For O_2 addition, n_F increases, n_{CF_2} is suppressed, and E_{SiO_2}, due to F atom etchants, increases with $\%O_2$ up to 30% O_2 addition. However, the etching is not selective over silicon. In contrast, with addition of H_2, n_F is suppressed and n_{CF_2} increases, leading to a mild increase in E_{SiO_2} and a strong decrease in E_{Si} (not shown), with SiO_2/Si selectivity as high as 15:1. This is the regime of anisotropic selective etching of SiO_2 over silicon. The SiO_2 etch rate is observed to abruptly fall to zero above roughly 20% H_2 addition. As will be seen below, this is due to polymer film formation on the SiO_2 surface.

Under the action of ion bombardment and high-incident CF_x radical flux, an $SiC_xF_yO_z$ layer as thick as 10–20 Å forms on the SiO_2 surface. For lower radical fluxes, the layer is thinner. Ion beam mixing of adsorbed CF_x radicals is believed to play an important role in formation of this layer. Under these conditions, the etch mechanisms are believed to be similar to those seen for F atom etching of silicon. The most important mechanism is probably the breaking and reforming of bonds within and on the surface of the $SiC_xF_yO_z$ layer due to the collision cascade produced when an energetic ion hits and penetrates the surface. This produces easily desorbable etch products that are weakly bound to the surface, such as SiF_4, SiF_2, CO, CO_2, COF_2, $SiOF_2$, and, possibly, O_2. A crucial point is that adsorbed C atoms can here act as etchants, removing oxygen from the surface by reactions such as

$$\text{ion} + \text{C(s)} + \text{SiO}_2\text{(s)} \longrightarrow \text{CO(g)} + \text{SiO(s)} + \text{ion}$$

$$\longrightarrow \text{CO}_2\text{(g)} + \text{Si(s)} + \text{ion}$$

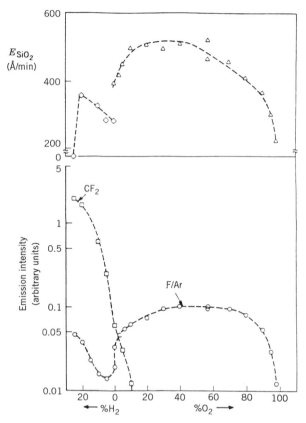

FIGURE 15.12. SiO_2 etch rate (top) and plasma-induced emission for CF_2 and F/Ar actino-metric emission ratio (bottom) versus $\%H_2$ and $\%O_2$ addition to a CF_4 parallel plate discharge (Flamm, 1989).

Hence the presence of oxygen in the lattice impedes carbon buildup, allowing the surface to be etched. For the same conditions, carbon-containing polymer films as thick as 100–200 Å are observed to form on silicon as well as on nonactive surfaces. The film on silicon inhibits the etch reaction there, leading to the high observed selectivities for SiO_2/Si under unsaturate rich conditions (see also Fig. 15.9). Even for discharge conditions that do not lead to carbon-containing film deposition on silicon, (15.3.17) shows that the silicon etch rate can be small in CF_x-rich discharges, thus leading to high selectivities in the absence of polymer buildup on silicon surfaces. On the other hand, for very high unsaturate concentrations ($> 20\%H_2$ in CF_4/H_2 mix), SiO_2 etching ceases due to formation of thick polymer films on SiO_2 surfaces also, as seen in Fig. 15.12.

Si_3N_4 Etching

Silicon nitride is commonly used as a mask material for patterned oxidation of silicon, and as a dielectric. There are two kinds of Si_3N_4 material: that produced by chemical vapor deposition (CVD) at high temperatures, and that grown under plasma-enhanced conditions (PECVD) at temperatures less than 400 °C. The latter material does not necessarily have 3:4 Si/N stochiometry and generally has a significant fraction of H atoms in the lattice. Etch rates for PECVD material are generally high compared to CVD Si_3N_4.

Pure chemical F-atom etching of Si_3N_4 can have selectivities of 5–10 over SiO_2, but is not selective over silicon. The etching is isotropic with an activation energy of order 0.17 V. Anisotropic ion energy driven etching of Si_3N_4 is performed using unsaturated fluorocarbon feedstocks. There is little selectivity over SiO_2, but fairly high selectivities over silicon and resist can be attained.

Aluminum Etching

Aluminum is commonly used as an interconnect material in integrated circuits because of its high electrical conductivity, excellent bondability and adherence to silicon and SiO_2, compatibility with CVD oxide and nitride, and ability to form both ohmic and Schottky contacts with silicon. Since AlF_3 is involatile, F atoms cannot be used to etch aluminum, and Cl_2 or Br_2 feedstocks are used instead. These vigorously and isotropically etch aluminum in the absence of ion bombardment. Molecular chlorine etches pure clean aluminum without a plasma; in fact, Cl_2 rather than Cl appears to be the primary etchant species for aluminum in etching discharges. The main etch products are $Al_2Cl_6(g)$ at low temperatures ($\lesssim 200$ °C) and $AlCl_3(g)$ at higher temperatures.

An ion-enhanced etch with inhibitor chemistry is needed to anisotropically etch aluminum. For Cl_2 feedstock, additives such as CCl_4, $CHCl_3$, $SiCl_4$, and BCl_3 are used. Cl_2 and Cl do not etch Al_2O_3, even in the presence of ion bombardment. Processes used to break through the native oxide (~ 30 Å thick) and initiate the aluminum etch include physical sputtering due to the ion bombardment and additives such as CCl_4, $SiCl_4$, and BCl_3. Unsaturated radicals produced from these feedstock additives etch Al_2O_3 at slow rates.

Water vapor interferes with aluminum etching and must be excluded from the system or scavenged using water-seeking additives such as BCl_3 and $SiCl_4$. Copper is often added to aluminum to harden the material and increase its resistance to electromigration and hillock formation. A material such as Al/5%Cu is hard to etch because copper chloride ($CuCl_3$) is nearly involatile at room temperature; its heat of vaporization is 15.4 kJ/mol and its vapor pressure is 1 Torr at 572 °C. This material can be etched at temperatures considerably exceeding room temperature or in systems having high ion-bombarding fluxes, with consequent high physical sputtering rates for the copper. Aluminum chloride products can react with photoresist mask materials during etching. To reduce the deterioration of mask materials, the substrate temperature must be kept below 100–150 °C. Post etch corrosion due to $AlCl_3$ or Al_2Cl_6 deposits on the wafer can be a problem. These react with water vapor to generate HCl, which can corrode the aluminum and other structures on the wafer. Hence a wet (HNO_3) or dry chemistry (fluorocarbon plasma) is used to remove any remaining aluminum chloride etch product from the surface. The dry process converts aluminum chlorides to unreactive aluminum fluorides.

Resist Etching

Photoresist mask materials are primarily long-chain organic polymers consisting mostly of carbon and hydrogen. Oxygen plasmas are used to isotropically etch ("strip") resist mask materials from wafers and are also used for anisotropic pattern transfer into these materials in so-called surface imaged dry development schemes for photoresists. An active etchant for both applications is O atoms. Pure chemical etching of resists using O atoms is isotropic and highly selective over silicon and SiO_2. It is generally characterized by an activation energy of 0.2–0.6 V. In some cases, the etch rate does not have a simple Arrhenius form, with two activation energies depending on whether the substrate temperature is above or below the "glass transition temperature" T_g of the resist. For example, polymethyl methacrylate (PMMA) has $T_g \sim 60$–90 °C, with $\mathcal{E}_a \approx 0.2$ V for $T < T_g$ and $\mathcal{E}_a \approx 0.4$ V for $T > T_g$.

Pure chemical etch rates for many resists are low but can be enhanced by addition of a few percent C_2F_6 or CF_4 to the feedstock mix. This may be due to F atom reactions with the resist to produce HF etch product, leaving unsaturated or radical sites on the polymer for subsequent O-atom attack. In some cases, the measured activation energy is lowered by a factor of as much as three with F-atom addition. Alternatively, it is known that small additions of fluorine atoms can increase the O-atom concentration in the discharge, thus increasing the etch rate. This may be due to a reduction in the O-atom recombination rate on the reactor walls because of F-atom chemisorption.

Ion-enhanced anisotropic etching in O_2 plasmas is used for dry development of surface-imaged photoresists, in which only a small fraction of the volume at the top of the photoresist layer is exposed to the light. There are a number of motivations for use of surface-imaged resists. As lateral feature sizes continue to decrease, optical wavelengths to expose the patterns must also decrease. For decreased wavelengths, the depth of focus in the resist is also reduced. Hence the pattern is not in accurate focus throughout the entire thickness of the resist. A second motivation is that optical reflections from the layer underlying the resist can be eliminated. These reflections can lead to photoresist exposure in regions not directly illuminated, with consequent pattern transfer that is not faithful to the original image. A final motivation is that resists become naturally opaque to light as the wavelength is reduced into the deep UV region below 2000–2200 Å; for wavelengths below these, surface imaging technology may be required.

A typical process flow is shown in Fig. 15.13. First, the top 0.2 μm of a 1.5-μm-thick layer of photoresist is optically exposed to a pattern. Second, the resist is silylated by exposure to a silicon-containing gas. The silicon is selectively absorbed into the exposed photoresist, but is not absorbed into the unexposed photoresist. Finally, the photoresist is anisotropically etched in an O_2 plasma. The O atoms initially react with the exposed, silicon containing surface layer to create an SiO_x mask that is impervious to subsequent ion-assisted O-atom etching. The unexposed,

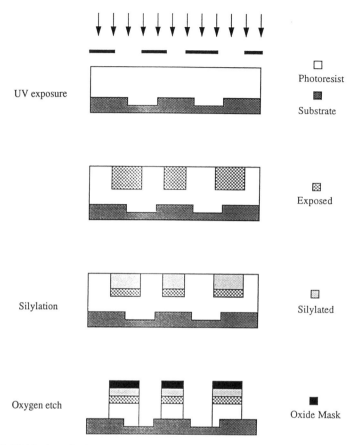

UV exposure

Photoresist

Substrate

Exposed

Silylation

Silylated

Oxygen etch

Oxide Mask

FIGURE 15.13. A typical process flow for a silylated surface imaged resist dry development scheme.

non-silicon-containing photoresist is anisotropically etched. The original surface image is therefore transferred into the entire thickness of the resist film. Clearly the unsilylated/silylated etch selectivity is a critical process parameter. Minimum selectivities of 10–20 are required.

The balance between ion bombardment and O-atom flux is delicate in this application. On the silylated areas, the O-atom flux must be large enough to oxidize the silylated layer to form the SiO_x mask, and the ion bombardment must be weak enough so that the mask is not physically sputtered away during the etch of the unsilylated areas. For the unsilylated areas, the ion energy and flux must be large enough to transfer the pattern with near vertical sidewalls into the resist. Any SiO_x sputtered onto the unsilylated areas may serve as a micro mask during the etch, leaving residues, often called "grass", on the unsilylated regions after the etch is completed. The requirement of anisotropic etching with low physical sputtering is severe and demands careful control of both ion energy and flux. Hence high-density discharges, rather than rf diode discharges, which lack this control, are generally used. Fluorocarbon feedstock additions cannot be used to enhance the etch rate of unsilylated resist, or to prevent formation of residues or remove them during etch, because such additions lead to greatly enhanced etch rates for the SiO_x mask, and consequent reductions in unsilylated/silylated selectivity below the required minimum. In fact, trace fluorine contaminants have been found to have adverse effects on the selectivity.

PROBLEMS

15.1. Sputter Etching   Estimate the maximum etch rate (Å/min) for physical sputtering of silicon using 600-V Ar^+ ions for an Ar^+ density n_{Ar^+} at the sheath edge of 10^{11} cm^{-3}. Use the data given in Table 9.2.

15.2. Capacitive Rf Discharge Scaling   Repeat the analysis leading to the scalings (15.2.20) and (15.2.29) of n_{is} and n_{OS} with $P_{abs}, n_g, l, T_e, \omega$, and A for a low pressure slab model of a symmetric capacitive rf discharge. Assume that the discharge is electropositive, that stochastic heating dominates ohmic heating, and that ion energy losses dominate electron collisional losses.

15.3. High-Pressure, High-Density Discharge Scaling   Repeat the analysis leading to the scalings (15.2.20) and (15.2.29) of n_{is} and n_{OS} with discharge parameters for a high-pressure slab model of a high-density discharge. Assume that the discharge is electropositive and that the ion flux to the wall is determined by an ambipolar diffusion coefficient $D_a \propto n_g^{-1}$.

15.4. Doping Effect for Cl Atom Silicon Etching   For Cl-atom etching of n-type silicon, there is a strong doping effect, with $E_{Si} \propto n_D^\gamma$, where γ is a coefficient given in Table 15.3. Consider a model for this, similar to that shown in Fig. 15.8, in which the voltage drop ΔV across the $SiCl_x$ dielectric layer is small compared to the voltage drop $\Delta V'$ due to depletion of free electron carriers n_e within the silicon near the $SiCl_x$–Si interface. The latter voltage drop was neglected in determining the surface charge density n'_- in (15.3.13). Here we make the opposite assumption. Assume that for a given n'_-, there is a uniform positive charge density n_D over a thickness δ' extending into the silicon from the $SiCl_x$-Si interface.

(a) Consider a voltage drop across the silicon having the form

$$\Delta V' = V'_0 + T \ln \left(\frac{n_D}{n_I} \right)$$

similar to (15.3.12), where V'_0 is a constant of order 1 V determined by the affinity level for formation of Cl^- ions. Use Poisson's equation to find an expression for δ' as a function of n_D and the dielectric constant $\epsilon_{Si} \approx 11.8 \, \epsilon_0$.

(b) Solve for n'_- as a function of n_D and V'_0, and show that $\gamma = 1/2$ in the limit $V'_0 \gg T \ln(n_D/n_I)$.

15.5. Free Radical Production in a CF_4 Discharge   Consider a simplified mechanism for F-atom production in a CF_4 discharge, consisting of reactions 2 and 3 in Table 15.1 and reactions 1 and 2 in Table 15.2. Assume that the electron density is specified to be $n_e = 10^{10}$ cm^{-3} and that there is no other generation or loss of CF_4, CF_3, CF_2, and F than given by these reactions; i.e., no surface losses, etc.. Assume that the rate constants for reactions 1 and 2 in Table 15.2 are second order; i.e., for the high-pressure limit in which the reactions are independent of the concentration of the third molecule M (here CF_4).

(a) Write the differential equations for the densities of the four species; e.g., $dn_{CF_4}/dt = \dots$,

(b) In the steady state, show that $n_F n_{CF_3}/n_{CF_4} = A(T)$ and $n_{CF_2}/n_{CF_3} = B(T)$. Obtain A and B in terms of K_4, K_5, K_1, K_2, and n_e.

(c) If the initial concentration of CF_4 is n_0 and all other initial concentrations are zero at time $t = 0$, then find the equilibrium concentration ($t \to \infty$) of F atoms in terms of n_0, A, and B.

(d) For a CF_4 pressure of 10 Torr at 300 K, and silicon etching due to a flux of F atoms only, use (15.3.1) to estimate the initial etch rate (Å/min) when a piece of silicon is inserted into the equilibrium gas mixture.

15.6. Surface Model for Silicon Etch in a CF_4 Discharge   Consider the following surface model for pure chemical silicon etch (no ion bombardment) in a CF_4 discharge. Let n_1 and n_2 be the gas-phase densities of CF_x radicals and F atoms near the surface, respectively, and let θ_1 and θ_2 be the fractions of the SiF_3 surface covered with CF_4 and SiF_4, respectively. Let K_{a1} and K_{a2} (cm^3/s) be the adsorption rate constants for CF_x radicals and F atoms, respectively, and let K_{d1} and K_{d2} be the desorption rate constants (s^{-1}) for $CF_4(g)$ and $SiF_4(g)$, respectively. Assume Langmuir kinetics with adsorption of CF_x and F on the SiF_3 surface only.

(a) In the steady state, give the two conservation equations for carbon and fluorine on the surface.

(b) Solve these to obtain the surfaces coverages θ_1 and θ_2.

(c) Find the silicon etch rate E_{Si} and plot the normalized etch rate per incident F atom, E_{Si}/n_2 (Å-cm^3/min) versus n_2/n_1 for $x = 3$, $K_{a1} = K_{a2} = 4 \times 10^{-14}$ cm^3/s, $n_0' = 7 \times 10^{14}$ cm^{-2}, $K_{d1} = K_{d2} = 10^{12}$ s^{-1}, and $n_{SiF_3} = 5 \times 10^{22}$ cm^{-3}. Assume that $n_1, n_2 \ll K_{d2}/K_{a1}$.

15.7. Comparison of Silicon and SiO_2 Loading Effects   For the same etch rates and discharge conditions, SiO_2 etching in CF_4 discharges exhibits a smaller loading effect (ratio of $n_{OS}^{(0)}/n_{OS}$ in (15.2.33)) than silicon etching. If the only etch products formed are CO_2 and SiF_4 for SiO_2 etching, and C_2F_6 and SiF_4 for silicon etching, then show that, for a fixed etchant generation rate and the same loading effect, the SiO_2 etch area is four times the silicon etch area. In both cases assume that $P_{rec} = 0$ on the walls and that $P_r = 1$ on the substrates.

CHAPTER 16

DEPOSITION AND IMPLANTATION

16.1 INTRODUCTION

Plasma-assisted deposition, implantation, and surface modification are important materials processes for producing films on surfaces and modifying their properties. For example, as described in Chapter 1, the cycle of film and mask deposition, mask patterning, implantation or other modification, etching, and mask stripping is repeated many times during the manufacture of modern integrated circuit devices. Because device structures are sensitive to temperature, high-temperature deposition processes cannot be used in many cases. Fortunately, due to the nonequilibrium nature of low-pressure processing discharges, high-temperature films can be deposited at low temperatures. Furthermore, films can be deposited with improved properties, nonequilibrium chemical compositions, and crystal morphologies that are unattainable under equilibrium deposition conditions at any temperature. Unique films not found in nature can be deposited, e.g., diamond.

Consider two examples for integrated circuit fabrication. Most aluminum thin films (i.e., actually Al/Cu or Al/Si) used for interconnection are deposited on the wafer by physical sputtering from an aluminum or alloy target; this is essentially a room-temperature process. Although thermal evaporation sources can be used, it is more difficult to control film uniformity and composition with these sources. Another example is the final insulating "capping" layer on many devices, silicon nitride, which is deposited by plasma-enhanced chemical vapor deposition (PECVD) at temperatures near 300 °C. An equivalent nonplasma chemical vapor deposition (CVD) would require temperatures near 900 °C, and therefore cannot be used because it would melt the aluminum, destroying the device. Furthermore, by varying the ion bombardment and other plasma parameters in PECVD of silicon nitride, the film composition, stress, and integrity can be controlled, greatly increasing its reliability as a capping layer. Let us note, however, that PECVD cannot replace CVD in some applications; e.g., most low-temperature PECVD films are amorphous and not crystalline, which can more easily be achieved with CVD. Where high temperatures are allowed, CVD can be the method of choice for deposition of metals, dielectrics, and semiconducting films.

Ion implantation is another important process for semiconductor doping, and has other uses, such as for surface hardening of materials. For silicon doping, ions such as boron, phosphorous, and arsenic are implanted. For surface hardening of metals, nitrogen or carbon are implanted. Conventional ion beam implanters are used for low-flux, high-energy implants. At high fluxes, particularly for low ion energies, and where mass-energy selection is not critical, plasma-immersion ion implantation (PIII) can be used to meet process requirements that are not attainable using conventional ion beam implanters. PIII processes have been developed for hardening medically implantable hip joints, for hardening tools and dies, and for doping semiconducting materials. Materials modifications through a combination of ion implantation and ion beam mixing of near surface layers are also under development. As was described in Chapter 15, ion beam mixing can also play a critical role in etch processes.

In this chapter, as in the previous, we focus on the area of integrated circuit processing to describe deposition, implantation, and other surface-modification processes. For a thorough review of plasma-assisted deposition and surface modification processes, the monograph of Konuma (1992), the collections of review articles edited by Vossen and Kern (1978, 1991), and references cited therein should be consulted.

The range of plasma-deposition processes is broadly divided into two areas: PECVD and sputter deposition. PECVD is described in Section 16.2, using the well-known example of amorphous silicon (a-Si) deposition to introduce the discharge regime, gas-phase chemistry, and surface-reaction model. While specific to a-Si deposition, the discussion is relevant to PECVD for many other materials. PECVD of SiO_2 is also described to introduce a more complicated surface chemistry and to treat the issues of anisotropic deposition and conformality of deposition over topography, e.g., deposition in trenches. Almost all the discharges described in previous chapters are widely used for PECVD, with the exception of dc discharges, although the bulk of the deposition is done commercially with some form of multi- or single-wafer capacitive rf discharge reactor.

Sputter deposition, which is discussed in Section 16.3, includes both physical sputtering and reactive sputtering. In the former, atoms are sputtered from a target material and are transported to and deposited on a substrate. The mechanism of physical sputtering is described in Section 9.3, and some data are given in Table 9.2. Some issues related to sputtering uniformity are also considered in Section 14.5. In Section 16.3 we describe the sputtered atom energy distributions and their influence on film properties. In reactive sputtering, a feedstock gas whose dissociation products chemically react with the target material is present in addition to the bombarding ions.

Hence the deposited film is a compound formed from the sputtered materials and the reactive gases. In contrast to physical sputtering, where a model for the generation and transport of sputtered atoms from target to substrate is relatively straightforward, a reactive sputtering model involves surface reactions at both target and substrate in addition to sputtering at the target and deposition at the substrate. A simple model for this process is given to conclude Section 16.3. Sputtering discharges for conducting films are generally dc discharges, usually dc planar magnetrons (see Section 14.5); for sputtering insulating films, capacitive rf discharges or rf driven planar magnetrons are commonly used.

Ion implantation using PIII is described in Section 16.4. The basic principles for the process are given, a simple model for the dynamic high-voltage sheath formation is developed, and some applications to integrated circuit and other processing are described. PIII must generally be done in low-pressure ($p \lesssim 1$ mTorr) processing discharges in which the ion mean free path is comparable to or larger than the high-voltage sheath width, but there are some applications where higher pressures are desirable.

Other plasma-enhanced surface modification processes, not treated in this text, include low-temperature oxidation of silicon, plasma polymerization, and additional (non-PIII) nitriding and carbiding techniques. For example, good quality thin SiO_2 films have been grown on single-crystal silicon in oxygen discharges at substrate temperatures of 250–400 °C (Carl et al., 1991). The process is called *plasma anodization* because the substrate is generally biased positive with respect to the plasma, drawing a net dc current through the film as it grows. Oxidation kinetics can be explained by O^- transport-limited growth at the $Si–SiO_2$ interface. Sputtering contamination during film growth is an issue, so microwave and other high-density discharges having low sheath voltages are generally used.

16.2   PLASMA-ENHANCED CHEMICAL VAPOR DEPOSITION

Chemical vapor deposition (CVD) consists of a thermally activated set of gas-phase and surface reactions that produce a solid product at a surface. In plasma-enhanced chemical vapor deposition (PECVD), the gas phase and often the surface reactions are controlled or strongly modified by the plasma properties. In place of thermal activation in CVD, the critical initial step in PECVD is electron impact dissociation of the feedstock gas. Since $T_e \sim 2$–5 V in a low-pressure discharge easily suffices for feedstock dissociation and since T_e is much greater than the substrate (and heavy particle) temperature, the deposition can be carried out at temperatures much lower than for CVD. Because chemical reactions between neutral gas-phase precursor components are often required for PECVD, the discharge pressures used are in the range 0.1–10 Torr, considerably higher than those used for plasma assisted etching. The neutral mean free paths are therefore small, of order 0.003–0.3 mm. The plasma densities are in the range 10^9–10^{11} cm^{-3}, and the fractional ionizations are low, of order 10^{-7}–10^{-4}. As for etching, the deposition is limited by either the feedstock gas flow rate and pressure or by the discharge power, depending on which is rate limiting.

Surface activation energies for PECVD are often small, occasionally negative. Hence deposition rates are usually not very sensitive to the substrate temperature T. However, film properties such as composition, stress, and morphology are generally strong functions of T. Consequently, T is usually optimized to achieve a desired set of film properties.

Deposited film uniformity is a critical issue for PECVD because of the high pressures, high flow rates, short mean free paths, high gas-phase reaction rates, and high surface sticking probabilities for some gas-phase deposition precursors (often, neutral radicals). This combination of factors makes it very difficult to achieve uniform precursor and ion fluxes across the substrate area. Hence, great care is required in design of the neutral transport system for flow of gases into and out of the reaction zone. Similarly, the variation of the power deposition per unit area in the discharge must be carefully controlled. For these reasons, rf-driven parallel-plate discharge geometries have been favored, although some depositions have been performed using high-density cylindrical discharges, such as ECRs, helicons, and rf inductive discharges (TCPs).

Amorphous Silicon

Amorphous silicon thin films are used in solar cells, for thin-film transistors for flat panel displays, and for exposure drums for xerography. Whereas epitaxial (crystalline) silicon has a density of 2.33 g/cm^3, PECVD amorphous silicon grown using silane (SiH$_4$) discharges has a lower density, ~ 2.2 g/cm^3, due to incorporation of 5–20% H atoms in the lattice. Hence this material is usually denoted as a-Si:H. The material is inexpensive to make and easily deposited over large areas on a wide variety of substrates including glasses, metals, polymers, and ceramics. The feedstock gas in a capacitive rf discharge is typically SiH$_4$ at pressures of order 0.2–1 Torr, although SiH$_4$/H$_2$/Ar mixes are sometimes used at somewhat higher pressures. Gas-phase additions such as B$_2$H$_6$ and PH$_3$ are used to grow p-type or n-type material, respectively. The rf power fluxes are typically 10–100 mW/cm^2, yielding deposition rates of 50–500 Å/min. The substrate temperatures are typically 25–400 °C, depending on the application. The activation energy for the deposition is low, 0.025–0.1 V, compared to 1.5 V for high-temperature CVD silicon deposition using SiH$_4$.

SiH$_4$ is a hazardous gas that reacts explosively with air or water vapor. The molecule is tetrahedral (symmetry group T_d, with the silicon atom in the center), having a heat of formation of 34.3 kJ/mol and a Si–H bond distance of 1.5 Å. The SiH$_3$–H bond energy is 3.9 V. The positive ion SiH$_4^+$ is unstable or weakly stable, and has not been observed under typical discharge conditions; SiH$_3^+$ is normally observed. Both SiH$_3$ and SiH$_2$ radicals have a positive electron affinity; hence, silane discharges can be electronegative. The SiH$_2$–H, SiH–H, and Si–H bond energies are 3.0, 3.4, and 3.0 V, respectively. Some rate constants for significant (mostly two-body) gas-phase reactions are given in Table 16.1. A relatively complete SiH$_4$/H$_2$/Ar gas-phase discharge model is described by Kushner (1988). This model includes over 35 electron impact reactions, 90 neutral–neutral reactions, 80 positive ion–neutral reactions, and a complete set of electron–ion and positive–negative ion recombination reactions.

TABLE 16.1. Selected Reaction Rate Constants for SiH$_4$ Discharges

Number	Reaction	Rate Constant (cm^3/s)	Source
1	$e + SiH_4 \rightarrow SiH_3 + H + e$	$1.5E\text{--}8 \exp(-10/T_e)$	a
2	$e + SiH_4 \rightarrow SiH_2 + 2H + e$	$1.8E\text{--}9 \exp(-10/T_e)$	a
3	$e + SiH_4 \rightarrow SiH_3^- + H$	$1.5E\text{--}11 \exp(-9/T_e)$	a
4	$e + SiH_4 \rightarrow SiH_2^- + H_2$	$9E\text{--}12 \exp(-9/T_e)$	a
5	$e + SiH_4 \rightarrow SiH_3^+ + H + 2e$	$3.3E\text{--}9 \exp(-12/T_e)$	a
6	$e + SiH_4 \rightarrow SiH_2^+ + H_2 + 2e$	$4.7E\text{--}9 \exp(-12/T_e)$	a
7	$SiH_4 + H \rightarrow SiH_3 + H_2$	$2.7E\text{--}12$	k
8	$SiH_4 + SiH_2 \rightarrow Si_2H_6^*$	$1E\text{--}11$	k
9	$Si_2H_6^* \rightarrow Si_2H_4 + H_2$	$5E6\ /s$	k
10	$Si_2H_6^* + SiH_4 \rightarrow Si_2H_6 + SiH_4$	$1E\text{--}10$	k
11	$SiH_4 + SiH_3 \rightarrow Si_2H_5 + H_2$	$1.8E\text{--}15$	k
12	$SiH_3 + H \rightarrow SiH_2 + H_2$	$1E\text{--}10$	k
13	$SiH_3 + SiH_3 \rightarrow SiH_2 + SiH_4$	$7E\text{--}12$	k
14	$e + SiH_n^+ \rightarrow SiH_{n-1} + H$	$2.5E\text{--}7\,T_e^{-1/2}$	a
15	$SiH_m^- + SiH_n^+ \rightarrow SiH_m + SiH_n$	$5E\text{--}7$	k

Note. T_e in volts and $T \sim 500\text{--}700$ K for ions and neutrals. The notation E–8 means 10^{-8}.
[a]Based on data in Kushner (1988).
[k]Kushner (1988).

There is considerable evidence (McCaughey and Kushner, 1989) that SiH$_3$ and SiH$_2$ radicals are important precursors for film growth, that SiH$_4$ also participates in surface reactions, and that ion (SiH$_3^+$) bombardment plays a critical role in film growth. A simple model of the surface, shown in Fig. 16.1, is that it consists of active sites, containing at least one dangling bond, and passive sites, containing either silicon or hydrogen atoms at all four bonds. The dangling bonds are created by ion bombardment, which also removes hydrogen from the surface. SiH$_2$ can insert itself into the lattice upon impact with the surface at either active or passive sites, leading to film growth in a manner similar to that of physical vapor deposition (i.e., as in (9.4.24)). Such films are generally of poor quality, having voids, undesired surface

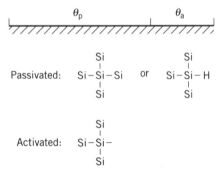

FIGURE 16.1. Surface coverage model for amorphous silicon deposition; θ_a and θ_p are the fractions of the surface that are active and passive, respectively.

roughness, and other surface defects. Adsorbed SiH_3 radicals can diffuse along the surface but can insert into the lattice only at active sites, filling in the surface roughness and contributing to growth of a smooth, high-quality film. SiH_4 adsorbed upon impact at active sites can lose an H atom, thus passivating the site. Based on these ideas, elaborated by McCaughey and Kushner, we let θ_a and θ_p be the fraction of the surface covered by active and passive sites, respectively, with $\theta_a + \theta_p = 1$. Then the surface reactions can be represented as

$$SiH_3{}^+ + \theta_p \xrightarrow{K_i} \theta_a + Y_iH(g) \tag{16.2.1}$$

$$SiH_2 + \theta_a \xrightarrow{K_2} \theta_a \tag{16.2.2}$$

$$SiH_2 + \theta_p \xrightarrow{K_{2p}} \theta_p \tag{16.2.3}$$

$$SiH_3 + \theta_a \xrightarrow{K_3} \theta_p \tag{16.2.4}$$

$$SiH_4 + \theta_a \xrightarrow{K_4} \theta_p + SiH_3(g) \tag{16.2.5}$$

where Y_i is the yield of H atoms removed per incident ion, $K_i \approx u_B/n_0'$, $K_2 \approx \frac{1}{4}s_2\bar{v}_2/n_0'$, $K_3 \approx \frac{1}{4}\bar{M}s_3\bar{v}_3/n_0'$, $K_4 \approx \frac{1}{4}s_4\bar{v}_4/n_0'$, and $K_{2p} \approx \frac{1}{4}s_{2p}\bar{v}_2/n_0'$ are the rate constants, with s_2, s_3, and s_4 the sticking coefficients on the activated surface for SiH_2, SiH_3, and SiH_4, respectively, s_{2p} is the sticking coefficient for SiH_2 on the passivated surface, n_0' is the area density of sites, and $\bar{M}$ is the mean number of sites visited by a surface-diffusing SiH_3 radical before desorption. In the steady state, the rate of creation of active sites is

$$\frac{d\theta_a}{dt} = Y_iK_in_{is}(1 - \theta_a) - K_3n_{3S}\theta_a - K_4n_{4S}\theta_a = 0 \tag{16.2.6}$$

Solving for θ_a, we obtain

$$\theta_a = \frac{Y_iK_in_{is}}{Y_iK_in_{is} + K_3n_{3S} + K_4n_{4S}} \tag{16.2.7}$$

Note that SiH_2 adsorption and reaction do not affect θ_a in this model. The deposition rate follows from reactions (16.2.2), (16.2.3), and (16.2.4), which each deposit one silicon atom:

$$D_{Si} = (K_3n_{3S}\theta_a + K_2n_{2S})\frac{n_0'}{n_{Si}} \tag{16.2.8}$$

For typical deposition processes, $Y_i \sim 5$–10, all s's are of order unity, $\bar{M} \sim 10$, and $K_4n_{4S} \gg Y_iK_in_{is} + K_3n_{3S}$. Under these conditions, (16.2.7) yields

$$\theta_a \approx \frac{Y_iK_in_{is}}{K_4n_{4S}} \tag{16.2.9}$$

For typical discharge parameters, $n_{is}/n_{4S} \sim 10^{-4}$ and, therefore, $\theta_a \sim 10^{-2}$. The overall sticking coefficient for SiH_3 to react with the entire surface (active and passive) is then $s_3 \bar{M} \theta_a \sim 0.1$. Under typical conditions for film deposition, $n_{2S} \sim 10^{-2} n_{3S}$, such that the first term in (16.2.8), responsible for the "good" film deposition, is roughly ten times larger than the second term, responsible for the "bad" film deposition. Clearly, from (16.2.8) and (16.2.9), good films are associated with high ion fluxes and energies, high SiH_3/SiH_2 ratios, and high SiH_3 surface diffusivities.

The preceding picture is oversimplified. For example, the reactions of H atoms at the surface,

$$H + \theta_a \longrightarrow \theta_p$$
$$H + \theta_p \longrightarrow \theta_a + H_2(g)$$

can modify the overall surface dynamics. A more complete model is presented by McCaughey and Kushner (1989).

Silicon Dioxide

SiO_2 can be grown by oxidation of bare silicon at 850–1100 °C using O_2 or H_2O gas. CVD oxide can also be deposited on substrates at 600–800 °C using SiH_4/O_2 or $TEOS/O_2$ feedstock gases, and can be grown at still lower temperatures, 100–300 °C, using PECVD with the same feedstocks. TEOS (tetraethoxysilane), $Si(OC_2H_5)_4$, has the chemical structure shown in Fig. 16.2, with C–O and Si–O bond energies of 3.7 and 4.7 V, respectively. In contrast to silane, which is an explosive gas at room temperature, TEOS is a relatively inert liquid. Gases such as N_2 or Ar are often used as carriers of the vapor. Highly dilute $TEOS/O_2$ feedstock mixtures are usually used; a 1%TEOS/99%O_2 mixture is typical. Under these conditions, much of the gas-phase kinetics is dominated by O_2, and the discharge can be modeled as if it were a pure O_2 discharge. Highly oxygen-rich mixtures are required for good quality films because TEOS contains carbon and hydrogen, which the O_2 burns to form $CO_2(g)$ and $H_2O(g)$ effluents. If this is not done efficiently, then the films can have a substantial carbon and/or hydrogen content.

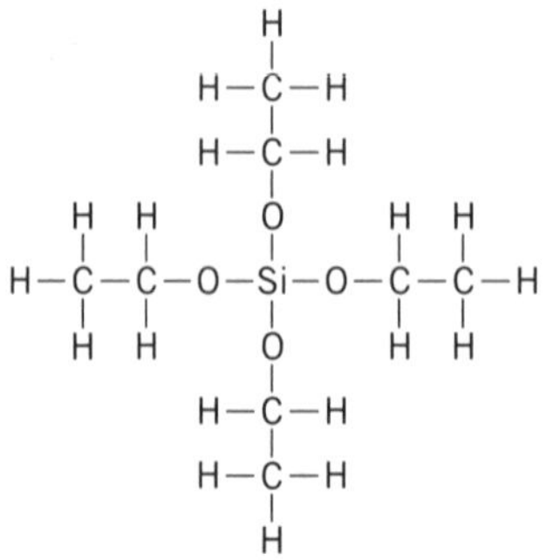

FIGURE 16.2.   Chemical structure of TEOS.

Oxide deposition using $SiH_4/Ar/N_2O$, $SiH_4/Ar/NO$, or $SiH_4/Ar/O_2$ gas mixtures can yield deposition rates of up to 2000 Å/min. The usual oxygen source is N_2O, as this produces copious oxygen atoms on dissociation and the best quality films. The deposition precursors are believed to be SiH_3, SiH_2, and O radicals created by electron impact dissociation of SiH_4 and the oxidant (N_2O, NO, or O_2). Initial steps in film formation consist of surface reactions such as

$$2SiH_3 + O(s) \longrightarrow (SiH_3)_2O$$
$$\longrightarrow SiH_3OH + H_2 \qquad (16.2.10)$$

Further oxygenation of the surface burns off most of the excess H atoms as $H_2O(g)$. The final film typically has 2–9% H atoms.

The sticking probabilities of the precursors SiH_3 and SiH_2 in silane discharges tend to be high, e.g., $s \sim 0.35$. This tends to lead to nonconformal deposition on topographical features such as in trenches; i.e., the deposition rates at various points on the trench surface are different. To understand this, consider a simple model of deposition on the sidewalls and bottom of a trench of initial width w and depth d, due to a uniform isotropic source of precursors at the top of the trench, as shown in Fig. 16.3a. Assume a unity sticking coefficient and ballistic transport of precursors within the trench; i.e., the mean free path for precursor collisions is much greater than w or d. Then it can be shown (Problem 16.2) that the deposition flux Γ_{SiO_2} on the sidewall is

$$\Gamma_{SiO_2} \propto 1 - \cos\theta_s \qquad (16.2.11)$$

where θ_s is the angle subtended by the trench opening as seen at a position along the sidewall. Note that θ_s is 90° near the top of the trench and falls monotonically with depth along the sidewall. Hence the maximum deposition rate is on the sidewall near the top of the trench. As deposition proceeds, as shown in Fig. 16.3b and c, this can lead to formation of a void or "keyhole" within the trench. This is undesirable for many applications. If the deposition is conformal, i.e., equal growth rates at all points within the trench, the keyhole is avoided and the trench completely fills with the insulating dielectric. To achieve conformal deposition, either the sticking probability should be small, leading to many precursor reflections within the feature, or precursors that stick with high probability should have high diffusion rates along the surface.

For SiO_2 deposition using silane-containing feedstocks, the sticking probabilities are high and the surface diffusion is not significant; consequently, the conformality of deposition is poor. Oxide deposition using highly dilute $TEOS/O_2$ feedstock at temperatures of 200–300 °C and pressures of 0.2–0.5 Torr leads to relatively low deposition rates, $\lesssim 500$ Å/min, compared to silane-based deposition. However, the sticking coefficients for TEOS deposition precursors seem to be almost an order of magnitude smaller than for the silane precursors; e.g., $s \sim 0.045$ for TEOS, leading to good deposition conformality. TEOS precursors are believed to be species such as $Si(OC_2H_5)_n(OH)_{4-n}$ or $Si(OC_2H_5)_nO_{4-n}$, $n = 0$–3. These can be formed by electron

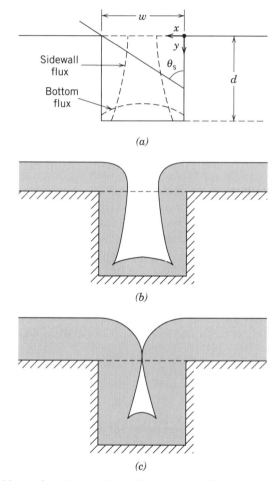

FIGURE 16.3. Nonconformal deposition within a trench, illustrating formation of a void as deposition proceeds: (a) before deposition, with the dashed lines giving the deposition flux incident on the sidewall and bottom; (b) midway during deposition; (c) just after the keyhole-shaped void has formed.

impact dissociation, e.g.,

$$e + Si(OC_2H_5)_n(OH)_{4-n} \longrightarrow Si(OC_2H_5)_{n-1}(OH)_{4-n+1} + C_2H_4 + e \qquad n = 1\text{--}4 \tag{16.2.12}$$

or by O-atom reactions with TEOS and its precursors, e.g.,

$$O + Si(OC_2H_5)_n(OH)_{4-n} \longrightarrow Si(OC_2H_5)_{n-1}(OH)_{4-n+1} + C_2H_4O \qquad n = 1\text{--}4 \tag{16.2.13}$$

In highly dilute TEOS/O_2 mixtures, the latter reactions predominate. It is not known which precursors are present in the highest concentrations. The precursors adsorb on the growing film surface where reactions with adsorbed O atoms further fragment the precursor and further oxidize the carbon and hydrogen. This oxidation process on the surface may be the rate-limiting step in the deposition. It is also known that there can be significant directionality in the deposition process. Presumably this is due to ion bombardment which enhances the vertical deposition rate. The measured TEOS deposition rates at moderate to high temperatures show a negative activation energy, $\mathcal{E}_a \sim -(0.1$–$0.2)$ V; i.e., the deposition rate increases as the substrate temperature is lowered. This can be interpreted in one of two ways: Either the desorption rate for TEOS precursors increases with increasing temperature, thus reducing the precursor coverage on the surface (Stout and Kushner, 1993), or there is increased surface recombination of O atoms as the temperature is increased, decreasing the gas-phase O-atom density (Cale et al., 1992).

The deposition chemistry and surface reactions for the former hypothesis can be described in a manner similar to that done previously for amorphous silicon deposition. The chemistry with the latter hypothesis can be described by the following three reactions:

1. O atoms oxidize TEOS precursors on the surface, leading to deposition. It is assumed that the TEOS fragments completely saturate the surface. Hence the reaction rate is independent of the precursor surface coverage. The deposition rate by this reaction is given as

$$D_{SiO_2}^{(1)} \approx \frac{0.9 n_{OS}}{n_{SiO_2}} \text{ cm/s} \qquad (16.2.14)$$

where the deposition rate constant 0.9 is determined by a fit to experimental data.

2. Oxygen ions also oxidize TEOS precursors, leading to a deposition rate

$$D_{SiO_2}^{(2)} \approx \frac{n_{O_2^+} u_B}{n_{SiO_2}} \text{ cm/s} \qquad (16.2.15)$$

3. Surface recombination of O atoms on wall (and deposition) surfaces to form nonreactive O_2 molecules reduces the available gas-phase O-atom concentration n_{OS} for reaction (16.2.14). The recombination probability $s_{rec}(T)$ on SiO_2 surfaces is activated but has a non-Arrhenius form (Greaves and Linnett, 1959), with $10^4 s_{rec} \approx 1.8, 2.7, 6.5,$ and 50 at $T = 20, 127, 200,$ and $394 \,^\circ C$, respectively. The flux of O atoms lost from the volume due to this process is

$$\Gamma_{rec} \approx 2 s_{rec}(T) \frac{1}{4} n_{OS} \bar{v}_O \qquad (16.2.16)$$

As T increases, Γ_{rec} increases, leading to a decrease in n_{OS} for a fixed generation rate, and hence a reduction in the deposition rate (16.2.14).

Silicon Nitride

Amorphous silicon nitride films were the first deposited on a large commercial scale using PECVD. They are used as a final encapsulating layer for integrated circuits because of their resistance to water vapor, salts, and other chemical contaminants. The usual feedstock mix is SiH_4/NH_3. The film precursors are probably SiH_3, SiH_2, and NH radicals, created by electron impact dissociation. Other possible precursors, such as Si_2H_6, $Si(NH_2)_4$, and $Si(NH_2)_3$, have been observed in SiH_4/NH_3 discharges by mass spectroscopy. The deposition is normally carried out at 0.25–3 Torr at 250–500 °C, yielding deposition rates of 200–500 Å/min. The activation energy for the deposition rate is small and can even be negative, depending on discharge conditions. The stoichiometry of the films is SiN_xH_y (sometimes called p-Sinh), with $x \sim 1$–1.2 and $y \sim 0.2$–0.6. The hydrogen atoms are bonded in the lattice, and low hydrogen content is associated with high temperatures and high rf power fluxes. Below 300 °C, the hydrogen content is relatively constant. A key step in film formation is thermal- or ion-induced desorption of H or H_2 from the growing film. The film characteristics depend strongly on the hydrogen content, with high hydrogen content yielding undesired films. The mechanical stress can be controlled by varying electrical properties of the discharge, such as the rf driving frequency. This variation is associated with the ion-bombarding energy, which decreases (at fixed rf power) with increasing frequency, as described in Chapter 11.

The source of most of the hydrogen in the films has been identified as NH_3, not SiH_4. This has motivated the use of other nitrogen sources such as N_2. Although SiH_4/N_2 can be used as a feedstock, the films are of poorer quality, the conformality is not as good, and the deposition rate is lower than with use of NH_3 as the source of nitrogen. On the other hand, the films have much less hydrogen and are richer in nitrogen than those grown using NH_3.

16.3   SPUTTER DEPOSITION

Physical Sputtering

In physical sputter deposition, ions incident on a target physically sputter target atoms, which ballistically flow to and are deposited on a substrate. Argon ions at 500–1000 V are usually used. Because sputter yields are of order unity for almost all target materials, a very wide variety of pure metals, alloys, and insulators can be deposited. Physical sputtering, especially of elemental targets, is a well-understood process, enabling sputtering systems for various applications to be relatively easily designed. Reasonable deposition rates with excellent film uniformity, good surface smoothness, and adhesion can be achieved over large areas. Refractory materials can also be easily sputtered. Sputter deposition is highly nonconformal, although redeposition techniques by ion bombardment of the deposited film can improve the conformality.

At first sight, it might seem that when a multicomponent target is sputtered, the deposited film will have a different composition than the target due to the difference in sputtering yields of the components. However, when multicomponent targets are sputtered, because of the difference in sputtering yields, an altered layer forms at the target surface having a different composition than the target. In the steady state, in the absence of diffusion of components between the layer and the bulk target, the flux of atoms sputtered from the layer has the stoichiometry of the original target material. If the sticking coefficients of the components on the substrate are all the same, then the deposited film will have the composition of the bulk target material. Thus, alloy targets can be sputter deposited on substrates. However, targets such as ceramics or oxides having high vapor pressure constituents, e.g., O atoms, usually cannot be physically sputter deposited.

In the area of metal film deposition, sputtering is commonly used to deposit electrode and interconnection material. For example, various films have been deposited such as aluminum in integrated circuit devices, transition metals films such as iron, cobalt, and nickel for magnetic coatings, superconducting films such as niobium, reflective optical films such as aluminum, silver, and gold, corrosion-resistant films such as chromium, and films such as chromium for decorative purposes.

Assuming that all the sputtered material is deposited on the substrate, the deposition rate for physical sputtering is

$$D_{\text{sput}} = \frac{\gamma_{\text{sput}}\Gamma_i}{n_f} \frac{A_t}{A_s} \text{ cm/s} \tag{16.3.1}$$

where Γ_i is the incident ion flux ($\text{cm}^{-2}\text{-s}^{-1}$), n_f is the density of the deposited film (cm^{-3}), A_t (cm^2) is the target area sputtered, A_s (cm^2) is the substrate area on which the film is deposited, and γ_{sput} is the sputtering yield. A scaling law for the sputtering yield in the linear cascade regime is given in (9.3.14), and some sputtering yields are tabulated in Table 9.2. For 1-kV argon ions with $A_t/A_s = 1$, $n_f = 5 \times 10^{22} /\text{cm}^{-3} \gamma_{\text{sput}} = 1$, and an ion current density of 1 mA/cm², ($\Gamma_i \approx 6.3 \times 10^{15} /\text{cm}^{-2}\text{-s}^{-1}$), the deposition rate is 750 Å/min. Rf- or dc-driven planar magnetron discharges are usually used for sputtering; the operating pressure is generally 10^{-3}– 10^{-2} Torr, which is low enough that the mean free path for sputtered atoms is larger than the separation between target and substrate.

Sputtered atoms are emitted with a cascade-type energy distribution (Winters and Coburn, 1992):

$$f(\mathcal{E}) \propto \frac{\mathcal{E}_t\mathcal{E}}{(\mathcal{E}_t + \mathcal{E})^3} \tag{16.3.2}$$

where $\mathcal{E}_t$ is the surface binding energy of the target material. The maximum of this distribution occurs at $\mathcal{E} = \mathcal{E}_t/2$. Since $\mathcal{E}_t \sim 1\text{–}4$ V, the characteristic sputtered atom energies are 0.5–2 V. Atoms striking the substrate with these energies can produce some mixing and diffusion between incoming atoms and substrate materials, leading to enhanced bonding and adhesion.

The morphology of sputtered films is primarily influenced by the substrate temperature, which is usually independently controlled, and secondarily, by the deposition pressure. The film morphology has been described by Thornton (1986) and is shown in Fig. 16.4. Letting T_m be the melting temperature of a sputter deposited metal film, then at low pressures and at very low temperatures, $T/T_m \lesssim 0.3$ (zone 1), the film consists of tapered columns with domed heads and significant voids between columns. These structures are formed by shadowing of atoms as they impinge on the growing film. The void fraction can be as high as 30%. For $0.3 \lesssim T/T_m \lesssim 0.5$ (zone T), the films have a fibrous structure in which crystallites grow perpendicular to the substrate plane without significant voids ($\lesssim$ 5% by volume). The film surface is relatively smooth and the film is almost as dense as that of the bulk material. These properties are a result of ion-bombardment-induced surface mobility of deposited atoms on the substrate. This film morphology is desirable for many applications. For $0.5 \lesssim T/T_m \lesssim 0.8$ (zone 2), thermally activated surface diffusion of deposited atoms leads to the appearance of columnar grains, which increase in diameter as T/T_m increases. For $0.8 \lesssim T/T_m \lesssim 1$ (zone 3), volume diffusion of atoms within the film leads to a smooth, randomly oriented polycrystalline film. All of these zones are used in various sputtering applications.

Ion and fast neutral bombardment of the substrate can strongly influence film properties and is responsible for the desirable properties in zone T. Positive ions with energies of 20–30 V arise from acceleration across the plasma–substrate sheath potential. The energy can be increased by applying a bias voltage (rf or, for conducting films, dc driven) to the substrate, and this is done in many commercial applications. Deposition pressure and ion bombardment also affect intrinsic film stress. High bombardment energies produce high compressive stresses due to recoil implantation. Generally, there is a transition from compressive to a generally more desirable tensile stress as the pressure is increased, with the transition pressure typically increasing with the atomic mass of the sputtered material (Konuma, 1992). This is believed to be due to a reduction in ion-bombarding energy as pressure is increased.

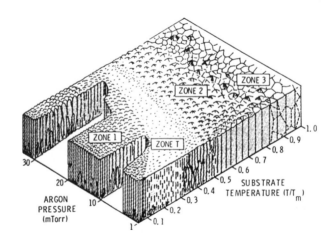

FIGURE 16.4.   Morphology of sputtered films (Thornton, 1986).

Reactive Sputtering

For reactive sputtering, a feedstock gas whose dissociation products chemically react with the target is present in addition to the bombarding ions. The deposited film is a compound formed from the sputtered target materials and the reactive gas. A common application is the sputter deposition of films whose components have strongly different vapor pressures, and hence, sticking probabilities on the substrate. For example, physical sputtering of a SiO_2 target in argon can lead to deposition of a silicon rich oxide film on the substrate. If O_2 gas is added to the system, then O atoms can be incorporated into the growing film to restore the 1:2 Si/O stoichiometry. A pure silicon target can also be used with O_2 gas to deposit SiO_2 films by reactive sputtering.

Reactive sputtering is widely used to deposit dielectrics such as oxides and nitrides, as well as carbides and silicides. Ceramics such as YBaCuO superconducting films can be sputter deposited from YBaCuO targets using O_2 as the reactive gas. Common reactive gases used for a wide variety of applications are O_2 and H_2O for O atoms, N_2 and NH_3 for N atoms, CH_4 and C_2H_2 for C atoms, and SiH_4 for Si atoms. Although ceramic or oxide targets can be used, they are not machinable and cannot handle high-power fluxes without cracking; hence metal targets are most commonly used where high deposition rates and controllable film stoichiometry are desired.

In reactive sputtering, chemical reactions occur at both target and substrate, in addition to sputtering at the target and deposition at the substrate. There are two "modes" of operation for reactive sputtering of a metal target to deposit a compound film. For low ion flux and high gas flux, the target is covered by the compound. For high ion flux and low gas flux, the target remains metallic. Higher deposition rates are achieved in the "metallic mode" than in the "covered mode." For fixed ion flux, as the reactive gas flux is varied, there is a transition between the covered and metallic modes exhibiting hysteresis; i.e., the transition flux for increasing the flux to pass from the metallic to the covered mode is higher than the transition flux for decreasing the flux to pass from the covered to the metallic mode.

A simple model of reactive sputtering described by Berg et al. (1989) makes it possible to understand the hysteresis and other properties of reactive sputter deposition. Let A_t and A_s be the target and substrate areas, θ_t and θ_s be the fractions of the target and substrate areas covered by the compound film, and γ_m and γ_c be the yields for sputtering the metal and the compound from the target. In the steady state, the compound formation rate on the target must be equal to the sputtering rate of the compound from the target. Letting Γ_i and Γ_r be the incident ion and reactive gas molecule fluxes, and letting s_r be the sticking coefficient of a reactive molecule on the metal part of the target, then

$$n_t' \frac{d\theta_t}{dt} = i\Gamma_r s_r(1 - \theta_t) - \Gamma_i \gamma_c \theta_t = 0 \qquad (16.3.3)$$

where i is the number of atoms per molecule of reactive gas (e.g., $i = 2$ for O_2 gas). Sputtered compound molecules and metal atoms are assumed to be evenly deposited over the substrate surface. The coverage θ_s of compound on the substrate increases because reactive gas molecules are incident on the metallic part $(1 - \theta_s)$, and because a fraction $(1 - \theta_s)$ of the compound flux sputtered from the target is deposited on the metallic part of the substrate. Similarly, θ_s decreases because a fraction θ_s of the metal-atom flux sputtered from the target is deposited on the compound part of the substrate. Hence, accounting for the ratio of target and substrate areas, we obtain

$$n_s' \frac{d\theta_s}{dt} = i\Gamma_r s_r (1 - \theta_s) + \Gamma_i \gamma_c \theta_t \left(\frac{A_t}{A_s}\right) (1 - \theta_s)$$

$$- \Gamma_i \gamma_m (1 - \theta_t) \left(\frac{A_t}{A_s}\right) \theta_s = 0 \qquad (16.3.4)$$

The total number of reactive gas molecules per second that are consumed to form the compound deposited on the substrate is

$$\frac{dN_r}{dt} = \Gamma_r s_r [(1 - \theta_t)A_t + (1 - \theta_s)A_s] \qquad (16.3.5)$$

and the target sputtering flux is

$$\Gamma_{\text{sput}} = \Gamma_i [\gamma_m (1 - \theta_t) + \gamma_c \theta_t] \qquad (16.3.6)$$

Equations (16.3.3) and (16.3.4) can be simultaneously solved to determine the compound coverages θ_t and θ_s on the target and substrate as a function of the fluxes, rate constants, and areas. Then dN_r/dt and Γ_{sput} can be evaluated, exhibiting the hysteresis (Problem 16.3).

For reactive sputter deposition of TiN films at 10 mTorr using a titanium target and an Ar/N_2 gas mixture, the optical intensity of a titanium emission line (proportional to Γ_{sput}) is plotted in Fig. 16.5 versus the input gas flow rate (equivalent to $dN_r/dt +$ const, where the constant is the number of reactive molecules per second removed by the pump). By controlling the input gas flow rate with a feedback system, the complete hysteresis curve could be traced out, as shown in Fig. 16.5a. In the absence of feedback control, which is more usual in reactive sputtering systems, the hysteresis yields jumps between the high and low deposition rate modes, as shown in Fig. 16.5b.

16.4 PLASMA-IMMERSION ION IMPLANTATION

Ion implantation is a process in which an energetic ion beam is injected into the surface of a solid material with the result that the atomic composition and structure of the near-surface region of the target material is changed, and thereby also the properties of the material surface are changed. The process is routine in semiconductor device fabrication. Metallurgical implantation is an emerging technology; in this application, new surface alloys are created with improved resistance to wear, corrosion, and fatigue.

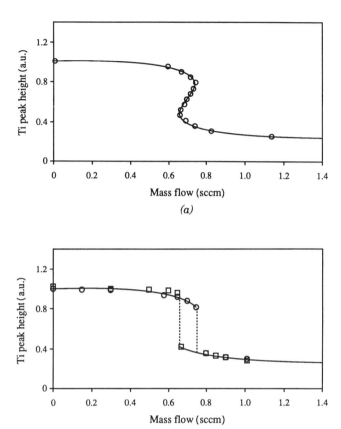

FIGURE 16.5. Reactive sputter deposition of TiN films, showing the optical emission signal for titanium versus the reactive gas flow rate (a) with and (b) without feedback control (Berg et al., 1989)

Conventional ion implantation is carried out in a vacuum environment in which an ion source is used to create an intense beam of ions of the species to be implanted. The ion beam is steered and accelerated through a potential of from tens to hundreds of kilovolts and transported to the target. Since the beam spot size is smaller than the wafer size, mechanical and electrostatic scanning are used to achieve dose uniformity. For some state-of-the-art semiconductor device structures, high angle tilting and rotation of the wafers are required to homogenize the dose uniformity on the sidewall area. This mechanical complexity significantly increases the physical size and cost of the implanter. The relatively low beam currents, limited by the source optics, lead to high costs for high-dose applications such as buried dielectric layers formed by implantation of oxygen (SIMOX), doping of thin-film transistors for active matrix flat panel displays, surface smoothing for optical coatings, ion beam mixing of thin films, and ion-assisted deposition. Lower-energy implantation (energies less than 5 kV) can also be limited by ion beam optics if high doses are required.

In plasma-immersion ion implantation (PIII), the intermediate stages of ion source, beam extraction, focusing, and scanning are omitted. The target is immersed in a plasma environment, and ions are extracted directly from the plasma and accelerated into the target by means of a series of negative high-voltage pulses applied to the target. Both metallurgical (Conrad et al., 1990) and semiconductor (Cheung, 1991) implantation processes have been demonstrated using PIII.

When a sudden negative voltage $-V_0$ is applied to the target, then, on the timescale of the inverse electron plasma frequency ω_{pe}^{-1}, electrons near the surface are driven away, leaving behind a uniform-density ion matrix sheath. The sheath thickness is a function of the applied voltage and the plasma density, as given in (6.3.4). Subsequently, on the timescale of the inverse ion plasma frequency, ions within the sheath are accelerated into the target. This, in turn, drives the sheath–plasma edge farther away, exposing new ions that are extracted. On a longer timescale, the system evolves toward a steady-state Child law sheath, with the sheath thickness given by (6.3.14). The Child law sheath is larger than the matrix sheath by a factor of order $(V_0/T_e)^{1/4}$, where T_e is the electron temperature. This steady state can be of interest in PIII for high-throughput implantations into conducting targets.

The matrix sheath and its time evolution determine the implantation current $J(t)$ and the energy distribution of implanted ions. The structures of the initial matrix sheath in one-dimensional planar, cylindrical, and spherical targets (Conrad, 1987) and two-dimensional wedge-shaped targets (Donnelly and Watterson, 1989) have been determined. In addition, analytical estimates of the sheath dynamics have been obtained (Lieberman, 1989c; Scheuer et al., 1990), and the self-consistent equations have been solved numerically to find the time evolution of the matrix sheath in planar geometry (Vahedi et al., 1991; Stewart and Lieberman, 1991; and references therein). In this section we first present some simple dynamical models for PIII sheath formation in the collisionless and collisional regimes, and then briefly describe some experimentally investigated applications of PIII for integrated circuit fabrication and metallurgical surface modification.

Collisionless Sheath Model

Figure 16.6 shows the PIII geometry. The planar target is immersed in a uniform plasma of density n_0. At time $t = 0$, a voltage pulse of amplitude $-V_0$ and time width t_p is applied to the target, and the plasma electrons are driven away to form the matrix sheath, with sheath edge at $x = s_0$. As time evolves (Fig. 16.7b), ions are implanted, the sheath edge recedes, and a nonuniform, time-varying sheath forms near the target. The model assumptions are as follows:

1. The ion flow is collisionless. This is valid for sufficiently low gas pressures.
2. The electron motion is inertialess. This follows because the characteristic implantation timescale much exceeds ω_{pe}^{-1}.
3. The applied voltage V_0 is much greater than the electron temperature T_e; hence the Debye length $\lambda_{De} \ll s_0$, and the sheath edge at s is abrupt.

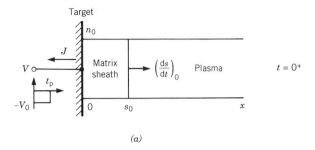

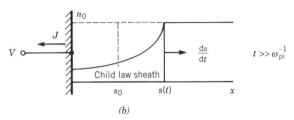

FIGURE 16.6 Planar PIII geometry (a) just after formation of the matrix sheath and (b) after evolution of the quasistatic Child law sheath.

4. During and after matrix sheath implantation, a quasistatic Child law sheath forms. The current demanded by this sheath is supplied by the uncovering of ions at the moving sheath edge and by the drift of ions toward the target at the Bohm (ion sound) speed $u_B = (eT_e/M)^{1/2}$. The assumption of Bohm flow will probably not be valid during the initial sheath expansion before the Bohm presheath has formed. The sheath dynamics using an alternative assumption, that the drift velocity of ions toward the target is zero, has been explored by Scheuer et al. (1990).

5. During the motion of an ion across the sheath, the electric field is frozen at its initial value, independent of time, except for the change in field due to the velocity of the moving sheath.

Assumptions 4 and 5 are approximations that permit an analytical solution to the sheath motion. These assumptions are justified post hoc by comparison with numerical results.

(a) Sheath Motion   We assume that after a short transient, the ion matrix sheath evolves into a Child law sheath with time-varying current density and sheath thickness. The Child law current density J_c for a voltage V_0 across a sheath of thickness s is given by (6.3.12):

$$J_c = \frac{4}{9}\epsilon_0 \left(\frac{2e}{M}\right)^{1/2} \frac{V_0^{3/2}}{s^2} \tag{16.4.1}$$

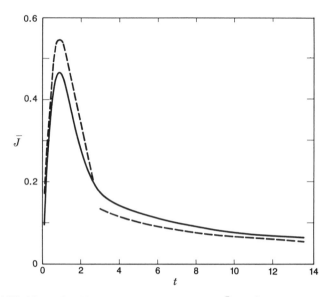

FIGURE 16.7. Normalized implantation current density $\bar{J} = J/(en_0u_0)$ versus normalized time $\bar{t} = \omega_{pi}t$. The dashed lines show the analytical solution for $\bar{t} < 2.7$ and $\bar{t} > 3.0$, and the solid line is the numerical solution of the fluid equations.

where ϵ_0 is the free space permittivity and e and M are the ion charge and mass. Equating J_c to the charge per unit time crossing the sheath boundary,

$$en_0 \left(\frac{ds}{dt} + u_B \right) = J_c \qquad (16.4.2)$$

we find the sheath velocity

$$\frac{ds}{dt} = \frac{2}{9} \frac{s_0^2 u_0}{s^2} - u_B \qquad (16.4.3)$$

where

$$s_0 = \left(\frac{2\epsilon_0 V_0}{en_0} \right)^{1/2} \qquad (16.4.4)$$

is the matrix sheath thickness and

$$u_0 = \left(\frac{2eV_0}{M} \right)^{1/2} \qquad (16.4.5)$$

is the characteristic ion velocity. Integrating (16.4.3), we obtain

$$\tanh^{-1}\left(\frac{s}{s_c}\right) - \frac{s}{s_c} = \frac{u_B t}{s_c} + \tanh^{-1}\left(\frac{s_0}{s_c}\right) - \frac{s_0}{s_c} \tag{16.4.6}$$

where

$$s_c = s_0 \left(\frac{2}{9}\frac{u_0}{u_B}\right)^{1/2} \tag{16.4.7}$$

is the steady-state Child law sheath thickness. Since $s_c \gg s_0$ and assuming $s_c \gg s$, we find by expanding (16.4.6), or by integrating (16.4.3) with $u_B = 0$, that

$$\frac{s^3}{s_0^3} = \frac{2}{3}\omega_{pi}t + 1 \tag{16.4.8}$$

where $\omega_{pi} = (e^2 n_0/\epsilon_0 M)^{1/2} = u_0/s_0$ is the ion plasma frequency in the matrix sheath. Substituting (16.4.7) into (16.4.8), we note that the timescale t_c for establishing the steady-state Child law sheath ($s = s_c$) is $t_c \approx (\sqrt{2}/9)\omega_{pi}^{-1}(2V_0/T_e)^{3/4}$; we assume that the pulse width $t_p \ll t_c$ in the development that follows.

(b) Matrix Sheath Implantation   Because the initial charge density in the matrix sheath is uniform, the initial electric field varies linearly with x: $E = (M/e)\omega_{pi}^2(x - s)$. Hence, the ion motion is

$$\frac{d^2 x}{dt^2} = \omega_{pi}^2(x - s) \tag{16.4.9}$$

where x is the particle position. Approximating $s = s_0 + (ds/dt)_0 t$ in (16.4.9) and using (16.4.3) with $s = s_0$ and $u_B \ll u_0$, we obtain

$$\frac{d^2 x}{dt^2} = \omega_{pi}^2(x - s_0) - \frac{2}{9}u_0\omega_{pi}^2 t \tag{16.4.10}$$

Integrating (16.4.10), we find

$$x - s_0 = (x_0 - s_0)\cosh\omega_{pi}t - \frac{2}{9}s_0\sinh\omega_{pi}t + \frac{2}{9}u_0 t \tag{16.4.11}$$

where we have let $x = x_0$ and $\dot{x} = 0$ at $t = 0$. (Choosing $\dot{x} \approx -u_B$, consistent with the sheath motion (16.4.3), yields a negligible correction to (16.4.11) because $u_B \ll u_0$.) Letting $x = 0$ in (16.4.11), we can obtain the ion flight time t from

$$s_0 = (s_0 - x_0)\cosh\omega_{pi}t + \frac{2}{9}s_0\sinh\omega_{pi}t - \frac{2}{9}u_0 t \tag{16.4.12}$$

In a time interval between t and $t + dt$, ions from the interval between x_0 and $x_0 + dx_0$ are implanted. Differentiating x_0 in (16.4.12) with respect to time, we find

$$\frac{dx_0}{dt} = \frac{\omega_{pi}(s_0 - x_0)\sinh \omega_{pi}t + \frac{2}{9}u_0(\cosh \omega_{pi}t - 1)}{\cosh \omega_{pi}t} \tag{16.4.13}$$

Using (16.4.12) to eliminate $s_0 - x_0$ in (16.4.13), we obtain the implantation current density $J = en_0 dx_0/dt$ as

$$\bar{J} = \frac{\sinh \bar{t}}{\cosh^2 \bar{t}} + \frac{2}{9}\frac{1 + \bar{t}\sinh \bar{t} - \cosh \bar{t}}{\cosh^2 \bar{t}} \tag{16.4.14}$$

where $\bar{J} = J/(en_0 u_0)$ is the normalized current density and $\bar{t} = \omega_{pi}t$ is the normalized time. Equation (16.4.14) gives the implantation current density versus time for those ions in the initial matrix sheath $0 \le x_0 \le s_0$. Setting $x_0 = s_0$ in (16.4.12), we obtain $\bar{t} \approx 2.7$. At this time, all matrix sheath ions are implanted; hence we take (16.4.14) to reasonably approximate the current for $0 \le \bar{t} \le 2.7$. The left dashed curve in Fig. 16.7 plots $\bar{J}$ versus $\bar{t}$. The maximum current density $\bar{J}_{max} \approx 0.55$ occurs at $\bar{t}_{max} \approx 0.95$, and $\bar{J}(2.7) \approx 0.19$.

(c) Child Law Sheath Implantation   Consider now the implanted ions having initial positions at $x_0 > s_0$. The time t_s for the initial sheath edge at s_0 to reach x_0 is found from (16.4.8):

$$\omega_{pi}t_s = \frac{3}{2}\frac{x_0^3}{s_0^3} - \frac{3}{2} \tag{16.4.15}$$

We expect this to be valid for $\bar{t} \gtrsim 2.7$. At time t_s, an ion at x_0 begins its flight across the sheath. The ion flight time is given by (Problem 6.2)

$$\omega_{pi}t' = \frac{3x_0}{s_0} \tag{16.4.16}$$

Hence, an ion at x_0 reaches the target at a time $t = t_s + t'$ given by

$$\bar{t} = \omega_{pi}t = \frac{3}{2}\frac{x_0^3}{s_0^3} - \frac{3}{2} + 3\frac{x_0}{s_0} \tag{16.4.17}$$

Differentiating (16.4.17), we obtain

$$\frac{dx_0}{dt} = \frac{u_0}{\frac{9}{2}(x_0^2/s_0^2) + 3} \tag{16.4.18}$$

The normalized implantation current density is thus

$$\bar{J} = \frac{1}{\frac{9}{2}(x_0^2/s_0^2) + 3} \tag{16.4.19}$$

Equations (16.4.17) and (16.4.19) give $\bar{J}(\bar{t})$ as a parametric function of x_0/s_0. If we set $x_0/s_0 = 1$, we find $\bar{t} = 3$ and $\bar{J}(3) = 2/15 \approx 0.133$. As $\bar{t} \to \infty$, $x_0 \to s_c \gg s_0$; hence $\bar{J}(\infty) \to (2/9)s_0^2/s_c^2$. Unnormalizing, we find $J(\infty) \to en_0 u_B$, which correctly gives the steady-state Child law current density. The right dashed curve in Fig. 16.7 shows the analytical results for $\bar{J}$ versus $\bar{t}$ for $\bar{t} \gtrsim 2.7$. We note that (16.4.14) and (16.4.19) do not smoothly join at $x_0 = s_0$, a consequence of the simplifying assumptions 4 and 5 that were used to solve for the sheath and ion motion.

The preceding analysis has been compared with numerical solutions of the nonlinear partial differential equations for the ion and electron motion (Lieberman, 1989c). The ion motion is collisionless, the electrons are in thermal equilibrium, and Poisson's equation relates the densities to the potential. Figure 16.7 shows a numerical solution for $V_0/T_e = 200$. We see that (16.4.14) for $\bar{t} \lesssim 2.7$ and (16.4.19) for $\bar{t} \gtrsim 2.7$ reasonably approximate the numerical results. The energy distribution of ions striking the target can be determined from the basic model. The analysis can also be applied to nonplanar geometries (Scheuer et al., 1990). The spatial structure and time evolution of the collisionless sheath has been measured by Cho et al. (1988) for planar targets at low voltages (~ 100 V) and by Shamim et al. (1991) for cylindrical and spherical targets at high voltages (~ 30 kV), obtaining good agreement with the collisionless model.

Collisional Sheath Model

Ion collisions within the sheath at high gas pressures lead to reduced implantation energies and finite width energy and angular distributions for ions that greatly affect their implantation over topography, i.e., within trenches. The energy and angular distributions have been determined analytically and compared with particle in cell, Monte Carlo collision (PIC-MCC) simulations (Vahedi et al., 1991, 1993). The collisionless model assumptions 2, 3, and 5 are retained, but 1 is replaced by the assumption that the ion motion within the sheath is highly collisional, with charge transfer the dominant source of ion–neutral collisions. It is also assumed that the ion charge density n_s in the sheath is uniform in space but slowly varying in time, with $n_s(t) < n_0$, the bulk plasma density. A uniform distribution is seen experimentally for similar sheaths, such as the cathode sheaths in dc glow discharges (see Chapter 14), and is also seen in PIC-MCC simulations of collisional PIII.

To determine the energy distribution of the bombarding ions, the Maxwell equation,

$$\frac{dE}{dx} = \frac{en_s}{\epsilon_0} \tag{16.4.20}$$

is integrated from the electrode surface at $x = 0$ to a position x within the sheath to obtain

$$E = \frac{en_s}{\epsilon_0}(s - x) \tag{16.4.21}$$

where the boundary condition $E = 0$ at $x = s$ has been used. Integrating again to determine the potential using $d\Phi/dx = -E$, we obtain

$$\Phi = -\frac{en_s}{2\epsilon_0}(s - x)^2 \tag{16.4.22}$$

where $\Phi = 0$ at $x = s$. Letting $\Phi = -V_0$ at $x = 0$, we obtain the matrix sheath result

$$n_s = \frac{2\epsilon_0 V_0}{es^2} \tag{16.4.23}$$

The equation of motion of an ion starting from rest at $x = x_0$, after a charge transfer collision in the sheath, is

$$\frac{d^2x}{dt^2} = \frac{eE}{M} = \frac{2eV_0}{Ms^2}(x - s) \tag{16.4.24}$$

Assuming that s varies slowly in time, this can be integrated to obtain the ion velocity $u(x)$,

$$u^2 = \frac{u_0^2[(x^2 - x_0^2) - 2s(x - x_0)]}{s^2} \tag{16.4.25}$$

where u_0 is given by (16.4.5). The ion velocity at the target is then

$$u_t^2 = \frac{u_0^2(2sx_0 - x_0^2)}{s^2} \tag{16.4.26}$$

The distribution of ion flux $f(u_t)$ is determined by applying conservation of particles to obtain

$$f(u_t)\,du_t = v_{cx}n_s\,e^{-x_0/\lambda_i}dx_0 \tag{16.4.27}$$

where $v_{cx} = \bar{u}_t/\lambda_i$ is the ion–neutral charge-transfer collision frequency, $\bar{u}_t$ is the mean ion velocity near the target, and the exponential factor gives the probability that an ion created by charge transfer at x_0 will hit the target before a subsequent ion–neutral collision. Differentiating (16.4.26) to determine dx_0/du_t and substituting this into (16.4.27), we obtain, for $\lambda_i \ll s$,

$$f(u_t) \propto \frac{u_t}{(1 - u_t^2/u_0^2)^{1/2}} \exp\left\{\frac{s}{\lambda_i}\left[\left(1 - \frac{u_t^2}{u_0^2}\right)^{1/2} - 1\right]\right\}, \qquad u_t < u_0 \tag{16.4.28}$$

Figure 16.8 compares the analytical theory (16.4.28) and the computer simulation over a range of pressures (mean free path regimes). The mean ion velocity near the

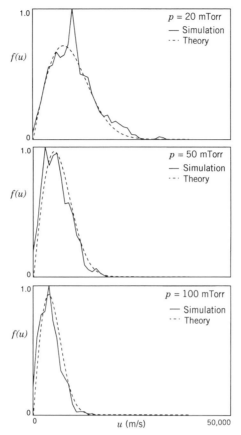

FIGURE 16.8. Ion velocity distribution at the target for a collisional sheath; the maximum velocity for collisionless acceleration to the target is roughly 5×10^4 m/s.

target is found from (16.4.28) to be

$$\bar{u}_t = \left(\frac{eV_0 \pi \lambda_i}{Ms} \right)^{1/2} \tag{16.4.29}$$

and the implantation current density is

$$J_t = en_s \bar{u}_t = \epsilon_0 \left(\frac{4\pi e \lambda_i}{M} \right)^{1/2} \frac{V_0^{3/2}}{s^{5/2}} \tag{16.4.30}$$

where the second equality follows by using (16.4.23) and (16.4.29). The scaling of the current density (16.4.30) is the same as found for the collisional rf sheath (11.2.54), showing the essential correspondence between the two sheaths.

Ion–neutral elastic scattering tends to isotropize the angular distribution of the impinging ions, leading to energetic ion bombardment of trench sidewalls for implantation over topography. Vahedi et al. (1993) have modeled this process to obtain the ratio of trench sidewall to bottom fluxes,

$$\frac{\Gamma_h}{\Gamma_v} \approx \frac{0.34}{0.93 + \sigma_{cx}/\sigma_{sc}} \tag{16.4.31}$$

which is valid in the limits $\lambda_i \ll s$ and $\sigma_{sc} \ll \sigma_{cx}$, where σ_{sc} and σ_{cx} are the ion–neutral elastic scattering and charge-transfer cross sections, respectively.

Applications of PIII to Materials Processing

(a) Semiconductor Processes   Figure 16.9 illustrates two PIII system configurations compatible with semiconductor thin-film processing requirements. To permit operation at pressures as low as 0.2 mTorr, ECR sources operating at 2.45 GHz provide the high ion density, $n_i \sim 10^{10}$–10^{11} cm^{-3}, to supply the required high implantation current. Hot filament sources, which are used for metallurgical implantation (see below), cannot be used because of contamination. The substrate is biased with a pulsed (2–30 kV, 1–3 μs), or, possibly, dc negative voltage to accelerate the ions toward the substrate surface.

With a diode configuration, shown in Fig. 16.9a, gaseous sources such as Ar, N$_2$, BF$_3$, H$_2$O, and O$_2$ can be used to provide the ionization medium and the implanting ions. The diode configuration is most convenient for doping applications such as shallow junction formation and conformal doping of nonplanar device structures because many dopant gaseous sources are available. When metal-containing gases are used, e.g., WF$_6$, the diode configuration can operate as an ion-assisted chemical vapor deposition system. By adding another negatively biased target controlled by a separate power supply to form a triode configuration, as shown in Fig. 16.9b, atoms from this target are sputtered into the plasma by the carrier gas plasma ions. Some of

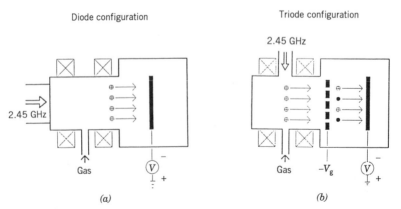

FIGURE 16.9. Schematic showing diode and triode configurations of PIII for semiconductor implantation.

the emitted target atoms are ionized in the plasma and subsequently implanted into the substrate. Secondary electron emission from the target has significant negative consequences at the high ion fluxes and energies of PIII. The secondary electron current can be 5–10 times as large as the ion implantation current at high voltages (30–100 kV) (Szapiro and Rocca, 1989). The secondary electrons from the target are accelerated across the sheath and subsequently impinge on the chamber surfaces, which can be a serious x-ray hazard. Also, the PIII power source must supply the power and current, which leads to poor power efficiencies.

PIII has been applied, experimentally, to a number of semiconductor processes. For sub-100-nm p^+/n junction formation where boron implantation is used, preamorphization of the crystalline silicon together with large doses of boron minimizes the source/drain resistance. The final junction depth of these ultra-shallow junctions is dominated by diffusion of dopants during thermal activation, which greatly modifies the implanted depth profile. Because of the high-flux capability of PIII at low implantation energies, it is well suited for ultra shallow junction formation. The silicon is preamorphized with a 4-kV SiF_4 PIII implantation prior to a 2-kV BF_3 PIII implantation. After annealing at 1060 °C for 1 s, an extremely shallow junction depth of 80 nm is obtained with a sheet resistance of 447Ω per square. Junctions with a total leakage current density at a reverse bias of -5 V lower than 30 nA/cm^2 have been fabricated (Pico et al., 1992), which compares to the state of the art using other technologies.

Another application has been to selective metal plating. Since copper has a very low electrical resistivity and good electromigration properties, it is an ideal conductor to replace aluminum for integrated circuit interconnects. However, plasma-assisted etching of copper has not been successful due to the lack of suitable volatile etch products. PIII has been used for selective and planarized plating of copper interconnects using palladium seeding, thus avoiding the need to etch the copper. A palladium sputtering target is immersed in the plasma and has an independently controlled negative bias to regulate the sputtering rate. The sputtered neutral palladium forms a continuous flux for deposition, while the Ar^+ and Pd^+ ions assist the penetration of deposited palladium into the substrate via ion beam mixing.

PIII has also been used to conformally dope silicon trenches. High packing densities of devices on silicon substrates are achievable by making use of vertical sidewalls for active transistor channels and as charge storage elements such as trench capacitors. Conventional implantation techniques have focused on multistep implants with collimated beams at controlled beam incidence angles. Taking advantage of the angular divergence of implanting ions in PIII at high gas pressures, conformal doping of high aspect ratio silicon trenches with BF_3 doping has been achieved. In these experiments, silicon trenches about 1 μm wide and 5 μm deep were implanted at -10 kV and a gas pressure of 5 mTorr, yielding a relatively uniform p^+/n junction depth on the top, bottom, and sidewalls of the trench.

(b) Metallurgical Processes PIII can also be used for metallurgical surface modification to improve wear, hardness, and corrosion resistance. In this context, the process has been called plasma source ion implantation (PSII). PSII can easily be

used to implant nonplanar targets, e.g., tools and dies, with minimum shadowing and sputtering of the target. The latter can limit the retained dose of the implanted ion species. Ions have been implanted under batch processing conditions, with acceptable dose uniformities to the depths and concentrations required for surface modification, resulting in dramatic improvement in the life of manufacturing tools under actual industrial conditions (Conrad et al., 1990; Redsten et al., 1992). In a typical PSII process, the target is immersed in a nitrogen plasma of density $n_0 \sim 5 \times 10^9$ cm^{-3}. A series of 50-kV, 10-μs pulses at 100 kHz are applied to the target for minutes to hours. For these conditions the initial matrix sheath thickness is 3 cm, and the Child law sheath thickness is 24 cm, but the pulse width is short enough that the Child law sheath does not have time to fully form. In the referenced work, the plasma is generated by a hot tungsten filament source, which is inserted into the chamber and biased at $-(100–300)$ V. The filament emits electrons that are accelerated across the filament sheath into the plasma, where they subsequently ionize the background gas, which is typically at a pressure of 10^{-4} Torr. The dynamics of hot filament plasma sources is fairly well understood (Leung et al., 1976). Multipole magnets are required on the surface of the implantation chamber to confine the primary electrons (see Section 5.6 for a description of multipole magnetic confinement).

PROBLEMS

16.1. Silane Discharge Model   Consider a simplified (uniform electron temperature) model for a high-pressure capacitive rf discharge in silane. Use the rate constants in Table 16.1. Assume that the silane density n_g is uniform and is much larger than all other densities, and ignore negative ions and all volume loss processes. The discharge parameters are $p = 200$ mTorr and $l = 3$ cm, and the gas is at room temperature.

(a) Assume that SiH$_3$ and SiH$_2$ are created by reactions 1 and 2 in Table 16.1, respectively, and that both species are lost to the electrode walls with unity sticking coefficient. Find the fluxes $\Gamma_{\mathrm{SiH_3}}$ and $\Gamma_{\mathrm{SiH_2}}$ at the electrodes, and find their ratio $\Gamma_{\mathrm{SiH_3}}/\Gamma_{\mathrm{SiH_2}}$, in terms of n_e, n_g, l, and the rate constants.

(b) By equating the volume rate of generation of positive ions (reactions 5 and 6) to the loss of ions to the discharge electrodes, determine T_e. Treat the SiH$_3{}^+$ and SiH$_2{}^+$ ions as identical, and assume an ion-neutral momentum transfer rate constant $K_{\mathrm{mi}} \approx 10^{-9}$ cm^3/s.

(c) Using your results in parts (a) and (b), determine values for $\Gamma_{\mathrm{SiH_3}}$, $\Gamma_{\mathrm{SiH_2}}$, $\Gamma_{\mathrm{SiH_4}}$, and Γ_i at the electrode for an ion (and electron) density $n_0 = 3 \times 10^{10}$ cm^{-3} in the center of the discharge.

(d) Assuming $V_{\mathrm{rf}} = 500$ V and collisional sheaths (see Section 11.2, and take $u_s = u_B$ in (11.2.53)), and assuming a reasonable value $\mathcal{E}_c = 100$ V for the collisional energy lost per electron–ion pair created, find the ion-bombarding energy $\mathcal{E}_{\mathrm{ic}}$, given by (11.2.57), and the absorbed power per unit area S_{abs}.

16.2. Deposition Rate Within a Trench Consider deposition within a trench of width w and depth h, as shown in Fig. 16.3a, due to an isotropic flux of precursors at the top of the trench having a sticking coefficient of unity. Assume ballistic transport within the trench; i.e., the mean free path for precursor collisions is much greater than w or h.

(a) Let dN be the number of precursor molecules incident on a differential width dy at a sidewall position y due to an isotropic flux emitted from a differential width dx at the top of the trench. Using precursor particle conservation, show that dN is proportional to $dx \cos \theta$ (emission by the source width) and to $dy \cos \theta'$ (reception by the sidewall width), and is inversely proportional to the distance r between the source and sidewall:

$$dN = A \frac{\cos \theta \, \cos \theta' \, dx \, dy}{r}$$

where $\cos \theta = x/r$, $\cos \theta' = y/r$, and A is a constant.

(b) Integrating the expression in part (a) from $x = 0$ to $x = w$, and assuming a uniform source at the top of the trench, show that $dN/dy \propto 1 - \cos \theta_s$, where $\cos \theta_s = y/(y^2 + w^2)^{1/2}$; note that θ_s is the angle subtended by the trench opening as seen at the position y along the sidewall.

(c) Using a similar procedure to that developed in parts (a) and (b), find an expression for the nonuniform deposition rate dN/dx versus x at the bottom of the trench.

16.3. Reactive Sputtering Deposition Consider the reactive sputtering model of Section 16.3 with equal target and substrate areas. Let $a = \gamma_m/\gamma_c$ and $Y = 2\Gamma_r s_r/\Gamma_i \gamma_c$ be the normalized sputtering coefficient and flux, respectively.

(a) Show using (16.3.3) and (16.3.4) that the surface coverages of the compound on the target and the substrate are given respectively by

$$\theta_t = \frac{Y}{1 + Y}$$

$$\theta_s = \frac{Y^2 + 2Y}{Y^2 + 2Y + a}$$

(b) Show using (16.3.5) and (16.3.6) that the reactive gas flow and the sputtering flux are given respectively by

$$\frac{dN_r}{dt} \propto Y \left(\frac{1}{1 + Y} + \frac{a}{Y^2 + 2Y + a} \right)$$

$$\Gamma_{\text{sput}} \propto \frac{a + Y}{1 + Y}$$

(c) For the limiting case $a \gg 1$, graph dN_r/dt versus Y and Γ_{sput} versus Y. From these graphs, sketch Γ_{sput} versus dN_r/dt and show that the curve exhibits hysteresis similar to that shown in Fig. 16.5. Is there hysteresis for the case $a \leq 1$? Prove your answer.

16.4. Collisionless PIII

(a) Derive the sheath motion (16.4.8) when a sudden negative voltage $-V_0$ is applied to the target by directly solving the differential equation (16.4.3) for the case $u_B \equiv 0$.

(b) Suppose a voltage $-V_0(t) = -\alpha t$ that varies linearly with time for $t > 0$ is applied to the target. Assuming a Child law sheath (16.4.1) and using (16.4.2) for the case $u_B \equiv 0$, determine the implantation current density $J_c(t)$ and sketch J_c versus t. Explain why the current density is singular at $t = 0+$ and suggest a value for the maximum current density at $t = 0+$.

16.5. Collisional PIII   Consider the collisional sheath model for high-voltage implantation given in Section 16.4.

(a) Using (16.4.30) for J_t, along with the basic relation (16.4.2) (with $u_B \equiv 0$), show that the collisional uniform density (matrix) sheath expands as

$$s(t) = s_0(1 + \omega_0 t)^{2/7}$$

and find an expression for ω_0.

(b) Find an expression for $n_s(t)$.

APPENDIX A

COLLISION DYNAMICS

The dynamics of a particle in a central force potential can be directly calculated. Using the center of mass coordinates, this corresponds to a collision between any two particles, in which their mutual interaction has a central force, e.g., the Coulomb and polarization potentials. The equation for the trajectory is straightforwardly calculated from the conservation of total energy and angular momentum. Referring to Fig. A.1, these equations can be written

$$\frac{1}{2}m_R \left(\dot{r}^2 + r^2 \dot{\theta}^2 \right) + U(r) = \frac{1}{2}m_R v_0^2 \tag{A.1}$$

and

$$-m_R r^2 \dot{\theta} = m_R b v_0 \tag{A.2}$$

where v_0 is the initial relative velocity, m_R is the reduced mass, and here θ is defined as shown in the figure. Substituting $\dot{\theta}$ from (A.2) in (A.1) we obtain an equation for $\dot{r}$ alone

$$\frac{1}{2}m_R \dot{r}^2 = \frac{1}{2}m_R v_0^2 - \left[U(r) + \frac{m_R b^2 v_0^2}{2r^2} \right] \tag{A.3}$$

where the term in brackets is an effective potential including the repulsive centrifugal potential $m_R b^2 v_0^2 / 2r^2$. For $U(r) \propto r^{-i}$, which includes the Coulomb and polarization potentials, this is readily solvable (see Goldstein, 1950). We form the trajectory

541

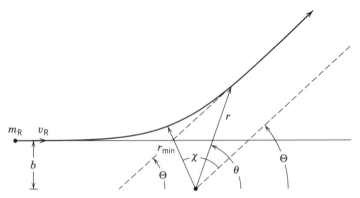

FIGURE A.1 Illustrating the exact classical calculation of the differential scattering cross section.

equation by solving (A.3) for $\dot{r}$ and then dividing by $\dot{\theta}$ from (A.2), to obtain

$$\frac{dr}{d\theta} = \frac{\pm\left\{v_0^2 - \dfrac{2}{m_R}\left[U(r) + \dfrac{m_R b^2 v_0^2}{2r^2}\right]\right\}^{1/2}}{bv_0/r^2} \tag{A.4}$$

where the minus sign is for the outward bound trajectory. The angle χ is then obtained by integrating (A.4),

$$\chi = \int_{r_{\min}}^{\infty} \frac{(bv_0/r^2)\,dr}{\left\{v_0^2 - \dfrac{2}{m_R}\left[U(r) + \dfrac{m_R b^2 v_0^2}{2r^2}\right]\right\}^{1/2}} \tag{A.5}$$

and from Fig. A.1

$$\Theta = \pi - 2\chi \tag{A.6}$$

We have used symmetry about the closest approach, to integrate (A.5) from the closest approach $r_{\min}$ to infinity, and then doubled the angle. The lower limit $r_{\min}$ is obtained from the condition that $\dot{r}$ changes sign, which from (A.3) is just

$$\frac{1}{2}m_R v_0^2 - \left[U(r_{\min}) + \frac{m_R b^2 v_0^2}{2r_{\min}^2}\right] = 0 \tag{A.7}$$

Although complete solutions of (A.5) are possible, they are not illuminating. However, for attractive potentials it is often useful to distinguish between two cases: the case for which the centrifugal force serves as a barrier to deep penetration and the case for which this repulsive potential is overcome by the attractive potential. From (A.3) the transition occurs when $U(r) = -Cr^{-i}$, with $i = 2$. For $i < 2$ the effective potential

$$U_{\text{eff}}(r) = U(r) + \frac{m_R b^2 v_0^2}{2r^2} \tag{A.8}$$

is always repelling at the origin. These two cases are illustrated in Fig. A.2, where $U_c(r)$ is the centrifugal potential. For $U(r) > 0$ the effective potential is, of course, always repulsive. An important characteristic of the potential in Fig. A.2b, with $i > 2$, is that a resonance phenomenon can occur for an energy close to the value for which the force disappears (W_0 in the figure).

Coulomb Cross Section

For Coulomb collisions, there is a straightforward solution to the trajectory equation (A.5). Substituting in the Coulomb potential for an electron–ion collision

$$U(r) = -\frac{Ze^2}{4\pi\epsilon_0 r}$$

and defining a new variable $\rho = b/r$, we recast (A.5) with (A.6) into the form

$$\Theta = 2 \int_0^{\rho_{\max}} \frac{d\rho}{\left(1 - \frac{2Ze^2}{4\pi\epsilon_0 m_R v_0^2 b}\rho - \rho^2\right)^{1/2}} - \pi \tag{A.9}$$

where $\rho_{\max}$ is obtained from the solution of (A.7). Integrating (A.9) we have

$$\Theta = -2\cos^{-1}\left\{\frac{Ze^2/4\pi\epsilon_0 m_R v_0^2 b}{\left[1 + \left(Ze^2/4\pi\epsilon_0 m_R v_0^2 b\right)^2\right]^{1/2}}\right\} + \pi \tag{A.10}$$

which can be rewritten

$$\frac{Ze^2/4\pi\epsilon_0 m_R v_0^2 b}{\left[1 + \left(Ze^2/4\pi\epsilon_0 m_R v_0^2 b\right)^2\right]^{1/2}} = \cos\left(\frac{\Theta}{2} - \frac{\pi}{2}\right) = \sin\frac{\Theta}{2} \tag{A.11}$$

Using a trigonometric identity we have

$$\frac{Ze^2}{4\pi\epsilon_0 m_R v_0^2 b} = \tan\frac{\Theta}{2} \tag{A.12a}$$

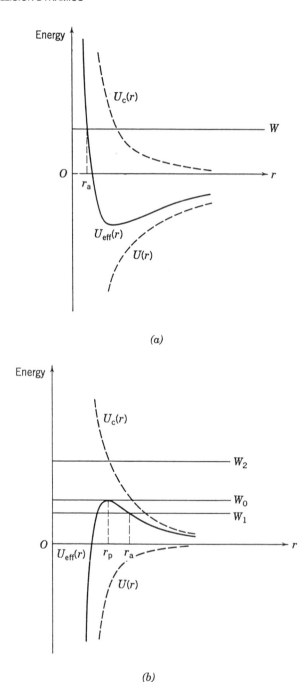

FIGURE A.2 The potential functions used for the calculation of elastic scattering in (a) an attractive inverse first power potential and (b) an attractive inverse third power potential.

or, solving for b,

$$b = \frac{Ze^2}{4\pi\epsilon_0 m_R v_0^2 \tan(\Theta/2)} \qquad (A.12b)$$

From the definition of the differential cross section in (3.1.13), we have, after taking a derivative of (A.12b) and performing a few trigonometric manipulations,

$$\sigma(\Theta) = \frac{b}{\sin\Theta}\left|\frac{db}{d\Theta}\right| = \frac{Z^2 e^4}{(8\pi\epsilon_0)^2 m_R^2 v_0^4 \sin^4(\Theta/2)} \qquad (A.13)$$

which is the well-known *Rutherford cross section* for Coulomb scattering. Because of the $\sin^4(\Theta/2)$ term in the denominator, the total scattering cross section is infinite, unless cut off by long-range shielding, as discussed in Sec. 3.3.

THE COLLISION INTEGRAL
AND KINETIC THEORY

Boltzmann Collision Integral

We now obtain the general form for the term $\partial f/\partial t|_c$, which occurs on the right-hand side (RHS) of the Boltzmann equation (2.3.3), known as the Boltzmann collision integral. Consider an elastic collision between incident and target particles having distributions f_1 and f_2 and velocities $\mathbf{v}_1$ and $\mathbf{v}_2$. The number of particles between $\mathbf{v}_1$ and $\mathbf{v}_1 + d\mathbf{v}_1$ is

$$f_1(\mathbf{r}, \mathbf{v}_1, t)\, d^3 v_1$$

and the number between $\mathbf{v}_2$ and $\mathbf{v}_2 + d\mathbf{v}_2$ is

$$f_2(\mathbf{r}, \mathbf{v}_2, t)\, d^3 v_2$$

The flux of incident particles in a coordinate system in which $\mathbf{v}_2$ is zero is

$$|\mathbf{v}_1 - \mathbf{v}_2|\, f_1\, d^3 v_1$$

The differential cross section for scattering through angle θ_1 is $I\left(|\mathbf{v}_1 - \mathbf{v}_2|, \theta_1\right)$. The number of particles per unit time scattered out of the differential volume $d^3 v_1\, d^3 v_2$ and into the solid angle $d\Omega$ is then

$$f_1 f_2\, |\mathbf{v}_1 - \mathbf{v}_2|\, d^3 v_1\, d^3 v_2\, I\, d\Omega$$

547

Dividing by $d^3 v_1$ and integrating over $d^3 v_2$ and $d\Omega$ we obtain all particles scattered out of the incident distribution f_1:

$$\frac{\partial f_1}{\partial t}\bigg|_{\text{out}} = \int \int f_1 f_2 \, |v_1 - v_2| \, d^3 v_2 \, I \, d\Omega \tag{B.1}$$

The particles from the distributions f_1 and f_2 having velocities v_1 and v_2 are scattered to primed velocities v_1' and v_2' in distributions $f_1' \equiv f_1(\mathbf{r}, \mathbf{v}_1', t)$ and $f_2' \equiv f_2(\mathbf{r}, \mathbf{v}_2', t)$. The rate of scattering into f_1, from the reversibility of the equations of motion, is then

$$\frac{\partial f_1}{\partial t}\bigg|_{\text{in}} = \int \int f_1' f_2' \, |v_1' - v_2'| \, d^3 v_2' \, I' \, d\Omega' \tag{B.2}$$

Finally, for elastic collisions, the relative velocity is conserved,

$$|v_1' - v_2'| = |v_1 - v_2| \tag{B.3}$$

and the differential cross sections $I \, d\Omega$ and $I' \, d\Omega'$ in the primed and unprimed coordinates can be identified. Substituting (B.3) in (B.2) and subtracting (B.1) from (B.2), we have

$$\frac{\partial f_1}{\partial t}\bigg|_{\text{c}} = \int d^3 v_2 \int_0^{2\pi} d\phi_1 \int_0^{\pi} \left(f_1' f_2' - f_1 f_2 \right) \, |v_1 - v_2| \, I \, \sin\theta_1 \, d\theta_1 \tag{B.4}$$

where we have written out $d\Omega$ explicitly and noted that the integral over θ_1 is done before the integral over v_2. The general form of (B.4) constitutes the RHS of (2.3.3).

It is not easy to evaluate (B.4) under the action of arbitrary forces on the LHS of (2.3.3). For small-angle Coulomb collisions an expansion of (B.4) is possible to obtain the *Fokker–Planck collision integral* described in many texts on fully ionized plasmas (e.g., see Schmidt, 1979). For large-angle collisions with neutrals a different expansion is usually used, which assumes a distribution close to equilibrium. We have already employed the resulting Krook collision operator in our formulation of the macroscopic equations in Sec. 2.3. We outline how this approximation is obtained below. First, however, we show that the general form (B.4) is satisfied by a Maxwellian distribution at equilibrium.

Maxwellian Distribution

At equilibrium the distribution is stationary, $\partial f / \partial t|_{\text{c}} = 0$, which is satisfied if

$$f_1' f_2' - f_1 f_2 = 0 \tag{B.5}$$

Taking the logarithm of (B.5) we have

$$\ln f_1' + \ln f_2' = \ln f_1 + \ln f_2 \tag{B.6}$$

Guessing a solution of the form

$$\ln f = -\xi^2 m v^2 + \ln C \tag{B.7}$$

for f_1, f_2, f_1' and f_2', and substituting into (B.6), we obtain

$$m_1 v_1'^2 + m_2 v_2'^2 = m_1 v_1^2 + m_2 v_2^2 \tag{B.8}$$

which expresses the conservation of energy in an elastic collision. Hence (B.7) is a solution of (B.5). Taking the antilog of (B.7), we obtain the equilibrium distribution

$$f_1(\mathbf{v}) = f_2(\mathbf{v}) = C\, e^{-\xi^2 m v^2} \tag{B.9}$$

which is the form assumed in (2.4.2). This gives, with the appropriate normalization, the Maxwellian distribution, with a common temperature for all the species, as in (2.4.7). We could also have included a function of the momentum in (B.7) and found a drifting Maxwellian at equilibrium. Equation (B.5) is clearly sufficient to satisfy $\partial f / \partial t \,|_c = 0$. It is also necessary, which can be shown by use of the *Boltzmann H Theorem*, which states that the time derivative of the function

$$H = \int f \ln f\, d^3 v \tag{B.10}$$

which measures the randomness of the distribution, is zero if and only if (B.5) is satisfied. A more detailed account of the above material, including a derivation of the H theorem, can be found in Holt and Haskell (1965, Chapter 5).

The Krook Collision Operator

We consider only elastic collisions between electrons and neutrals, such that the collision integral can be written, from (B.4),

$$\frac{\partial f_e}{\partial t}\bigg|_c = \int d^3 v_g \int_0^{2\pi} d\phi_1 \int_0^{\pi} (f_e' f_g - f_e f_g)\, v\, I(v, \theta_1)\, \sin\theta_1\, d\theta_1 \tag{B.11}$$

where we assume the neutrals are infinitely massive, so that $|\mathbf{v} - \mathbf{v}_g| = v = v'$, the electron speed, and $f_g' = f_g$. We expand the electron distribution function to first order in the deviation from isotropy, which we take to be cylindrically symmetric along the v_z direction:

$$f_e(z, v, \psi, t) \approx f_{e0}(z, v, t) + \frac{v_z}{v} f_{e1}(z, v, t) \tag{B.12}$$

where $v_z / v = \cos\psi$, with ψ the spherical polar angle in velocity space, and where f_{e1} is not a function of ψ. Because $v' = v$, we have $f_{e1}' = f_{e1}$ and $f_{e0}' = f_{e0}$. Substituting the expansion (B.12) into the factor in parentheses in (B.11), we have

$$(f_e' f_g - f_e f_g)v = (f_{e0} f_g - f_{e0} f_g)v + f_{e1} f_g v_z' - f_{e1} f_g v_z \tag{B.13}$$

where the first term on the right-hand side is zero. For infinitely massive neutrals, the scattering process yields

$$v_z' = v_x \cos \phi_1 \sin \theta_1 + v_y \sin \phi_1 \sin \theta_1 + v_z \cos \theta_1$$

Substituting this and (B.13) into (B.11), we perform the ϕ_1 integration to obtain

$$\left. \frac{\partial f_e}{\partial t} \right|_c = 2\pi \int f_g \, d^3 v_g \int_0^\pi f_{e1} \, \cos \psi \, (\cos \theta_1 - 1) \, v \, I(v, \theta_1) \, \sin \theta_1 \, d\theta_1 \tag{B.14}$$

where $\int f_g \, d^3 v_g = n_g$, the neutral gas density.

Kinetic Equations for Electrons

Writing the Boltzmann equation (2.3.3) in one spatial dimension, we have

$$\frac{\partial f_e}{\partial t} + v_z \frac{\partial f_e}{\partial z} - \frac{e}{m} E_z \frac{\partial f_e}{\partial v_z} = \left. \frac{\partial f_e}{\partial t} \right|_c \tag{B.15}$$

Using (B.12), we expand (B.15) to obtain

$$\frac{\partial f_{e0}}{\partial t} + \cos \psi \frac{\partial f_{e1}}{\partial t} + v \cos \psi \frac{\partial f_{e0}}{\partial z} + v \cos^2 \psi \frac{\partial f_{e1}}{\partial z}$$
$$- \frac{e}{m} E_z \cos \psi \frac{\partial f_{e0}}{\partial v} - \frac{e}{m} E_z \left[\frac{f_{e1}}{v} + v \frac{\partial}{\partial v} \left(\frac{f_{e1}}{v} \right) \cos^2 \psi \right] = \left. \frac{\partial f_e}{\partial t} \right|_c \tag{B.16}$$

where $\partial f_e / \partial t|_c$ is given by (B.14). Multiplying (B.16) by $\sin \psi$ and integrating over ψ from 0 to π, we obtain, after collecting terms,

$$\frac{\partial f_{e0}}{\partial t} + \frac{v}{3} \frac{\partial f_{e1}}{\partial z} - \frac{e}{m} E_z \frac{1}{3v^2} \frac{\partial}{\partial v} (v^2 f_{e1}) = 0 \tag{B.17}$$

Equation (B.17) gives the time rate of change of the isotropic part of the distribution, given the anisotropic part, and does not directly depend on the collisions. Multiplying (B.16) by $\sin \psi \cos \psi$ and integrating, as before, we obtain

$$\frac{\partial f_{e1}}{\partial t} + v \frac{\partial f_{e0}}{\partial z} - \frac{e}{m} E_z \frac{\partial f_{e0}}{\partial v} = -\nu_m(v) f_{e1} \tag{B.18}$$

where

$$\nu_{\mathrm{m}}(v) = n_{\mathrm{g}} v \, 2\pi \int_0^{\pi} (1 - \cos \theta_1) I(v, \theta_1) \sin \theta_1 \, d\theta_1 \tag{B.19}$$

is the momentum transfer collision frequency. Given f_{e0}, we find the time variation of the anisotropic part of the distribution function from the collision frequency ν_{m}. We see that ν_{m} is defined in the usual way:

$$\nu_{\mathrm{m}} = n_{\mathrm{g}} \sigma_{\mathrm{m}}(v) v \tag{B.20}$$

where n_{g} is the neutral density and σ_{m} is the momentum transfer cross section.

We note that the form of (B.17) follows because the elastic collisions of the electrons are with infinitely massive neutrals. If the neutrals have a Maxwellian distribution and are not infinitely massive, then for elastic collisions, a collision term appears on the right-hand side of (B.17) (Holt and Haskell, 1965, Chapter 10; Smirnov, 1981, p. 66):

$$\frac{\partial f_{\mathrm{e0}}}{\partial t} + \frac{v}{3} \frac{\partial f_{\mathrm{e1}}}{\partial z} - \frac{e}{m} E_z \frac{1}{3v^2} \frac{\partial}{\partial v}(v^2 f_{\mathrm{e1}})$$
$$= \frac{m}{m+M} \frac{1}{v^2} \frac{\partial}{\partial v} \left[v^3 \nu_{\mathrm{m}}(v) \left(f_{\mathrm{e0}} + \frac{eT}{mv} \frac{\partial f_{\mathrm{e0}}}{\partial v} \right) \right] \tag{B.21}$$

where T is the neutral gas temperature. The first term in parentheses on the right-hand side accounts for elastic scattering energy losses, while the second term accounts for the non-zero gas temperature. Equations (B.18) and (B.17) or (B.21) are the fundamental kinetic equations for the electron distribution function (B.12), in the limit that the anisotropy is small, $|f_{\mathrm{e1}}| \ll |f_{\mathrm{e0}}|$, and that Coulomb collisions are negligible. The equations can be modified to include the effect of a dc magnetic field force (Holt and Haskell, 1965), but we do not introduce this complication here. If there are also energy losses due to inelastic collisions, then an additional term such as $-n_{\mathrm{g}} \sigma_{\mathrm{inel}}(v) v f_{\mathrm{e0}}$ can be added to the right-hand side of (B.21). For a steady-state distribution with no spatial gradients or electric field, the left-hand side of (B.21) vanishes. We can then set the right-hand side equal to zero, to find that f_{e0} is a Maxwellian distribution at temperature T; i.e., the electrons and neutrals have equilibrated. However, this is rarely the situation in gas discharges.

Diffusion and Mobility

Consider now a steady state with nonzero density gradient and dc electric field. Solving (B.18) for f_{e1} yields

$$f_{\mathrm{e1}} = -\frac{1}{\nu_{\mathrm{m}}} \left(v \frac{\partial f_{\mathrm{e0}}}{\partial z} - \frac{e}{m} E_z \frac{\partial f_{\mathrm{e0}}}{\partial v} \right) \tag{B.22}$$

Introducing the particle flux

$$\mathbf{\Gamma}_e = \int \mathbf{v} \left(f_{e0} + \frac{v_z}{v} f_{e1} \right) d^3 v \tag{B.23}$$

and using spherical coordinates in velocity space, we see that the isotropic part of f_e does not contribute to the flux. From the anisotropic part, we obtain only a z component:

$$\Gamma_{ez} = 2\pi \int_0^\pi \sin \psi \, d\psi \, \cos^2 \psi \int_0^\infty v f_{e1} v^2 \, dv$$

$$= \frac{4\pi}{3} \int_0^\infty v^3 f_{e1} \, dv \tag{B.24}$$

Inserting (B.22) into (B.24), we obtain

$$\Gamma_{ez} = -D_e \frac{dn_e}{dz} - \mu_e n_e E_z \tag{B.25}$$

where

$$D_e = \frac{4\pi}{3n_e} \int_0^\infty \frac{v^4}{\nu_m(v)} f_{e0} \, dv \tag{B.26}$$

is the diffusion coefficient, and

$$\mu_e = -\frac{4\pi e}{3mn_e} \int_0^\infty \frac{v^3}{\nu_m(v)} \frac{df_{e0}}{dv} \, dv \tag{B.27}$$

is the mobility. For a Maxwellian distribution, D_e and μ_e are related by the Einstein relation (5.1.9), as can be shown directly from (B.26) and (B.27). Equations (B.26) and (B.27) are important because they give the proper prescription for averaging over $\nu_m(v)$ to determine D_e and μ_e.

Druyvesteyn Distribution

Consider now the steady-state electron distribution function in a uniform steady electric field and with elastic collisions between electrons and neutral gas atoms. From (B.18), we have

$$f_{e1} = \frac{eE_z}{m\nu_m} \frac{df_{e0}}{dv} \tag{B.28}$$

For ease of analysis, we take the gas temperature T to be negligible and take $m \ll M$ such that (B.21) becomes

$$\frac{eE_z}{3m} \frac{\mathrm{d}(v^2 f_{e1})}{\mathrm{d}v} = -\frac{m}{M} \frac{\mathrm{d}(v^3 \nu_m f_{e0})}{\mathrm{d}v} \tag{B.29}$$

Integrating (B.29), we obtain

$$f_{e1} = -\frac{3m^2}{eE_z M} v \nu_m f_{e0} \tag{B.30}$$

Equating (B.28) and (B.30) and integrating, we find

$$f_{e0} = A \exp\left[-\frac{3m^3}{e^2 E_z^2 M} \int_0^v v' \nu_m^2(v') \, \mathrm{d}v' \right] \tag{B.31}$$

where A is a normalization constant determined by $\int f_{e0} \mathrm{d}^3 v = n_e$. For a constant collision frequency, $\nu_m(v) = $ const, we obtain a Maxwellian distribution. However, for constant cross section (hard-sphere) collisions, $\sigma_m = $ const, with $\nu_m = n_g \sigma_m v$, we find that $f_{e0} \propto e^{-\alpha v^4}$, which is known as the *Druyvesteyn distribution*. Many electron–neutral cross sections behave as hard-sphere interactions at low energies (see Fig. 3.9).

Electron Distribution in an Rf Field

We consider now the electron distribution in an rf electric field $E_z(t) = \mathrm{Re}\, E_0 e^{j\omega t}$ due to elastic collisions of electrons with neutral gas atoms, in the frequency regime $\omega \gg (m/M)\nu_m$. In this case, the energy transferred by electrons to gas atoms over one rf period is small, and f_{e0} is independent of time. Introducing $f_{e1}(t) = \mathrm{Re}\, \tilde{f}_{e1} e^{j\omega t}$ into (B.18), we obtain

$$\tilde{f}_{e1} = \frac{eE_0}{m(j\omega + \nu_m)} \frac{\mathrm{d}f_{e0}}{\mathrm{d}v} \tag{B.31}$$

Substituting $E_z(t)$ and $f_{e1}(t)$ into (B.21) and time averaging the resulting equation over an rf period, we obtain, in analogy to (B.29), and with the same assumptions $(T \to 0, m/M \ll 1)$,

$$\frac{1}{2} \mathrm{Re} \frac{eE_0}{3m} \frac{\mathrm{d}(v^2 \tilde{f}_{e1})}{\mathrm{d}v} = -\frac{m}{M} \frac{\mathrm{d}(v^3 \nu_m f_{e0})}{\mathrm{d}v} \tag{B.32}$$

Integrating (B.32), we find

$$\mathrm{Re}\, \tilde{f}_{e1} = -\frac{6m^2}{eE_0 M} v \nu_m f_{e0} \tag{B.33}$$

Equating the real part of (B.31) to (B.33), we obtain

$$\frac{eE_0 \nu_m}{m(\omega^2 + \nu_m^2)} \frac{df_{e0}}{dv} = -\frac{6m^2}{eE_0 M} v\nu_m f_{e0} \tag{B.34}$$

Equation (B.34) can be integrated to obtain

$$f_{e0} = A \exp\left\{ -\frac{6m^3}{e^2 E_0^2 M} \int_0^v v'[\omega^2 + \nu_m^2(v')] \, dv' \right\} \tag{B.35}$$

where A is the normalization constant. We note that at high frequencies or low pressures, such that $\omega \gg \nu_m$, f_{e0} reduces to a Maxwellian distribution. Substituting (B.35) into (B.31) determines f_{e1}, the oscillating anisotropic part of f_e. For further consideration of the application of kinetic theory to determine the electron distribution, see Smirnov (1981, Chapter 2).

APPENDIX C

STOCHASTIC HEATING IN AN INDUCTIVE DISCHARGE

The theory of anomalous collisions in a thin electric field layer (Ichimaru, 1973; Alexandrov et al., 1984), originally developed by Pippard (1949) to describe the high-frequency skin resistance of metals at low temperatures, has been applied to collisionless heating in a low-pressure inductive discharge (Turner, 1993; Godyak et al., 1994). To determine the heating, we consider an electron from the bulk plasma incident on the rf electric field within a skin depth layer in slab geometry. We assume that the transverse electric field within the slab decays exponentially with distance x from the edge into the slab:

$$E_y(x,t) = E_0 e^{-|x|/\delta} \cos(\omega t + \phi) \tag{C.1}$$

We assume that $\nu_m \ll \bar{v}_e/2\delta$; hence there are no electron collisions within the slab. Because there are no x-directed forces, we can write

$$\begin{aligned} x(t) &= -v_x t & t < 0 \\ &= v_x t & t > 0 \end{aligned} \tag{C.2}$$

where the electron reflects from the surface at $t = 0$. Substituting (C.2) into (C.1) yields the transverse electric field seen by the electron,

$$\begin{aligned} E_y(t) &= \operatorname{Re} E_0 e^{(j\omega + v_x/\delta)t + j\phi} & t < 0 \\ &= \operatorname{Re} E_0 e^{(j\omega - v_x/\delta)t + j\phi} & t > 0 \end{aligned} \tag{C.3}$$

The transverse velocity impulse,

$$\Delta v_y = -\int_{-\infty}^{\infty} dt \frac{eE_y(t)}{m} \tag{C.4}$$

is calculated by substituting (C.3) into (C.4) and integrating to obtain

$$\Delta v_y = \frac{2eE_0\delta}{m} \frac{v_x}{v_x^2 + \omega^2\delta^2} \cos\phi \tag{C.5}$$

The energy change ΔW, averaged over a uniform distribution of initial electron phases ϕ, is then

$$\begin{aligned}
\Delta W &= \frac{1}{2}m\langle(\Delta v_y)^2\rangle_\phi \\
&= \frac{1}{4}m\left(\frac{2eE_0\delta}{m}\right)^2 \frac{v_x^2}{(v_x^2 + \omega^2\delta^2)^2}
\end{aligned} \tag{C.6}$$

which can be integrated over the particle flux to obtain the stochastic heating power

$$S_{\text{stoc}} = \int_{-\infty}^{\infty} dv_y \int_{-\infty}^{\infty} dv_z \int_0^{\infty} dv_x f_e v_x \Delta W(v_x) \tag{C.7}$$

For a Maxwellian electron distribution f_e, the integrals over v_y and v_z are easily done. Substituting $\zeta = mv_x^2/2eT_e$ to evaluate the v_x integral, we obtain

$$S_{\text{stoc}} = \frac{mn_s}{\bar{v}_e}\left(\frac{eE_0\delta}{m}\right)^2 \mathcal{I} \tag{C.8}$$

where

$$\mathcal{I}(\alpha) = \frac{1}{\pi}\int_0^{\infty} d\zeta\, e^{-\zeta} \frac{\zeta}{(\zeta + \alpha)^2} \tag{C.9}$$

with

$$\alpha = \frac{4\omega^2\delta^2}{\pi\bar{v}_e^2} \tag{C.10}$$

Letting $\zeta' = \zeta + \alpha$, then $\mathcal{I}$ can be expressed as

$$\mathcal{I} = \frac{1}{\pi}\left[e^\alpha(1 + \alpha)E_1(\alpha) - 1\right] \tag{C.11}$$

where the exponential integral,

$$E_1(\alpha) = \int_\alpha^\infty d\zeta \, \frac{e^{-\zeta}}{\zeta} \tag{C.12}$$

is tabulated in the literature. For $\alpha \ll 1$, we find

$$\mathcal{I}(\alpha) \approx \frac{1}{\pi} \left[\ln\left(\frac{1}{\alpha}\right) - 1.58 \right] \tag{C.13}$$

and for $\alpha \gg 1$, we find

$$\mathcal{I}(\alpha) \approx \frac{1}{\pi\alpha^2} \tag{C.14}$$

A graph of $\mathcal{I}$ versus α is given in Fig. C.1.

We can introduce an effective collision frequency ν_{stoc} by equating the stochastic heating (C.8) to an effective collisional heating power flux:

$$
\begin{aligned}
S_{\text{ohm}} &= \frac{1}{2} \int_0^\infty dx \, \left(E_0 e^{-x/\delta} \right)^2 \frac{e^2 n_s}{m} \frac{\nu_{\text{stoc}}}{\nu_{\text{stoc}}^2 + \omega^2} \\
&= \frac{1}{4} \frac{e^2 n_s \delta}{m} \frac{\nu_{\text{stoc}}}{\nu_{\text{stoc}}^2 + \omega^2} E_0^2
\end{aligned}
\tag{C.15}
$$

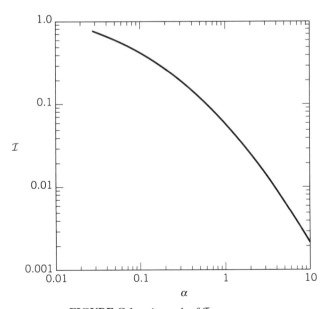

FIGURE C.1. A graph of $\mathcal{I}$ versus α.

There are two frequency regimes. For low frequency, $\omega \ll \nu_{\text{stoc}}$, we have

$$S_{\text{ohm}} = \frac{1}{4} \frac{e^2 n_s \delta}{m \nu_{\text{stoc}}} E_0^2$$

and equating this to (C.8), we determine

$$\nu_{\text{stoc}} = \frac{C_e \bar{v}_e}{\delta} \tag{C.16}$$

where $C_e = \frac{1}{4} \mathcal{I}(\alpha)$. In this regime, with $\alpha \ll 1$, using Fig. C.1, we find $C_e \sim 1$, implying $\nu_{\text{stoc}} \sim \bar{v}_e / \delta$. In the opposite limit of high frequency, $\omega \gg \nu_{\text{stoc}}$, we have

$$S_{\text{ohm}} = \frac{1}{4} \frac{e^2 n_s \delta}{m} \frac{\nu_{\text{stoc}}}{\omega^2} E_0^2$$

and we determine

$$\nu_{\text{stoc}} = \frac{4\mathcal{I}(\alpha)\omega^2 \delta}{\bar{v}_e} \tag{C.17}$$

In this regime, with $\alpha \gg 1$, using (C.14), we find

$$\nu_{\text{stoc}} \sim \frac{\bar{v}_e^3}{\delta^3 \omega^2} \tag{C.18}$$

REFERENCES

Alexandrovich, A. F., L. S. Bogdankevich, and A.A. Rukhadze (1984), *Principles of Plasma Electrodynamics*, Springer, New York.

Allis, W. P., S. J. Buchsbaum, and A. Bers (1963), *Waves in Anisotropic Plasmas*, MIT Press, Cambridge, MA.

Atkins, P. W. (1986), *Physical Chemistry*, 3d ed., Freeman, New York.

Barnett, C. F. (1989), in *A Physicist's Desk Reference*, H.L. Anderson, ed., American Institute of Physics, New York.

Barone, M. E., and D. B. Graves (1994), to appear in *J. Appl. Phys.*

Berg, S., H. O. Blom, M. Moradi, C. Nender, and T. Larsson (1989), *J. Vac. Sci. Technol.* **A7**, 1225.

Book, D. L. (1987), *NRL Plasma Formulary (Revised)*, Naval Research Laboratory, Washington, DC.

Born, M., and E. Wolf (1980), *Principles of Optics*, 6th ed., Pergamon, New York.

Boswell, R. W. (1970), *Plasma Phys. Controlled Fusion* **26**, 1147.

Boyd, R. L. F., and J. B. Thompson (1959), *Proc. R. Soc.* **A252**, 102.

Bransden, B. H., and C. J. Joachain (1983), *Physics of Atoms and Molecules*, Wiley, New York.

Bransden, B. H., and M. R. C. McDowell (1992), *Charge Exchange and the Theory of Ion-Atom Collisions*, Clarendon, Oxford, UK.

Brown, S. C. (1959), *Basic Data of Plasma Physics*, Technology Press and Wiley, New York.

Budden, K. G. (1966), *Radio Waves in the Ionosphere*, Cambridge University Press, Cambridge, UK.

Cale, T. S., G. B. Raupp, and T. H. Gandy (1992), *J. Vac. Sci. Technol.* **A10**, 1128.

Carl, D. A., D. W. Hess, M. A. Lieberman, T. D. Nguyen, and R. Gronsky (1991), *J. Appl. Phys.* **70**, 3301.

Chen, F. F. (1965), in *Plasma Diagnostic Techniques*, R.H. Huddlestone and S.L. Leonard, eds., Academic, New York.

Chen, F. F. (1984), *Introduction to Plasma Physics and Controlled Fusion*, 2d ed., Plenum, New York.

Chen, F. F. (1991), *Plasma Phys. Controlled Fusion* **33**, 339.

Chen, F. F. (1992), *J. Vac. Sci. Technol.* **A10**, 1389.

Cheung, N. W. (1991), *Nucl. Instrum. Methods* **55**, 811.

Cho, M. H., N. Hershkowitz, and T. Intrator (1988), *J. Vac. Sci. Technol.* **A6**, 2978.

Cobine, J. D. (1958), *Gaseous Conductors*, Dover, New York.

Coburn, J. W., and H. F. Winters (1979), *J. Vac. Sci. Technol.* **16**, 391.

Conrad, J. R. (1987), *J. Appl. Phys.* **62**, 777.

Conrad, J. R., R. A. Dodd, S. Han, M. Madapura, J. Scheuer, K. Sridharam, and F. J. Worzala (1990), *J. Vac. Sci. Technol.* **A8**, 3146.

Cook, J. M., D. E. Ibbotson, P. D. Foo, and D. L. Flamm (1990), *J. Vac. Sci. Technol.* **A8**, 1820.

Donnelly, I. J., and P. A. Watterson (1989), *J. Phys. D.* **22**, 90.

Donnelly, V. M. (1989), in *Plasma Diagnostics*, Vol. 1, O. Auciello and D. L. Flamm, eds., Academic, New York.

Eckert, H. U. (1986), *Proc. 2nd Int. Conf. on Plasma Chem. and Technol.*, H. Boening, ed., Technomic Publ., Lancaster, PA.

Eliasson, B., and U. Kogelschatz (1986), Basic data for modelling of electrical discharges in gases: oxygen, *Report KLR 86–11C*, Brown Boveri Konzernforschung, CH-5405 Baden.

Feldman, L. C., and J. W. Mayer (1986), *Fundamentals of Surface and Thin Film Analysis*, North-Holland, New York.

Flamm, D. L., (1989), in *Plasma Etching: An Introduction*, D. M. Manos and D. L. Flamm, eds., Academic, New York.

Flamm, D. L. (1990), Mechanisms of silicon etching in fluorine- and chlorine-containing plasmas, *Report UCB/ERL M90/41*, College of Engineering, University of California, Berkeley.

Flamm, D. L., and G. K. Herb (1989), in *Plasma Etching: An Introduction*, D.M. Manos and D.L. Flamm, eds., Academic, New York.

Franklin, R. N. (1976), *Plasma Phenomena in Gas Discharges*, Clarendon, Oxford, UK.

Gilmore, F. R. (1964), Potential energy curves for N_2, NO, O_2 and corresponding ions, *RAND Corporation Memorandum R-4034-PR*.

Ginzburg, V. L. (1964), *The Propagation of Electromagnetic Waves in Plasma*, Pergamon, Oxford, UK.

Godyak, V. A. (1986), *Soviet Radio Frequency Discharge Research*, Delphic Associates, Falls Church, VA.

Godyak, V. A. (1990a), in *Plasma–Surface Interactions and Processing of Materials*, O. Auciello et al., eds., Kluwer Academic, Boston.

Godyak, V. A. (1990b), Private communication.

Godyak, V. A., and R. B. Piejak (1990a), *Phys. Rev. Lett.* **65**, 996.

Godyak, V. A., and R. B. Piejak (1990b) *J. Vac. Sci. Technol.* **A8**, 3833.

Godyak, V. A. and N. Sternberg (1990a), *IEEE Trans. Plasma Sci.* **18**, 159.

Godyak, V. A. and N. Sternberg (1990b), *Phys. Rev. A.* **42**, 2299.

Godyak, V. A., R. B. Piejak, and B. M. Alexandrovich (1991) *IEEE Trans. Plasma Sci.* **19**, 660.

Godyak, V. A., R. B. Piejak, and B. M. Alexandrovich (1992) *Plasma Sources Sci. Technol.* **1**, 36.

Godyak, V. A., R. B. Piejak, and B. M. Alexandrovich (1993), *J. Appl. Phys.* **73**, 3657.

Godyak, V. A., R. B. Piejak, and B. M. Alexandrovich (1994), *Plasma Sources Sci. Technol.*, to be published.

Goldstein, H. (1950), *Classical Mechanics*, Addison-Wesley, Cambridge, MA.

Gray, D. C., I. Tepermeister, and H. H. Sawin (1993), *J. Vac. Sci. Technol.* **B11**, 1243.

Greaves, J. C., and W. Linnett (1959) *Trans. Faraday Soc.* **55**, 1355.

Harrington, R. F. (1961), *Time-Harmonic Electromagnetic Fields*, McGraw-Hill, New York.

Haugsjaa, P. O., and R. C. Amme (1970), *J. Chem. Phys.* **52**, 4874.

Hayashi, M. (1987), in *Swarm Studies and Inelastic Electron–Molecule Collisions*, L.C. Pitchford, B. V. McKoy, A. Chutjian, and S. Trajmar, eds., Springer, New York.

Heald, M. A., and C. B. Wharton (1965), *Plasma Diagnostics with Microwaves*, Wiley, New York.

Hershkowitz, N. (1989), in *Plasma Diagnostics*, Vol. 1, O. Auciello and D.L. Flamm, eds., Academic, New York.

Hershkowitz, N., K. N. Leung, and T. Romesser (1975) *Phys. Rev. Lett.* **35**, 277.

Herzberg, G. (1971), *The Spectra and Structures of Simple Free Radicals*, Dover, New York.

Hittorf, W. (1884), *Wiedemanns Ann Phys.* **21**, 90.

Holt, H. E., and R. E. Haskell (1965), *Plasma Dynamics*, Macmillan, New York.

Hopwood, J., C. R. Guarnieri, S. J. Whitehair, and J. J. Cuomo (1993a), *J. Vac. Sci. Technol.* **A11**, 152.

Hopwood, J., C. R. Guarnieri, S. J. Whitehair, and J. J. Cuomo (1993b), *J. Vac. Sci. Technol.* **A11**, 147.

Huddlestone, R. H., and S. L. Leonard, eds. (1965), *Plasma Diagnostic Techniques*, Academic, New York.

Hussein, M. A., and G. A. Emmert (1990), *Phys. Fluids* **B2**, 218.

Ichimaru, I. (1973), *Basic Principles of Plasma Physics: A Statistical Approach*, Benjamin, Reading MA.

Jackson, J.D. (1975), *Classical Electrodynamics*, 2d ed., Wiley, New York.

Jaeger, F., A. J. Lichtenberg, and M. A. Lieberman (1972), *Plasma Phys.* **14**, 1073.

Joyce, B. A., and C. T. Foxon (1984), in *Simple Processes at the Gas–Solid Interface*, C. H. Bamford, C. F. H. Tipper, and R. G. Compton, eds., Elsevier, Amsterdam.

Kagan, Y. M., and V. I. Perel (1964), *Sov. Phy. Usp.* **6**, 767.

Keller, J. H., J. C. Forster, and M. S. Barnes (1993), *J. Vac. Sci. Technol.* **A11** 2487.

Komachi, K. (1992), *J. Vac. Sci. Technol.* **A11**, 164.

Komori, A., T. Shoji, K. Miyamoto, J. Kawai, and Y. Kawai (1991), *Phys. Fluids* **B3**, 893.

Konuma, M. (1992), *Film Deposition by Plasma Techniques*, Springer, New York.

Kossyi, I. A., A. Y. Kostinsky, A. A. Matveyev, and V. P. Silakov (1992), *Plasma Sources Sci. Technol.* **1**, 207.

Krall, N. A., and A. W. Trivelpiece (1973), *Principles of Plasma Physics*, McGraw-Hill, New York.

Kushner, M. J. (1988), *J. Appl. Phys.* **63**, 2532.

Laframboise, J. G. (1966), Theory of spherical and cylindrical langmuir probes in a collision-less, Maxwellian plasma at rest, *UTIAS Report No. 100*, University of Toronto.

Lawton, S. A., and A. V. Phelps (1978), *J. Chem. Phys.* **69**, 1055.

Lee, C., D. B. Graves, M. A. Lieberman, and D.W. Hess (1994) *J. Electrochem. Soc.*, **141**, 1546.

Leung, K. N., T. K. Samec, and A. Lamm (1975), *Phys. Lett.* **51A**, 490.

Leung, K. N., G. R. Taylor, J. M. Barrick, S. L. Paul and R. E. Kribel (1976), *Phys. Lett.* **57A**, 145.

Lichtenberg, A. J., V. Vahedi, M. A. Lieberman, and T. Rognlien (1994), *J. Appl. Phys.*, **75**, 2339.

Lieberman, M. A. (1988), *IEEE Trans. Plasma Sci.* **16**, 638.

Lieberman, M. A. (1989a), *IEEE Trans. Plasma Sci.* **17**, 338.

Lieberman, M. A. (1989b), *J. Appl. Phys.* **65**, 4168.

Lieberman, M. A. (1989c), *J. Appl. Phys.* **66**, 2926.

Lieberman, M. A., and R. A. Gottscho (1994), Design of high-density plasma sources for materials processing, in *Physics of Thin Films*, Vol. 18, M. H. Francombe and J. L. Vossen, eds., Academic, New York.

Lieberman, M. A., and S. E. Savas (1990), *J. Vac. Sci. Technol.* **A8**, 1632.

Lieberman, M. A., A. J. Lichtenberg, and S. E. Savas (1991), *IEEE Trans. Plasma Sci.* **19**, 189.

Manos, D. M., and H. F. Dylla (1989), in *Plasma Etching: An Introduction*, D. M. Manos and D. L. Flamm, eds., Academic, New York.

Manos, D. M., and D. L. Flamm, eds. (1989), *Plasma Etching: An Introduction*, Academic, New York.

Massey, H. S. W, E. H. S. Burhop, and H. B. Gilbody (1969–74), *Electron and Ion Impact Phenomena*, 2d ed., Clarendon, Oxford, UK.

Matsuoka, M., and K. Ono (1988), *J. Vac. Sci. Technol.* **A6**, 25.

Matthieussent, G., and J. Pelletier (1992), in *Microwave Excited Plasmas*, M. Moissan and J. Pelletier, eds., Elsevier, Amsterdam.

McCaughey, M. J., and M. J. Kushner (1989), *J. Appl. Phys.* **65**, 186.

McDaniel, E. W. (1964), *Collision Phenomena in Ionized Gases*, Wiley, New York.

McDaniel, E. W. (1989), *Atomic Collisions: Electron and Photon Projectiles*, Wiley, New York.

McDaniel, E. W., J. B. A. Mitchell, and M. E. Rudd (1993), *Atomic Collisions: Heavy Particle Projectiles*, Wiley, New York.

Misium, G. R., A. J. Lichtenberg, and M.A. Lieberman (1989), *J. Vac. Sci. Technol.* **A7**, 1007.

Mogab, C. J., A. C. Adams, and D. L. Flamm (1979) *J. Appl. Phys.* **49**, 3796.

Moisan, M., and Z. Zakrzewski (1991), *J. Phys. D: Appl. Phys.* **24**, 1025.

Moroney, R. M., A. J. Lichtenberg, and M. A. Lieberman (1989), *J. Appl Phys.* **66**, 1618.

Morris, M. A., M. Bowker, and D. A. King (1984), in *Simple Processes at the Gas–Solid Interface*, C. H. Bamford, C. F. H. Tipper, and R. G. Compton, eds., Elsevier, Amsterdam.

Mott-Smith, H.M., and I. Langmuir (1926), *Phys. Rev.* **28**, 727.

Musil, J., and F. Zacek (1970), *Plasma Phys.* **12**, 17.

Musil, J., and F. Zacek (1971) *Plasma Phys.* **13**, 471.

Niazi, K., A. J. Lichtenberg, M. A. Lieberman, and D.L. Flamm (1994), *Plasma Sources Sci. Technol.*, to be published.

Ogryzlo, E. A., D. E. Ibbotson, D. L. Flamm, and J. A. Mucha (1990), *J. Appl. Phys.* **67**, 3115.

Parker, J. V., J. C. Nickel, and R. W. Gould (1964), *Phys. Fluids* **7**, 1489.

Perry, A. J., D. Vender, and R. W. Boswell (1991) *J. Vac. Sci. Technol.* **B9**, 310.

Phelps, A. V. (1985), Tabulations of cross sections and calculated transport and reaction coefficients for electron collisions with O_2, *JILA Information Center Report*, University of Colorado, Boulder.

Pico, C. A., M. A. Lieberman, and N. W. Cheung (1992), *J. Electron. Mater.* **21**, 75.

Piejak, R. B., V. A. Godyak, and B. M. Alexandrovich (1992), *Plasma Sources Sci. Technol.* **1**, 179.

Pippard, A. B. (1949), *Physica* **15**, 45.

Plumb, I. C., and K. R. Ryan (1986), *Plasma Chem. Plasma Process.* **6**, 205.

Popov, O. A., and V. A. Godyak (1985), *J. Appl. Phys.* **57**, 53.

Porteous, R. K., H. M. Wu, and D. B. Graves (1994), *Plasma Sources Sci. Technol.* **3**, 25.

Raizer, Y. P. (1991), *Gas Discharge Physics*, Springer, New York.

Ramo, S., J. R. Whinnery, and T. Van Duzer (1984), *Fields and Waves in Communication Electronics*, 2d ed., Wiley, New York.

Rapp, D., and D. Briglia (1965), *J. Chem. Phys.* **43**, 1480.

Rapp, D., and W. E. Francis (1962), *J. Chem Phys.* **37**, 2631.

Redsten, A. M., K. Sridharan, F. J. Worzala, and J. R. Conrad (1992), *J. Mater. Process. Technol.* **30**, 253.

Riemann, K-U. (1991), *J. Phys. D: Appl. Phys.* **24**, 493.

Rogoff, G. L. (1985), *J. Phys. D: Appl Phys.* **18**, 1533.

Scheuer, J. T., M. Shamim, and J. R. Conrad (1990), *J. Appl. Phys.* **67**, 1241.

Schmidt, G. (1979), *Physics of High Temperature Plasmas*, 2d ed., Academic, New York.

Schwarz, S. E., and W. G. Oldham (1984), *Electrical Engineering: An Introduction*, Holt, Rinehart & Winston, New York.

Selwyn, G. S. (1993), *Optical Diagnostic Techniques for Plasma Processing*, Monograph M11, American Vacuum Society Press, New York.

Shamim, M., J. T. Scheuer, and J. R. Conrad (1991), *J. Appl. Phys.* **69**, 2904.

Sharp, T. (1971), *Atomic Data*, Vol. 2, Academic, New York

Sheridan, T. E., M. J. Goeckner, and J. Goree (1991), *J. Vac. Sci. Technol.* **A9**, 688.

Simon, A. (1959), *An Introduction to Thermonuclear Research*, Pergamon, New York.

Smirnov, B. M. (1977), *Introduction to Plasma Physics*, Mir, Moscow.

Smirnov, B. M. (1981), *Physics of Weakly Ionized Gases*, Mir, Moscow.

Smirnov, B. M. (1982), *Negative Ions*, McGraw-Hill, New York.

Smullin, L. D., and P. Chorney (1958), *Proc. IRE* **46**, 360.

Spitzer, L. (1956), *Physics of Fully Ionized Gases*, Interscience, New York.

Stevens, J. E., Y. C. Huang, R. L. Jarecki, and J.L. Cecchi (1992), *J. Vac. Sci. Technol.* **A10**, 1270.

Stewart, R. A., and M. A. Lieberman (1991), *J. Appl. Phys.* **70**, 3481.

Stix, T. H. (1992), *Waves in Plasmas*, American Institute of Physics, New York.

Stout, P. J., and M. J. Kushner (1993), *J. Vac. Sci. Technol.* **A11**, 2562.

Su, C. H., and S. H. Lam (1963), *Phys. Fluids*, **6**, 1479.

Su, T., and M. T. Bowers (1973), *Int. J. Mass Spectrom. Ion Phys.* **12**, 347.

Szapiro, B., and J. J. Rocca (1989), *J. Appl. Phys.* **65**, 3713.

Thompson, J. B. (1959), *Proc. Phys. Soc.* **73**, 818.

Thomson, J. J. (1912), *Philos. Mag.* **23**, 449.

Thomson, J. J. (1924), *Philos. Mag.* **47**, 337.

Thomson, J. J. (1927), *Philos. Mag.* **4**, 1128.

Thorne, A. P. (1988), *Spectrophysics*, Chapman & Hall, London.

Thornton, J. A. (1986), *J. Vac. Sci. Technol.* **A4**, 3059.

Thornton, J. A., and A. S. Penfold (1978), in *Thin Film Processes*, J. L. Vossen and W. Kern, eds., Academic, New York.

Tonks, L., and I. Langmuir (1929), *Phys. Rev.* **34**, 876.

Trivelpiece, A. W. and R. W. Gould (1959), *J. Appl. Phys.* **30**, 1784.

Turner, M. M. (1993), *Phys. Rev. Lett.* **71**, 1844.

Vahedi, V. (1993), *Modeling and Simulation of Rf Discharges Used for Plasma Processing*, Thesis, University of California, Berkeley.

Vahedi, V., M. A. Lieberman, M. V. Alves, J. P. Verboncoeur, and C. K. Birdsall (1991), *J. Appl. Phys.* **69**, 2008.

Vahedi, V., R. A. Stewart, and M. A. Lieberman (1993), *J. Vac. Sci. Technol.* **A11**, 1275.

Vahedi, V., C. K. Birdsall, M. A. Lieberman, G. DiPeso, and T. D. Rognlien (1994), *Plasma Sources Sci. Technol.*, **2**, 273.

Vender, D., and R. W. Boswell (1990), *IEEE Trans. Plasma Sci.* **18**, 725.

Vossen, J. L., and W. Kern, eds. (1978), *Thin Film Processes*, Academic, New York.

Vossen, J. L., and W. Kern, eds. (1991), *Thin Film Processes II*, Academic, New York.

Waits, R. K. (1978), in *Thin Film Processes*, J. L. Vossen and W. Kern, eds., Academic, New York.

Walkup, R. E., K. L. Saenger, and G. S. Selwyn (1986), *J. Chem. Phys.* **84**, 2668.

Wendt, A. E. (1993), *2nd Workshop on High Density Plasmas and Applications*, AVS Topical Conference, August 5–6, 1993, San Francisco, CA.

Wendt, A. E., and M. A. Lieberman (1990), *J. Vac. Sci. Technol.* **A8**, 902.

Williamson, M. C., A. J. Lichtenberg, and M.A. Lieberman (1992), *J. Appl. Phys.* **72**, 3924.

Winters, H. F., and J. W. Coburn (1992), *Surface Sci. Rep.* **14**, 161.

Winters, H. F., and D. Haarer (1987), *Phys. Rev. B* **36**, 6613;(1988) **37**, 10379.

Wood, B. P. (1991) *Sheath Heating in Low Pressure Capacitive Radio Frequency Discharges*, Thesis, University of California, Berkeley.

Wu, H-M., D. B. Graves, and R. K. Porteous (1994), *Plasma Sources Sci. Technol.*, to be published.

Zangwill, A. (1988), *Physics at Surfaces*, Cambridge University Press, Cambridge, UK.

INDEX